MODELING AND ANALYSIS OF DYNAMIC SYSTEMS

Third Edition

THIRD EDITION

MODELING AND ANALYSIS OF DYNAMIC SYSTEMS

CHARLES M. CLOSE
Rensselaer Polytechnic Institute

DEAN K. FREDERICK
Unified Technologies, Inc.

JONATHAN C. NEWELL
Rensselaer Polytechnic Institute

JOHN WILEY & SONS, INC.

New York / Chichester / Weinheim / Brisbane / Singapore / Toronto

ACQUISITIONS EDITOR   *Joseph Hayton*
EDITORIAL ASSISTANT   *Steve Peterson*
MARKETING MANAGER   *Katherine Hepburn*
SENIOR PRODUCTION EDITOR   *Norine M. Pigliucci*
SENIOR DESIGNER   *Kevin Murphy*

Production Management Services: Publication Services

Cover photo courtesy of NASA.

This book was set in Times Roman by Publication Services and printed and bound by Hamilton Printing Company. The cover was printed by Lehigh Press.

This book was printed on acid-free paper. ∞

Library of Congress Cataloging in Publication Data:
Close, Charles M.
    Modeling and analysis of dynamic systems / Charles M. Close and Dean K. Frederick
    and Jonathan C. Newell—3rd ed.
        p. cm.
    Includes bibliographical references.
    ISBN 0-471-39442-4 (cloth : alk. paper)
        1. System analysis. I. Frederick, Dean K., 1934– II. Newell, Jonathan C. III. Title.

QA402.C53 2001
003–dc21

                                                                                          2001033010

Printed in the United States of America

10 9 8 7 6 5 4 3 2 1

To my wife, Margo,

and to our children and grandchildren

CMC

To my mother,

Elizabeth Dean Frederick,

and to the memory of my father,

Charles Elder Frederick

DKF

To my wife, Sigrin, and my sons, Andrew and Raymond

JCN

PREFACE

The primary purpose of this edition remains the same as in the previous editions: to provide an introductory treatment of dynamic systems suitable for all engineering students regardless of discipline. We have, however, made significant changes as a result of experiences with our many students, comments from numerous professors around the country, and the increasing educational use of computer packages. We have maintained flexibility in the selection and ordering of material.

The book can be adapted to several types of courses. One such use is for students who need a detailed treatment of modeling mechanical and electrical systems and of obtaining analytical and computer solutions before proceeding to more advanced levels. Such courses can serve as a foundation for subsequent courses in vehicular dynamics, vibrations, circuits and electronics, chemical process control, linear systems, feedback systems, nuclear reactor control, and biocontrol systems.

The book also covers such general topics as transfer functions, state variables, the linearization of nonlinear models, block diagrams, and feedback systems. Hence it is suitable for a general dynamic systems course for students who have completed a disciplinary course such as machine dynamics, electrical circuits, or chemical process dynamics.

This text can also be used for students with significant modeling and analysis experience who wish to emphasize computer techniques and feedback control systems. Topics include computer solutions for both linear and nonlinear models, as well as root-locus diagrams, Bode plots, block diagrams, and operational amplifiers. We explain some of the practical design criteria for control systems and illustrate the use of analytical and computer methods to meet those criteria. Finally, the book can provide a general introduction to dynamic systems for students in broad-based engineering programs or in programs such as biomedical and materials engineering who may have limited time for this subject.

We assume that the reader has had differential and integral calculus and basic college physics, including mechanics and electrical phenomena. Many students will have had a course in differential equations at least concurrently. We have been careful to present the mathematical results precisely (although without the rigorous proofs required for a mathematics book), so that the concepts learned will remain valid in subsequent courses. For example, the impulse has been treated in a manner that is consistent with distribution theory but is no more difficult to grasp than the usual approach taken in introductory engineering books.

Approach

The book reflects the approach we have used for many years in teaching basic courses in dynamic systems. Whether for a particular discipline or for a general engineering course, we have found it valuable to include systems from at least two disciplines in some depth. This illustrates the commonality of the modeling and analysis techniques, encourages students to avoid compartmentalizing their knowledge, and prepares them to work on projects as part of an interdisciplinary team.

Mechanical systems are examined first because they are easily visualized and because most students have had previous experience with them. The basic procedures for obtaining models for analytical and computer solutions are developed in terms of translational systems. Models include those in state-variable, input-output, and matrix form, as well as block diagrams. The techniques are quickly extended to rotational systems, and there are also chapters on electrical, electromechanical, thermal, and fluid systems. Each type of system is modeled in terms of its own fundamental laws and nomenclature.

After introducing block diagrams in Chapter 4, we show how to use Simulink and MATLAB to obtain responses of simple systems. We introduce Laplace transforms fairly early and use them as the primary means of finding analytical solutions. We emphasize the transfer function as a unifying theme.

We treat both linear and nonlinear models, although the book allows nonlinear systems to be deemphasized if desired. Students should realize that inherent nonlinearities generally cannot be ignored in the formulation of an accurate model. Techniques are introduced for approximating nonlinear systems by linear models, as well as for obtaining computer solutions for nonlinear models.

We believe that obtaining and interpreting computer solutions of both linear and nonlinear models constitute an important part of any course on dynamic systems. We introduce, early on, MATLAB and Simulink, two computer packages that are widely used in both educational and industrial settings. Although it is possible to use the book without detailed computer work, the inclusion of such methods enhances the understanding of important concepts, permits more interesting examples, allows the early use of computer projects, and prepares the students for real-life work.

An important feature is providing motivation and guidance for the reader. Each chapter except Chapter 1 has an introduction and summary. There are approximately 200 examples to reinforce new concepts as soon as they are introduced. Before the examples, there are explicit statements about the points to be illustrated. Where appropriate, comments about the significance of the results follow the examples. There are over 400 end-of-chapter problems. The answers to selected problems are contained in Appendix G.

Organization

The majority of the material can be covered in a one semester course, but the book can also be used as the basis for a two quarter or year long course. A number of chapters (including Chapters 10, 11, 12, 14, and 15) and a number of individual sections (including Sections 3.3, 6.7, 8.6, and 9.4) can be omitted or abbreviated without any loss of continuity. The chapters can be grouped into the following four blocks:

1. **Modeling of mechanical and electrical systems:** Chapters 1 through 6. The sections of Chapter 4 on Simulink and MATLAB can be deferred. However, we believe it is beneficial to encourage students to prepare simulations and to note typical features of the responses even before complete analytical explanations are presented.

2. **Analytical solutions for linear systems and linearizing nonlinear models:** Chapters 7 through 9. The Laplace transform is used as the principal tool for linear systems, although the classical approach is also summarized in Appendix D. Chapter 9 presents procedures for linearizing nonlinear models and compares some of the results with computer solutions for the original models.

3. **Modeling other types of systems:** Chapters 10 through 12.

4. **Block diagrams, feedback systems, and design tools:** Chapters 13 through 15.

A core sequence for students without a previous course in dynamic systems might include Chapters 1 through 9 plus Chapter 13. Readers with prior modeling experience might find parts of the first six chapters to be review, although they should be sure to understand the state-variable and input-output models in Chapter 3. As an illustration of how to extend the modeling and analysis techniques to other types of systems, we suggest including one or more of Chapters 10, 11, and 12.

Chapters 13 through 15 can be used as an introduction to the modeling, analysis, and design of feedback control systems. In Chapter 14, we present important features of MATLAB's Control System Toolbox, including root-locus and Bode diagrams. Rather than provide lengthy descriptions of how to draw such diagrams by hand, we emphasize interpreting computer-generated plots and using them in an iterative process to meet some practical design criteria.

The following table of prerequisites should be helpful in selecting material from the later chapters:

Topic	Chapter	Prerequisite material
Linearization	9	Chapter 5[1]
Electromechanical systems	10	Chapters 5–8
Thermal systems	11	Chapters 8, 9
Fluid systems	12	Chapters 8, 9
Block diagrams	13	Chapter 8
Feedback systems	14	Chapter 13

[1] Section 9.2 also requires Chapter 6.

New to This Edition

1. Computer methods are introduced in a completely new Chapter 4 and used throughout the rest of the book. Among the many advantages of doing this is the instructor's ability to add features to an earlier computer problem. For example, the simulation in Chapter 4 of a jumper with a parachute is extended in Chapter 9.

2. The Laplace transform is used earlier for the analytical solution of linear models, rather than being preceded by a lengthy discussion of classical time-domain techniques. There is a brief comparison of different analytical techniques, and the classical approach is summarized in Appendix D. We illustrate major concepts (such as transient, steady-state, zero-input, and zero-state responses; step, impulse, and ramp responses; time constants, damping ratios, and undamped natural frequencies; complex-plane plots; and mode functions and frequency-response functions) in the context of transform solutions. There is also a heightened emphasis on the central role of the transfer function.

3. The material leading to the analysis and design of feedback systems has been reorganized around the use of computer tools. We point out where design choices are required and how to use these tools to help make such choices.

4. We have removed material that instructors did not often use and have streamlined some of the other treatments. An example is the elimination of the detailed pencil-and-paper solutions of matrix equations. At the same time, we have added material, such as Appendix C on complex algebra, where some students seem weak.

Solutions Manual and Web Site

A solutions manual containing detailed solutions to all the problems in the text can be obtained by contacting John Wiley & Sons. At the beginning of each chapter in the manual, a table lists the following information for each of the major topics: the section of the text in which it is covered, the examples in which it is illustrated, and the problems that relate to it. Additional information related to the examples and problems can be found by visiting http://www.wiley.com/college/close.

Acknowledgments

We are indebted to our colleagues at Rensselaer who provided support and helpful comments, including faculty members from a number of departments who have taught dynamic systems. We especially thank Kim Stelson of the University of Minnesota for participating in discussions with us and for his useful suggestions. We benefited from the feedback of 50 faculty members across the country who responded to a survey. We would also like to thank those who reviewed the revision plan, and parts or all of the new edition. They include E. Harry Law, Clemson University; Raghu Echempati, Kettering University; Muhammad Ali Rob Sharif, University of Alabama; Subhash C. Sinha, Auburn University; and Horn S. Tzou, University of Kentucky.

Harriet Borton of Rensselaer's Information Technology Services gave valuable advice about the use of LaTeX. It was a pleasure to work with Joseph Hayton and Norine Pigliucci at John Wiley & Sons and with Jan Fisher at Publication Services in the planning and production of the book. Finally and most importantly, we sincerely appreciate the patience, support, and encouragement of our wives throughout this project.

CMC
DKF
JCN

CONTENTS

CHAPTER

1

INTRODUCTION

In this chapter we present the rationale for the book, define several terms that will be used throughout, and describe various types of systems. The chapter concludes with a description of the particular types of systems to be considered and a summary of the techniques that the reader should be able to apply after finishing the book.

▶ 1.1 RATIONALE

The importance of understanding and being able to determine the dynamic response of physical systems has long been recognized. It has been traditional in engineering education to have separate courses in dynamic mechanical systems, circuit theory, chemical-process dynamics, and other areas. Such courses develop techniques of modeling, analysis, and design for the particular physical systems that are relevant to that specific discipline, even though many of the techniques taught in these courses have much in common. This approach tends to reinforce the student's view of such courses as isolated entities with little in common, and to foster a reluctance to apply what has been learned in one course to a new situation.

Another justification for considering a wide variety of different types of systems in an introductory book is that the majority of systems that are of practical interest contain components of more than one type. In the design of electronic circuits, for example, attention must also be paid to mechanical structure and to dissipation of the heat generated. Hydraulic motors and pneumatic process controllers are other examples of useful combinations of different types of elements. Furthermore, the techniques in this book can be applied not only to pneumatic, acoustical, and other traditional areas but also to systems that are quite different, such as sociological, physiological, economic, and transportation systems.

Because of the universal need for engineers to understand dynamic systems and because there is a common methodology applicable to such systems regardless of their physical origin, it makes sense to present them all together. This book considers both the problem of obtaining a mathematical description of a physical system and the various analysis techniques that are widely used.

▶ 1.2 ANALYSIS OF DYNAMIC SYSTEMS

Because the most frequently used key word in the text is likely to be *system*, it is appropriate to define it at the outset. A **system** is any collection of interacting elements for which there are cause-and-effect relationships among the variables. This definition is necessarily general, because it must encompass a broad range of systems. The most important feature of the definition is that it tells us we must take interactions among the variables into account in system modeling and analysis, rather than treating individual elements separately.

Our study will be devoted to **dynamic systems**, for which the variables are time-dependent. In nearly all our examples, not only will the excitations and responses vary with time, but at any instant the derivatives of one or more variables will depend on the values of the system variables at that instant. The system's response will normally depend on initial conditions, such as stored energy, in addition to any external excitations.

In the process of analyzing a system, two tasks must be performed: modeling the system and solving for the model's response. The combination of these steps is referred to as **system analysis**.

Modeling the System

A **mathematical model**, or **model** for short, is a description of a system in terms of equations. The basis for constructing a model of a system are the physical laws (such as the conservation of energy and Newton's laws) that the system elements and their interconnections are known to obey.

The type of model sought will depend on both the objective of the engineer and the tools for analysis. If a pencil-and-paper analysis with parameters expressed in literal rather than numerical form is to be performed, a relatively simple model will be needed. To achieve this simplicity, the engineer should be prepared to neglect elements that do not play a dominant role in the system.

On the other hand, if a computer is available for carrying out simulations of specific cases with parameters expressed in numerical form, a comprehensive mathematical model that includes descriptions of both primary and secondary effects might be appropriate. In short, a variety of mathematical models are possible for a system, and the engineer must be prepared to decide what form and complexity are most consistent with the objectives and the available resources.

One example of a dynamic system that is familiar to everyone is the automobile. In order to limit the complexity of any model we wish to make of this system, we must omit some of the system's features. And, in fact, many of the parameters may be relatively unimportant for the objective of a particular study. Among many possible concerns are ease of handling on the straightaway or while turning a corner; comfort of the driver; fuel efficiency; stopping ability; crash resistance; and the effects of wind gusts, potholes, and other obstacles.

Suppose that we limit our concern to the forces on the driver when the vehicle is traveling over a rough road. Some of the key characteristics of the system are represented in Figure 1.1(a) by masses, springs, and shock absorbers. The chassis has by far the largest mass, but other masses that may be significant are the front axles, rear axles, wheels, and driver. Suspension systems between the chassis and the axles are designed to minimize the vertical motion of the chassis when the tires undergo a sudden change in motion because of the road surface. The tires themselves have some elasticity, which is represented in Figure 1.1(a) by additional springs between the wheels and the road. The driver is somewhat cushioned from the chassis motion by the characteristics of the seat, and there is also some friction between the driver and the seat back.

We assume that the vehicle is traveling at a constant speed and that the horizontal motion of the chassis does not concern us. We must certainly allow for the vertical motion caused by the uneven road surface. We may also wish to consider the pitching effect when the front tires hit a bump or depression, causing the front of the chassis to move up or down before the rear. This would require us to consider not only the vertical motion of the chassis but also rotation about its center of mass.

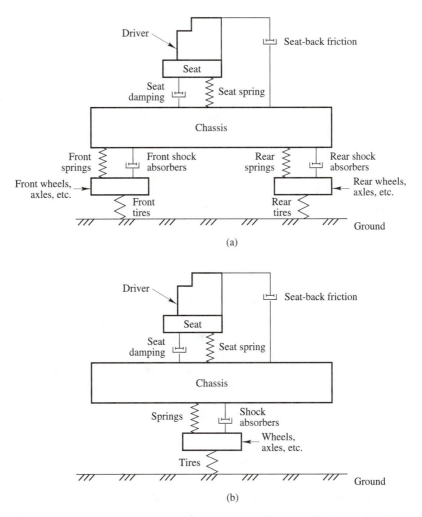

Figure 1.1 (a) One representation of an automobile. (b) A simplified representation.

This figure is adapted from a drawing in Chapter 42 of *The Shock and Vibration Handbook*, third edition (1988), edited by Cyril M. Harris. It is used with the permission of the publisher, McGraw-Hill, Inc. Part (a) of the figure also appears in the fourth edition (1996) of that book.

The complexity of a system model is sometimes measured by the number of independent energy-storing elements. For Figure 1.1(a), energy can be stored in four different masses and in five different springs. If we ignore the pitching effect, we might simplify the analysis by combining the front and rear axles into a single mass, as shown in part (b) of the figure, which has only three masses and three springs.

In the initial phase of the analysis, we might make other simplifying assumptions. Perhaps some of the elements remaining in Figure 1.1(b) could be omitted. Perhaps we would use a mathematical description of the individual elements that is simpler than that required for the final analysis.

On the other hand, for a more thorough study of the effect of a bumpy road on the driver, it might be necessary to add other characteristics to those represented in Figure 1.1(a). When one of the two wheels on the front axle encounters a bump or depression, the displacement and forces on it are different from those on its mate. Thus we might want to consider each of the four wheels as a separate mass and to allow for side-to-side rotation of the chassis, in addition to the vertical and pitching motions.

When devising models at various stages in the design process, engineers usually give considerable thought to how detailed the representation of the system's characteristics should be. Many of the remarks we have made about the automobile can be extended to airplanes, boats, rockets, motorcycles, and other vehicles. In the next few chapters, we shall show how to describe the important characteristics by sets of equations.

Solving the Model

The process of using the mathematical model to determine certain features of the system's cause-and-effect relationships is referred to as **solving the model**. For example, the responses to specific excitations may be desired for a range of parameter values, as guides in selecting design values for those parameters. As described in the discussion of modeling, this phase may include the analytical solution of simple models and the computer solution of more complex ones.

The type of equation involved in the model has a strong influence on the extent to which analytical methods can be used. For example, nonlinear differential equations can seldom be solved in closed form, and the solution of partial differential equations is far more laborious than that of ordinary differential equations. Computers can be used to generate the responses to specific numerical cases for complex models. However, using a computer to solve a complex model has its limitations. Models used for computer studies should be chosen with the approximations encountered in numerical integration in mind, and should be relatively insensitive to system parameters whose values are uncertain or subject to change. Furthermore, it may be difficult to generalize results based only on computer solutions that must be run for *specific* parameter values, excitations, and initial conditions.

The engineer must not forget that the model being analyzed is only an approximate mathematical description of the system, not the physical system itself. Conclusions based on equations that required a variety of assumptions and simplifications in their development may or may not apply to the actual system. Unfortunately, the more faithful a model is in describing the actual system, the more difficult it is to obtain general results.

One procedure is to use a simple model for analytical results and design, and then to use a different model to verify the design by means of computer simulation. In very complex systems, it may be feasible to incorporate actual hardware components into the simulation as they become available, thereby eliminating the corresponding parts of the mathematical model.

▶ 1.3 CLASSIFICATION OF VARIABLES

A system is often represented by a box (traditionally called a black box), as shown in Figure 1.2. The system may have several **inputs**, or **excitations**, each of which is a function

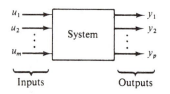

Figure 1.2 Black-box representation of a system.

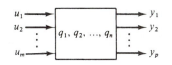

Figure 1.3 General system representation showing inputs, state variables, and outputs.

of time. Typical inputs include a force applied to a mass, a voltage source applied to an electrical circuit, and a heat source applied to a vessel filled with a liquid. For the model of an automobile in Figure 1.1(a), the inputs might be the vertical displacements of the bottoms of the springs representing the tires, as the tires move over the bumps in the road. In general discussions that are not related to specific systems, we shall use the symbols $u_1(t), u_2(t), \ldots, u_m(t)$ to denote the m inputs, shown by the arrows directed into the box.

Outputs are variables that are to be calculated or measured. Typical outputs include the velocity of a mass, the voltage across a resistor, and the rate at which a liquid flows through a pipe. For the model in Figure 1.1(a), one of the outputs might be the vertical acceleration of the driver. The p outputs are represented in Figure 1.2 by the arrows pointing away from the box representing the system. They are denoted by the symbols $y_1(t), y_2(t), \ldots, y_p(t)$. There is a cause-and-effect relationship between the outputs and inputs. To calculate any one of the outputs for all $t \geq t_0$, we must know the inputs for $t \geq t_0$ and also the accumulated effect of any previous inputs. One approach to constructing a mathematical model is to find equations that relate the outputs directly to the inputs by eliminating all the other variables that are internal to the system. If we are interested only in the input-output relationships, eliminating extraneous variables may seem appealing. However, by deleting information from the model, we may lose potentially important aspects of the system's behavior.

Another modeling technique is to introduce a set of **state variables**, which generally differs from the set of outputs but may include one or more of them. The state variables must be chosen so that a knowledge of their values at any reference time t_0 and a knowledge of the inputs for all $t \geq t_0$ is sufficient to determine the outputs and state variables for all $t \geq t_0$. An additional requirement is that the state variables be independent; that is, it must not be possible to express one state variable as an algebraic function of the others. This approach is particularly convenient for working with multi-input, multi-output systems and for obtaining computer solutions. In Figure 1.3, the representation of the system has been modified to include the state variables denoted by the symbols $q_1(t), q_2(t), \ldots, q_n(t)$ within the box. The state variables can account for all important aspects of the system's internal behavior, regardless of the choice of output variables. Equations for the outputs are then written as algebraic functions of the state variables, the inputs, and time.

Whenever it is appropriate to indicate units for the variables and parameters, we shall use the International System of Units (abbreviated SI, from the French *Système International d'Unités*). A list of the units used in this book appears in Appendix A.

▶ 1.4 CLASSIFICATION OF SYSTEMS

Systems are grouped according to the types of equations that are used in their mathematical models. Examples include partial differential equations with time-varying coefficients, ordinary differential equations with constant coefficients, and difference equations. In this

TABLE 1.1   Criteria for Classifying Systems

Criterion	Classification
Spatial characteristics	Lumped
	Distributed
Continuity of the time variable	Continuous
	Discrete-time
	Hybrid
Quantization of the dependent variable	Nonquantized
	Quantized
Parameter variation	Fixed
	Time-varying
Superposition property	Linear
	Nonlinear

section we define and briefly discuss ways of classifying the models, and in the next section we indicate those categories that will be treated in this book. The classifications that we use are listed in Table 1.1.

Spatial Characteristics

A **distributed system** does not have a finite number of points at which state variables can be defined. In contrast, a **lumped system** can be described by a finite number of state variables.

To illustrate these two types of systems, consider the flexible shaft shown in Figure 1.4(a), with one end embedded in a wall and with a torque applied to the other end. The angle through which a point on the surface of the shaft is twisted depends on both its distance from the wall and the applied torque. Hence the shaft is inherently distributed and would be modeled by a partial differential equation. However, if we are interested only in the angle of twist at the right end of the shaft, we may account for the flexibility of the shaft by a rotational spring constant K, and represent the effect of the distributed mass by the single moment of inertia J. Making these approximations results in the lumped system shown in Figure 1.4(b), which has the important property that its model is an ordinary differential equation. Because ordinary differential equations are far easier to solve than partial differential equations, converting from a distributed system to a lumped approximation is often essential if the resulting model is to be solved with the resources available.

Another example of a distributed system is an inductor that consists of a wire wound around a core, as shown in Figure 1.5(a). If an electrical excitation is applied across the terminals of the coil, then different values of voltage exist at all points along the coil, char-

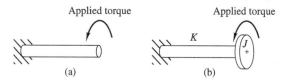

Figure 1.4   (a) A torsional shaft. (b) Its lumped approximation.

Figure 1.5 (a) An inductor. (b) Its lumped approximation.

acteristic of a distributed system. To develop a lumped circuit whose behavior as calculated at the terminals closely approximates that of the distributed device, we might account for the resistance of the wire by a lumped resistance R, and for the inductive effect related to the magnetic field by a single inductance L. The resulting lumped circuit is shown in Figure 1.5(b). Note that in these two examples (though not in all cases), the two elements in the lumped model do not correspond to separate physical parts of the actual system. The stiffness and moment of inertia of the flexible shaft cannot be separated into two physical pieces, nor can the resistance and inductance of the coil.

Continuity of the Time Variable

A second basis for classifying dynamic systems is the independent variable time. A **continuous system** is one for which the inputs, state variables, and outputs are defined over some continuous range of time (although the signals may have discontinuities in their waveshapes and not be continuous functions in the mathematical sense). A **discrete-time system** has variables that are determined at distinct instants of time, and that are either not defined or not of interest between those instants. Continuous systems are described by differential equations, discrete-time systems are described by difference equations.

Examples of the variables associated with continuous and discrete-time systems are shown in Figure 1.6. In fact, the discrete-time variable $f_2(kT)$ shown in Figure 1.6(b) is the sequence of numbers obtained by taking the values of the continuous variable $f_1(t)$ at instants separated by T units of time. Hence $f_2(kT) = f_1(t)|_{t=kT}$, where k takes on integer values. In practice, a discrete-time variable may be composed of pulses of very short duration (much less than T) or numbers that reside in digital circuitry. In either case, the variable is assumed to be represented by a sequence of numbers, as indicated by the dots in Figure 1.6(b). There is no requirement that their spacing with respect to time be uniform, although this is often the case.

A system that contains both discrete-time and continuous subsystems is referred to as a **hybrid system**. Many modern control and communication systems contain a digital computer as a subsystem. In such cases, those variables that are associated with the computer

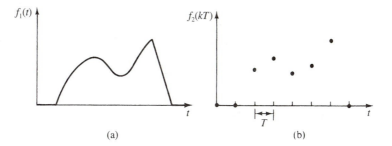

Figure 1.6 Sample variables. (a) Continuous. (b) Discrete-time.

are discrete in time, whereas variables elsewhere in the system are continuous. In such systems, sampling equipment is used to form discrete-time versions of continuous variables, and signal reconstruction equipment is used to generate continuous variables from discrete-time variables.

Quantization of the Dependent Variable

In addition to possible restrictions on the values of the independent variable time, the system variables may be restricted to certain distinct values. If within some finite range a variable may take on only a finite number of different values, it is said to be **quantized**. A variable that may have any value within some continuous range is **nonquantized**. Quantized variables may arise naturally, or they may be created by rounding or truncating the values of a nonquantized variable to the nearest quantization level.

The variables shown in Figure 1.7(a) and Figure 1.7(b) are both nonquantized, whereas those in the remaining parts of the figure are quantized. Although the variable in Figure 1.7(b) is restricted to the interval $-1 \leq f_b \leq 1$, it is nonquantized because it can take on a continuous set of values within that interval. The variable f_c shown in Figure 1.7(c)

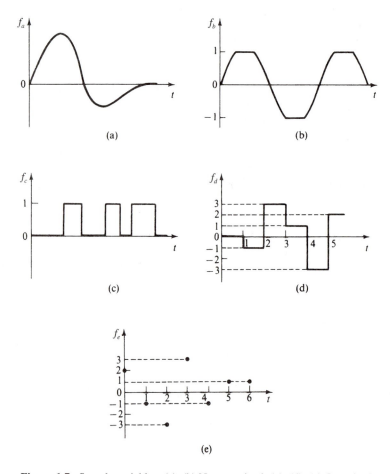

Figure 1.7 Sample variables. (a), (b) Nonquantized. (c), (d), (e) Quantized.

is restricted to the two values 0 and 1 and is representative of the signals found in devices that perform logical operations. The variable f_d is restricted to integer values and thus is quantized, although there need not be any other restriction on the magnitude of its values. The variable f_e is a discrete-time variable that is also quantized.

Because variables that are both discrete in time and quantized in amplitude (such as f_e) occur within a digital computer, they are referred to as **digital variables**. In contrast, the variable f_a in Figure 1.7(a) is both continuous and nonquantized and is representative of a signal within an analog computer. Hence continuous, nonquantized variables are often referred to as **analog variables**.

Parameter Variation

Systems may be classified according to properties of their parameters as well as of their variables. **Time-varying systems** are systems whose characteristics (such as the value of a mass or a resistance) change with time. Element values may change because of environmental factors such as temperature and radiation. Other examples of time-varying elements include the mass of a rocket, which decreases as fuel is burned, and the inductance of a coil, which increases as an iron slug is inserted into the core. In the differential equations describing time-varying systems, some of the coefficients are functions of time. Delaying the input to a time-varying system affects the size and shape of the response.

For **fixed** or **time-invariant systems**, whose characteristics do not change with time, the system model that describes the relationships between the inputs, state variables, and outputs is independent of time. If such a system is initially at rest, delaying the input by t_d units of time just delays the output by t_d units, without any change in its size or waveshape.

Superposition Property

A system can also be classified in terms of whether it obeys the **superposition property**, which requires that the following two tests be satisfied when the system is initially at rest with zero energy. (1) Multiplying the inputs by any constant α must multiply the outputs by α. (2) The response to several inputs applied simultaneously must be the sum of the individual responses to each input applied separately. **Linear systems** are those that satisfy the conditions for superposition; **nonlinear systems** are those for which superposition does not hold. For a linear system, the coefficients in the differential equations that make up the system model do not depend on the size of the excitation, whereas for nonlinear systems, at least some of the coefficients do. For a linear system initially at rest, multiplying all the inputs by a constant multiplies the output by the same constant. Likewise, replacing all the original inputs by their derivatives (or integrals) gives outputs that are the derivatives (or integrals) of the original outputs.

Nearly all systems are inherently nonlinear if no restrictions at all are placed on the allowable values of the inputs. If the values of the inputs are confined to a sufficiently small range, the original nonlinear model of a system may often be replaced by a linear model whose response closely approximates that of the nonlinear model. This type of approximation is desirable because analytical solutions to linear models are more easily obtained.

In other applications, the nonlinear nature of an element may be an essential feature of the system and should not be avoided in the model. Examples include mechanical valves or electrical diodes designed to give completely different types of responses for positive and negative inputs. Devices used to produce constant-amplitude oscillations generally have the amplitude of the response determined by nonlinear elements in the system.

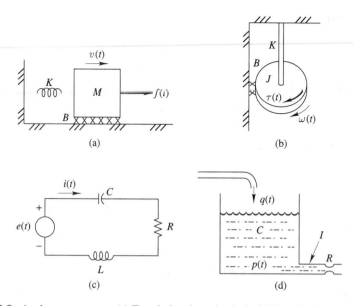

Figure 1.8 Analogous systems. (a) Translational mechanical. (b) Rotational mechanical. (c) Electrical. (d) Hydraulic.

To illustrate the difference between linear and nonlinear models, consider a system with the single input $u(t)$ and the single output $y(t)$. If the input and output are related by the differential equation

$$a_1 \frac{dy}{dt} + a_0 y(t) = b_0 u(t)$$

where a_0, a_1, and b_0 may be functions of time but do not depend on $u(t)$ or $y(t)$ in any way, then the system is linear. However, if one or more of the coefficients is a function of the input or output, as in

$$\frac{dy}{dt} + u(t)y(t) = u(t)$$

or

$$\frac{dy}{dt} + |y(t)|y(t) = u(t)$$

then the system is nonlinear.

Analogous Systems

Different systems that are described by equations that are identical except for the use of different symbols are called **analogs**. Consider the four simple systems depicted in Figure 1.8. In later chapters, we shall examine mechanical, electrical, and hydraulic systems in detail. For the systems in this figure, we shall give at this time only a very brief explanation of the variables, the symbols used, and the equations describing the systems.

Each type of system has two basic variables, both of which are functions of time t. For the translational mechanical system, the variables used in the figure are force $f(t)$ and velocity $v(t)$; for the rotational mechanical system, torque $\tau(t)$ and angular velocity $\omega(t)$; for the electrical system, voltage $e(t)$ and current $i(t)$; and for the hydraulic system, flow rate $q(t)$ and pressure difference $p(t)$. Instead of velocity, angular velocity, and current, the alternative variables of displacement, angular displacement, and charge could be used just as well. For simplicity, we assume for Figure 1.8 that all the elements are linear and that no energy is stored within the system before the input is applied.

In the translational system shown in part (a) of Figure 1.8, an external force $f(t)$ is applied to a mass M, whose motion is restrained by a spring K and a friction element B. For the rotational system in part (b), a torque $\tau(t)$ is exerted on a disk whose moment of inertia is J, which is restrained by the torsional bar K and the friction B. The electrical system in part (c) consists of an inductor L, a resistor R, and a capacitor C excited by a voltage source $e(t)$. In part (d), fluid with a known flow rate $q(t)$ enters a vessel that has hydraulic capacitance C. The orifice in the outlet pipe is approximated by the hydraulic resistance R. The inertia effect of the fluid mass can usually be neglected, but in order to complete the analogy, we represent it here by the inertance I.

The following equations, which describe the four systems in Figure 1.8, are identical except for the symbols used. The inputs are on the right sides of the equations, the output variables on the left.

$$M\frac{dv}{dt} + Bv(t) + K\int_0^t v(\lambda)d\lambda = f(t)$$

$$J\frac{d\omega}{dt} + B\omega(t) + K\int_0^t \omega(\lambda)d\lambda = \tau(t)$$

$$L\frac{di}{dt} + Ri(t) + \frac{1}{C}\int_0^t i(\lambda)d\lambda = e(t)$$

$$C\frac{dp}{dt} + \frac{1}{R}p(t) + \frac{1}{I}\int_0^t p(\lambda)d\lambda = q(t)$$

If the inputs are identical, then the respective responses will have the same form. The expressions for the power supplied by each of the inputs in the four parts of the figure are $f(t)v(t), \tau(t)\omega(t), e(t)i(t)$, and $p(t)q(t)$, respectively.

Other analogous components could have been included in our discussion. Levers, gears, transformers, and double-headed pistons constitute one set of analogous components. Brakes, diodes, and overflow valves are another. Our treatment of analogies could also be extended to other types of systems, such as thermal, pneumatic, and acoustical systems. We shall not overemphasize analogies, because it is generally better to treat each type of system separately. However, it is important to realize that the modeling and analysis tools that we shall develop are applicable to a very wide range of physical systems.

▶ 1.5 COMPUTER TOOLS

Modern systems analysis makes use of digital computers to facilitate the modeling of complex systems and to solve the resulting modeling equations. Software packages are available that provide a user-friendly interface to powerful numerical techniques for simulating arbitrary input functions, solving differential equations, and plotting and analyzing the results. This allows the engineer to develop models of complex dynamic systems and to study their responses. The effects of selecting different component values can be easily seen during the design process. In addition, powerful tools are available to design controllers applicable to the systems being studied.

This book uses a popular software package consisting of MATLAB and Simulink.[1] This is a package for technical computation, data analysis, and visualization widely used in education and industry. MATLAB supports the modeling, simulation, and analysis of linear and nonlinear dynamic systems. Simulink provides a graphical user interface (GUI) for building models as block diagrams, using click-and-drag mouse operations. It contains an extensive block library of sinks, sources, linear and nonlinear components, and connectors. Other software packages are available to accomplish these tasks, and the reader who is familiar with them should be able to translate the examples shown here into those packages.

We introduce these computer tools in Chapter 4 with a discussion of block-diagram representations of system models. This leads directly to the Simulink diagram, and its use in simulating the responses to desired inputs and initial conditions. As different types of systems are introduced in Chapters 5, 6, 10, 11, and 12, examples are solved using Simulink. In Chapter 9, computer techniques are used to simulate nonlinear systems and to compare the results with linearized models.

We return in Chapter 13 to more advanced topics in block diagram and simulation techniques. Chapters 14 and 15 use MATLAB and Simulink for more advanced work in analysis and control system design, and introduce the Control System Toolbox. This toolbox consists of a collection of functions for modeling, analyzing, and designing automatic control systems written in MATLAB.

MATLAB is available on many different computer platforms, including Windows, Unix, and Macintosh.[2] Full professional versions usually have many toolboxes and can handle very large models. A Student Version of MATLAB and Simulink is available for students at degree-granting institutions. This version is limited to 300 Simulink blocks and does not contain the Control Systems Toolbox, although that is available at added cost. All the material in the first 13 chapters of this book is supported by the Student Version of MATLAB and Simulink. Some of the functions used in Chapters 14 and 15 require the Control System Toolbox.

▶ 1.6 SCOPE AND OBJECTIVES

This book is restricted to lumped, continuous, nonquantized systems that can be described by sets of ordinary differential equations. Because well-developed analytical techniques are available for solving linear ordinary differential equations with constant coefficients, we shall emphasize such techniques. The majority of our examples will involve systems that are both fixed and linear. A method for approximating a nonlinear system by a fixed linear model will be developed. For time-varying or nonlinear systems that cannot be approximated by a fixed linear model, one can resort to computer solutions.

We list as **objectives** the following things that the reader should be able to do after finishing this book. These objectives are grouped in the two general categories of modeling and solving for the response.

After finishing this book, the reader should be able to do the following for dynamic systems composed of mechanical, electrical, thermal, and hydraulic components:

1. Given a description of the system, construct a simplified version using idealized elements and define a suitable set of variables.

2. Use the appropriate element and interconnection laws to obtain a mathematical model generally consisting of ordinary differential equations.

---

[1] MATLAB and Simulink are registered trademarks of The MathWorks, Inc.

[2] Windows is a registered trademark of Microsoft Corporation. Unix is a registered trademark licensed exclusively through X/Open Company, Limited. Macintosh is a registered trademark of Apple Computer, Inc.

3. If the model is nonlinear, determine the equilibrium conditions and, where appropriate, obtain a linearized model in terms of incremental variables.

4. Arrange the equations that make up the model in a form suitable for solution, and use them to construct and simplify block diagrams.

When a linear mathematical model has been determined or is given, the reader should be able to do the following:

1. For a first-order system, solve directly for the time-domain response without transforming the functions of time into functions of other variables.

2. For a model of up to fourth order, use the Laplace transform to
   (a) Find the complete time response.
   (b) Determine the transfer function and its poles and zeros.
   (c) Analyze stability and, where appropriate, evaluate time constants, damping ratios, and undamped natural frequencies.

3. Find from the transformed expression the steady-state response to a constant or sinusoidal input without requiring a general solution.

4. Use a suitable software package to obtain the response of a system to initial stored energy and to arbitrary inputs, typically including steps, impulses, and sinusoids.

5. Use computer tools to study the influence of changing system parameters on the system response, describing and, where possible, predicting the response.

6. Use block diagrams, root-locus plots, and Bode diagrams as aids in analyzing and designing feedback systems.

This book investigates all the foregoing procedures in detail. We illustrate the basic modeling approaches and introduce state variables first in the context of mechanical systems and later for electrical, electromechanical, fluid, and thermal systems. The numerical solution of modeling equations is introduced in Chapter 4, where block diagram representations are used to form Simulink diagrams, which are solved to give the dynamic responses.

There are two chapters dealing with the analytical solution of mathematical models. In Chapter 7 we introduce the solution of ordinary differential equations using the Laplace transform. The Laplace transform provides the basis for transfer functions, upon which a variety of methods for the analysis and design of linear systems are based.

Chapter 9 gives procedures for approximating a nonlinear model by a linear one. The usefulness and versatility of the MATLAB and Simulink tools are demonstrated with examples of several nonlinear systems in this chapter. Chapters 10 through 12 extend the procedures of the earlier chapters to electromechanical, thermal, and hydraulic systems.

In Chapter 13, we extend the treatment of block diagrams and numerical solutions to several new topics, and provide additional theoretical background beyond that covered in Chapter 4. In Chapter 14, a number of computer tools useful for analyzing and designing feedback control systems are introduced. The final chapter discusses the issue of stability covered in Chapter 8, and design guidelines for feedback systems, before concluding with several examples of control system design.

A total of seven appendices provide supporting material, including selected readings and answers to a number of the end-of-chapter problems. A solutions manual is available to instructors, upon request from the publisher. Users of this book may wish to look on the publisher's Web site for additional information, such as computer assignments and examples that have been developed for use at Rensselaer.

CHAPTER

2

TRANSLATIONAL MECHANICAL SYSTEMS

The modeling techniques for translational mechanical systems are discussed in this and the following chapter. Rotational systems are treated in Chapter 5. Procedures for solving the mathematical models are developed in later chapters.

After introducing the variables to be used, we discuss the laws for the individual elements, the laws governing the interconnections of the elements, and the use of free-body diagrams as an aid in formulating the equations of the model. Inputs will consist of either the application of a known external force or the moving of some body with a known displacement.

The first examples will be systems that can have only horizontal or only vertical motion. For masses that can move vertically, the gravitational forces must be considered. If the system contains an ideal pulley, then some parts can move horizontally and other parts vertically. Special situations, such as free-body diagrams for massless junctions, and rules for the series or parallel combination of similar elements, will be treated later in the chapter.

▶ 2.1 VARIABLES

The symbols for the basic variables used to describe the dynamic behavior of translational mechanical systems are

> x, displacement in meters (m)
>
> v, velocity in meters per second (m /s)
>
> a, acceleration in meters per second per second (m /s^2)
>
> f, force in newtons (N)

All these variables are functions of time. In general, however, we shall add a t in parentheses immediately after the symbol only when it denotes an input or when we find doing so useful for clarity or emphasis.

Displacements are measured with respect to some reference condition, which is often the equilibrium position of the body or point in question. Velocities are normally expressed as the derivatives of the corresponding displacements. If the reference condition of a displacement is not indicated because it is not of interest, then the reference condition for the velocity needs to be given.

Two conventions used to define displacements are illustrated in Figure 2.1(a) and Figure 2.1(b). In Figure 2.1(a), the variable x represents the displacement of the left side of the body from the fixed vertical wall, whereas in Figure 2.1(b), the reference position corresponding to $x = 0$ is not specifically shown.

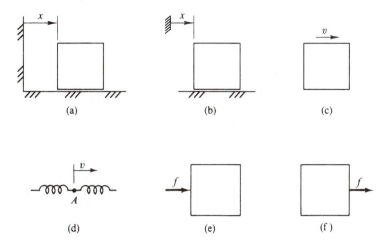

Figure 2.1 Conventions for designating variables.

Generally, the reference position will correspond to a condition of equilibrium for which the system inputs are constants and in which the net force on the body being considered is zero. Figure 2.1(c) and Figure 2.1(d) indicate two methods of defining a velocity. All points on the body in Figure 2.1(c) must move with the same velocity, so there is no possible ambiguity about which point has the velocity v. In Figure 2.1(d), the vertical line at the base of the arrow indicates that v is the velocity of the point labeled A. Forces can be represented by arrows pointing either into or away from a body, as depicted in Figure 2.1(e) and Figure 2.1(f), which are equivalent to one another.

Remember that the arrows only indicate an assumed positive sense for the displacement, velocity, or force being considered, and by themselves do not imply anything about the actual direction of the motion or of the force at a given instant. If, for example, in Figure 2.1(e) and Figure 2.1(f), the force acting on the body is $f(t) = \sin t$, the force acts to the right for $0 < t < \pi$ and to the left for $\pi < t < 2\pi$, and it continues to change direction every π seconds. Note that an alternative way of describing the identical situation is to draw the arrow pointing to the left and then write $f(t) = -\sin t$. Reversing a reference arrow is equivalent to reversing the sign of the algebraic expression associated with it. There is no unique way of choosing reference directions on a diagram, but the equations must be consistent with whatever choice is made for the arrows.

Reference arrows for the displacement, velocity, and acceleration of a given point are invariably drawn in the same direction so that the equations

$$v = \frac{dx}{dt}$$

and

$$a = \frac{dv}{dt} = \frac{d^2x}{dt^2}$$

can be used. With this understanding, a reference arrow for acceleration is not shown explicitly on the diagrams, and for the same reason only the reference arrow for either the displacement or the velocity of a point (but not both) is shown in many examples.

Variables in addition to those defined at the beginning of this section include

w, energy in joules (J)

p, power in watts (W)

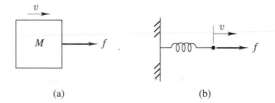

(a) (b)

Figure 2.2 Reference arrows for (1).

where 1 joule = 1 newton-meter and 1 watt = 1 joule per second. Because the arrows defining the positive senses of the velocity and the force point in the same direction, the power supplied to the mass in Figure 2.2(a) and to the spring in Figure 2.2(b) is

$$p = fv \tag{1}$$

Because power is defined to be the rate at which energy is supplied or dissipated, it follows that

$$p = \frac{dw}{dt} \tag{2}$$

and the energy supplied between time t_0 and t_1 is

$$\int_{t_0}^{t_1} p(t)\,dt$$

If $w(t_0)$ denotes the energy supplied up to time t_0, then the total energy supplied up to any later time t is

$$w(t) = w(t_0) + \int_{t_0}^{t} p(\lambda)\,d\lambda \tag{3}$$

In the last integrand, t has been replaced by the dummy variable λ in order to avoid confusion between the upper limit and the variable of integration.

▶ 2.2 ELEMENT LAWS

Physical devices are represented by one or more idealized elements that obey laws involving the variables associated with the elements. As we mentioned in Chapter 1, some degree of approximation is required in selecting the elements to represent a device, and the behavior of the combined elements may not correspond exactly to the behavior of the device. The elements that we include in translational systems are mass, friction, stiffness, and the lever. The element laws for the first three relate the external force to the acceleration, velocity, or displacement associated with the element. The lever is considered in Chapter 5.

Mass

Figure 2.2(a) shows a **mass** M, which has units of kilograms (kg), subjected to a force f. Newton's second law states that the sum of the forces acting on a body is equal to the time rate of change of the momentum:

$$\frac{d}{dt}(Mv) = f \tag{4}$$

which, for a constant mass, can be written as

$$M\frac{dv}{dt} = f \tag{5}$$

For (4) and (5) to hold, the momentum and acceleration must be measured with respect to an inertial reference frame. For ordinary systems at or near the surface of the earth, the earth's surface is a very close approximation to an inertial reference frame, so it is the one we use. The momentum, acceleration, and force are really vector quantities, but in this chapter the mass is constrained to move in a single direction, so we can write scalar equations.

We shall restrict our attention to constant masses and neglect relativistic effects so that we can use (5). Hence a mass can be modeled by an algebraic relationship between the acceleration dv/dt and the external force f. For (5) to hold, the positive senses of both dv/dt and f must be the same, because the force will cause the velocity to increase in the direction in which the force is acting.

Energy in a mass is stored as kinetic energy if the mass is in motion, and as potential energy if the mass has a vertical displacement relative to its reference position. The kinetic energy is

$$w_k = \tfrac{1}{2}Mv^2 \tag{6}$$

and the potential energy, assuming a uniform gravitational field, is

$$w_p = Mgh \tag{7}$$

where g is the gravitational constant (approximately 9.807 m/s^2 at the surface of the earth) and h is the height of the mass above its reference position. In order to determine the response for $t \geq t_0$ of a dynamic system containing a mass, we must know its initial velocity $v(t_0)$ and, if vertical motion is possible, its initial height $h(t_0)$.

Friction

Forces that are algebraic functions of the relative velocity between two bodies are modeled by friction elements. A mass sliding on an oil film that has laminar flow, as depicted in Figure 2.3(a), is subject to **viscous friction** and obeys the linear relationship

$$f = B\,\Delta v \tag{8}$$

where B has units of newton-seconds per meter (N·s/m) and where $\Delta v = v_2 - v_1$. The direction of a frictional force will be such as to oppose the motion of the mass. For (8) to apply to Figure 2.3(a), the force f exerted on the mass M by the oil film is to the left. (By Newton's third law, the mass exerts an equal force f to the right on the oil film.)

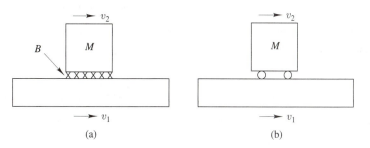

(a)   (b)

Figure 2.3 (a) Friction described by (8) with $\Delta v = v_2 - v_1$. (b) Adjacent bodies with negligible friction.

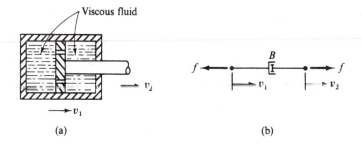

Figure 2.4 (a) A dashpot. (b) Its representation.

The friction coefficient B is proportional to the contact area and to the viscosity of the oil, and inversely proportional to the thickness of the film. A heavier mass would further compress the oil film, making it thinner and increasing the value of B.

Sometimes the frictional forces on adjacent bodies that have relative motion are small enough to be neglected. This might be the situation, for example, if the bodies are separated by bearings. The diagrams for such cases often show small wheels between the two bodies, as illustrated in Figure 2.3(b), in order to emphasize the lack of frictional forces.

Viscous friction also may be used to model a dashpot, such as the shock absorbers on an automobile. As indicated in Figure 2.4(a), a piston moves through an oil-filled cylinder, and there are small holes in the face of the piston through which the oil passes as the parts move relative to each other. The symbol often used for a dashpot is shown in Figure 2.4(b). Many dashpot devices involve high rates of fluid flow through the orifices and have nonlinear characteristics. If the flow is laminar, then the element is again described by (8). If the lower block in Figure 2.3(a) or the cylinder of the dashpot in Figure 2.4(a) is stationary, then $v_1 = 0$ and the element law reduces to $f = Bv_2$.

If the dashpot or oil film is assumed to be massless and if the accelerations are to remain finite, then when a force f is applied to one side, a retarding force of equal magnitude must be exerted on the other side (either by a wall or by some other component) as shown in Figure 2.4(b), again with $f = B(v_2 - v_1)$. This means that in the system shown in Figure 2.5, the force f is transmitted through the dashpot and exerted directly on the mass M.

The viscous friction described by (8) is a linear element, for which the plot of f versus Δv is a straight line passing through the origin, as shown in Figure 2.6(a). Examples of friction that obey nonlinear relationships are **dry friction** and **drag friction**. The former is modeled by a force that is independent of the magnitude of the relative velocity, as indicated in Figure 2.6(b), and that can be described by the equation

$$f = \begin{cases} -A & \text{for } \Delta v < 0 \\ A & \text{for } \Delta v > 0 \end{cases}$$

Drag friction is caused by resistance to a body moving through a fluid (such as wind resistance) and can often be described by an equation of the form $f = D|\Delta v|\Delta v$, as depicted in Figure 2.6(c). Various other nonlinearities may be encountered in friction elements.

The power dissipated by friction is the product of the force exerted and the relative velocity of the two ends of the element. This power is immediately converted to heat and

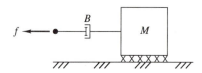

Figure 2.5 Force transmitted through a dashpot.

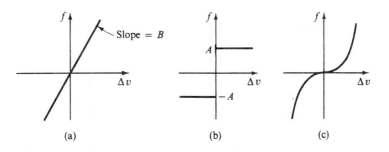

Figure 2.6 Friction characteristics. (a) Linear. (b) Dry. (c) Drag.

thus cannot be returned to the rest of the mechanical system at a later time. Accordingly, we do not usually need to know the initial velocities of the friction elements in order to solve the model of a system.

Stiffness

Any mechanical element that undergoes a change in shape when subjected to a force can be characterized by a **stiffness element**, provided only that an algebraic relationship exists between the elongation and the force. The most common stiffness element is the spring, although most mechanical elements undergo some deflection when stressed. For the spring sketched in Figure 2.7(a), we define d_0 to be the length of the spring when no force is applied and x to be the elongation caused by the force f. Then the total length at any instant is $d(t) = d_0 + x$, and the stiffness property refers to the algebraic relationship between x and f, as depicted in Figure 2.7(b). Because x has been defined as an elongation and the plot shows that f and x always have the same sign, it follows that the positive sense of f must be to the right in Figure 2.7(a); that is, f represents a tensile rather than a compressive force. For a linear spring, the curve in Figure 2.7(b) is a straight line and $f = Kx$, where K is a constant with units of newtons per meter (N/m).

Figure 2.7(c) shows a spring whose ends are displaced by the amounts x_1 and x_2 relative to their respective reference positions. If $x_1 = x_2 = 0$ corresponds to a condition when no force is applied to the spring, then the elongation at any instant is $x_2 - x_1$. For a linear spring,

$$f = K \Delta x \tag{9}$$

where $\Delta x = x_2 - x_1$. For small elongations of a structural shaft, K is proportional to the cross-sectional area and to Young's modulus and is inversely proportional to the length.

When a force f is applied to one side of a stiffness element that is assumed to have no mass, a force equal in magnitude but of opposite direction must be exerted on the other

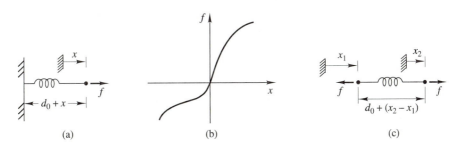

Figure 2.7 Characteristics of a spring.

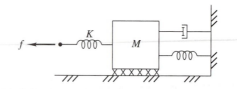

Figure 2.8 Force transmitted through a spring.

side. Thus for the system shown in Figure 2.8, the force f passes through the first spring and is exerted directly on the mass M. Of course, all physical devices have some mass, but to obtain a lumped model we assume either that it is negligible or that it is represented by a separate element.

Potential energy is stored in a spring that has been stretched or compressed, and for a linear spring that energy is given by

$$w_p = \tfrac{1}{2}K(\Delta x)^2 \tag{10}$$

This energy may be returned to the rest of the mechanical system at some time in the future. Therefore, the initial elongation $\Delta x(t_0)$ is one of the initial conditions we need in order to find the complete response of a system.

▶ 2.3 INTERCONNECTION LAWS

Having identified the individual elements in translational systems and having given equations describing their behavior, we next present the laws that describe the manner in which the elements are interconnected. These include D'Alembert's law, the law of reaction forces, and the law for displacement variables.

D'Alembert's Law

D'Alembert's law is just a restatement of Newton's second law governing the rate of change of momentum. For a constant mass, we can write

$$\sum_i (f_{\text{ext}})_i = M\frac{dv}{dt} \tag{11}$$

where the summation over the index i includes all the external forces $(f_{\text{ext}})_i$ acting on the body. The forces and velocity are in general vector quantities, but they can be treated as scalars provided that the motion is constrained to be in a fixed direction. Rewriting (11) as

$$\sum_i (f_{\text{ext}})_i - M\frac{dv}{dt} = 0 \tag{12}$$

suggests that the mass in question can be considered to be in equilibrium—that is, the sum of the forces is zero—provided that the term $-M\,dv/dt$ is thought of as an additional force. This fictitious force is called the **inertial force** or **D'Alembert force**, and including it along with the external forces allows us to write the force equation as one of equilibrium:

$$\sum_i f_i = 0 \tag{13}$$

This equation is known as **D'Alembert's law**. The minus sign associated with the inertial force in (12) indicates that when $dv/dt > 0$, the force acts in the negative direction.

Many readers are more used to writing modeling equations using Newton's second law in (11) rather than D'Alembert's law in (13). The two are completely equivalent to one another. Frequently, the Newtonian formulation is used in beginning courses, and the D'Alembert procedure in later courses. However, it usually takes only a few minutes to become comfortable with the latter method. Using D'Alembert's law, where all forces including the inertial force are shown on the modeling diagrams, people are less likely to inadvertently leave out a term or have a sign error. Furthermore, this procedure produces equations that have the same form as those for other types of systems, including electrical and electromechanical ones. This makes it easier to talk about analogies.

In addition to applying (13) to a mass, we can apply it to any point in the system, such as the junction between components. Because a junction is considered massless, the inertial force is zero in such a case.

The Law of Reaction Forces

In order to relate the forces exerted by the elements of friction and stiffness to the forces acting on a mass or junction point, we need Newton's third law regarding reaction forces. Accompanying any force of one element on another, there is a **reaction force** on the first element of equal magnitude and opposite direction.

In Figure 2.9(a), for example, let f_k denote the force exerted by the mass on the right end of the spring, with the positive sense defined to be to the right. Newton's third law tells us that there acts on the mass a reaction force f_k of equal magnitude with its positive sense to the left, as indicated in Figure 2.9(b). Likewise, at the left end, the fixed surface exerts a force f_k on the spring with the positive sense to the left, while the spring exerts an equal and opposite force on the surface.

The Law for Displacements

If the ends of two elements are connected, those ends are forced to move with the same displacement and velocity. For example, because the dashpot and spring in Figure 2.10(a) are both connected between the wall and the mass, the right ends of both elements have the same displacement x and move with the same velocity v. In Figure 2.10(b), where B_2 and K are connected between two moving masses, the elongation of both elements is $x_2 - x_1$. An equivalent statement is that if we go from M_1 to M_2 and record the elongation of the dashpot B_2, and then return to M_1 and subtract the elongation of the spring K, the result is zero. In effect, we are saying that the difference between the displacements of any two points is the same regardless of which elements we are examining between those points. This statement is really a consequence of our being able to uniquely define points in space.

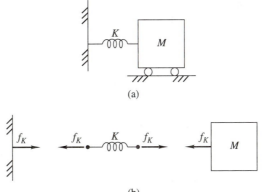

Figure 2.9   Example of reaction forces.

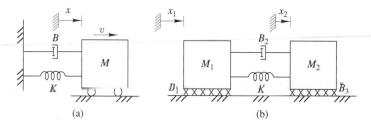

(a)   (b)

Figure 2.10 Two elements connected between the same endpoints. (a) One endpoint fixed. (b) Both endpoints movable.

The discussion in the previous paragraph can be summarized by saying that at any instant the algebraic sum of the elongations around any closed path is zero; that is

$$\sum_i (\Delta x)_i = 0 \ \text{ around any closed path} \tag{14}$$

It is understood that the left side of (14) is the algebraic sum of the elongations with signs that take into account the direction in which the path is being traversed. Furthermore, it is understood that for two elements connected between the same two points, such as B_2 and K in Figure 2.10(b), the elongations of both elements must be measured with respect to the same references. If for some reason the two elongations were measured with respect to different references, then the algebraic sum of the elongations around the closed path would be a constant, but not zero.

In Figure 2.11, let x_1 and x_2 denote displacements measured with respect to reference positions that correspond to a single equilibrium condition of the system. Then the respective elongations of B_1, B_2, and K are x_1, $x_2 - x_1$, and x_2. When the elongations are summed going from the fixed surface to the mass by way of the friction elements B_1 and B_2 and going back by way of the spring K, (14) gives

$$x_1 + (x_2 - x_1) - x_2 = 0$$

This equation can be regarded as a justification for the statement that the elongation of B_2 is $x_2 - x_1$ and that an additional symbol for this elongation is not needed.

In the analysis of mechanical systems, (14) is normally used implicitly and automatically in the process of labeling the system diagram. For example, we use the same symbol for two elongations that are forced to be equal by the element interconnections, and we avoid using different symbols for displacements that are known to be identical.

Differentiating (14) would lead to a similar equation in terms of relative velocities. However, we shall use only a single symbol for the velocities of two points that are constrained to move together. It will therefore not be necessary to invoke (14) in a formal way.

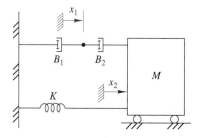

Figure 2.11 Illustration for the displacement law.

▶ 2.4 OBTAINING THE SYSTEM MODEL

The system model must incorporate both the element laws and the interconnection laws. The element laws involve displacements, velocities, and accelerations. Because the acceleration of a point is the derivative of the velocity, which in turn is the derivative of the displacement, we could write all the element laws in terms of x and its derivatives or in terms of x, v, and dv/dt. It is important to indicate the assumed positive directions for displacements, velocities, and accelerations. We shall always choose the assumed positive directions for a, v, and x to be the same, so it will not be necessary to indicate all three positive directions on the diagram. Throughout this book, dots over the variables are used to denote derivatives with respect to time. For example, $\dot{x} = dx/dt$ and $\ddot{y} = d^2y/dt^2$.

Free-Body Diagrams

We normally need to apply D'Alembert's law, given by (13), to each mass or junction point in the system that moves with a velocity that is unknown beforehand. To do so, it is useful to draw a free-body diagram for each such mass or point, showing all external forces and the inertial force by arrows that define their positive senses. The element laws are used to express all forces except inputs in terms of displacements, velocities, and accelerations. We must be sure that the signs of these expressions are consistent with the directions of the reference arrows. After the free-body diagram is completed, we can apply (13) by summing the forces indicated on the diagram, again taking into account their assumed positive senses. Normally all forces must be added as vectors, but in our examples the forces in the free-body diagram will be collinear and can be summed by scalar equations. The following two examples illustrate the procedure in some detail. The first system contains a single mass; the second has two masses that can move with different velocities.

▶ EXAMPLE 2.1

Draw the free-body diagram and apply D'Alembert's law to write a modeling equation for the system shown in Figure 2.12(a). The mass is assumed to move horizontally on frictionless bearings, and the spring and dashpot are linear.

SOLUTION The free-body diagram for the mass is shown in Figure 2.12(b). The vertical forces on the mass (the weight Mg and the upward forces exerted by the frictionless bearings) have been omitted because these forces are perpendicular to the direction of motion. The horizontal forces, which are included in the free-body diagram, are

f_K, the force exerted by the spring

f_B, the force exerted by the dashpot

f_I, the inertial force

$f_a(t)$, the applied force

The choice of directions for the arrows representing f_K, f_B, and f_I is arbitrary and does not affect the final result. However, the expressions for these individual forces must agree with the choice of arrows. The use of a dashed arrow for the inertial force f_I emphasizes that it is not an external force like the other three.

We next use the element laws to express the forces f_K, f_B, and f_I in terms of the element values K, B, and M and the system variables x and v. In Figure 2.12(a), the positive

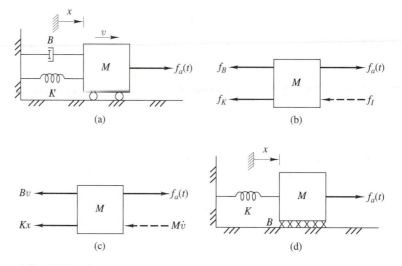

Figure 2.12 (a) Translational system for Example 2.1. (b) Free-body diagram. (c) Free-body diagram including element laws. (d) Additional system described by (15).

direction of x and v is defined to be to the right, so the spring is stretched when x is positive, and compressed when x is negative. If the spring undergoes an elongation x, then there must be a tensile force Kx on the right end of the spring directed to the right and a reaction force $f_K = Kx$ on the mass directed to the left. In other words, if x is positive, the spring is stretched, and it therefore pulls mass M to the left, as seen in Figure 2.12(c). Because the arrow for f_K does point to the left in Figure 2.12(b), we may relabel this force as Kx in Figure 2.12(c). Note that if x is negative at some instant of time, the spring will be compressed and will exert a force to the right on the mass. Under these conditions, Kx will be negative and the free-body diagram will show a negative force on the mass to the left, which is equivalent to a positive force to the right. Although the result is the same either way, it is customary to assume that all displacements are in the assumed positive directions when determining the proper expressions for the forces.

Similarly, when the right end of the dashpot moves to the right with velocity v, a force $f_B = Bv$ is exerted on the mass to the left. Finally, because of (12), the inertial force $f_I = M\dot{v}$ must have its positive direction opposite to that of dv/dt. After trying a few examples, the reader should be able to draw a free-body diagram such as the one in Figure 2.12(c) without first having to show explicitly the diagram in Figure 2.12(b).

D'Alembert's law can now be applied to the free-body diagram in Figure 2.12(c), with due regard for the assumed arrow directions. If forces acting to the right are regarded as positive, the law yields

$$f_a(t) - (M\dot{v} + Bv + Kx) = 0$$

Replacing v by $\dot{x}$ and $\dot{v}$ by $\ddot{x}$, and rearranging the terms, we can rewrite this equation as

$$M\ddot{x} + B\dot{x} + Kx = f_a(t) \tag{15}$$

In place of the dashpot between the mass and the wall in part (a) of Figure 2.12, part (d) shows viscous friction between the mass and the horizontal support. The free-body diagram for this system would remain the same as in Figure 2.12(c), and D'Alembert's law would again yield (15).

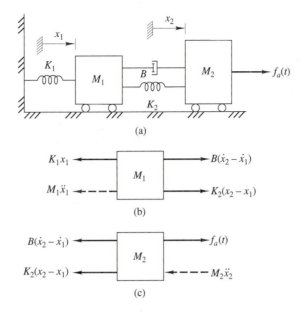

Figure 2.13 (a) Translational system for Example 2.2. (b), (c) Free-body diagrams.

EXAMPLE 2.2

Draw the free-body diagrams and use D'Alembert's law to write the two modeling equations for the two-mass system shown in Figure 2.13(a).

SOLUTION Because there are two masses that can move with different unknown velocities, a separate free-body diagram is drawn for each one. This is done in Figure 2.13(b) and Figure 2.13(c). In Figure 2.13(b), the forces $K_1 x_1$ and $M_1 \ddot{x}_1$ are similar to those in Example 2.1. As indicated in our earlier discussion of displacements, the net elongation of the spring and dashpot connecting the two masses is $x_2 - x_1$. Hence a positive value of $x_2 - x_1$ results in a reaction force by the spring to the right on M_1 and to the left on M_2, as indicated in Figure 2.13. Of course, the force on either free-body diagram could be labeled $K_2(x_1 - x_2)$, provided that the corresponding reference arrow were reversed. For a positive value of $\dot{x}_2 - \dot{x}_1$, the reaction force of the middle dashpot is to the right on M_1 and to the left on M_2. As always, the inertial forces $M_1 \ddot{x}_1$ and $M_2 \ddot{x}_2$ are opposite to the positive directions of the accelerations.

Summing the forces on each free-body diagram separately and taking into account the directions of the reference arrows give the following pair of differential equations:

$$B(\dot{x}_2 - \dot{x}_1) + K_2(x_2 - x_1) - M_1 \ddot{x}_1 - K_1 x_1 = 0$$
$$f_a(t) - M_2 \ddot{x}_2 - B(\dot{x}_2 - \dot{x}_1) - K_2(x_2 - x_1) = 0$$

Rearranging, we have

$$M_1 \ddot{x}_1 + B\dot{x}_1 + (K_1 + K_2)x_1 - B\dot{x}_2 - K_2 x_2 = 0 \tag{16a}$$
$$-B\dot{x}_1 - K_2 x_1 + M_2 \ddot{x}_2 + B\dot{x}_2 + K_2 x_2 = f_a(t) \tag{16b}$$

Equations (16a) and (16b) constitute a pair of coupled second-order differential equations. In the next chapter, we shall discuss two alternative methods of presenting the information contained in such a set of equations.

In the force equation (16a) for the mass M_1 in the last example, note that all the terms involving the displacement x_1 and its derivatives have the same sign. Similarly, in (16b) for the mass M_2, all the terms with x_2 and its derivatives have the same sign. This is generally true for systems where the only permanent energy sources are associated with the external inputs. The reason for this will become apparent from the discussion of stability in Section 8.2, but the reader may wish to use this fact now as a check on the work.

Suppose that D'Alembert's law is applied to any mass M_i and that terms involving corresponding variables are collected together. Then all the terms involving the displacement x_i of the mass M_i should be expected to have the same sign. If they do not, the engineer should suspect that an error has been made and should check the steps leading to that equation. In the simplified equation for M_i no general statement can be made about the signs of terms involving displacements other than x_i and its derivatives, because the other signs depend on the reference directions used for the definition of the variables.

Inputs are variables that are specified functions of time. These functions are completely known and do not depend on the values of the system's components. Inputs for translational mechanical systems may be either forces or displacements. A displacement input exists when one part of a system is moved in a predetermined way. We assume that the mechanism that provides the specified displacement has a source of energy sufficient to carry out the motion regardless of any retarding forces that might come from the system components. Similarly, a force input is assumed to be available over whatever range of displacements may result.

▶ **EXAMPLE 2.3**

The system shown in Figure 2.14 is the same as that in Figure 2.13(a), except that the excitation is the displacement input $x_2(t)$ instead of an applied force. Write the equation governing the motion.

SOLUTION   We know the motion of mass M_2 in advance, so there is no need to draw a free-body diagram for it. The free-body diagram for M_1 is still the one shown in Figure 2.13(b), and D'Alembert's law again gives

$$M_1\ddot{x}_1 + B\dot{x}_1 + (K_1 + K_2)x_1 - B\dot{x}_2 - K_2x_2(t) = 0 \tag{17}$$

Because $x_2(t)$ and $\dot{x}_2$ are known functions of time, x_1 is the only unknown variable in (17) and a second equation is not needed.

If we wished to determine the force f_2 that needs to be applied in order to move M_2 with the prescribed displacement, we could also draw a free-body diagram for M_2. This would be the same as the one in Figure 2.13(c), except that $f_a(t)$ would be replaced by the unknown force f_2, with its positive sense to the right. The corresponding equation is

$$f_2 = -B\dot{x}_1 - K_2x_1 + M_2\ddot{x}_2 + B\dot{x}_2 + K_2x_2(t) \tag{18}$$

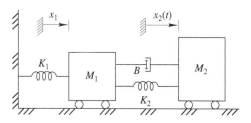

Figure 2.14 Translational system with displacement input for Example 2.3.

The quantities $x_2(t)$, $\dot{x}_2$, and $\ddot{x}_2$ are known. Once we have solved (17) for the unknown displacement $x_1(t)$ as a function of time, we can insert that result into (18) in order to calculate f_2.

A D'Alembert equation is not usually written for a body whose motion is known in advance. If, however, a free-body diagram is drawn for such a body, it is important to show the force associated with the displacement input, even though the force is unknown. The quantity f_2 in (18) would then be regarded as one of the outputs in the system model.

Relative Displacements

In the previous examples, the displacement variable associated with each mass was expressed with respect to its own fixed reference position. However, the position of a mass is sometimes measured with respect to some other moving body, rather than from a fixed reference. Such a **relative displacement** is used as one of the variables in the following example.

▶ **EXAMPLE 2.4**

For the two-mass system shown in Figure 2.15(a), x denotes the position of mass M_1 with respect to a fixed reference, and z denotes the relative displacement of mass M_2 with respect to M_1. The positive direction for both displacements is to the right. Assume that the two springs are neither stretched nor compressed when $x = z = 0$. Find the equations describing the system.

SOLUTION The free-body diagrams for the two masses are shown in parts (b) and (c) of Figure 2.15. The force on mass M_1 through the viscous friction element B_2 is proportional to the relative velocity $\dot{z}$ of the two masses. If M_2 is moving to the right faster than M_1, so that $\dot{z}$ is positive, then the force on M_1 through B_2 tends to pull the mass to the right, as indicated in Figure 2.15(b).

To draw the free-body diagram for M_2, we note that the elongation of the spring K_2 is $x + z$. Furthermore, the inertial force is always proportional to the *absolute* acceleration $\ddot{x} + \ddot{z}$, not to the *relative* acceleration with respect to some other moving body. Thus only the force exerted through the friction element B_2 is expressed in terms of the *relative* motion of the two masses. Summing the forces on each of the free-body diagrams gives

$$B_2\dot{z} - K_1 x - M_1\ddot{x} - B_1\dot{x} - B_3\dot{x} = 0$$
$$f_a(t) - M_2(\ddot{x} + \ddot{z}) - K_2(x + z) - B_2\dot{z} = 0$$

or, after we rearrange the terms,

$$M_1\ddot{x} + (B_1 + B_3)\dot{x} + K_1 x - B_2\dot{z} = 0 \tag{19a}$$
$$M_2\ddot{x} + K_2 x + M_2\ddot{z} + B_2\dot{z} + K_2 z = f_a(t) \tag{19b}$$

The comments made after Example 2.2 about the signs in the force equations for M_1 and M_2 still apply when a relative displacement variable is used. In the force equation (19a) for M_1, all the terms with the displacement x of the mass M_1 and its derivatives have the same sign. Similarly in (19b), all the terms involving the relative displacement z of M_2 with respect to M_1 have the same sign.

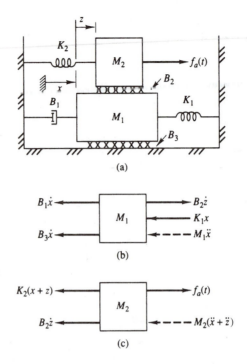

Figure 2.15 (a) Translational system for Example 2.4. (b), (c) Free-body diagrams.

The reader is encouraged to repeat this example when the displacement of each mass is expressed with respect to its own fixed reference position. If x_1 and x_2 denote the displacements of M_1 and M_2, respectively, with the positive senses to the right, we find that

$$M_1\ddot{x}_1 + (B_1 + B_2 + B_3)\dot{x}_1 + K_1 x_1 - B_2\dot{x}_2 = 0$$
$$-B_2\dot{x}_1 + M_2\ddot{x}_2 + B_2\dot{x}_2 + K_2 x_2 = f_a(t)$$

(20)

When x_1 is replaced by x, and x_2 is replaced by $x + z$, this pair of equations reduces to those in (19). However, it is useful to be able to obtain (19) directly from the free-body diagrams in Figure 2.15.

Displacement variables must be expressed with respect to some reference position. These references are commonly chosen such that any springs are neither stretched nor compressed when the values of the displacement variables are zero. The following two examples show why this may not necessarily be the case for systems with vertical motion.

▶ **EXAMPLE 2.5**

Draw the free-body diagram, including the effect of gravity, and find the differential equation describing the motion of the mass shown in Figure 2.16(a).

SOLUTION   Assume that x is the displacement from the position corresponding to a spring that is neither stretched nor compressed. The gravitational force on the mass is Mg, and we include it in the free-body diagram shown in Figure 2.16(b) because the mass moves vertically. By summing the forces on the free-body diagram, we obtain

$$M\ddot{x} + B\dot{x} + Kx = f_a(t) + Mg$$

(21)

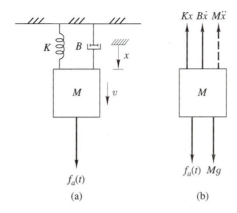

Figure 2.16 (a) Translational system with vertical motion. (b) Free-body diagram.

Suppose that the applied force $f_a(t)$ is zero and that the mass is not moving. Then $x = x_0$, where x_0 is the constant displacement caused by the gravitational force. Because $\dot{x}_0 = \ddot{x}_0 = 0$, the foregoing differential equation reduces to the algebraic equation

$$Kx_0 = Mg \tag{22}$$

We also can see this directly from the free-body diagram by noting that all but two of the five forces vanish under these conditions.

We now reconsider the case where $f_a(t)$ is nonzero and where the mass is moving. Let

$$x = x_0 + z \tag{23}$$

This equation defines z as the displacement caused by the input $f_a(t)$, namely the additional displacement beyond that resulting from the constant weight Mg. Substituting (23) into (21) and again noting that $\dot{x}_0 = \ddot{x}_0 = 0$, we have

$$M\ddot{z} + B\dot{z} + K(x_0 + z) = f_a(t) + Mg$$

or, by using (22),

$$M\ddot{z} + B\dot{z} + Kz = f_a(t) \tag{24}$$

Comparison of (21) and (24) indicates that we can ignore the gravitational force Mg when drawing the free-body diagram and when writing the system equation, provided that the displacement is defined to be the displacement from the static position corresponding to no inputs except gravity.

This conclusion is valid only when masses are suspended vertically by one or more linear springs. Under static-equilibrium conditions, there are no inertial or friction forces, and the force exerted by the spring has the form

$$f_K(t) = K(x_0 + z)$$
$$= Kx_0 + Kz$$

which is just the superposition of the static force caused by gravity and the force caused by additional inputs. For nonlinear springs, however, this conclusion is not valid because superposition does not hold.

Normally, a new symbol such as z is not introduced in problems involving gravitational forces. Instead, the symbol x can be redefined to be the additional displacement from the static position.

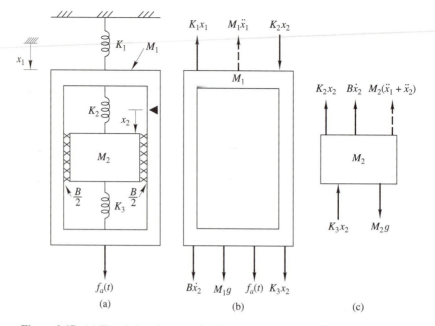

Figure 2.17 (a) Translational system for Example 2.6. (b), (c) Free-body diagrams.

▶ **EXAMPLE 2.6**

For the system shown in Figure 2.17(a), x_1 and x_2 denote the elongations of K_1 and K_2, respectively. Note that x_1 is the displacement of mass M_1 with respect to a fixed reference but that x_2 is the relative displacement of M_2 with respect to M_1. When $x_1 = x_2 = 0$, all three springs shown in the figure are neither stretched nor compressed. Draw the free-body diagram for each mass, including the effect of gravity, and find the differential equations describing the system's behavior. Determine the values of x_1 and x_2 that correspond to the static-equilibrium position, when $f_a(t) = 0$ and when the masses are motionless.

SOLUTION The free-body diagrams are shown in parts (b) and (c) of the figure. Many of the comments made in Example 2.4 also apply to this problem. If x_1 and x_2 are positive, then K_1 and K_2 are stretched and K_3 is compressed. Under these circumstances, K_1 exerts an upward force on M_1, and K_2 and K_3 exert downward forces on M_1. The relative velocity of M_2 with respect to M_1 is $\dot{x}_2$, so frictional forces of $B\dot{x}_2$ are exerted downward on M_1 and upward on M_2. The inertial force on M_2 is proportional to its absolute acceleration, which is $\ddot{x}_1 + \ddot{x}_2$. Summing the forces on each of the free-body diagrams gives

$$M_1\ddot{x}_1 + K_1x_1 - B\dot{x}_2 - (K_2 + K_3)x_2 = M_1g + f_a(t)$$
$$M_2\ddot{x}_1 + M_2\ddot{x}_2 + B\dot{x}_2 + (K_2 + K_3)x_2 = M_2g \tag{25}$$

This pair of coupled equations has two unknown variables. If the element values and $f_a(t)$ are known, and if the necessary initial conditions are given, then (25) can be solved for x_1 and x_2 as functions of time by the methods discussed in later chapters. Because x_2 is a relative displacement, the total displacement of M_2 is $x_1 + x_2$.

To find the displacements x_{1_0} and x_{2_0} that correspond to the static-equilibrium position, we replace $f_a(t)$ and all the displacement derivatives by zero. Then (25) reduces to

$$K_1 x_{1_0} - (K_2 + K_3) x_{2_0} = M_1 g$$
$$(K_2 + K_3) x_{2_0} = M_2 g \tag{26}$$

from which

$$x_{1_0} = \frac{(M_1 + M_2)g}{K_1}$$
$$x_{2_0} = \frac{M_2 g}{K_2 + K_3} \tag{27}$$

If we want the differential equations in terms of displacements z_1 and z_2 measured with respect to the equilibrium conditions given by (27), then we can write $x_1 = x_{1_0} + z_1$ and $x_2 = x_{2_0} + z_2$. Substituting these expressions into (25) and using (26), we find that

$$M_1 \ddot{z}_1 + K_1 z_1 - B \dot{z}_2 - (K_2 + K_3) z_2 = f_a(t)$$
$$M_2 \ddot{z}_1 + M_2 \ddot{z}_2 + B \dot{z}_2 + (K_2 + K_3) z_2 = 0$$

As expected, these equations are similar to (25) except for the absence of the gravitational forces.

The last two examples illustrate the choice that we have when choosing references for linear systems with vertical motion. We will generally define displacements from the positions corresponding to unstressed springs. This means that we must show the Mg force on the free-body diagram for any mass that can move vertically. If, however, we were to define displacements from the static-equilibrium positions caused by the gravitational forces, then we would not show the Mg forces on our diagrams. The modeling equations will have the same form as for the first method, except that the Mg terms will be missing. The first method can be used even for nonlinear springs; however, the second cannot.

The Ideal Pulley

A pulley can be used to change the direction of motion in a translational mechanical system. Frequently, part of the system then moves horizontally and the rest moves vertically. The basic pulley consists of a cylinder that can rotate about its center and that has a cable resting on its surface. An ideal pulley has no mass and no friction. We assume that there is no slippage between the cable and the surface of the cylinder—that is, that they both move with the same velocity. We also assume that the cable is always in tension but that it cannot stretch. If we need to consider a cable that can stretch, we can approximate that effect by showing a separate spring leading to an ideal cable. If the pulley is not ideal, then its mass and any frictional effects must be considered, as will be discussed in Chapter 5. The action of an ideal pulley is illustrated in the following example.

▷ **EXAMPLE 2.7**

Find and compare the equations describing the systems shown in Figure 2.18(a) and Figure 2.19(a). Let $x_1 = x_2 = 0$ correspond to the condition when the springs are neither stretched nor compressed.

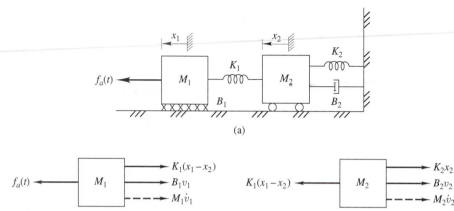

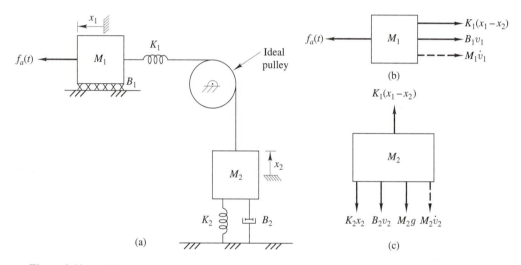

Figure 2.18 (a) Translational system for Example 2.7. (b), (c) Free-body diagrams.

Figure 2.19 (a) Translational system with an ideal pulley added. (b), (c) Free-body diagrams.

SOLUTION   Free-body diagrams for the system in part (a) of Figure 2.18 are shown in parts (b) and (c). Note that the spring K_1 exerts equal but opposite forces on M_1 and M_2. Summing the forces shown in the diagrams gives

$$M_1\dot{v}_1 + B_1 v_1 + K_1(x_1 - x_2) = f_a(t)$$
$$M_2\dot{v}_2 + B_2 v_2 + K_2 x_2 = K_1(x_1 - x_2) \tag{28}$$

When an ideal pulley is added to give the system in Figure 2.19(a), the free-body diagram for M_1, which is repeated in part (b) of the figure, remains unchanged. To draw the diagram for M_2, we note that both ends of the cable move with the same motion because the cable cannot stretch. The force exerted by K_1 passes through the cable and is exerted directly on M_2. Hence the cable does not affect the magnitude of the forces exerted by K_1 but merely changes the direction of the force on M_2 to be upward. Because M_2 moves vertically, we also include the gravitational force in its free-body diagram, which is shown in Figure 2.19(c). Summing the forces shown in parts (b) and (c) of the figure gives

$$M_1\dot{v}_1 + B_1 v_1 + K_1(x_1 - x_2) = f_a(t)$$
$$M_2\dot{v}_2 + B_2 v_2 + K_2 x_2 + M_2 g = K_1(x_1 - x_2) \tag{29}$$

As expected, (28) and (29) are identical except for the presence of the gravitational force $M_2 g$.

It is important to be aware of a significant difference between the models given by (28) and (29) for the last example. The difference involves their validity. Positive or negative values of the quantity $x_1 - x_2$ correspond to the elongation or compression, respectively, of the spring K_1. Equations (28) are valid for Figure 2.18(a) for all values of x_1 and x_2. For the system with the pulley, however, the free-body diagram in Figure 2.19(c) indicates that the cable would exert a downward force on M_2 for a negative value of $x_1 - x_2$. This implies that the cable would tend to push apart the bodies attached to its ends, which is not physically possible. In such a case the cable would buckle, become detached from the pulley, and exert no force at all on M_2. Thus (29) can be used only when $x_1 - x_2 \geq 0$. Although this condition is not normally written next to the corresponding equations, the engineer should always examine the analytical or computer solution to be sure that the results correspond to conditions for which the modeling equations are valid.

Parallel Combinations

In some cases, two or more springs or dashpots can be replaced by a single equivalent element. Two springs or dashpots are said to be in **parallel** if the first end of each is attached to the same body and if the remaining ends are also attached to a common body. We shall consider a specific example before formulating a general rule.

▷ **EXAMPLE 2.8**

The system shown in Figure 2.20(a) includes two linear springs between the wall and the mass M. Write the differential equation describing the motion of the mass. Find the spring constant K_{eq} for a single spring that could replace K_1 and K_2. Assume first that the springs have the same unstretched length. Then repeat the problem when the unstretched lengths of K_1 and K_2 are d_1 and d_2, respectively.

SOLUTION   If the unstretched lengths of the two springs are identical, then they will have the same elongation, denoted by x, when the mass is in motion. The free-body diagram is shown in Figure 2.20(b). Summing the forces gives

$$M\ddot{x} + B\dot{x} + (K_1 + K_2)x = f_a(t) \tag{30}$$

If the combination of K_1 and K_2 is replaced by a single equivalent spring, then the system reduces to that shown in Figure 2.12(d), which is described by (15). Comparing (30) and (15) reveals that

$$K_{eq} = K_1 + K_2 \tag{31}$$

Next suppose that the springs have different unstretched lengths, and let $d(t)$ denote the distance from the wall to the left side of the mass. The elongations of K_1 and K_2, which are no longer the same, are $d(t) - d_1$ and $d(t) - d_2$, respectively. The mass has velocity $\dot{d}$ and acceleration $\ddot{d}$. The new free-body diagram, with all the retarding forces expressed in terms of $d(t)$, is shown in Figure 2.20(c). By D'Alembert's law,

$$M\ddot{d} + B\dot{d} + K_1[d(t) - d_1] + K_2[d(t) - d_2] = f_a(t) \tag{32}$$

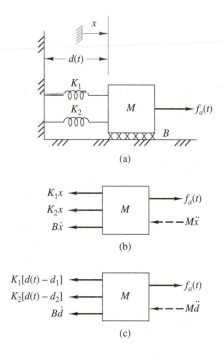

Figure 2.20 (a) Translational system for Example 2.8. (b) Free-body diagram when the springs have the same unstretched lengths. (c) Free-body diagram when the springs have different unstretched lengths.

Let d_0 be the distance from the wall to the mass when $f_a(t) = 0$ and when the mass is motionless in a position of static equilibrium. The only forces on the mass will be those exerted by the two springs, and those forces will be equal and opposite. From (32), or directly from Figure 2.20(c),

$$K_1(d_0 - d_1) + K_2(d_0 - d_2) = 0 \tag{33}$$

from which

$$d_0 = \frac{K_1 d_1 + K_2 d_2}{K_1 + K_2} \tag{34}$$

The respective static elongations of K_1 and K_2 are

$$d_0 - d_1 = \frac{K_2(d_2 - d_1)}{K_1 + K_2}$$

$$d_0 - d_2 = \frac{K_1(d_1 - d_2)}{K_1 + K_2}$$

If $d_1 \neq d_2$, one of these elongations will be negative. This means that one spring will be stretched and the other compressed, so that the net static force exerted by the two springs on the mass can be zero.

It is now reasonable to define the variable z as the distance of the mass beyond d_0 when a force input is applied—that is, as the displacement with respect to the reference position given by (34). Then

$$d(t) = d_0 + z$$

Substituting this expression into (32) and noting that $\dot{d} = \dot{z}$ and $\ddot{d} = \ddot{z}$, we have

$$M\ddot{z} + B\dot{z} + K_1(d_0 + z - d_1) + K_2(d_0 + z - d_2) = f_a(t)$$

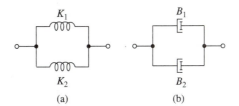

Figure 2.21 Parallel combinations. (a) $K_{eq} = K_1 + K_2$. (b) $B_{eq} = B_1 + B_2$.

Using (33) to cancel some of the terms in this equation, we obtain

$$M\ddot{z} + B\dot{z} + (K_1 + K_2)z = f_a(t)$$

which is identical to (30) except for the use of z in place of x. Thus (31) is valid for the equivalent spring constant even if the unstretched lengths of the springs are different, provided that x is interpreted as the displacement beyond the static-equilibrium position given by (34).

Two parallel springs or dashpots have their respective ends joined, as shown in Figure 2.21. From the last example, we see that for the parallel combination of two springs,

$$K_{eq} = K_1 + K_2 \tag{35}$$

Similarly, it can be shown that for two dashpots in parallel, as in part (b) of Figure 2.21,

$$B_{eq} = B_1 + B_2 \tag{36}$$

The formulas for parallel stiffness or friction elements can be extended to situations that may seem somewhat different from those in Figure 2.21. The key requirement for parallel elements is that respective ends move with the same displacement. The individual ends need not be tied directly together in the figure depicting the system. Although one pair of ends may sometimes be connected to a fixed surface, in other cases both pairs of ends may be free to move.

▶ **EXAMPLE 2.9**

Find the equation describing the motion of the mass in the translational system shown in Figure 2.22(a). Show that the two springs can be replaced by a single equivalent spring, and the three friction elements by an equivalent element.

SOLUTION Note that each of the springs has one side attached to the mass and the other attached to the fixed surface on the perimeter of the diagram. When x is positive, K_1 is stretched and K_2 compressed. Thus the springs will exert forces of $K_1 x$ and $K_2 x$ on the mass to the left, as shown on the free-body diagram in Figure 2.22(b).

For each of the three friction elements, one side moves with the velocity of the mass and the other side is stationary. When $\dot{x}$ is positive, each of these elements exerts a retarding force on the mass to the left. Summing the forces shown on the free-body diagram, we have

$$M\ddot{x} + (B_1 + B_2 + B_3)\dot{x} + (K_1 + K_2)x = f_a(t)$$

With $K_{eq} = K_1 + K_2$ and $B_{eq} = B_1 + B_2 + B_3$, this equation becomes

$$M\ddot{x} + B_{eq}\dot{x} + K_{eq}x = f_a(t)$$

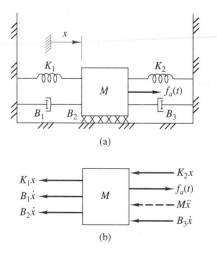

Figure 2.22 (a) Translational system with parallel stiffness and friction elements. (b) Free-body diagram.

When the parallel combinations of stiffness and friction elements in Figure 2.22(a) are replaced by equivalent elements, the system reduces to that shown in Figure 2.12(d), which is described by (15).

Series Combinations

Two springs or dashpots are said to be in **series** if they are joined at only one end of each element and if there is no other element connected to their common junction. The following example has a series combination of two springs and also illustrates the application of D'Alembert's law to a massless junction.

▶ **EXAMPLE 2.10**

When $x_1 = x_2 = 0$, the two springs shown in Figure 2.23(a) are neither stretched nor compressed. Draw free-body diagrams for the mass M and for the massless junction A, and then write the equations describing the system. Show that the motion of point A is not independent of that of the mass M and that x_1 and x_2 are directly proportional to one another. Finally, find K_{eq} for a single spring that could replace the combination of K_1 and K_2.

SOLUTION The free-body diagrams are shown in parts (b) and (c) of the figure. Because there is no mass at point A, there is no inertial force in its free-body diagram. Summing the forces for each diagram gives

$$M\ddot{x}_1 + B\dot{x}_1 + K_1(x_1 - x_2) = f_a(t)$$
$$K_2 x_2 = K_1(x_1 - x_2)$$

Solving the second equation for x_2 in terms of x_1 gives

$$x_2 = \left(\frac{K_1}{K_1 + K_2}\right) x_1$$

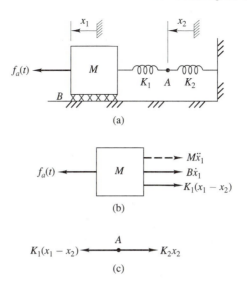

Figure 2.23 (a) Translational system with a massless junction. (b), (c) Free-body diagrams.

which shows that the two displacements are proportional to one another. Substituting this expression back into the first equation, we have

$$M\ddot{x}_1 + B\dot{x}_1 + K_1 \left[1 - \frac{K_1}{K_1 + K_2} \right] x_1 = f_a(t)$$

from which

$$M\ddot{x}_1 + B\dot{x}_1 + \frac{K_1 K_2}{K_1 + K_2} x_1 = f_a(t)$$

This equation describes the system formed when the two springs in Figure 2.23(a) are replaced by a single spring for which

$$K_{eq} = \frac{K_1 K_2}{K_1 + K_2} \tag{37}$$

Series combinations of stiffness and friction elements are shown in Figure 2.24. It is assumed that no other element is connected to the common junctions. For the two springs in part (a) of the figure, the equivalent spring constant is given by (37). For two dashpots in series, as in part (b) of the figure, it can be shown that

$$B_{eq} = \frac{B_1 B_2}{B_1 + B_2} \tag{38}$$

In order to reduce certain combinations of springs or dashpots to a single equivalent element, we may have to use the rules for both parallel and series combinations. The following example illustrates the procedure for two different combinations of dashpots.

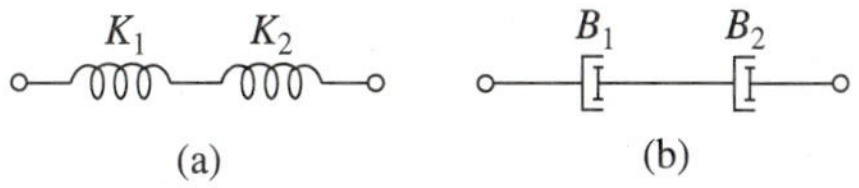

Figure 2.24 Series combinations. (a) $K_{eq} = K_1 K_2 / (K_1 + K_2)$. (b) $B_{eq} = B_1 B_2 / (B_1 + B_2)$.

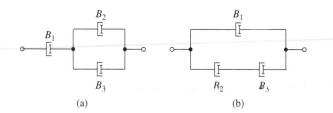

Figure 2.25 Parallel-series combinations.

▶ **EXAMPLE 2.11**

Find B_{eq} for the single friction element that can replace the three dashpots in each part of Figure 2.25.

SOLUTION In part (a) of the figure, the parallel combination of B_2 and B_3 can be replaced by a single element whose viscous friction coefficient is $B_2 + B_3$. This is then in series with B_1, so that for the overall combination,

$$B_{eq} = \frac{B_1(B_2 + B_3)}{B_1 + B_2 + B_3}$$

In Figure 2.25(b), the series combination of B_2 and B_3 is in parallel with B_1. Thus

$$B_{eq} = B_1 + \frac{B_2 B_3}{B_2 + B_3}$$

▶ SUMMARY

In this chapter we have introduced the variables, element laws, and interconnection laws for linear, lumped-element translational mechanical systems. Either force or displacement inputs can be applied to any part of the system. An applied force is a known function of time, but the motion of the body to which it is applied is not known at the outset of a problem. Conversely, a displacement input moves some part of the system with a known specified motion, but the force exerted by the external mechanism moving that part is normally not known.

Displacements may be measured with respect to fixed reference positions or with respect to some other moving body. When relative displacements are used, it is important to keep in mind that the inertial force of a mass is always proportional to its absolute acceleration, not to its relative acceleration.

To model a system, we draw a free-body diagram and sum the forces for every mass or other junction point whose motion is unknown. The free-body diagram for a massless junction is drawn in the usual way, except that there is no inertial force. The modeling process can sometimes be simplified by replacing a series-parallel combination of stiffness or friction elements with a single equivalent element.

Special attention was given to linear systems that involve vertical motion. If displacements are measured from positions where the springs are neither stretched nor compressed, the gravitational forces must be included in the free-body diagrams for any masses that can move vertically. If, however, the displacements are measured with respect to the static-equilibrium positions when the system is motionless and when no other external inputs are applied, then the gravitational forces do not appear in the final equations of motion. Systems containing pulleys can have some parts of the system moving horizontally and other parts moving vertically.

Our basic task in this chapter was to draw free-body diagrams and to write the corresponding equations describing the system. In the next chapter, we shall discuss how to present the information contained in these equations in ways that will facilitate the development of both computer and analytical solutions.

▶ PROBLEMS

Throughout this book, answers to problems marked with an asterisk are given in Appendix G.

*2.1. For the system shown in Figure P2.1, the springs are undeflected when $x_1 = x_2 = 0$. The input is $f_a(t)$. Draw free-body diagrams and write the modeling equations.

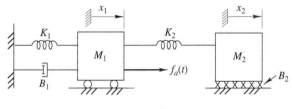

Figure P2.1

2.2. Repeat Problem 2.1 for the system shown in Figure P2.2.

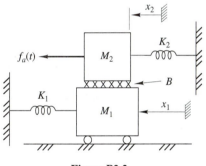

Figure P2.2

2.3. Repeat Problem 2.1 for the system shown in Figure P2.3.

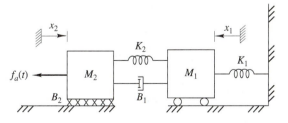

Figure P2.3

*2.4. For the system shown in Figure P2.4, draw the free-body diagram for each mass and write the differential equations describing the system.

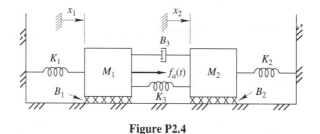

Figure P2.4

2.5. Repeat Problem 2.4 for the system shown in Figure P2.5.

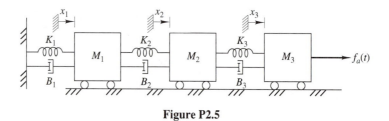

Figure P2.5

2.6. In the mechanical system shown in Figure P2.6, the spring forces are zero when $x_1 = x_2 = x_3 = 0$. Let the base be stationary so that $x_3(t) = 0$ for all values of t. Draw free-body diagrams and write a pair of coupled differential equations that govern the motion when the only input is $f_a(t)$.

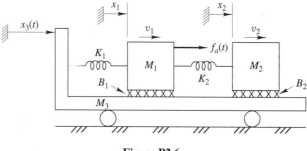

Figure P2.6

*2.7. For the system shown in Figure P2.7, the springs are undeflected when $x_1 = x_2 = 0$. The input is $x_2(t)$, the displacement of the left edge of M_2.

    a.   Write the equation governing the motion of M_1.

    b.   Write an expression for the force f_2, positive sense to the right, that must be applied to M_2 in order to achieve the displacement $x_2(t)$.

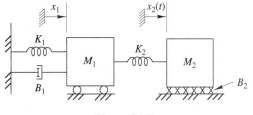

Figure P2.7

2.8. Repeat Problem 2.7 for the system shown in Figure P2.8.

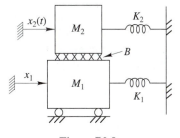

Figure P2.8

2.9. Repeat Problem 2.6 with the following changes:

    a.   The displacement $x_3(t)$ is the input and the applied force $f_a(t)$ is zero for all time.

    b.   Write an expression for the force f_3, positive sense to the right, that must be applied to M_3 in order to move M_3 with the specified displacement $x_3(t)$.

*2.10. For the system shown in Figure P2.10, the distance between masses M_1 and M_2 is $A + x_2$, where A is a constant. The springs are undeflected when $x_1 = x_2 = 0$. Draw free-body diagrams and write the differential equations for the system.

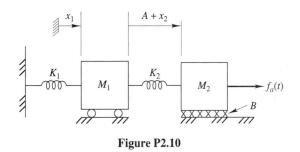

Figure P2.10

2.11. For the system shown in Figure P2.11, the input is the displacement $x_1(t)$. The springs are undeflected when $x_1 = x_2 = x_3 = 0$. The variable x_2 represents the displacement of M_2 with respect to M_1. Write the mathematical model to describe the motion of the masses as a set of coupled differential equations. Include appropriate free-body diagrams.

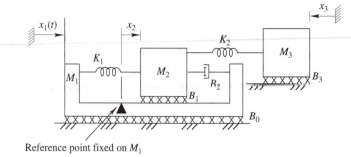

Reference point fixed on M_1

Figure P2.11

*2.12. The input to the translational mechanical system shown in Figure P2.12 is the displacement $x_3(t)$ of the right end of the spring K_1. The displacement of M_2 relative to M_1 is x_2. The forces exerted by the springs are zero when $x_1 = x_2 = x_3 = 0$. Draw the free-body diagrams and write the modeling equations.

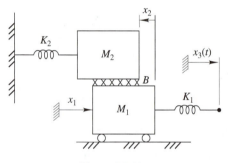

Figure P2.12

2.13. For the system shown in Figure P2.13, the displacement of M_1 relative to M_2 is x_1. The spring force is zero when $x_1 = x_2 = 0$. Draw free-body diagrams and write the modeling equations.

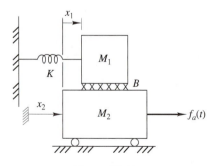

Figure P2.13

2.14. For the system shown in Figure P2.14, the displacements x_1 and x_2 are measured relative to M_3. Both springs are undeflected when $x_1 = x_2 = 0$. Draw free-body diagrams and write the modeling equations.

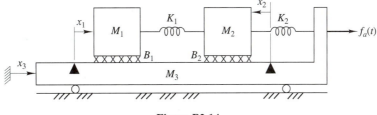

Figure P2.14

*2.15. The mechanical system shown in Figure P2.15 is driven by the applied force $f_a(t)$. When $x_1 = x_2 = 0$, the springs are neither stretched nor compressed.

    a.   Draw the free-body diagrams and write the differential equations of motion for the two masses in terms of x_1 and x_2.

    b.   Find x_{1_0} and x_{2_0}, the constant displacements of the masses caused by the gravitational forces when $f_a(t) = 0$ and when the system is in static equilibrium.

    c.   Rewrite the system equations in terms of z_1 and z_2, the relative displacements of the masses with respect to the static-equilibrium positions found in part (b).

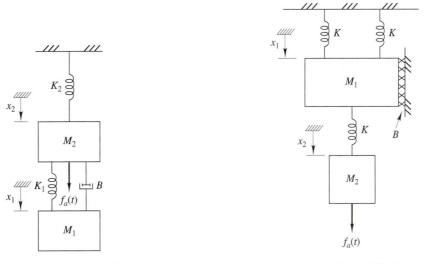

Figure P2.15

Figure P2.16

2.16. Repeat all three parts of Problem 2.15 for the system shown in Figure P2.16. Each of the three springs has the same spring constant K.

2.17. All the springs in Figure P2.17 are identical, each with spring constant K. The spring forces are zero when $x_1 = x_2 = x_3 = 0$.

  a. Draw the free-body diagrams, including the gravitational forces, and write the differential equations describing the system.

  b. Determine the constant elongation of each spring caused by the gravitational forces when the masses are stationary in a position of static equilibrium and when $f_a(t) = 0$.

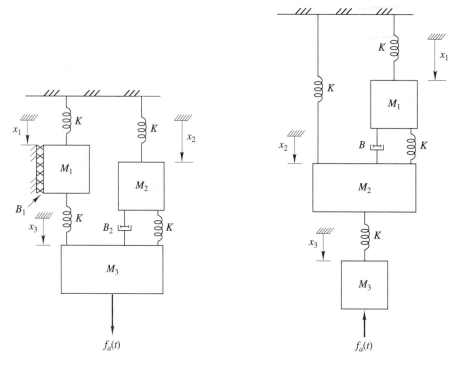

Figure P2.17   Figure P2.18

*2.18. Repeat Problem 2.17 for the system shown in Figure P2.18.

2.19. The system shown in Figure P2.19 has a nonlinear spring that obeys the expression $f_K = x^3$.

  a. Write the differential equation describing the system in terms of the displacement x.

  b. Let $x = x_0 + z$, where x_0 is the constant displacement caused by the gravitational force when the system is in static equilibrium. Rewrite the differential equation in terms of the variable z, canceling the gravitational term.

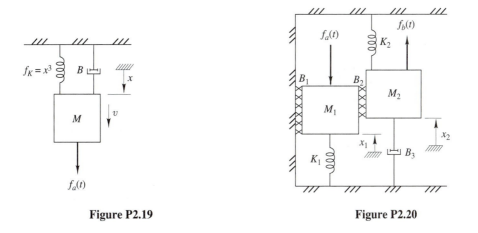

Figure P2.19   **Figure P2.20**

c. Comparison of (21) and (24) in Example 2.5 for a linear spring showed that the differential equation in z was the same as the one in x except for the deletion of the Mg term. Compare the results of parts (a) and (b) to show that such is not the case in this problem.

2.20. For the system shown in Figure P2.20, x_1 and x_2 are displacements relative to the undeflected spring lengths. The inputs are $f_a(t)$, $f_b(t)$, and gravity. Draw free-body diagrams and write the modeling equations.

2.21. Repeat Problem 2.20 for the system shown in Figure P2.21.

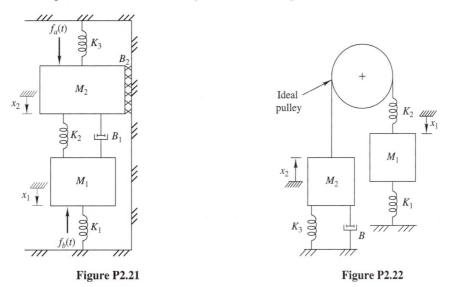

Figure P2.21   **Figure P2.22**

*2.22.** The pulley shown in Figure P2.22 is ideal. Draw free-body diagrams and write the modeling equations.

2.23. Repeat Problem 2.22 for the system shown in Figure P2.23.

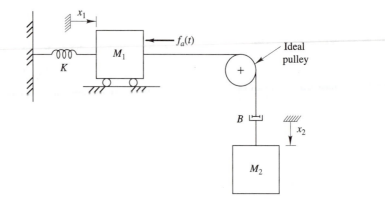

Figure P2.23

2.24. Repeat Problem 2.22 for the system shown in Figure P2.24.

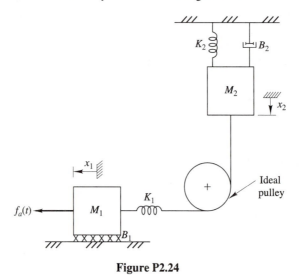

Figure P2.24

*2.25. For the system shown in Figure 2.11, draw the free-body diagrams for M and the junction of the dashpots. Apply D'Alembert's law and show that $\dot{x}_1$ is proportional to $\dot{x}_2$. Determine an equivalent coefficient B_{eq} for the combination of B_1 and B_2.

2.26. For the system shown in Figure P2.26, draw the free-body diagram and write a single differential equation. Determine an equivalent coefficient B_{eq} for the combination of B_1 and B_3.

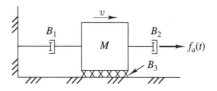

Figure P2.26

*2.27. The input for the system shown in Figure P2.27 is the displacement $x_3(t)$. Draw the free-body diagrams for the mass M and for the massless point A. Write the differential equations describing the system.

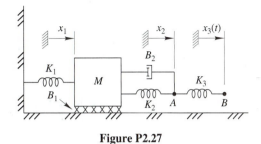

Figure P2.27

2.28. Repeat Problem 2.27 when the input is the force $f_a(t)$ applied to point B, with its positive sense to the right.

2.29. For the system shown in Figure P2.29, $x_1 = 0$ when the spring is undeflected. Draw free-body diagrams and write the modeling equations.

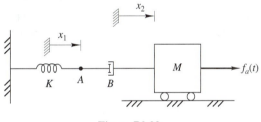

Figure P2.29

2.30. For the series combination of elements shown in Figure P2.30,

   a. Write the equations describing part (a) of the figure.

   b. Find expressions for K_{eq} and B_{eq} in part (b) of the figure such that the motions of the ends of the combination are the same as those in part (a).

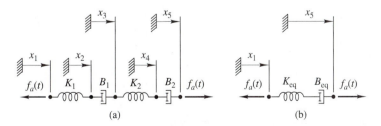

Figure P2.30

3

STANDARD FORMS FOR SYSTEM MODELS

In Chapter 2, we introduced the element and interconnection laws for translational systems and the procedure for drawing free-body diagrams and applying D'Alembert's law. These are essential steps regardless of the final form of the equations.

Not only should the complete mathematical model contain as many independent equations as unknown variables, but it should also have a form that is convenient for its solution. The two most common forms for the model when a direct solution is contemplated are discussed, illustrated, and compared in this chapter. Although translational mechanical systems are used as examples, the same techniques are applied in subsequent chapters to other types of systems. Other important forms of the model, based on Laplace transform procedures, will be found later in the book.

3.1 STATE-VARIABLE EQUATIONS

One procedure for formulating the system model is to begin by selecting a set of **state variables**. This set of variables must completely describe the effect of the past history of the system on its response in the future. Then, knowing the values of the state variables at a reference time t_0 and the values of the inputs for all $t \geq t_0$ is sufficient for evaluating the state variables and outputs for all $t \geq t_0$. It is further understood that the state variables must be independent; that is, it must be impossible to express any state variable as an algebraic function of the remaining state variables and the inputs.

Although the choice of state variables is not unique, the state variables for mechanical systems are usually related to the energy stored in each of the system's energy-storing elements. Because any energy that is initially stored in these elements can affect the response of the system at a later time, one state variable is normally associated with each of the energy-storing elements. Generally, this adequately summarizes the effect of the past history of the system. In some systems, the number of state variables is different from the number of energy-storing elements because a particular interconnection of elements causes redundant variables or because there is need for a state variable that is not related to the storage of energy.

Suppose the system has n state variables, m inputs, and p outputs. Let $q_1, q_2, \ldots, q_n$ denote the state variables, let $u_1, u_2, \ldots, u_m$ be the inputs, and let $y_1, y_2, \ldots, y_p$ be the outputs. We must find a set of first-order differential equations having a particular form. Each equation must express the derivative of one of the state variables as an *algebraic* function of the state variables and inputs. For a linear system, these algebraic functions will consist of a sum of terms, each of which is just a state variable multiplied by a coefficient or an input multiplied by a coefficient. If, for example, there are three state variables and two inputs, the state-variable equations have the form

$$\dot{q}_1 = a_{11}q_1 + a_{12}q_2 + a_{13}q_3 + b_{11}u_1 + b_{12}u_2$$
$$\dot{q}_2 = a_{21}q_1 + a_{22}q_2 + a_{23}q_3 + b_{21}u_1 + b_{22}u_2 \qquad (1)$$
$$\dot{q}_3 = a_{31}q_1 + a_{32}q_2 + a_{33}q_3 + b_{31}u_1 + b_{32}u_2$$

The outputs of interest are then expressed as algebraic functions of the state variables and inputs. If, for example, the linear system described by (1) happens to have two outputs, then the output equations normally have the form

$$y_1 = c_{11}q_1 + c_{12}q_2 + c_{13}q_3 + d_{11}u_1 + d_{12}u_2$$
$$y_2 = c_{21}q_1 + c_{22}q_2 + c_{23}q_3 + d_{21}u_1 + d_{22}u_2 \qquad (2)$$

For fixed linear systems, all of the coefficients given by a_{ij}, b_{ij}, c_{ij}, and d_{ij} are constants. For linear systems whose parameters vary with time, some of these coefficients are functions of time. Only for nonlinear elements could some of the coefficients themselves become functions of the state variables or inputs. In the most general case of time-varying, nonlinear systems, the right sides of the state-variable and output equations can be quite complicated functions of the state variables, the inputs, and time. Even then, however, the right sides of these equations should still be algebraic functions, without any derivative or integral terms. An uncommon exception wherein derivatives of the input must appear in (2) will be illustrated in one of the examples.

In this chapter we shall deal almost entirely with fixed linear systems. The form of the equations will then be generally similar to (1) and (2), although there may be more state variables, inputs, or outputs. An output variable may be identical to a state variable, in which case (2) may contain one or more equations such as $y_i = q_i$. When all the outputs are identical to some of the state variables, (2) becomes trivial and can be omitted.

Solving (1) for $t \geq t_0$ requires knowledge of the inputs for $t \geq t_0$ and also of the initial values of the state variables, namely $q_1(t_0)$, $q_2(t_0)$, and $q_3(t_0)$. In general, we cannot solve the individual state-variable equations separately but must solve them as a group. For example, the equation $\dot{q}_1 = -q_1 + q_2$ cannot be solved for q_1 unless another equation exists that can be solved for q_2.

The energy stored in translational mechanical systems must be associated with the masses or springs. From Equation (2.6)[1] the kinetic energy of a mass is $\frac{1}{2}Mv^2$, and from Equation (2.10) the potential energy of a spring is $\frac{1}{2}K(\Delta x)^2$. Thus it is logical to consider as possible state variables the velocities of the masses and the elongations of the springs. In most problems, we can express the elongations of the springs in terms of the displacements of the masses.

When drawing free-body diagrams, we try to express the forces in terms of state variables and inputs, in a way that avoids unnecessary derivatives. Thus we normally label inertial and friction forces in the form $M\dot{v}$ and Bv, rather than $M\ddot{x}$ and $B\dot{x}$.

In the remainder of this section, a variety of examples illustrate the technique of deriving the mathematical model in state-variable form. The general approach is as follows:

1. Identify the state variables and write those state-variable equations that do not require a free-body diagram, such as equations of the form $\dot{x} = v$.

2. Draw free-body diagrams for each independent mass and junction point that can move with an unknown motion. Sum the forces on each free-body diagram separately to obtain a set of differential equations.

---

[1] This denotes Equation (6) in Chapter 2.

3. Manipulate the equations into state-variable form. For each of the state variables, there must be an equation that expresses its derivative as an algebraic function of the state variables, the inputs, and possibly time.

4. Express the output variables as algebraic functions of the state variables, the inputs, and possibly time. In some unusual cases, it may be necessary to have derivatives of the input on the right side of the output equations. However, input derivatives should be avoided wherever possible.

▷ **EXAMPLE 3.1**

Find the state-variable model for the system shown in Figure 3.1(a), which is identical to Figure 2.12(a). The outputs of interest are the tensile force for the spring K and the velocity and acceleration of the mass M.

SOLUTION   We choose as state variables the elongation x of the spring, which is related to its potential energy, and the velocity v of the mass, which is related to its kinetic energy. Having made this choice, we can write the first of the two state-variable equations by inspection:

$$\dot{x} = v$$

Summing the forces on the free-body diagram, which was drawn for Example 2.1 and which is repeated in Figure 3.1(b), gives

$$M\dot{v} + Bv + Kx = f_a(t)$$

Solving this equation for $\dot{v}$ results in the second state-variable equation, where the right side is a function of only the state variables x and v and the input $f_a(t)$, as required. Thus the state variable equations are

$$\dot{x} = v$$

$$\dot{v} = \frac{1}{M}[-Kx - Bv + f_a(t)]$$

Computer solutions of such a set of equations are treated in Chapter 4. In order to find the state variables x and v as functions of time for all $t \geq 0$, we would need to know the input $f_a(t)$ for $t \geq 0$ and also the initial values of the state variables at $t = 0$. Any outputs that are not also state variables are expressed as algebraic functions of the state variables and inputs, so they can be evaluated as soon as the state variables have been found.

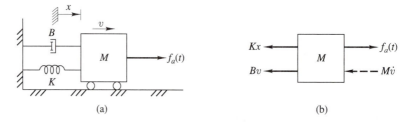

(a)   (b)

Figure 3.1   (a) Translational system for Example 3.1. (b) Free-body diagram.

The velocity of the mass is one of the state variables. The output equations for the tensile force f_K in the spring and the acceleration a_M of the mass are

$$f_K = Kx$$
$$a_M = \frac{1}{M}\left[-Kx - Bv + f_a(t)\right]$$

The system considered in the following three examples has four energy-storing elements and four state variables. The modeling is carried out for two different choices of state variables in order to illustrate that there are often several satisfactory choices. The examples also show that finding the state-variable equations is not made significantly more difficult when some of the system's elements are nonlinear.

▶ **EXAMPLE 3.2**

Find the state-variable equations for the system shown in Figure 3.2(a), which is identical to Figure 2.13(a). Write output equations for the tensile force f_{K_2} in the spring K_2 and for the total momentum m_T of the masses.

SOLUTION An appropriate choice of state variables is x_1, v_1, x_2, and v_2, because we can express the velocity of each mass and the elongation of each spring in terms of these four variables and because none of these variables can be expressed in terms of the other three. Because $\dot{x}_1 = v_1$ and $\dot{x}_2 = v_2$, two of the four state-variable equations are available immediately.

The free-body diagrams for the two masses are repeated from Example 2.2 in Figure 3.2(b) and Figure 3.2(c), with all forces labeled in terms of the state variables and the input. By D'Alembert's law,

$$M_1\dot{v}_1 + K_1 x_1 - K_2(x_2 - x_1) - B(v_2 - v_1) = 0$$
$$M_2\dot{v}_2 + K_2(x_2 - x_1) + B(v_2 - v_1) = f_a(t)$$

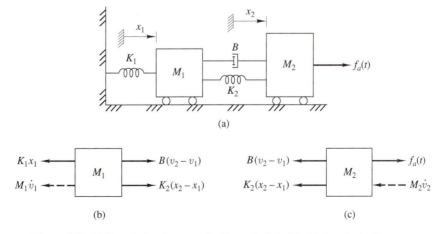

(a)

(b)   (c)

Figure 3.2 (a) Translational system for Example 3.2. (b), (c) Free-body diagrams.

which may be solved for $\dot{v}_1$ and $\dot{v}_2$, respectively. The state-variable equations are

$$\dot{x}_1 = v_1 \tag{3a}$$

$$\dot{v}_1 = \frac{1}{M_1}[-(K_1+K_2)x_1 - Bv_1 + K_2x_2 + Bv_2] \tag{3b}$$

$$\dot{x}_2 = v_2 \tag{3c}$$

$$\dot{v}_2 = \frac{1}{M_2}[K_2x_1 + Bv_1 - K_2x_2 - Bv_2 + f_a(t)] \tag{3d}$$

If we know the element values, the input $f_a(t)$ for $t \geq 0$, and the initial conditions $x_1(0)$, $v_1(0)$, $x_2(0)$, and $v_2(0)$, then we can solve this set of simultaneous first-order differential equations for x_1, v_1, x_2, and v_2 for all $t \geq 0$. The output equations are

$$f_{K_2} = K_2(x_2 - x_1)$$
$$m_T = M_1v_1 + M_2v_2 \tag{4}$$

▷ **EXAMPLE 3.3**

As an alternative choice of state variables for the system in Figure 3.2(a), use the relative displacement x_R and the relative velocity v_R of mass M_2 with respect to M_1, in place of x_2 and v_2. Again find the state-variable and output equations.

SOLUTION   We choose the four state variables x_1, v_1, x_R, and v_R, where

$$x_R = x_2 - x_1$$
$$v_R = v_2 - v_1$$

The free-body diagrams when the forces are labeled in terms of these variables are shown in Figure 3.3. Remember that the inertial forces depend on the absolute accelerations of the respective masses. By D'Alembert's law,

$$M_1\dot{v}_1 + K_1x_1 - K_2x_R - Bv_R = 0$$
$$M_2\dot{v}_1 + M_2\dot{v}_R + Bv_R + K_2x_R = f_a(t)$$

from which

$$\dot{v}_1 = \frac{1}{M_1}(-K_1x_1 + K_2x_R + Bv_R) \tag{5a}$$

$$\dot{v}_R = \frac{1}{M_2}[-M_2\dot{v}_1 - Bv_R - K_2x_R + f_a(t)] \tag{5b}$$

Although (5a) is a satisfactory state-variable equation, (5b) is not because of the derivative on the right side. Thus we must substitute (5a) into (5b) in order to get an appropriate expression for $\dot{v}_R$. When this is done, we obtain for the complete set of state-variable

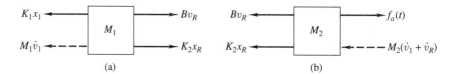

(a)   (b)

Figure 3.3   Free-body diagrams for Example 3.3 using relative displacement and relative velocity.

equations

$$\dot{x}_1 = v_1$$
$$\dot{v}_1 = \frac{1}{M_1}(-K_1x_1 + K_2x_R + Bv_R)$$
$$\dot{x}_R = v_R$$
$$\dot{v}_R = \frac{1}{M_1M_2}[K_1M_2x_1 - K_2(M_1 + M_2)x_R - B(M_1 + M_2)v_R + M_1f_a(t)]$$

(6)

To write the output equations in terms of the new state variables, we note that x_R is the elongation of K_2 and that the momentum of a mass depends on its absolute velocity. Thus

$$f_{K_2} = K_2x_R$$
$$m_T = M_1v_1 + M_2(v_1 + v_R)$$
$$= (M_1 + M_2)v_1 + M_2v_R$$

(7)

▶ **EXAMPLE 3.4**

For the system shown in Figure 3.2(a), the forces exerted by the linear spring K_1 and the linear dashpot B were K_1x_1 and Bv_R, respectively. Now assume that these two elements are replaced by nonlinear elements. The expression relating the force f_{K_1} to the displacement x_1 is denoted by $f_{K_1}(x_1)$, whereas the force on the dashpot is denoted by $f_B(v_R)$. Also take as an additional output the energy stored in the linear spring K_2.

SOLUTION   We again choose the state variables to be x_1, v_1, x_R, and v_R. The free-body diagrams are the same as those in Figure 3.3 except that K_1x_1 is replaced by $f_{K_1}(x_1)$, and Bv_R by $f_B(v_R)$, respectively. The resulting state-variable equations are identical to (6) except for the contributions of the two nonlinear elements:

$$\dot{x}_1 = v_1$$
$$\dot{v}_1 = \frac{1}{M_1}[-f_{K_1}(x_1) + K_2x_R + f_B(v_R)]$$
$$\dot{x}_R = v_R$$
$$\dot{v}_R = \frac{1}{M_1M_2}[M_2f_{K_1}(x_1) - K_2(M_1 + M_2)x_R - (M_1 + M_2)f_B(v_R) + M_1f_a(t)]$$

Note that the procedure for drawing the free-body diagrams and writing the state-variable equations is no more complicated than for the linear case. However, solving the differential equations analytically would be much more difficult, if not impossible.

The output equations for f_{K_2} and m_T remain the same as those in (7). The additional output equation for the energy stored in the linear spring K_2 is

$$w_{K_2} = \frac{1}{2}K_2x_R^2$$

which has a squared variable on the right side. Even in the case of linear components, some possible outputs may require output equations that are not just linear combinations of the state variables and inputs. Even then, however, the output equation should normally be an algebraic function of the state variables and inputs.

Next we consider two modifications of Figure 3.2(a) that illustrate circumstances in which the number of state variables might be different from the number of energy-storing elements. In the first of these, the spring K_1 is removed; in the second, an additional spring is added between the mass M_2 and a wall at the right.

▶ **EXAMPLE 3.5**

Find the state-variable model for the linear system shown in Figure 3.2(a) when the spring K_1 is removed. Let the outputs again be the tensile force f_{K_2} in the spring K_2 and the total momentum m_T of the masses. Then reconsider the problem when the displacement x_1 of the mass M_1 is an additional output of interest.

SOLUTION   The free-body diagrams will be the same as those in parts (b) and (c) of Figure 3.2, except that the force K_1x_1 will be missing. Thus (3) and (4) will still be valid if we just let $K_1 = 0$.

Because there are only three energy-storing elements in the modified system, however, we would expect to need only three state variables, rather than the four that appear in (3). Two of the three state variables can be chosen to be v_1 and v_2, which are related to the kinetic energy stored in the masses. We choose the third to be the elongation of the spring K_2, which is related to the potential energy in that element and which is $x_R = x_2 - x_1$. One of the state-variable equations is $\dot{x}_R = v_2 - v_1$. The other two follow from (3b) and (3d), with $K_1 = 0$ and with $x_2 - x_1$ replaced by x_R:

$$\dot{x}_R = v_2 - v_1$$
$$\dot{v}_1 = \frac{1}{M_1}[K_2 x_R - B v_1 + B v_2] \tag{8}$$
$$\dot{v}_2 = \frac{1}{M_2}[-K_2 x_R + B v_1 - B v_2 + f_a(t)]$$

The only variables on the right side of these equations are state variables and the input $f_a(t)$, as is required. The output equations are

$$f_{K_2} = K_2 x_R \tag{9}$$
$$m_T = M_1 v_1 + M_2 v_2$$

An alternative form of the state-variable model can be found by letting $K_1 = 0$ in (6) and (7). Because the only state variables that are needed are v_1, x_R, and v_R, we can omit the first equation in (6). Then the state-variable and output equations become

$$\dot{v}_1 = \frac{1}{M_1}(K_2 x_R + B v_R)$$
$$\dot{x}_R = v_R \tag{10}$$
$$\dot{v}_R = \frac{1}{M_1 M_2}[-K_2(M_1 + M_2)x_R - B(M_1 + M_2)v_R + M_1 f_a(t)]$$

and

$$f_{K_2} = K_2 x_R \tag{11}$$
$$m_T = (M_1 + M_2)v_1 + M_2 v_R$$

Finally, suppose that one of the outputs of interest is x_1. It is not possible to express x_1 as an algebraic function of only x_R, v_1, v_2, and $f_a(t)$. Thus we cannot construct a suitable output equation based on (8). Similarly, because x_1 cannot be expressed as an algebraic function of only v_1, x_R, v_R, and $f_a(t)$, we cannot write a suitable output equation based on (10). Hence, if x_1 is specified as one of the outputs of interest, three state variables are not sufficient. We would need to consider x_1 as an additional state variable and add the equation $\dot{x}_1 = v_1$ to either (8) or (10).

The previous example shows why an additional state variable that is not related to energy storage is sometimes needed. The following example illustrates how the number of state variables can be less than the number of energy-storing elements when the system's stored energy can be expressed by a reduced number of variables.

▶ **EXAMPLE 3.6**

Find a state-variable model for the linear system shown in Figure 3.4(a).

SOLUTION   Although there are three springs, their elongations are not all independent and can be specified in terms of the two displacement variables x_1 and x_2. The elongations of K_1, K_2, and K_3 are $x_1, x_2 - x_1$, and $-x_2$, respectively. The free-body diagrams for the two masses, which are shown in Figures 3.4(b) and 3.4(c), are the same as those in Figure 3.2 except for the additional force associated with K_3. By D'Alembert's law,

$$M_1\dot{v}_1 + (K_1 + K_2)x_1 + Bv_1 - K_2x_2 - Bv_2 = 0$$
$$M_2\dot{v}_2 - K_2x_1 - Bv_1 + (K_2 + K_3)x_2 + Bv_2 = f_a(t)$$

By solving these equations for $\dot{v}_1$ and $\dot{v}_2$, we can write the following four state-variable equations:

$$\dot{x}_1 = v_1$$

$$\dot{v}_1 = \frac{1}{M_1}[-(K_1 + K_2)x_1 - Bv_1 + K_2x_2 + Bv_2]$$

$$\dot{x}_2 = v_2 \qquad\qquad (12)$$

$$\dot{v}_2 = \frac{1}{M_2}[K_2x_1 + Bv_1 - (K_2 + K_3)x_2 - Bv_2 + f_a(t)]$$

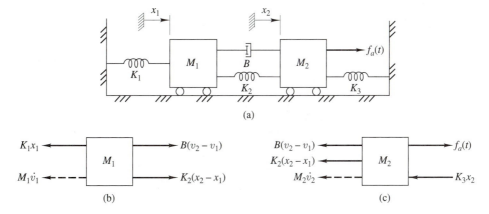

(a)

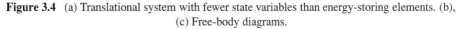

(b)   (c)

Figure 3.4   (a) Translational system with fewer state variables than energy-storing elements. (b), (c) Free-body diagrams.

If the outputs of interest are the tensile force f_{K_2} and the total momentum m_T, then the output equations are again given by (4).

___

In the following example, two springs and a dashpot are attached to a massless junction. The system contains three energy-storing elements and normally requires three state variables. However, when the dashpot is removed from the massless junction, the number of state variables is reduced.

▶ **EXAMPLE 3.7**

Find the state-variable model for the system shown in Figure 3.5(a). The input is the force $f_a(t)$, and the output is the displacement x_2 of the massless junction A. Repeat the problem when the dashpot B_2 is removed.

SOLUTION A satisfactory choice of state variables is x_1, v_1, and x_2, because these three variables determine the elongations of the springs and the velocity of the mass. One of the three state-variable equations is $\dot{x}_1 = v_1$.

To obtain the other two equations, we draw free-body diagrams for both the mass and the junction point, as shown in Figure 3.5(b) and Figure 3.5(c). Because the point A is massless, no inertial force is present in its free-body diagram. Summing the forces shown in these diagrams gives

$$M\dot{v}_1 + B_1 v_1 + K_1(x_1 - x_2) = f_a(t) \tag{13a}$$

$$B_2 \dot{x}_2 + K_2 x_2 + K_1(x_2 - x_1) = 0 \tag{13b}$$

Solving (13) for $\dot{v}_1$ and $\dot{x}_2$, we arrive at the state-variable equations

$$\dot{x}_1 = v_1 \tag{14a}$$

$$\dot{v}_1 = \frac{1}{M}[-K_1 x_1 - B_1 v_1 + K_1 x_2 + f_a(t)] \tag{14b}$$

$$\dot{x}_2 = \frac{1}{B_2}[K_1 x_1 - (K_1 + K_2)x_2] \tag{14c}$$

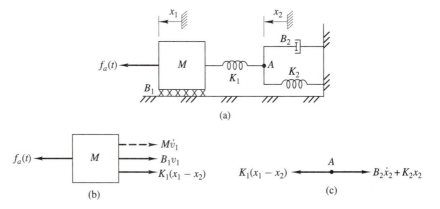

(a)

(b)

(c)

Figure 3.5 (a) Translational system containing a massless junction. (b), (c) Free-body diagrams.

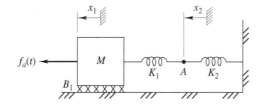

Figure 3.6 Translational system with springs in series.

Because the only output is one of the state variables, we shall not write a separate output equation.

Removing the dashpot corresponds to setting $B_2 = 0$. We cannot make this substitution in (14c), because division by zero is an invalid mathematical operation. However, replacing B_2 by zero in (13) gives

$$M\dot{v}_1 + B_1 v_1 + K_1(x_1 - x_2) = f_a(t) \tag{15a}$$

$$K_2 x_2 + K_1(x_2 - x_1) = 0 \tag{15b}$$

The second of these equations is purely algebraic and can be solved for x_2 in terms of x_1:

$$x_2 = \left(\frac{K_1}{K_1 + K_2}\right) x_1 \tag{16}$$

The displacements x_1 and x_2 are now proportional to each other, so they cannot both be state variables. If we choose x_1 and v_1 as the state variables, we still have $\dot{x}_1 = v_1$. To find the second state-variable equation, we substitute (16) into (15a) and solve for $\dot{v}_1$. Thus

$$\dot{x}_1 = v_1$$

$$\dot{v}_1 = \frac{1}{M}\left[-\left(\frac{K_1 K_2}{K_1 + K_2}\right) x_1 - B_1 v_1 + f_a(t)\right] \tag{17}$$

As required, the variables appearing on the right sides of these equations are either state variables or the input.

The system for $B_2 = 0$ is shown in Figure 3.6. From the discussion associated with Example 2.10 and Figure 2.23(a), we see that the two springs are now in series and can be replaced by a single equivalent spring for which $K_{eq} = K_1 K_2/(K_1 + K_2)$. This is reflected by the corresponding term in (17) and by the fact that only two state variables are needed. Because the output x_2 is no longer a state variable, we need a separate output equation to accompany (17). It is given by (16).

Sometimes the proposed state-variable equations that are initially found contain the derivative of an input on the right side. It might be argued that because the input is presumably known completely, its derivative could be found. However, it is important for the standard methods of solution (whether the solution is to be carried out analytically or with a computer) to eliminate any derivatives on the right side of the state-variable equations.

A basic method for doing this is illustrated in the following example, which also has a displacement input rather than an applied force as the specified input. An input derivative on the right side of the state-variable equations is removed by redefining one of the state variables.

It is also desirable to avoid input derivatives on the right side of the output equations when possible. In the following example, however, one of the output equations has unavoidable derivatives on the right side.

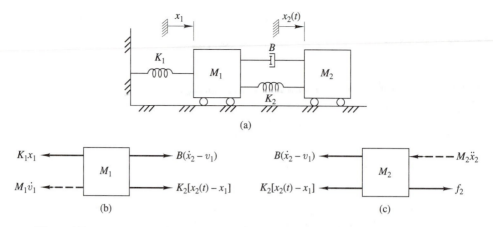

Figure 3.7 (a) Translational system with displacement input. (b), (c) Free-body diagrams.

▶ **EXAMPLE 3.8**

Find the state-variable model for the system shown in Figure 3.7(a), which is identical to Figure 2.14. The displacement $x_2(t)$ is a prescribed function of time that is the input of the system. The primary outputs of interest are the displacement and velocity of M_1. Consider as an additional output the force f_2 that must be applied to M_2 in order to move it with the prescribed displacement.

SOLUTION We know the motion of M_2 in advance, so there is no need to draw a free-body diagram for it unless it is needed for one of the output equations. The free-body diagram for M_1 is shown in Figure 3.7(b), and the corresponding force equation is

$$M_1 \dot{v}_1 + B v_1 + (K_1 + K_2)x_1 = B_2 \dot{x}_2 + K_2 x_2(t)$$

We may choose x_1 and v_1 as the state variables, write $\dot{x}_1 = v_1$, and solve the last equation for $\dot{v}_1$, which yields

$$\dot{x}_1 = v_1 \tag{18a}$$

$$\dot{v}_1 = \frac{1}{M_1}[-(K_1 + K_2)x_1 - B v_1 + B \dot{x}_2 + K_2 x_2(t)] \tag{18b}$$

Equation (18b) does not fit the required form for state-variable equations given by (1) because its right side contains $\dot{x}_2$, the derivative of the input. In order to eliminate $\dot{x}_2$ from the initial state-variable equations, we introduce a new state variable q to be used in place of v_1. To determine how q must be defined, we rewrite (18b) with the derivative of x_2 moved to the left side:

$$\dot{v}_1 - \frac{B}{M_1}\dot{x}_2 = \frac{1}{M_1}[-(K_1 + K_2)x_1 - B v_1 + K_2 x_2(t)] \tag{19}$$

If we select the new state variable q as

$$q = v_1 - \frac{B}{M_1}x_2(t) \tag{20}$$

then the left side of (19) can be written as $\dot{q}$. Rearranging (20), we see that the old state variable v_1 is given by

$$v_1 = q + \frac{B}{M_1}x_2(t) \tag{21}$$

Finally, we insert (21) into the right sides of (18a) and (19) to obtain

$$\dot{x}_1 = q + \frac{B}{M_1} x_2(t)$$
$$\dot{q} = \frac{1}{M_1}\left[-(K_1 + K_2)x_1 - Bq + \left(K_2 - \frac{B^2}{M_1}\right)x_2(t)\right]$$
(22)

which is in state-variable form. The output x_1 is a state variable, and the output v_1 is given by (21), which fits the standard form for an output equation.

To obtain an expression for the force f_2 applied to M_2, we draw the free-body diagram shown in Figure 3.7(c). By D'Alembert's law,

$$f_2 = -K_2 x_1 - Bv_1 + M_2 \ddot{x}_2 + B\dot{x}_2 + K_2 x_2(t)$$

Using (21) for v_1, we rewrite this equation as

$$f_2 = -K_2 x_1 - Bq + M_2 \ddot{x}_2 + B\dot{x}_2 + \left(K_2 - \frac{B^2}{M_1}\right)x_2(t)$$

which contains derivatives of the input in addition to the input $x_2(t)$ and the state variables x_1 and q. For this particular output, however, there is no way to avoid input derivatives on the right side.

Systems may have several inputs. In the final state-variable example, a time-varying force $f_a(t)$ is applied to M_1, and there is a gravitational force on M_2.

▶ **EXAMPLE 3.9**

Find a state-variable model for the system that was shown in Figure 2.19(a) and considered in Example 2.7. The two inputs are $f_a(t)$ and the gravitational constant g. The outputs are x_{K_1} and f_{B_2}, defined as the elongation of K_1 and the upward force on the top of B_2, respectively.

SOLUTION The free-body diagrams for the two masses were shown in parts (b) and (c) of Figure 2.19. The four state variables can be chosen to be the displacements x_1 and x_2 and the velocities v_1 and v_2. In Example 2.7, the force equations were shown to be

$$M_1 \dot{v}_1 + B_1 v_1 + K_1(x_1 - x_2) = f_a(t)$$
$$M_2 \dot{v}_2 + B_2 v_2 + K_2 x_2 + M_2 g = K_1(x_1 - x_2)$$

We solve the first equation for $\dot{v}_1$, solve the second equation for $\dot{v}_2$, and then write the following four state-variable equations:

$$\dot{x}_1 = v_1$$
$$\dot{v}_1 = \frac{1}{M_1}[-K_1 x_1 - B_1 v_1 + K_1 x_2 + f_a(t)]$$
$$\dot{x}_2 = v_2$$
$$\dot{v}_2 = \frac{1}{M_2}[K_1 x_1 - (K_1 + K_2)x_2 - B_2 v_2 - M_2 g]$$
(23)

The output equations are

$$x_{K_1} = x_1 - x_2$$
$$f_{B_2} = B_2 v_2$$
(24)

The right sides of both sets of equations are algebraic functions of the four state variables and the two inputs, as required.

▶ ## 3.2 INPUT-OUTPUT EQUATIONS

In this section, we develop the system models in the form of input-output differential equations by eliminating all variables except the inputs and outputs and their derivatives. For a system with one input $u(t)$ and one output y, the input-output equation has the general form

$$a_n y^{(n)} + \cdots + a_2 \ddot{y} + a_1 \dot{y} + a_0 y = b_m u^{(m)} + \cdots + b_1 \dot{u} + b_0 u(t) \tag{25}$$

where $y^{(n)} = d^n y / dt^n$ and $u^{(m)} = d^m u / dt^m$, and where for systems of practical interest $m \leq n$. For fixed linear systems, all the coefficients in (25) are constants. In order to solve such an equation for y for all $t \geq t_0$, we need to know not only the input $u(t)$ for $t \geq t_0$ but also the n initial conditions $y(t_0), \dot{y}(t_0), \ddot{y}(t_0), \ldots, y^{(n-1)}(t_0)$. Finding these initial conditions may be a difficult task.

For systems with more than one input, the right side of (25) will include additional input terms. If there are several outputs, we need a separate equation similar to (25) for each output. For example, the pair of equations

$$\ddot{y}_1 + 2\dot{y}_1 + 2y_1 = 3\dot{u}_1 + 2u_1(t) + u_2(t) + 3\dot{u}_3$$
$$\ddot{y}_2 + 2\dot{y}_2 + 2y_2 = u_1(t) + 2u_2(t) + u_3(t)$$

corresponds to a system with two outputs and three inputs. In the general case, each of the input-output equations involves only one unknown variable and its derivatives. Thus, unlike state-variable equations, each equation can be solved independently of the others.

An input-output equation can be constructed by combining the equations in a state-variable model, if such a model has already been found. A more direct method is to label the forces in the free-body diagrams in terms of the output variables and the minimum number of additional variables. For a translational system in which the outputs are displacements, for example, we would normally write the inertial and friction forces in the form $M\ddot{x}$ and $B\dot{x}$, rather than as $M\dot{v}$ and Bv.

▶ ### EXAMPLE 3.10

Write the input-output equation for the system shown in Figure 3.1(a) when the output is the displacement x.

SOLUTION   In Example 3.1, the state-variable equations were found to be

$$\dot{x} - v$$

$$\dot{v} = \frac{1}{M}[-Kx - Bv + f_a(t)]$$

We can combine these equations by replacing v and $\dot{v}$ in the second equation by $\dot{x}$ and $\ddot{x}$, respectively, to obtain

$$M\ddot{x} + B\dot{x} + Kx = f_a(t)$$

The direct approach, which does not make use of the state-variable model, would be to label all the forces on the free-body diagram except $f_a(t)$ in terms of x and its derivatives. Then the foregoing input-output equation follows directly from D'Alembert's law.

► **EXAMPLE 3.11**

Find the input-output equations that relate the outputs x_1 and x_2 to the input $f_a(t)$ for the system shown in Figure 3.5(a).

SOLUTION The free-body diagrams for the mass M and the junction point A were shown in parts (b) and (c) of Figure 3.5, and the corresponding force equations were written in (13). When all forces except the input are expressed in terms of x_1 and x_2, these equations become

$$M\ddot{x}_1 + B_1\dot{x}_1 + K_1x_1 - K_1x_2 = f_a(t) \tag{26a}$$

$$B_2\dot{x}_2 + (K_1 + K_2)x_2 - K_1x_1 = 0 \tag{26b}$$

To obtain a single differential equation relating x_1 to $f_a(t)$ from this pair of coupled equations, we must eliminate x_2 and $\dot{x}_2$. If one of the equations contains an unwanted variable but none of its derivatives, we can solve for it in terms of the remaining variables and their derivatives. Then we can eliminate the unwanted variable from the model by substitution. Thus we rewrite (26a) as

$$x_2 = \frac{1}{K_1}[M\ddot{x}_1 + B_1\dot{x}_1 + K_1x_1 - f_a(t)]$$

and, by differentiating once, we obtain

$$\dot{x}_2 = \frac{1}{K_1}(M\dddot{x}_1 + B_1\ddot{x}_1 + K_1\dot{x}_1 - \dot{f}_a)$$

Substituting these expressions for x_2 and $\dot{x}_2$ into (26b) gives

$$\frac{B_2}{K_1}(M\dddot{x}_1 + B_1\ddot{x}_1 + K_1\dot{x}_1 - \dot{f}_a) + \frac{K_1 + K_2}{K_1}[M\ddot{x}_1 + B_1\dot{x}_1 + K_1x_1 - f_a(t)] - K_1x_1 = 0$$

or

$$MB_2\dddot{x}_1 + (B_1B_2 + K_1M + K_2M)\ddot{x}_1 + (B_2K_1 + B_1K_1 + B_1K_2)\dot{x}_1 + K_1K_2x_1$$
$$= B_2\dot{f}_a + (K_1 + K_2)f_a(t) \tag{27}$$

which is the input-output equation for x_1. To obtain a differential equation with x_2 as the only unknown variable, we use (26b) to get an expression for x_1, which we then substitute into (26a). The result is

$$MB_2\dddot{x}_2 + (B_1B_2 + K_1M + K_2M)\ddot{x}_2 + (B_2K_1 + B_1K_1 + B_1K_2)\dot{x}_2 + K_1K_2x_2 = K_1f_a(t) \tag{28}$$

Note that the coefficients on the left sides of (27) and (28) are the same. Note also that the input-output equations are third order and that the system has three state variables.

For the special case where $B_2 = 0$, which corresponds to removing the dashpot, (27) and (28) reduce to

$$M\ddot{x}_1 + B_1\dot{x}_1 + \left(\frac{K_1K_2}{K_1 + K_2}\right)x_1 = f_a(t)$$

$$M\ddot{x}_2 + B_1\dot{x}_2 + \left(\frac{K_1K_2}{K_1 + K_2}\right)x_2 = \left(\frac{K_1}{K_1 + K_2}\right)f_a(t)$$

The input-output equation for x_1 is now similar to that in Example 3.10, with K replaced by an equivalent constant $K_{eq} = K_1K_2/(K_1 + K_2)$ for the two springs that are now in series. We also see that $x_2 = [K_1/(K_1 + K_2)]x_1$ and that our results are consistent with those in Example 3.7. Because the two springs can be replaced by a single equivalent spring, the input-output equations are only second order.

Reduction of Simultaneous Differential Equations

As we have seen, it is often necessary to combine a set of differential equations involving more than one dependent variable into a single differential equation with a single dependent variable. We did this in Example 3.11 by straightforward substitution. However, when it is not obvious how to eliminate the unwanted variable easily, the following procedure is recommended.

Let p denote the differentiation operator d/dt such that $py = \dot{y}$, $p^2y = \ddot{y}$, etc. Then, for example,

$$(p+2)y = \dot{y} + 2y$$
$$[(p+1)(p+2)]y = (p^2 + 3p + 2)y$$
$$= \ddot{y} + 3\dot{y} + 2y$$
$$(a_n p^n + \cdots + a_2 p^2 + a_1 p + a_0)y = a_n y^{(n)} + \cdots + a_2 \ddot{y} + a_1 \dot{y} + a_0 y$$

where $y^{(k)} = d^k y/dt^k$ for any positive integer value of k. Remember that p must operate on the variable or expression that follows it and that it is not a variable or algebraic quantity itself.

Suppose that we have the pair of equations

$$\dot{y}_1 + 2y_1 + y_2 = 3u(t)$$
$$2\dot{y}_1 + 5y_1 - 2\dot{y}_2 + 2y_2 = 0 \tag{29}$$

and want to find a single differential equation involving only the variables y_2 and $u(t)$. In terms of the p operator, we can rewrite (29) as

$$(p+2)y_1 + y_2 = 3u(t)$$
$$(2p+5)y_1 + (-2p+2)y_2 = 0$$

We now premultiply the first equation by $(2p+5)$, premultiply the second equation by $(p+2)$, and then subtract the new second equation from the first. The result is

$$[(2p+5) - (p+2)(-2p+2)]y_2 = (2p+5)3u(t)$$

from which

$$(2p^2 + 4p + 1)y_2 = (6p+15)u(t) \tag{30}$$

To return to a differential equation relating y_2 and $u(t)$, we observe that (30) is the operator form of

$$2\ddot{y}_2 + 4\dot{y}_2 + y_2 = 6\dot{u} + 15u(t)$$

which is indeed the correct input-output equation. This algebraic procedure provides a useful means of manipulating sets of differential equations with constant coefficients.

▶ **EXAMPLE 3.12**

For the system in Figure 3.2(a), which is identical to Figure 2.13(a), find the input-output equation relating x_1 and $f_a(t)$.

SOLUTION   The free-body diagrams for the two masses were shown in Figure 2.13 (with all forces labeled in terms of displacements and the input) and in Figure 3.2 (with forces labeled in terms of state variables and the input). Repeating (2.16), where the force equations were written in terms of displacements, we have the pair of simultaneous second-order differential equations

$$M_1\ddot{x}_1 + B\dot{x}_1 + (K_1 + K_2)x_1 - B\dot{x}_2 - K_2x_2 = 0$$
$$-B\dot{x}_1 - K_2x_1 + M_2\ddot{x}_2 + B\dot{x}_2 + K_2x_2 = f_a(t)$$

In terms of the p operator, these equations become

$$[\,p^2M_1 + pB + (K_1 + K_2)]\,x_1 - (pB + K_2)x_2 = 0$$
$$-(pB + K_2)x_1 + (p^2M_2 + pB + K_2)x_2 = f_a(t)$$

When we combine this pair of operator equations algebraically to eliminate x_2, we find that

$$\{M_1M_2p^4 + (M_1 + M_2)Bp^3 + [M_1K_2 + M_2(K_1 + K_2)]p^2 + BK_1p + K_1K_2\}x_1 = (pB + K_2)f_a(t)$$

which is the operator form of the differential equation

$$M_1M_2x_1^{(iv)} + (M_1 + M_2)Bx_1^{(iii)} + [M_1K_2 + M_2(K_1 + K_2)]\ddot{x}_1 + BK_1\dot{x}_1 + K_1K_2x_1 = B\dot{f}_a + K_2f_a(t)$$
$$\tag{31}$$

The symbols $x^{(iv)}$ and $x^{(iii)}$ are defined as $x^{(iv)} = d^4x/dt^4$ and $x^{(iii)} = d^3x/dt^3$. As expected for a system that has four state variables, the input-output differential equation is of order four.

Comparison with the State-Variable Method

The order of the input-output differential equation describing a system is usually the same as the number of state variables. Occasionally one or more of the state variables may have no effect on the output, in which case the order of the input-output equation is less than the number of state variables.

For a first-order system, both forms of the system model involve a single first-order differential equation and are essentially identical. For higher-order systems they are quite different. We must solve a set of n first-order differential equations in state-variable form as a group, and we must know the initial value of each state variable to solve the set of n equations. An input-output equation of order n contains only one dependent variable, but we need to know the initial values of that variable and its first $n - 1$ derivatives. In practice, finding the input-output equation and the associated initial conditions may require more effort than finding the information needed for a state-variable solution.

Using the state-variable equations has significant computational advantages when a computer solution is to be found, as is discussed in Chapter 4. In fact, standard methods for solving a high-order, possibly nonlinear input-output differential equation on a computer usually require decomposition into a set of simultaneous first-order equations anyway. The analytical solution of various models for fixed linear systems is considered in Chapters 7 and 8.

State-variable equations are particularly convenient for complex multi-input, multi-output systems. They are often written in matrix form, and, in addition to their computational advantages, they can be used to obtain considerable insight into system behavior. The state-variable concept has formed the basis for many of the recent theoretical developments in system analysis.

▶ 3.3 MATRIX FORMULATION OF STATE-VARIABLE EQUATIONS

One of the characteristics of state-variable models is that they are suitable for matrix notation and for the techniques of linear algebra. For example, we can represent any number of first-order state-variable equations by a single matrix differential equation merely by making the appropriate definitions. The data for computer solutions are frequently entered in matrix notation. The important commercial program MATLAB, which is discussed in considerable detail in Chapters 4 and 14, manipulates matrix equations and presents some of its output displays in matrix form.

Standard procedures exist for solving a set of matrix equations involving inputs and outputs that are functions of time. We can apply many of the theoretical properties of matrices that are taught in introductory linear algebra courses to the study of dynamic systems once we have put their models into matrix form. Such an approach leads to important concepts and a fuller understanding of system behavior, especially when dealing with complex multi-input, multi-output systems.

Although we shall not discuss the analytical solution of matrix state-variable equations in this book, some references are given in Appendix F, Selected Reading. We do, however, need to know how to put the modeling equations into matrix form. Basic matrix operations such as addition and multiplication are summarized in Appendix B.

The required form for the state-variable model was explained in Section 3.1. For a fixed linear system in which $q_1, q_2, \ldots, q_n$ are the state variables, where $u_1, u_2, \ldots, u_m$ are the inputs, and where $y_1, y_2, \ldots, y_p$ are the outputs, the state-variables equations are

$$
\begin{aligned}
\dot{q}_1 &= a_{11}q_1 + a_{12}q_2 + \cdots + a_{1n}q_n + b_{11}u_1 + \cdots + b_{1m}u_m \\
\dot{q}_2 &= a_{21}q_1 + a_{22}q_2 + \cdots + a_{2n}q_n + b_{21}u_1 + \cdots + b_{2m}u_m \\
&\vdots \\
\dot{q}_n &= a_{n1}q_1 + a_{n2}q_2 + \cdots + a_{nn}q_n + b_{n1}u_1 + \cdots + b_{nm}u_m
\end{aligned}
\tag{32}
$$

and the output equations have the form

$$
\begin{aligned}
y_1 &= c_{11}q_1 + c_{12}q_2 + \cdots + c_{1n}q_n + d_{11}u_1 + \cdots + d_{1m}u_m \\
y_2 &= c_{21}q_1 + c_{22}q_2 + \cdots + c_{2n}q_n + d_{21}u_1 + \cdots + d_{2m}u_m \\
&\vdots \\
y_p &= c_{p1}q_1 + c_{p2}q_2 + \cdots + c_{pn}q_n + d_{p1}u_1 + \cdots + d_{pm}u_m
\end{aligned}
\tag{33}
$$

where the coefficients a_{ij}, b_{ij}, c_{ij}, and d_{ij} are constants. The initial conditions associated with (32) are the initial values of the state variables, $q_1(0), q_2(0), \ldots, q_n(0)$.

When dealing with the set of n state variables $q_1, q_2, \ldots, q_n$, we shall use the symbol $\mathbf{q}$ to denote the entire set.[2] Thus

$$
\mathbf{q} =
\begin{bmatrix}
q_1 \\
q_2 \\
\vdots \\
q_n
\end{bmatrix}
$$

which is a matrix having n rows and a single column, with each of its elements being one of the state variables. As such, its elements are functions of time. Matrices having a single column are commonly referred to as **column vectors** or, for short, **vectors**. Hence the symbol $\mathbf{q}$ will be called the **state vector**, and it is understood that its ith element is the state

---

[2]Boldface symbols are used to denote matrices. The symbols are generally capitals, but lowercase letters may be used for vectors. Variables written in italic type are scalars.

variable q_i. The initial value of the state vector **q** is the vector

$$\mathbf{q}(0) = \begin{bmatrix} q_1(0) \\ q_2(0) \\ \vdots \\ q_n(0) \end{bmatrix}$$

The vector $\dot{\mathbf{q}}$ is an n-element vector whose elements are the derivatives of the corresponding elements in **q**:

$$\dot{\mathbf{q}} = \begin{bmatrix} \dot{q}_1 \\ \dot{q}_2 \\ \vdots \\ \dot{q}_n \end{bmatrix}$$

The m inputs $u_1, u_2, \ldots, u_m$ will be represented by **u**, the **input vector**, which is defined as

$$\mathbf{u} = \begin{bmatrix} u_1 \\ u_2 \\ \vdots \\ u_m \end{bmatrix}$$

The p outputs $y_1, y_2, \ldots, y_p$ will be elements in **y**, the **output vector**, which is given by

$$\mathbf{y} = \begin{bmatrix} y_1 \\ y_2 \\ \vdots \\ y_p \end{bmatrix}$$

The coefficients are put into the following four matrices. We define the $n \times n$ matrix

$$\mathbf{A} = \begin{bmatrix} a_{11} & a_{12} & \cdots & a_{1n} \\ a_{21} & a_{22} & \cdots & a_{2n} \\ \vdots & \vdots & & \vdots \\ a_{n1} & a_{n2} & \cdots & a_{nn} \end{bmatrix}$$

the $n \times m$ matrix

$$\mathbf{B} = \begin{bmatrix} b_{11} & b_{12} & \cdots & b_{1m} \\ b_{21} & b_{22} & \cdots & b_{2m} \\ \vdots & \vdots & & \vdots \\ b_{n1} & b_{n2} & \cdots & b_{nm} \end{bmatrix}$$

the $p \times n$ matrix

$$\mathbf{C} = \begin{bmatrix} c_{11} & c_{12} & \cdots & c_{1n} \\ c_{21} & c_{22} & \cdots & c_{2n} \\ \vdots & \vdots & & \vdots \\ c_{p1} & c_{p2} & \cdots & c_{pn} \end{bmatrix}$$

and the $p \times m$ matrix

$$\mathbf{D} = \begin{bmatrix} d_{11} & d_{12} & \cdots & d_{1m} \\ d_{21} & d_{22} & \cdots & d_{2m} \\ \vdots & \vdots & & \vdots \\ d_{p1} & d_{p2} & \cdots & d_{pm} \end{bmatrix}$$

Then we can write the state-variable and output equations for the fixed linear system described by (32) and (33) as

$$\dot{\mathbf{q}} = \mathbf{Aq} + \mathbf{Bu}$$
$$\mathbf{y} = \mathbf{Cq} + \mathbf{Du} \tag{34}$$

Reflecting for a moment, we note that $\dot{q}_i$ is the ith element of $\dot{\mathbf{q}}$ and is obtained by multiplying the ith row of $\mathbf{A}$ by the corresponding elements in the column vector $\mathbf{q}$ and adding to that product the product of the ith row of $\mathbf{B}$ and the column vector $\mathbf{u}$. Carrying out these operations gives

$$\dot{q}_i = a_{i1}q_1 + a_{i2}q_2 + \cdots + a_{in}q_n + b_{i1}u_1 + \cdots + b_{im}u_m$$

Likewise, we find the kth element of $\mathbf{y}$ by multiplying the kth row of $\mathbf{C}$ by $\mathbf{q}$ and adding to that product the product of the kth row of $\mathbf{D}$ and the vector $\mathbf{u}$, obtaining

$$y_k = c_{k1}q_1 + c_{k2}q_2 + \cdots + c_{kn}q_n + d_{k1}u_1 + \cdots + d_{km}u_m$$

▶ **EXAMPLE 3.13**

For the system shown in Figure 3.1(a) and modeled in Example 3.1, write the state-variable model in the form of (34) and identify the coefficient matrices $\mathbf{A}$, $\mathbf{B}$, $\mathbf{C}$, and $\mathbf{D}$.

SOLUTION   From Example 3.1, the state-variable equations are

$$\dot{x} = v$$
$$\dot{v} = \frac{1}{M}[-Kx - Bv + f_a(t)] \tag{35}$$

and the output equations are

$$f_K = Kx$$
$$v = v$$
$$a_M = \frac{1}{M}[-Kx - Bv + f_a(t)] \tag{36}$$

Note that v is both a specified output and a state variable. Although the corresponding output equation may seem trivial, we should include it when writing the model in matrix form.

Because there are two state variables, one input, and three outputs, the state vector $\mathbf{q}$, the input vector $\mathbf{u}$, and the output vector $\mathbf{y}$ will have two, one, and three elements respectively. Let

$$\mathbf{q} = \begin{bmatrix} x_1 \\ v_1 \end{bmatrix}, \quad \mathbf{u} = [f_a(t)], \quad \text{and} \quad \mathbf{y} = \begin{bmatrix} f_K \\ v \\ a_M \end{bmatrix}$$

When a matrix contains only a single element, as is the case with $\mathbf{u}$, the brackets around the matrix are usually omitted. With the above definitions of $\mathbf{q}$, $\mathbf{u}$, and $\mathbf{y}$, we can write (35) and (36) as

$$\begin{bmatrix} \dot{x}_1 \\ \dot{v}_1 \end{bmatrix} = \underbrace{\begin{bmatrix} 0 & 1 \\ -K/M & -B/M \end{bmatrix}}_{\mathbf{A}} \begin{bmatrix} x \\ v \end{bmatrix} + \underbrace{\begin{bmatrix} 0 \\ 1/M \end{bmatrix}}_{\mathbf{B}} f_a(t)$$

and

$$\begin{bmatrix} f_K \\ v \\ a_M \end{bmatrix} = \underbrace{\begin{bmatrix} K & 0 \\ 0 & 1 \\ -K/M & -B/M \end{bmatrix}}_{\mathbf{C}} \begin{bmatrix} x \\ v \end{bmatrix} + \underbrace{\begin{bmatrix} 0 \\ 0 \\ 1/M \end{bmatrix}}_{\mathbf{D}} f_a(t)$$

which have the form of (34) and where the coefficient matrices $\mathbf{A}$, $\mathbf{B}$, $\mathbf{C}$, and $\mathbf{D}$ are specifically identified. Because there is only one input, $\mathbf{B}$ and $\mathbf{D}$ have only one column. The reader should verify that carrying out the matrix operations in these equations gives (35) and (36).

▶ **EXAMPLE 3.14**

Write the state-variable model in matrix form for the system shown in Figure 3.5(a) and considered in Example 3.7. Let the only output be the tensile force f_{K_1} in the spring K_1.

SOLUTION   From (14), the state-variable equations are

$$\dot{x}_1 = v_1$$

$$\dot{v}_1 = \frac{1}{M}[-K_1 x_1 - B_1 v_1 + K_1 x_2 + f_a(t)]$$

$$\dot{x}_2 = \frac{1}{B_2}[K_1 x_1 - (K_1 + K_2)x_2]$$

and the output equation is $f_{K_1} = K_1(x_1 - x_2)$. We take the state vector $\mathbf{q}$, the input vector $\mathbf{u}$, and the output vector $\mathbf{y}$ to be

$$\mathbf{q} = \begin{bmatrix} x_1 \\ v_1 \\ x_2 \end{bmatrix}, \mathbf{u} = [f_a(t)], \text{ and } \mathbf{y} = [f_{K_1}]$$

Then

$$\dot{\mathbf{q}} = \begin{bmatrix} 0 & 1 & 0 \\ -\dfrac{K_1}{M} & -\dfrac{B_1}{M} & \dfrac{K_1}{M} \\ \dfrac{K_1}{B_2} & 0 & -\left(\dfrac{K_1 + K_2}{B_2}\right) \end{bmatrix} \mathbf{q} + \begin{bmatrix} 0 \\ \dfrac{1}{M} \\ 0 \end{bmatrix} \mathbf{u}$$

and

$$\mathbf{y} = [K_1 \quad 0 \quad -K_1]\mathbf{q} + [0]\mathbf{u}$$

which have the form of (34). Because there is only one input, $\mathbf{B}$ and $\mathbf{D}$ have only one column. Because there is only one output, $\mathbf{C}$ and $\mathbf{D}$ have only one row. Thus $\mathbf{B}$ is a column vector, $\mathbf{C}$ a row vector, and $\mathbf{D}$ a scalar. Such systems are called single-input, single-output systems to distinguish them from the more general class of multi-input, multi-output systems.

▶ **EXAMPLE 3.15**

Rewrite in matrix form the state-variable and output equations for the system shown in Figure 2.19(a) and modeled in Example 3.9.

SOLUTION   From Example 3.9, the state-variable equations are

$$\dot{x}_1 = v_1$$

$$\dot{v}_1 = \frac{1}{M_1}[-K_1x_1 - B_1v_1 + K_1x_2 + f_a(t)]$$

$$\dot{x}_2 = v_2$$

$$\dot{v}_2 = \frac{1}{M_2}[K_1x_1 - (K_1 + K_2)x_2 - B_2v_2 - M_2g]$$

and the output equations are

$$x_{K_1} = x_1 - x_2$$

$$f_{B_2} = B_2v_2$$

The state vector has four elements and is taken to be

$$\mathbf{q} = \begin{bmatrix} x_1 \\ v_1 \\ x_2 \\ v_2 \end{bmatrix}$$

We take the input vector **u** and the output vector **y** to be

$$\mathbf{u} = \begin{bmatrix} f_a(t) \\ g \end{bmatrix}$$

and

$$\mathbf{y} = \begin{bmatrix} x_{K_1} \\ f_{B_2} \end{bmatrix}$$

With these definitions, we can write the model in the form of (34) as

$$\dot{\mathbf{q}} = \begin{bmatrix} 0 & 1 & 0 & 0 \\ -\dfrac{K_1}{M_1} & -\dfrac{B_1}{M_1} & \dfrac{K_1}{M_1} & 0 \\ 0 & 0 & 0 & 1 \\ \dfrac{K_1}{M_2} & 0 & -\dfrac{(K_1 + K_2)}{M_2} & -\dfrac{B_2}{M_2} \end{bmatrix} \mathbf{q} + \begin{bmatrix} 0 & 0 \\ \dfrac{1}{M_1} & 0 \\ 0 & 0 \\ 0 & -1 \end{bmatrix} \mathbf{u}$$

and

$$\mathbf{y} = \begin{bmatrix} 1 & 0 & -1 & 0 \\ 0 & 0 & 0 & B_2 \end{bmatrix} \mathbf{q} + \begin{bmatrix} 0 & 0 \\ 0 & 0 \end{bmatrix} \mathbf{u}$$

For any fixed linear system, the elements in the coefficients matrices **A**, **B**, **C**, and **D** will all be constants. Equations (34) would also be valid for a time-varying linear system, with the understanding that some of the elements in **A**, **B**, **C**, and **D** would then be functions of time.

For the case of systems that can be time-varying and nonlinear, a more comprehensive form for the equations is necessary. For a general nth-order system having m inputs, the individual state-variable equations would have the form

$$\begin{aligned} \dot{q}_1 &= f_1(q_1, q_2, \ldots, q_n, u_1, \ldots, u_m, t) \\ \dot{q}_2 &= f_2(q_1, q_2, \ldots, q_n, u_1, \ldots, u_m, t) \\ &\vdots \\ \dot{q}_n &= f_n(q_1, q_2, \ldots, q_n, u_1, \ldots, u_m, t) \end{aligned} \qquad (37)$$

where $q_1, q_2, \ldots, q_n$ are the state variables and $u_1, u_2, \ldots, u_m$ are the inputs. The algebraic functions $f_1, f_2, \ldots, f_n$ express the state-variable derivatives $\dot{q}_1, \dot{q}_2, \ldots, \dot{q}_n$ in terms of the state variables, the inputs, and possibly the time t.[3] The initial conditions associated with (37) are the initial values of the state variables, $q_1(0), q_2(0), \ldots, q_n(0)$. If the system has p outputs, the individual output equations have the general form

$$
\begin{aligned}
y_1 &= g_1(q_1, q_2, \ldots, q_n, u_1, u_2, \ldots, u_m, t) \\
y_2 &= g_2(q_1, q_2, \ldots, q_n, u_1, u_2, \ldots, u_m, t) \\
&\vdots \\
y_p &= g_p(q_1, q_2, \ldots, q_n, u_1, u_2, \ldots, u_m, t)
\end{aligned}
\tag{38}
$$

where $g_1, g_2, \ldots, g_p$ are again algebraic functions of the state variables, the inputs, and possibly time. The sets of equations in (37) and (38) can be represented by the two vector equations

$$
\begin{aligned}
\dot{\mathbf{q}} &= \mathbf{f}(\mathbf{q}, \mathbf{u}, t) \\
\mathbf{y} &= \mathbf{g}(\mathbf{q}, \mathbf{u}, t)
\end{aligned}
\tag{39}
$$

where $\mathbf{f}$ and $\mathbf{g}$ are vector functions having n and p elements, respectively.

▷ SUMMARY

In this chapter we introduced the concept of state variables and showed how to write the mathematical model of a dynamic system in terms of these variables. Such a model consists of a set of coupled first-order differential equations, whose right-hand sides are algebraic functions of the state variables and inputs. Once the state variables have been found as known functions of time, the system outputs can be found easily as algebraic functions of the state variables and inputs.

In Section 3.2 we used a very different approach, eliminating from our equations all variables except the inputs and outputs and their derivatives. For an nth-order system, we obtain an nth-order differential equation for each output. These equations, which are not coupled together and which can be solved independently of each other, have only the inputs and their derivatives as variables on the right-hand side.

The final section showed how to write the state-variable equations in matrix form. In Chapter 14 we show how these matrices can be used to construct computer implementations of linear dynamic systems in very compact form using MATLAB and its Control System Toolbox. In a similar fashion, models in input-output form can be implemented on the computer once we have developed transfer functions, which is done in Chapter 8. Computer solutions obtained with simulation techniques are discussed next, in Chapter 4.

▷ PROBLEMS

3.1. Find the state-variable model for the system shown in Figure 2.16(a) and considered in Example 2.5:

    a.   When x and $v = \dot{x}$ are state variables

    b.   When z and $w = \dot{z}$ are state variables

The inputs are $f_a(t)$ and g. Take the energy stored in the spring as the output.

---

[3]Note that the use of the symbol f in this very general context is not related to the use of f to denote forces in mechanical systems.

3.2. The mathematical model of a linear system with input $f_a(t)$ and output y is

$$\ddot{x} + \alpha\ddot{x} + \beta\dot{x} + \gamma x = f_a(t)$$
$$y = \ddot{x} + \dot{x}$$

Write the model in state-variable form, where the state variables are $x, v = \dot{x}$, and $a = \ddot{x}$.

*3.3. The following pair of equations corresponds to a linear dynamic system having $f_a(t)$ as its input and y as its output:

$$\ddot{y} + 4\dot{y} + 2y = x$$
$$\dot{x} + x + y = f_a(t)$$

Derive the state-variable form of the model. Define any new symbols.

3.4. A mechanical system having input $f_a(t)$ and output $y = \dot{x}_1 - \dot{x}_2$ obeys the pair of differential equations

$$M_1\ddot{x}_1 + B(\dot{x}_1 - \dot{x}_2) + K_1 x_1 = 0$$
$$M_2\ddot{x}_2 + B(\dot{x}_2 - \dot{x}_1) + K_2 x_2 = f_a(t)$$

Write the model in state-variable form when the state variables are $x_1, v_1 = \dot{x}_1, x_2$, and $v_2 = \dot{x}_2$.

*3.5. Write state-variable equations for the mechanical system shown in Figure 2.15(a) and modeled in Example 2.4 by Equation (2.20). Take the state variables to be the absolute displacements x_1 and x_2 and their derivatives v_1 and v_2. Include algebraic output equations for y_1, the viscous force of M_2 on M_1 (positive to the right), and y_2, the tensile force in spring K_2.

3.6. Repeat Problem 3.5 using Equation (2.19). Take the state variables to be the absolute displacement x, its derivative $v_x = \dot{x}$, the relative displacement z, and its derivative $v_z = \dot{z}$.

3.7. Write state-variable equations for the mechanical system shown in Figure 2.17(a) and modeled in Example 2.6 by Equation (2.25). The inputs are the applied force $f_a(t)$ and the gravitational constant g. Include algebraic output equations for the tensile force in each of the three springs.

3.8. Find a set of state-variable equations for the system shown in Figure P2.5. The output is the total momentum of the system.

*3.9. Find the state-variable model for the system shown in Figure P2.15 when the outputs are the elongation y_1 of the spring K_1 and the acceleration a_1 of the mass M_1. The inputs are $f_a(t)$ and the gravitational constant g.

3.10. Repeat Problem 3.5 when the spring K_1 in Figure 2.15(a) is removed. Keep the same outputs, but use x_2, v_1, and v_2 as the state variables. Explain why it is not necessary to have x_1 as a state variable.

*3.11. Find the state-variable model for the system shown in Figure P2.16. The output is the elongation of the lower spring.

3.12. Find the state-variable model for the system shown in Figure P2.21. The output is the increase in the separation between M_1 and M_2.

*3.13. Write the state-variable equations for the system shown in Figure P2.27, where the input is the displacement $x_3(t)$. The output is the force exerted by the dashpot B_2 on point A, with the positive sense to the left.

3.14. Repeat Problem 3.13 when $B_2 = 0$. Let the output be the displacement x_2 instead of the force exerted by the dashpot.

*3.15. The model for a certain dynamic system is

$$\dot{x}_1 = -3x_1 + 2x_2 + u_1(t) + 2\dot{u}_2$$
$$\dot{x}_2 = 2x_1 + x_2 + \dot{u}_1$$
$$y = x_1 - x_2 + u_2(t)$$

where x_1 and x_2 are state variables, $u_1(t)$ and $u_2(t)$ the inputs, and y the output. Rewrite the model in state-variable form by avoiding derivatives on the right side of the equations. Define any new symbols and identify the state variables.

3.16. The input to the system shown in Figure P3.16 is the displacement $x_2(t)$ of M_2. Write a set of state-variable equations. The springs are relaxed when $x_1 = x_2 = 0$. Include an output equation for the downward force f_2 that must be applied to M_2 to achieve the specified motion.

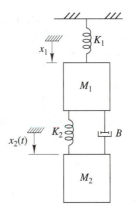

Figure P3.16

3.17. For the system shown in Figure P2.2, replace the force input $f_a(t)$ by the displacement $x_2(t)$ of M_2. The springs are relaxed when $x_1 = x_2 = 0$. Write a set of state-variable equations. Include an output equation for the force f_2 that must be applied to M_2 to achieve the specified motion, with the positive sense to the left.

*3.18. Find a set of state-variable equations for the system shown in Figure P2.6 when the inputs are $x_3(t)$ and $f_a(t)$. The outputs are the velocities v_1 and v_2.

3.19. A linear dynamic system with input $u(t)$, output y, and state variables x_1 and x_2 is characterized by the equations

$$\dot{x}_1 + 2\dot{x}_2 = 3x_1 + 4x_2 - 5u(t)$$
$$\dot{x}_1 - \dot{x}_2 = 2x_1 + x_2 + u(t)$$
$$y = \dot{x}_1 + 2x_2$$

Find the state-variable and output equations.

3.20. Find the state-variable model for the system shown in Figure P2.22, when the energy stored in each spring is a separate output.

*3.21. Find the state-variable model for the system shown in Figure P2.21. The outputs are the momentum of each mass and the tensile force in the cable.

3.22. Find the state-variable model for the system shown in Figure P2.24 when the elongation of each spring is a separate output.

3.23. Redo Example 3.11 by using the p operator to obtain
    a. The input-output equation for x_1
    b. The input-output equation for x_2

*3.24. Use the p operator to obtain the input-output equation relating x_1 and $x_2(t)$ for the system shown in Figure 3.7(a), starting with
    a. The state-variable equations for x_1 and v_1 given in (18)
    b. The state-variable equations for x_1 and q given in (22)

3.25.

    a. Starting with Equation (2.19) and using the p operator, obtain an input-output differential equation for x in the system model developed in Example 2.4. To simplify the calculations, let all parameters have values of unity in the appropriate units.
    b. Repeat part (a) by starting with Equation (2.20) to find the input-output equation for x_1. Compare the answer to that which you obtained in part (a).

*3.26. Set all of the parameters to unity in Equation (2.25) and then use the p operator to obtain the input-output equation relating x_1 to the inputs $f_a(t)$ and g for the system shown in Figure 2.17(a).

3.27.

    a. Repeat Problem 3.26 for Equation (2.29), describing the system shown in Figure 2.19(a).
    b. Obtain the input-output equation for x_2.

3.28. Write in matrix form the state-variable model developed in Example 3.8, with the input $x_2(t)$ and the outputs x_1 and v_1. Identify the matrices $\mathbf{A}$, $\mathbf{B}$, $\mathbf{C}$, and $\mathbf{D}$, the state vector $\mathbf{q}$, the input vector $\mathbf{u}$, and the output vector $\mathbf{y}$. Also give the values of n, m, and p.

*3.29. Repeat Problem 3.28 for the system modeled by (8) and (9) in Example 3.5, with the input $f_a(t)$ and the outputs f_{K_2} and m_T.

3.30. Repeat Problem 3.29 using (10) and (11) in place of (8) and (9).

3.31. Repeat Problem 3.28 for the system modeled by (3) and (4) in Example 3.2, with the input $f_a(t)$ and the outputs f_{K_2} and m_T.

3.32. Repeat Problem 3.28 for the state-variable model developed in Example 3.6, with the input $f_a(t)$ and the outputs f_{K_2}, m_T, and v_2.

4

BLOCK DIAGRAMS AND COMPUTER SIMULATION

In Chapter 3 we presented the two major forms of modeling equations: state-variable and input-output. In this chapter, we introduce a powerful tool for solving these equations and obtaining the response of a system to different inputs. This tool is the digital computer, using numerical simulation software. Both linear and nonlinear differential equations can be solved numerically with high precision and speed, allowing system responses to be calculated and displayed for many input functions. An easy way to make an interface between a system's modeling equations and the digital computer is to use block diagrams that can be drawn from the system's differential equations. With the aid of appropriate software, the diagrams can then be used to produce a computer program that solves the original differential equations.

Inspecting the block diagram of a system may provide new insight into the system's structure and behavior beyond that available from the differential equations themselves. In this chapter we restrict the discussion to fixed linear systems, but the use of block diagrams is valid also for nonlinear systems, as we will see in Chapter 9.

We begin by defining the components to be used in our diagrams. We then show how these blocks can be interconnected to represent differential equations. Finally, we introduce a computational language called MATLAB and its graphical user interface, Simulink, to obtain solutions from these block diagrams.

The operations that we generally use in block diagrams are summation, gain, and integration. Other blocks, including nonlinear elements such as multiplication, square root, exponential, logarithmic, and other functions, are available. Provisions are also included for supplying input functions, using a signal generator block, and for displaying results, using a scope block.

▶ 4.1 DIAGRAM BLOCKS

A **block diagram** is an interconnection of blocks representing basic mathematical operations in such a way that the overall diagram is equivalent to the system's mathematical model. In such a diagram, the lines interconnecting the blocks represent the variables describing the system behavior. These may be inputs, outputs, state variables, or other related variables. The blocks represent operations or functions that use one or more of these variables to calculate other variables. For example, the force f produced by a spring can be calculated from its displacement by multiplying the displacement x by the spring constant K. In block-diagram form, this can be shown as in Figure 4.1.

Figure 4.1 A block diagram representing $f = Kx$.

Figure 4.2 A summer representing $y = x_1 + x_2 - x_3$.

Summer

The addition and subtraction of variables is represented by a **summer**, or summing junction. A summer is usually drawn as a circle or rectangle that has any number of arrows directed toward it (the inputs) and a single arrow directed away from it (the output). Associated with each entering arrowhead is a plus or minus symbol indicating the sign of the associated term. The output variable, appearing as the single arrow leaving the circle or rectangle, is defined to be the sum of all the incoming variables, with the associated signs taken into account. For example, a summer that produces $y = x_1 + x_2 - x_3$ is shown in Figure 4.2.

Gain

The multiplication of a single variable by a constant is represented by a **gain** block. We place no restriction on the value of the gain, which may be positive, zero, or negative. Figure 4.1 shows a gain block that obeys the relationship $f = Kx$.

Integrator

Integration with respect to time is performed in an **integrator** block, as shown in Figure 4.3(a). The output of an integrator is given by

$$y(t) = y(0) + \int_0^t u(\lambda)\, d\lambda$$

so that

$$\dot{y}(t) = u(t)$$

If the input to this block is the derivative of y with respect to time, its output must be $y(t)$. This is shown explicitly in Figure 4.3(b). The initial condition, $y(0)$, is not usually shown explicitly in block diagrams, but it is nevertheless important and must be specified or assumed to be zero. For now, we will assume the initial conditions to be zero.

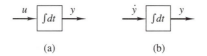

(a)   (b)

Figure 4.3 The block diagram for an integrator.

Figure 4.4 A constant block.

Constant

The **constant** block in Figure 4.4 has no input, and its output never changes. As shown, it obeys the relationship $y = c$.

▶ 4.2 COMBINING BLOCKS TO SOLVE MODELING EQUATIONS

We now show how to combine these elements into a block diagram that represents the solution to a differential equation. Consider the modeling equation

$$\dot{x} = f_a(t) - Ax \tag{1}$$

where A is a known constant, $f_a(t)$ is a known input function, and x is the output. Before using the single modeling equation in (1), we must take advantage of the relationship between x and $\dot{x}$, so that the diagram has only one unknown variable. We do this by making $\dot{x}$ the input to an integrator, whose output must then be x, as shown in Figure 4.5(a).

Because the equation for $\dot{x}$ is the difference of two terms, we show it as the output of a summer. The inputs to the summer are $f_a(t)$ and Ax, with appropriate signs, as shown in part (b) of the figure. Finally, we complete the diagram by using a gain block to form the signal Ax, as in part (c) of the figure. Notice that the only external signal coming into the final diagram is the known input $f_a(t)$.

Diagrams for Input-Output Equations

The process of writing block diagrams for models in input-output form can be developed from this simple beginning. For a system having a single output and no input derivatives, the steps are as follows:

1. Solve the given equation for the highest derivative of the unknown output variable.

2. Connect one or more integrator blocks in series to integrate that derivative successively as many times as necessary to produce the output variable.

3. Use the result of step 1 to form the highest output derivative as the output of a summer and a gain block. The signals entering the summer usually consist of the output and its lower-order derivatives, each multiplied by a constant, and with the appropriate sign. There will also generally be a term representing the input.

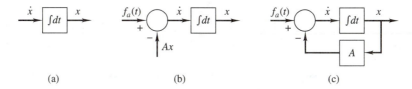

(a)   (b)   (c)

Figure 4.5 The block diagram representation of (1). (a), (b) Partial diagrams. (c) Complete diagram.

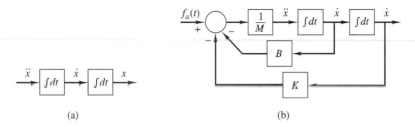

Figure 4.6 (a)

Figure 4.6 The block diagram representation of (2). (a) Two integrators. (b) Complete diagram.

▶ **EXAMPLE 4.1**

Construct a block diagram for the system shown in Example 2.1, whose input-output model was given by Equation (2.15) as

$$M\ddot{x} + B\dot{x} + Kx = f_a(t) \qquad (2)$$

SOLUTION   Solving for the highest derivative of x, namely $\ddot{x}$, we obtain

$$\ddot{x} = \frac{1}{M}[f_a(t) - B\dot{x} - Kx]$$

We use two integrators in series to relate $\ddot{x}$, $\dot{x}$, and x to one another, as shown in Figure 4.6(a). Because the equation for $\ddot{x}$ is a constant times the sum of three terms, we form this quantity by means of a gain block preceded by a summer. The signals entering the summer are the three terms inside the brackets: $f_a(t)$, $B\dot{x}$, and Kx, with appropriate signs. The last two of these signals are formed by picking off $\dot{x}$ and x from the diagram and then multiplying them by the gain constants B and K, respectively. The complete block diagram is shown in part (b) of Figure 4.6.

▶ **EXAMPLE 4.2**

Construct a block diagram from the input-output equation for the system shown in Figure 2.13(a), which has two masses and two springs and a force input $f_a(t)$. Let $B = 0$ and take the output to be x_1.

SOLUTION   We found the input-output equation in Example 3.12. From Equation (3.31) with $B = 0$,

$$M_1 M_2 x_1^{(iv)} + [M_1 K_2 + M_2(K_1 + K_2)]\ddot{x}_1 + K_1 K_2 x_1 = K_2 f_a(t)$$

so

$$x_1^{(iv)} = \frac{1}{M_1 M_2}[-A\ddot{x}_1 - K_1 K_2 x_1 + K_2 f_a(t)] \qquad (3)$$

where $A = M_1 K_2 + M_2(K_1 + K_2)$. We start with a chain of four integrators to relate x_1 and its four derivatives to one another. Looking at the equation for the fourth derivative of x_1, we form it as the output of a gain block preceded by a summer. The signals entering the summer are the three terms inside the brackets in that equation. The complete diagram is shown in Figure 4.7.

We see from the last two examples that for an nth-order input-output differential equation, we would start with a chain of n integrators in series. Notice, however, that if $B \neq 0$

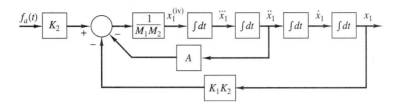

Figure 4.7 The block diagram representation of (3).

in Example 4.2, the input-output equation will contain a derivative of the input. It might be argued that because $f_a(t)$ is a known function, we should be able to obtain its derivative as another known function, perhaps by defining and using a differentiator block in our diagram. We do not normally do this, however, because that could lead to numerical inaccuracies when we use the diagram to obtain a computer solution. Instead, Chapter 13 will contain a more complete treatment, which will include representing input-output equations that contain derivatives of the input on the right side.

Diagrams for the Original Modeling Equations

Diagrams based on the input-output equations of complex systems may become quite cluttered and, more importantly, may not convey as much understanding of the physical system as we might wish. Rather than trying to first find an input-output model, we may start with the original modeling equations. If we have several free-body diagrams, then the overall model will probably consist of a set of coupled second-order equations. It is sensible to draw the block diagram directly from these coupled equations. Because the variables associated with a single mass are its position, velocity, and acceleration, the corresponding portion of the diagram generally consists of a chain of two integrators with the variables being fed back to a summer. The complete diagram then becomes an interconnection of subdiagrams that contain two-integrator chains, each with a summing junction. These ideas are illustrated in the following example.

► **EXAMPLE 4.3**

Consider again the fourth-order mechanical system studied in Examples 2.2 and 4.2. Draw a block diagram representing the original modeling equations that were found in Equations (2.16a) and (2.16b).

SOLUTION In these modeling equations there are two unknown displacements, x_1 and x_2, and their highest derivatives are $\ddot{x}_1$ and $\ddot{x}_2$. As step 1, we solve for $\ddot{x}_1$ and $\ddot{x}_2$ to get

$$\ddot{x}_1 = \frac{1}{M_1}\left[-B\dot{x}_1 - (K_1 + K_2)x_1 + B\dot{x}_2 + K_2x_2\right]$$

$$\ddot{x}_2 = \frac{1}{M_2}\left[f_a(t) - B\dot{x}_2 + B\dot{x}_1 - K_2x_2 + K_2x_1\right] \tag{4}$$

As step 2, for each of these two equations we connect two integrators in series in order to relate the unknown displacement to its two derivatives. In step 3, we see from (4) that each second derivative can be formed as the output of a gain block, preceded by a summer whose inputs are the terms inside the brackets. Except for the external input $f_a(t)$, each of the signals entering the summers can be obtained by picking off x_1, $\dot{x}_1$, x_2, or $\dot{x}_2$ from one of the integrator chains as the input to a gain block. The complete diagram is in Figure 4.8.

This diagram represents two coupled second-order subsystems. The two rows containing the integrators represent the two subsystems, and the connections between the

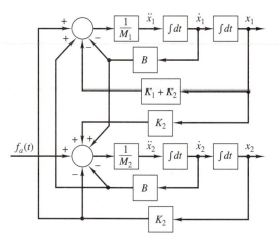

Figure 4.8 A block-diagram representation of (4).

subsystems show the effects of the forces in the springs and dashpot connecting the masses. The short arrows labelled x_1 and x_2 pointing to the right are shown at the right side of the diagram to designate these variables as outputs.

Diagrams for State-Variable Models

To construct a diagram for a state-variable model, we first draw an integrator for each state variable. The inputs to these integrators are the first derivatives of the corresponding state variables. We then use the state-variable equations to form each of these derivatives in terms of state variables and inputs. In general, each derivative will be the output of a gain block preceded by a summer, although some derivatives may be formed in a simpler way. The steps can be summarized as follows.

1. Draw an integrator for each state variable.
2. Assemble the terms needed to form the derivative of each state variable, as specified in the state-variable equations. These will be functions of the state variables and inputs.
3. Assemble the blocks needed for any output equations, using the state variables and inputs as required.

We now illustrate this process by constructing a diagram for a fourth-order system.

▶ **EXAMPLE 4.4**

Draw a block diagram to represent the state-variable model of the system shown in Figure P2.4. It can be shown that for the four state variables x_1, v_1, x_2, and v_2, the state-variable equations are

$$\dot{x}_1 = v_1 \tag{5a}$$

$$\dot{v}_1 = \frac{1}{M_1}[f_a(t) - K_1 x_1 - B_1 v_1 - B_3(v_1 - v_2) - K_3(x_1 - x_2)] \tag{5b}$$

$$\dot{x}_2 = v_2 \tag{5c}$$

$$\dot{v}_2 = \frac{1}{M_2}[-K_2 x_2 - B_2 v_2 + B_3(v_1 - v_2) + K_3(x_1 - x_2)] \tag{5d}$$

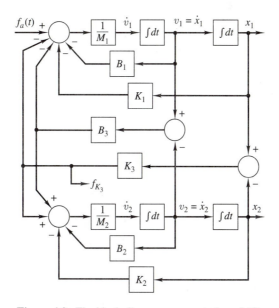

Figure 4.9 The block-diagram representation of (5).

SOLUTION Figure 4.9 is a block diagram that represents these equations. It contains an integrator for each state variable, plus a number of summers and gain blocks to produce the terms required for the derivative of each state variable. The integrators that produce the velocities have inputs, $\dot{v}_1$ and $\dot{v}_2$, that are formed by summing the appropriate terms in (5b) and (5d). The system input $f_a(t)$ is shown here as a line entering the uppermost of these summing junctions from the left.

As presented, this example does not have an output equation. However, if we wanted to study the compressive force in spring K_3, which obeys the relationship $f_{K3} = K_3(x_1 - x_2)$, we observe that this value exists as the output of the gain block K_3 near the center of the diagram. Hence, we need only to add a short line with an arrow labelled f_{K3} to represent this force as an output.

Similarly, the forces in either of the springs K_1 or K_2, or the viscous drag forces in the dashpot B_3 or the sliding surfaces B_1 and B_2, are all available in the diagram at the left sides of the blocks labeled K_1, K_2, B_3, B_1, and B_2, respectively.

Structure of Block Diagrams

The form of the block diagram in Figure 4.8 is different from that of Figure 4.9 in an important detail. In Figure4.8, there has been an algebraic simplification in (4), in which the coefficients of each variable were collected together. In this example, the coefficients of x_1, namely K_1 and K_2, were added together. This simplifies the block diagram by minimizing the total number of blocks required. In (5) and Figure 4.9, on the other hand, terms are grouped as they occur in the free-body diagram. In the block diagram in Figure 4.9, the differences in displacement and velocity are formed explicitly and then multiplied by the appropriate coefficient, K or B. This results in an interesting and useful isomorphism between the block diagram in Figure 4.9 and the physical description. In particular, the spring K_1 and the damping B_1 are located between M_1 and a fixed frame. As a result, the gain blocks representing K_1 and B_1 connect only to the variables x_1 and v_1, respectively.

Furthermore, the spring K_3 and the dashpot B_3 are placed between M_1 and M_2. Consequently, in the block diagram, the relative velocity and relative displacement are formed by two summers and then multiplied by B_3 and K_3, respectively. The signs of the resulting forces on M_1 are directed oppositely from those on M_2. Finally, the spring K_2 and damping B_2 are located between M_2 and a fixed reference frame. Again, in the block diagram, these gains are associated only with the inputs to the integrator for v_2, and do not affect the inputs to the integrator for v_1.

A further desirable feature of the structure used in Figure 4.9 is that it naturally produces output equations for many physical variables in the system. Forces in all the components are calculated explicitly before they are applied to the masses upon which they act.

▶ 4.3 RUNNING SIMULINK WITH MATLAB

In the previous section, we showed how to draw block diagrams to represent modeling equations in both input-output and state-variable form. A principal reason for doing so was to prepare to solve these equations numerically. We now describe a system of numerical computation called **MATLAB**, which is a product of The MathWorks available on all commonly used computer platforms and operating systems. A Student Version is available for the PC. This powerful system for simulating and displaying complex systems is widely used in academic and industrial settings. Only a small subset of the functions of MATLAB will be considered here. The examples chosen can all be solved using the Student Version of these programs. Other simulation and modeling programs are available, and everything that is covered in this book can be translated into other programs.

We use MATLAB with its companion package **Simulink**, which provides a graphical user interface (GUI) for building system models and executing the simulation. These models are constructed by drawing interconnected blocks that represent the algebraic and differential equations that describe the system behavior, similar to a block diagram. MATLAB is used in a supporting role to initialize parameter values and to produce plots of the system response.

Simulink provides access to an extensive set of blocks that accomplish a wide range of functions useful for the simulation and analysis of dynamic systems. The blocks are grouped into libraries, by general classes of functions. Mathematical functions such as summers and gains are in the `Math` library. Integrators are in the `Continuous` library. Constants, common input functions, and a clock can all be found in the `Sources` library.[1] Simulink is a graphical interface that allows the user to create programs that are actually run in MATLAB. When these programs run, they create arrays of the variables defined in Simulink that can be made available to MATLAB for analysis and/or plotting. The variables to be used in MATLAB must be identified by Simulink using a `ToWorkspace` block, which is found in the `Sinks` library. (When using this block, open its dialog box and specify that the save format should be `Matrix`, rather than the default, which is called `Structure`.) The `Sinks` library also contains a `Scope`, which allows variables to be displayed as the simulated system responds to an input. This is most useful when studying responses to repetitive inputs, as discussed in Section 4.4.

Blocks

Simulink uses **blocks** that are similar to the ones in Section 4.1. The summer block can be represented in two ways in Simulink: by a circle or by a rectangle. Both choices are shown

---

[1]Throughout the remainder of this book, we will use `courier type` to designate formal Simulink or MATLAB commands or names.

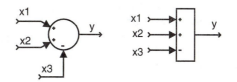

Figure 4.10 Two Simulink blocks for a summer representing $y = x_1 + x_2 - x_3$.

Gain

Figure 4.11 The Simulink block for a gain of K.

in Figure 4.10. The configuration of the summer, allowing any number of inputs and the sign of each, is established by a dialog box obtained by double-clicking on the block.

A gain block is shown by a triangular symbol, with the gain expression written inside if it will fit. If not, the symbol -k- is used. This block is shown in Figure 4.11. The value used in each gain block is established in a dialog box that appears if the user double-clicks on its block.

The block for an integrator as shown in Figure 4.12 looks unusual. The quantity $1/s$ comes from the Laplace transform expression for integration, and will seem more natural after we have covered the Laplace transform in Chapter 7. If the user double-clicks on the symbol for an integrator, a dialog box appears allowing the initial condition for that integrator to be specified. It may be implicit, and not shown on the block, as in Figure 4.12(a). Alternatively, a second input to the block can be displayed to supply the initial condition explicitly, as in part (b) of Figure 4.12. Initial conditions may be specific numerical values, literal variables, or algebraic expressions. The user is encouraged to consult the help function and other documentation supplied by The MathWorks for details.

Constants are created by the `Constant` block, which closely resembles Figure 4.4. Double-clicking on the symbol opens a dialog box to establish the constant's value. It can be a number or an algebraic expression using constants whose values are defined in the workspace and are therefore known to MATLAB.

Two additional blocks will be needed if we wish to use MATLAB to plot the responses versus time. These are the `Clock` and the `ToWorkspace` blocks. The clock produces the variable "time" that is associated with the integrators as MATLAB calculates a numerical (digital) solution to a model of a continuous system. The result is a string of sample values of each of the output variables. These samples are not necessarily at uniform time increments, so it is necessary to have the variable "time" that contains the time corresponding to each sample point. Then MATLAB can make plots versus "time." The clock output could be given any arbitrary name; we use "t" in most of the examples.

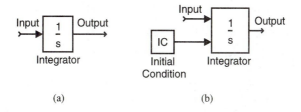

(a) (b)

Figure 4.12 Two forms of the Simulink block for an integrator. (a) Implicit initial condition. (b) Initial condition shown explicitly.

The `ToWorkspace` block is used to return the results of a simulation to the MATLAB workspace, where they can be analyzed and/or plotted. Any variable in a Simulink diagram can be connected to a `ToWorkspace` block. In our examples, all of the state variables and the input variables are usually returned to the workspace. In addition, the result of any output equation that may be simulated would usually be sent to the workspace.

Simulink provides scores of other blocks with different functions. Users are encouraged to browse the Simulink libraries and consult the online `Help` facility provided with MATLAB.

Inputs

A system responds to one or more inputs or initial conditions. It may have an input shown explicitly, such as the force $f_a(t)$ in (5b). Simulink provides several blocks that can produce such inputs. It is also possible that no input is applied, but that the system's response to some set of initial conditions is desired. If this is the case, the initial condition can be specified for any integrator by double-clicking on its block. The intial conditions can be given as numerical or literal constants or algebraic expressions of constants.

A Simulink block is provided for a `Step` input, a signal that changes (usually from zero) to a specified new, constant level at a specified time. These levels and time are specified by double-clicking on the `Step` block, which is found in the `Sources` library. Another block designates the `Signal Generator`, which can produce a sinusoidally varying input, a series of square pulses, a sawtooth wave, or a random signal.

Running the Simulation

To create a simulation in Simulink, the user should follow these steps:

- Start MATLAB.
- Start Simulink.
- Open the libraries that contain the blocks you will need. These usually will include the `Sources`, `Sinks`, `Math`, and `Continuous` libraries, and possibly others.
- Open a new Simulink window.
- Drag the needed blocks from their library folders to that window. The `Math` library, for example, contains the `Gain` and `Sum` blocks.
- Arrange the blocks in the window in an orderly way that corresponds to the equations to be solved.
- Interconnect the blocks by dragging the cursor from the output of one block to the input of another block. Interconnecting branches can be made by right-clicking on an existing branch.
- Double-click on any block having parameters that must be established, and set these parameters. For example, the gain of all `Gain` blocks must be set. The number and signs of the inputs to a `Sum` block must be established. The parameters of any source blocks should also be set in this way.

The details of these procedures can be found in the MATLAB and Simulink documentation. Users are particularly encouraged to read the Student Versions of *Learning MATLAB* and *Learning Simulink*.

It is necessary to specify a `Stop time` for the solution. This is done by clicking on the `Simulation > Parameters` entry on the Simulink toolbar. In order to be able to produce MATLAB plots versus time, the output of a `Clock` block must be sent to MATLAB by using a `ToWorkspace` block. At the `Simulation > Parameters` entry, other parameters can be selected in this dialog box, but the default values of all of

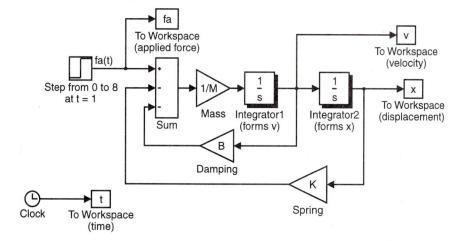

Figure 4.13 A Simulink diagram to solve (2) for the system in Figures 4.6 and 2.12.

them should be adequate for almost all of the examples used in this book. If the response before time zero is needed, it can be obtained by setting the Start time to a negative value. It may be necessary in some problems to reduce the maximum integration step size used by the numerical algorithm. If the plots of the results of a simulation appear "choppy" or composed of straight line segments when they should be smooth, this problem can be solved by reducing the Max step size permitted.

▶ **EXAMPLE 4.5**

Construct a Simulink diagram to calculate the response of the system shown in Example 4.1 to an input force that increases from 0 to 8 N at $t = 1$ s. The parameter values are $M = 2$ kg, $K = 16$ N/m, and $B = 4$ N·s/m.

SOLUTION   We wrote the modeling equation, solved for the highest derivative of the output, and drew a block diagram to represent this equation in Example 4.1. The corresponding Simulink diagram is shown in Figure 4.13. In this diagram, we have emphasized the importance of the integrator by drawing its block with a drop shadow, which appears to give the block depth. This is accomplished in the Simulink window by highlighting the integrator block, then clicking on the Show Drop Shadow option on the Format drop-down menu. We use the ToWorkspace blocks for $t, f_a(t), x,$ and v in order to allow MATLAB to plot the desired responses.

Note that the Step block has been used to provide the input $f_a(t)$. We have set the initial and final values and the time at which the step occurs. For clarity, we have also written these values underneath the symbol.

We next cause MATLAB to solve the equations represented by this Simulink diagram. The duration of the simulation is selected to be 10 seconds from the Simulation > Parameters entry on the toolbar. The simulation is run by clicking the Start entry. When the computer makes a short chirp sound, the simulation is complete, and the variables that have been sent to the workspace are available there for analysis and/or plotting. If we enter the command plot(t,x) in the main MATLAB window, the desired plots are displayed. Several functions are available to make plots more informative. Among these are: xlabel, ylabel, title, gtext, text, grid, axis, and subplot. Figure 4.14 shows the complete response of this system.

Some properties of this solution can be checked manually, a practice that is recommended whenever a simulation is employed. For example, the steady-state value of x is

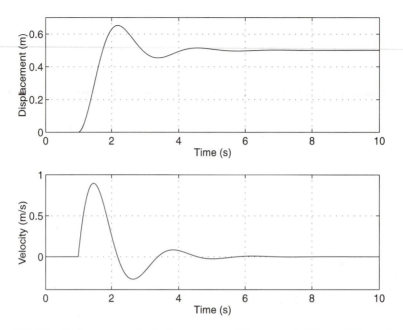

Figure 4.14 The displacement and velocity responses of the system in Figure 4.13 to a change in $f_a(t)$ from 0 to 8 N at $t = 1$ s.

0.5 m. This can be calculated by setting the derivatives equal to zero in (2) and solving for x when $f_a = 8$. This steady-state value is seen in Figure 4.14. Note that the responses of both variables are oscillations that decay exponentially to steady-state values. The displacement shown in the upper plot shows an overshoot starting at $t = 1.8$ s that peaks at $t = 2.2$ s. There is a smaller undershoot of the displacement between $t = 2.9$ s and $t = 4.0$ s.

Another important feature of a numerical simulation is the ease with which parameters can be varied and the results observed directly. In Example 4.5, the effect of changing B is to alter the amount of overshoot or undershoot. These are related to a term called the damping ratio, a topic we will examine in detail in Chapter 8. To illustrate, begin the simulation in Example 4.5 with $B = 4$ N·s/m, but with the input applied at $t = 0$, and plot the result. Then rerun it with $B = 8$ N·s/m. Hold the first plot active, by the command hold on, and reissue the plot command. The second plot is superimposed on the first. This can be repeated for $B = 12$ N·s/m and $B = 25$ N·s/m, with the result shown in Figure 4.15.

Another convenient feature of MATLAB, or other numerical simulation tools, is the ability to introduce initial conditions on any integrator. This process allows energy stored in the system at $t = 0$ to be represented explicitly. Double-clicking on an integrator block opens a dialog box where the initial condition on the output of that integrator can be set.

▶ **EXAMPLE 4.6**

Find the response of the system in Example 4.5 when there is no input for $t \geq 0$, but when the initial value of the displacement x is zero and the initial velocity $\dot{x}(0)$ is 1 m/s.

SOLUTION The Simulink diagram shown in Figure 4.13 is used, with the size of the input step set to zero and with the initial condition on Integrator1 set to 1.0. The resulting displacement and velocity are shown in Figure 4.16.

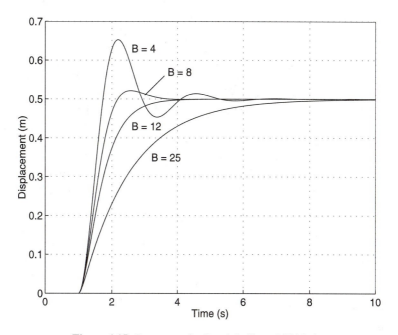

Figure 4.15   Responses for $B = 4, 8, 12$, and 25 N·s/m.

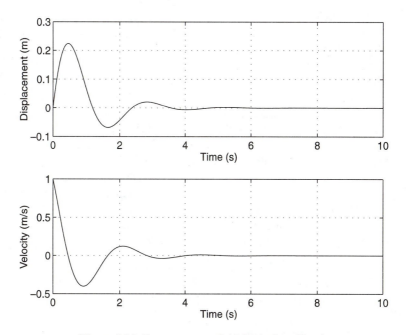

Figure 4.16   Responses to an initial velocity of 1 m/s.

M-files

MATLAB has a feature called M-files that can simplify its operational use. These are files containing MATLAB statements, which can be saved and reexecuted by a single MATLAB command. Three types of statements are particularly useful with Simulink. The values of the parameters of a model can be saved in an M-file and changed by editing this file if needed. Commands to execute a model can be included, so that the results of a simulation are created. Finally, plot commands and other postsimulation data analysis and display commands can be used to cause the production, printing, or storage of the desired product. M-files are produced with a text editor; there is one built into MATLAB that is accessible from the toolbar or through the edit command.

A file that will set up the MATLAB workspace by establishing the values of the parameters needed for the Simulink simulation of Example 4.5 is shown below.

```
% This file is named ex4_5_setup.m.
% Everything after a % sign on a line is a comment that
% is ignored by MATLAB. This file establishes the
% parameter values for ex4_5_model.mdl.
%
M = 2;          % kg
K = 16;         % N/m
B = 4;          % Ns/m
fafinal = 8;    % N
```

The next M-file can be run to produce the upper plot in Figure 4.14.

```
% This file is named ex4_5_plot.m.
% It makes a plot of the data produced by ex4_5_model.mdl.
plot(t,x)    ;    grid    % Plots x for the case with B=4.
xlabel('Time (s)');
ylabel('Displacement (m)');
```

A semicolon in a physical line ends the logical line, and anything after it is treated as if it were on a new physical line. A semicolon at the end of a line that generates output to the command window suppresses the printing of that output. These files can be used by first entering the command ex4_5_setup in the command window. Then open the Simulink model and start the simulation by clicking on the toolbar entry Simulation > Start. Finally, enter the command ex4_5_plot in the command window to make the plot.

When two or more variables are being studied simultaneously, it is frequently desirable to plot them one above the other on separate axes, as is done for displacement and velocity in Figure 4.14. This is accomplished with the subplot command. The following M-file uses this command to produce both parts of Figure 4.14.

```
% This file is named ex4_5_plot2.m.
% It makes both parts of Figure 4.14.
% Execute ex4_5_setup.m first.
%
subplot(2,1,1);
plot(t,x);    grid      % Plots x for the case with B=4.
xlabel('Time (s)');
```

```
ylabel('Displacement (m)');
subplot(2,1,2);
plot(t,v);    grid    % Plots v for the case with B=4.
xlabel('Time (s)');
ylabel('Velocity (m/s)');
```

If a complex plot is desired, in which several runs are needed with different parameters, this can all be accomplished by executing ex4_5_setup followed by this M-file:

```
% This file is named ex4_5_plots.m.
% It plots the data produced by ex4_5_model.mdl for
% several values of B. Execute ex4_5_setup.m first.
%
sim('ex4_5_model')  % Has the same effect as clicking on
                    % Start on the toolbar.
plot(t,x)           % Plots the initial run with B=4.
hold on             % Plots later results on the same axes
                    % as the first.
B = 8;              % New value of B; other parameter values
                    % stay the same.
sim('ex4_5_model')  % Rerun the simulation with new B value.
plot(t,x)           % Plots new x on original axes.
B = 12;  sim('ex4_5_model');  plot(t,x)
B = 25;  sim('ex4_5_model');  plot(t,x)
hold off
```

Entering the command ex4_5_plots in the command window results in a figure similar to Figure 4.15.

▷ EXAMPLE 4.7

A person wearing a parachute jumps out of an airplane. Assume that there is no wind and that the parachute provides viscous damping relative to a fixed reference frame. The parachute has a relatively small mass M_p and a large damping coefficient B_p. The jumper has a larger mass M_j and a smaller drag coefficient B_j. The cords attaching the parachute to the jumper are called risers and are assumed to be quite springy. The elastic effect of the risers is represented by the spring constant K_R. The deformation of the parachute itself can also be included in the value of K_R. A simple model of the system is shown in Figure 4.17. A more comprehensive model will be discussed in Chapter 9.

Write the modeling equations describing the motions of the jumper and the parachute. Choose the references such that the displacements x_j of the jumper and x_p of the parachute are both zero at $t = 0$. Take the positive direction of these displacements to be downward, as shown in Figure 4.17. Draw a Simulink diagram to solve these equations. Use the following parameter values: $M_p = 10$ kg, $M_j = 60$ kg, $B_p = 100$ N·s/m, $B_j = 10$ N·s/m, and $K_R = 400$ N/m.

Simulate the following experiment: The jumper jumps out of the plane, and the parachute opens. Let $t = 0$ be the instant the parachute opens. At that time, the risers are fully extended, so they are neither slack nor stretched. At $t = 0$ the parachute and the jumper are falling at 20 m/s. Run the simulation for the first 3 seconds after the parachute opens.

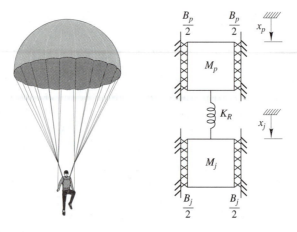

Figure 4.17 A parachute jumper and a mechanical model for the jumper.

SOLUTION When we draw the free-body diagrams for M_p and M_j, sum the forces, and solve for the highest derivatives of the displacements, we obtain

$$\ddot{x}_p = \frac{1}{M_p}\left[-B_p\dot{x}_p + K_R(x_j - x_p) + M_p g\right]$$
$$\ddot{x}_j = \frac{1}{M_j}\left[M_j g - B_j\dot{x}_j - K_R(x_j - x_p)\right]$$

(6)

which are represented in the Simulink diagram in Figure 4.18.

The following M-file can be executed to set up and run this model.

```
%file is ex4_7_run.m
Mp = 10;   Mj = 60;
Bp = 100;  Bj = 10;
KR = 400;
sim('ex4_7_model')
subplot(4,1,1)
plot(t,xj,t,xp,'--')
subplot(4,1,2)
plot(t,vj,t,vp,'--')
subplot(4,1,3)
plot(t,aj)
subplot(4,1,4)
plot(t,delx)
```

The resulting MATLAB plots are similar to Figure 4.19. They have been labeled with letters A through E to help the reader identify the following features:

A. The rapid slowing of the parachute as its drag becomes active but the jumper's weight has not yet extended the risers to pull down hard on the parachute.

B. The increase in velocity of the jumper due to gravity, before the parachute and risers begin to take effect.

C. The increase in length of the risers as the weight of the jumper is supported by the parachute.

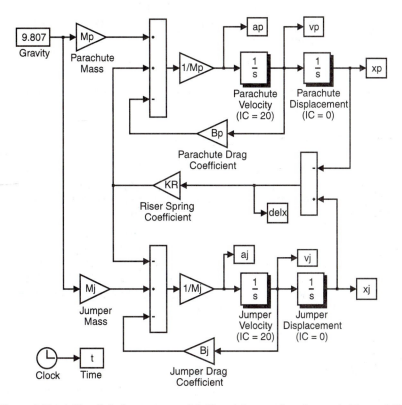

Figure 4.18 A Simulink diagram to model (6) and the parachute jumper in Figure 4.17.

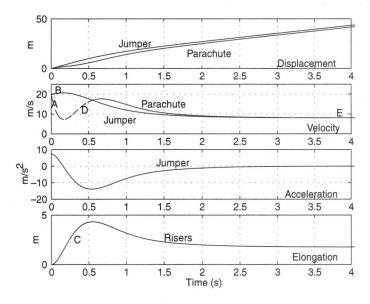

Figure 4.19 Responses for Example 4.7.

    D.   The increase in parachute velocity due to the weight of the jumper.

    E.   The terminal velocity of the jumper and parachute. The reader may show that this velocity is $(M_p + M_j)g/(B_p + B_j)$, which is 6.24 m/s for the stated parameter values.

▶ 4.4 REPETITIVE INPUTS

In the previous sections, we have been concerned with the responses to systems when the input is a function that only occurs once, such as a step. Many systems are designed to operate with **repetitive inputs** such as a sinusoid or a train of pulses. Other systems may be tested this way, even if they are to be used with nonrepeating inputs. Simulink provides a source for such input signals and a convenient way to display the resulting responses. By setting the final time for a simulation to a very large number, we can observe the simulation continuously, and study the effect of changes in the parameters of the model or the input. This ability to simulate a system and study it while it runs is analogous to building the system and studying it in a laboratory. We demonstrate this ability by providing a train of pulses as the input to the model described in Example 4.5.

Signal Generator

One source of repetitive signals in Simulink is called the `Signal Generator`. This block, which is located in the `Sources` library, is placed on the Simulink diagram in place of the `Step` input. Double-clicking on the `Signal Generator` block opens a dialog box, where a sine wave, a square wave, a ramp (sawtooth), or a random waveform can be chosen. In addition, the amplitude and frequency of the signal may be specified. The signals produced have a mean value of zero. The repetition frequency can be given in Hertz (Hz), which is the same as cycles per second, or in radians/second.

Scope

When the signals to be examined are repetitive, and the startup transient is not of interest, the system response can be examined graphically, as the simulation runs, using the `Scope` block found in the `Sinks` library. This name is derived from the electronic instrument called an oscilloscope, which performs a similar function with electronic signals. Any of the variables in a Simulink diagram can be connected to the `Scope` block, and when the simulation is started, that variable is displayed. It is possible to include several `Scope` blocks. The `Scope` normally chooses its scales automatically to best display the data. When using the scope, it is sometimes necessary to slow down the execution of the simulation in order to make the display occur at a rate the user can see. Particularly on fast computers, the scope may flash the entire answer too quickly for the user to see. One method to prevent this is to open the `Simulation > Parameters` dialog box on the Simulink task bar and set the maximum step size to a small number. This forces the numerical algorithm to make many small steps, which requires more execution time, so the `Scope` runs more slowly. A disadvantage of this solution is that if data are also being sent to the Workspace, the data arrays generated will become large. See the MATLAB and Simulink user documentation for further details, including how to display more than one variable at a time.

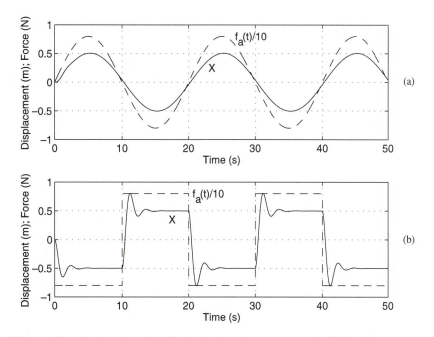

Figure 4.20 The function $f_a(t)/10$ (dashed) and the response x (solid) for the system in Figure 4.13. (a) Sinusoidal input. (b) Pulse input.

▶ **EXAMPLE 4.8**

Prepare a Simulink implementation of the system shown in Figure 4.13, with the same element values as in Example 4.5, and use it to simulate and plot the response to the sinusoidal input

$$f_a(t) = 8\sin\omega t$$

where $\omega = 2\pi(0.05)$. This specified input is a sinusoid having an amplitude of 8 N (16 N peak-to-peak) and a frequency of 0.05 Hz. The period is $1/(0.05) = 20$ s. Then generate the response of the same system to a series of square waves of amplitude 8 N and period 20 s, and compare the two responses by plotting them one above the other over the interval $0 \le t \le 50$ s.

SOLUTION The Simulink diagram needed to solve this problem is Figure 4.13, with the `Step` input at the upper left corner replaced by the `Signal Generator`. The `Signal Generator` is set to produce sine waves of amplitude 8 at a frequency of 0.05 Hz. The response is plotted in the top half of Figure 4.20. This figure shows two variables: the input $f_a(t)/10$ as a dashed line, and the response x as a solid line. This plot could be made by running the model, then giving the command `plot(t,fa/10,'--',t,x)`.

If one wishes to compare directly the responses to the square wave and the sinusoidal inputs, it is convenient to plot their results on adjacent, similar axes. To accomplish this, precede the first plot command with `subplot(2,1,1)`. This will cause two plots to appear, one above the other. The first plot will appear in the upper half of the window. To continue the comparison, change the `Signal Generator` to produce square waves and rerun the simulation. Give the command `subplot(2,1,2)` to make the lower half of the window active and then repeat the plot command. The two plots now appear similar, but are not yet on the same scale. To make the axes identical, give the command `axis([0 50 -1 1])` to each subplot.

The reader should consult the MATLAB documentation for the complete syntax for the `subplot` and `axis` commands. The result of these commands is shown in Figure 4.20. The inputs are represented by dashed curves, and the outputs by solid curves. In order to be able to use a single scale for the vertical axes, we plot $f_a(t)/10$ instead of $f_a(t)$. The reader may wish to compare the lower part of the figure to the curve in Figure 4.14. For the first half-cycle, the output curve shows the same overshoot as in Figure 4.14. For the remaining half-cycles, the overshoot is twice as great, because the height of the input steps is 16 N rather than 8 N.

▶ SUMMARY

In this chapter we showed how to construct block diagrams that are equivalent to the modeling equations developed in Chapters 2 and 3. Blocks representing elementary operations such as addition, gain, or integration were assembled to represent the system equations. The model may have been written directly from free-body diagrams, or may have been presented in input-output or state-variable form. We introduced Simulink, a graphical user interface to the analysis software package MATLAB. Simulink diagrams, which are similar to block diagrams, are used to simulate the response to a specified input or set of initial conditions. Complex systems whose solutions would be difficult or impossible to find analytically can be simulated by assembling simple elements. The responses of these systems can then be studied quantitatively. The techniques of modeling and computer analysis introduced in this chapter allow the digital computer to replace prototyping and testing in many practical engineering applications. We will use these techniques and tools throughout this book as we model and study the responses of electrical, electromechanical, thermal, and hydraulic systems.

▶ PROBLEMS

4.1. Draw block diagrams for each of the following sets of state-variable equations.

   a. $\dot{x} = -4x + 6y + 2u(t)$
       $\dot{y} = -2x - 3y$

   b. $\dot{x}_1 = -3x_1 + 5x_2 + 3u(t)$
       $\dot{x}_2 = 4x_1 - 6x_2 - u(t)$

   c. $\dot{\theta} = \omega$
       $\dot{\omega} = -8\theta - 4\omega + 2x$
       $\dot{x} = v$
       $\dot{v} = 6\theta - 3x + u(t)$

In Problems 4.2 through 4.6, draw block diagrams for the systems that were modeled in the examples indicated, using the equations cited.

*4.2. Equations (2.19) in Example 2.4

4.3. Equations (2.25) in Example 2.6

4.4. Equations (3.8) and (3.9) in Example 3.5

*4.5. Equations (3.21) and (3.22) in Example 3.8

4.6. Equations (3.23) and (3.24) in Example 3.9

4.7. Draw a Simulink diagram to represent the system in Example 4.1. Plot the first 20 seconds of the response when the applied force $f_a(t)$ is zero for $t < 2$ and has a constant value of 2 N for $t \geq 2$ s. Let $M = 10$ kg, $B = 10$ N·s/m, and $K = 10$ N/m.

*4.8. Draw a Simulink diagram to represent the system shown in Figure P2.7. Use the following parameter values: $M_1 = 5$ kg, $B_1 = 20$ N·s/m, and $K_1 = K_2 = 100$ N/m. The input $x_2(t)$ increases from zero to 0.1 m at $t = 1$ s. Plot the displacement x_1 as a function of time for 5 seconds.

4.9. Draw a Simulink diagram to represent the system shown in Example 4.3. Plot x_1 and x_2 for the first 50 seconds when the applied force $f_a(t)$ increases from 0 to 10 N at $t = 1$ s. The parameter values are $M_1 = M_2 = 10$ kg, $B = 20$ N·s/m, and $K_1 = K_2 = 10$ N/m.

*4.10. Draw a Simulink diagram to represent the system shown in Example 4.4. Plot the first 10 seconds of the response when the applied force $f_a(t)$ increases from 0 to 10 N at $t = 1$ s. Let $M_1 = M_2 = 10$ kg, $B_1 = B_2 = B_3 = 20$ N·s/m, and $K_1 = K_2 = K_3 = 10$ N/m. Plot the displacements x_1 and x_2 on the same axes.

4.11. Draw a Simulink diagram to represent the system shown in Figure 2.15(a). Plot the first 10 seconds of the response when the applied force $f_a(t)$ increases from 0 to 100 N at $t = 1$ s. The parameter values are: $M_1 = M_2 = 5$ kg, $B_1 = B_2 = 20$ N·s/m, $B_3 = 50$ N·s/m, and $K_1 = K_2 = 100$ N/m. Plot both x and z on the same axes.

4.12. Draw a Simulink diagram to represent the system shown in Figure P2.14. Plot the first 10 seconds of the response of the system. The input $f_a(t)$ is a force that increases from 0 to 100 N at $t = 1$ s, and then returns to zero at $t = 5$ s. The parameter values are: $M_1 = M_2 = M_3 = 5$ kg, $B_1 = B_2 = 20$ N·s/m, and $K_1 = K_2 = 100$ N/m. Plot the two displacements x_1 and x_2 on the same axes. Plot the displacement x_3 separately. Also plot the velocities of these displacements on a third set of axes.

*4.13. Repeat Problem 4.7 with the initial value of the displacement $x(0) = 0.5$ m, and with the input $f_a(t) = 0$ for all $t \geq 0$. Keep the initial value of the velocity at zero. Let $M = 10$ kg, $B = 20$ N·s/m, and $K = 10$ N/m.

4.14. Repeat Problem 4.9 with the initial values of the displacements $x_1(0) = 0.5$ m and $x_2(0) = 0$, and with the input $f_a(t) = 0$ for all $t \geq 0$. Keep the initial values of the velocities at zero. The parameter values are: $M_1 = 2$ kg, $M_2 = 10$ kg, $B = 20$ N·s/m, $K_1 = 50$ N/m, and $K_2 = 5$ N/m. Plot x_1 and x_2 for the first 10 seconds.

*4.15. Draw a Simulink diagram for the system shown in Figure P4.15. Because this is a first-order system, its model has only one state variable and only one integrator is required. Use the state-variable approach, and take the velocity as the state variable. The displacement is not of interest, as might be expected, since there is no connection to a reference frame in this mobile system. For each of the four sets of values for the mass and damping coefficient given below, plot the first 5 seconds of the response to an initial condition of $v(0) = 1$ m/s when $f_a(t) = 0$ for all $t \geq 0$.

    a. $M = 10$ kg and $B = 10$ N·s/m
    b. $M = 10$ kg and $B = 25$ N·s/m
    c. $M = 25$ kg and $B = 10$ N·s/m
    d. $M = 25$ kg and $B = 25$ N·s/m
    e.   How does increasing the value of M or B affect the rate at which $v(t)$ decays to zero?

4.16. Construct the Simulink diagram described in Example 4.8, using the parameter values given in that example. For a sinusoidal input of amplitude 8, record the amplitudes of the responses to frequencies of 0.1, 0.5, 1, and 5 Hz. Plot the responses for $0 \leq t \leq 100$ s, and record the amplitude for the last complete cycle shown on these plots.

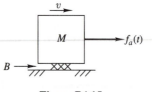

Figure P4.15

4.17. Repeat Problem 4.12 with the same parameter values, except for $B_1 = 50$ N·s/m. Give an explanation in physical terms for the difference in the response compared to that of Problem 4.12.

4.18. Draw a Simulink diagram for the system shown in Figure 2.22(a), with $B_1 = B_2 = B_3 = 10$ N·s/m and $K_1 = K_2 = 10$ N/m. The applied force $f_a(t)$ is zero for all $t \leq 1$ s and is 5 N for all $t > 1$ s.

    a. Plot the displacement x for $0 \leq t \leq 8$ when $M = 4$ kg.

    b. Repeat part (a) when $M = 20$ kg.

    c. How does the value of M affect the steady-state response?

    d. By trial and error, determine the range of values of M for which the displacement x never overshoots the steady-state value.

4.19. Change the value of B_1 to 2 N·s/m, and repeat Problem 4.12. Give an explanation in physical terms for the change in the response from the response of Problem 4.12.

4.20. Repeat Problem 4.11 with the same parameter values, except for $B_2 = 30$ N·s/m, $B_3 = 20$ N·s/m, and $K_1 = 20$ N/m. Give an explanation in physical terms for the difference in the response compared to that of Problem 4.11.

5

ROTATIONAL MECHANICAL SYSTEMS

In Chapter 2 we presented the laws governing translational systems and introduced the use of free-body diagrams as an aid in writing equations describing the motion. In Chapter 3 we showed how to rearrange the equations and develop state-variable and input-output models.

Extending these procedures to rotational systems requires little in the way of new concepts. We first introduce the three rotational elements that are analogs of mass, friction, and stiffness in translational systems. Two other elements, levers and gears, are characterized in a somewhat different way. The use of interconnection laws and free-body diagrams for rotational systems is very similar to their use for translational systems. In the examples, we seek models consisting of sets of state-variable and output equations, or input-output equations that contain only a single unknown variable. We include examples of combined translational and rotational systems, followed by a computer simulation.

▶ 5.1 VARIABLES

For rotational mechanical systems, the symbols used for the variables are

θ, angular displacement in radians (rad)

ω, angular velocity in radians per second (rad/s)

α, angular acceleration in radians per second per second (rad/s^2)

τ, torque in newton-meters (N·m)

all of which are functions of time. Angular displacements are measured with respect to some specified reference angle, often the equilibrium orientation of the body or point in question. We shall always choose the reference arrows for the angular displacement, velocity, and acceleration of a body to be in the same direction so that the relationships

$$\omega = \dot{\theta}$$
$$\alpha = \dot{\omega} = \ddot{\theta}$$

hold. The conventions used are illustrated in Figure 5.1, where τ denotes an external torque applied to the rotating body by means of some unspecified mechanism, such as by a gear on the supporting shaft. Because of the convention that the assumed positive directions for θ, ω, and α are the same, it is not necessary to show all three reference arrows explicitly.

The power supplied to the rotating body in Figure 5.1 is

$$p = \tau\omega \tag{1}$$

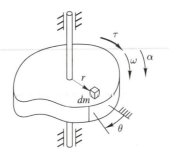

Figure 5.1 Conventions for designating rotational variables.

The power is the derivative of the energy w, and the energy supplied to the body up to time t is

$$w(t) = w(t_0) + \int_{t_0}^{t} p(\lambda)d\lambda$$

▶ 5.2 ELEMENT LAWS

The elements used to represent physical devices in rotational systems are moment of inertia, friction, stiffness, levers, and gears. We shall restrict our consideration to elements that rotate about fixed axes in an inertial reference frame.

Moment of Inertia

When Newton's second law is applied to the differential mass element dm in Figure 5.1 and the result is integrated over the entire body, we obtain

$$\frac{d}{dt}(J\omega) = \tau \qquad (2)$$

where $J\omega$ is the angular momentum of the body and where τ denotes the net torque applied about the fixed axis of rotation. The symbol J denotes the **moment of inertia** in kilogram-meters^2 (kg·m^2). We can obtain it by carrying out the integration of $r^2\,dm$ over the entire body.

The moment of inertia for a body whose mass M can be considered to be concentrated at a point is ML^2, where L is the distance from the point to the axis of rotation. Figure 5.2 shows a slender bar and a solid disk, each of which has a total mass M that is uniformly distributed throughout the body. In parts (a) and (b) of the figure, expressions for J are given for the case where the axis of rotation passes through the center of mass. The results for other common shapes can be found in basic physics and mechanics books. If we have an axis that does not pass through the center of mass, we can use the **parallel-axis theorem**. Let J_0 denote the moment of inertia about the parallel axis that passes through the center of mass, and let a be the distance between the two axes. Then the desired moment of inertia is

$$J = J_0 + Ma^2 \qquad (3)$$

For the slender bar shown in Figure 5.2(c) with the axis of rotation at one end, (3) gives

$$J = \frac{1}{12}ML^2 + M\left(\frac{L}{2}\right)^2 = \frac{1}{3}ML^2 \qquad (4)$$

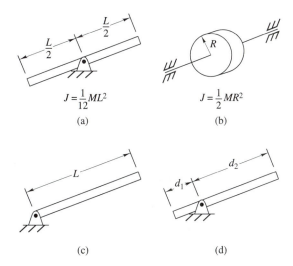

Figure 5.2 Moments of inertia. (a) Slender bar. (b) Disk. (c), (d) Slender bar where axis of rotation does not pass through the center of mass.

For two or more components rotating about the same axis, we can find the total moment of inertia by summing the individual contributions. If the uniform bar shown in Figure 5.2(d) has a total mass M, then the masses for the sections of length d_1 and d_2 will be $M_1 = d_1M/(d_1 + d_2)$ and $M_2 = d_2M/(d_1 + d_2)$. Using (4), we see that the total moment of inertia is

$$J = \frac{1}{3}M_1d_1^2 + \frac{1}{3}M_2d_2^2 = \frac{M(d_1^3 + d_2^3)}{3(d_1 + d_2)} \tag{5}$$

We consider only nonrelativistic systems and constant moments of inertia, so (2) reduces to

$$J\dot{\omega} = \tau \tag{6}$$

where $\dot{\omega}$ is the angular acceleration. As is the case for a mass having translational motion, a rotating body can store energy in both kinetic and potential forms. The kinetic energy is

$$w_k = \tfrac{1}{2}J\omega^2 \tag{7}$$

and for a uniform gravitational field, the potential energy is

$$w_p = Mgh \tag{8}$$

where M is the mass, g the gravitational constant, and h the height of the center of mass above its reference position. If the fixed axis of rotation is vertical or passes through the center of mass, there is no change in the potential energy as the body rotates, and (8) is not needed. To find the complete response of a dynamic system containing a rotating body, we must know its initial angular velocity $\omega(t_0)$. If its potential energy can vary or if we want to find $\theta(t)$, then we must also know $\theta(t_0)$.

Friction

A **rotational friction** element is one for which there is an algebraic relationship between the torque and the relative angular velocity between two surfaces. **Rotational viscous friction** arises when two rotating bodies are separated by a film of oil. Figure 5.3(a) shows two concentric rotating cylinders separated by a thin film of oil, where the angular velocities of the cylinders are ω_1 and ω_2 and the relative angular velocity is $\Delta\omega = \omega_2 - \omega_1$. The torque

$$\tau = B\Delta\omega \tag{9}$$

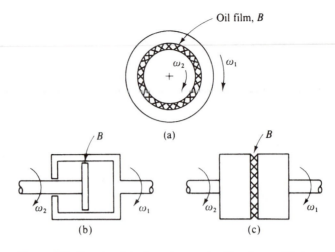

Figure 5.3 Rotational devices characterized by viscous friction.

will be exerted on each cylinder, in directions that tend to reduce the relative angular veloc-ity $\Delta\omega$. Hence the positive sense of the frictional torque must be counterclockwise on the inner cylinder and clockwise on the outer cylinder. The friction coefficient B has units of newton-meter-seconds. Note that the same symbol is used for translational viscous friction, where it has units of newton-seconds per meter.

Equation (9) also applies to a rotational dashpot that might be used in modeling a fluid drive system, such as that shown in Figure 5.3(b) or Figure 5.3(c). The inertia of the parts is assumed to be negligible or else is accounted for in the mathematical model by separate moments of inertia. If the rotational friction element is assumed to have no inertia, then when a torque τ is applied to one side, a torque of equal magnitude but opposite direction must be exerted on the other side (by a wall, the air, or some other component), as shown in Figure 5.4(a), where $\tau = B(\omega_2 - \omega_1)$. Thus in Figure 5.4(b), the torque τ passes through the first friction element and is exerted directly on the moment of inertia J.

Other types of friction, such as the damping vanes shown in Figure 5.5(a), may exert a retarding torque that is not directly proportional to the angular velocity but that may be described by a curve of τ versus $\Delta\omega$, as shown in Figure 5.5(b). For a linear element, the curve must be a straight line passing through the origin. The power supplied to the friction element, $\tau \Delta\omega$, is immediately lost from the mechanical system in the form of heat.

Stiffness

Rotational stiffness is usually associated with a torsional spring, such as the mainspring of a clock, or with a relatively thin shaft. It is an element for which there is an algebraic

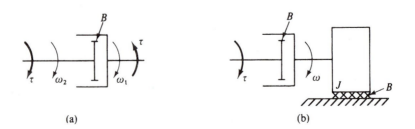

(a)   (b)

Figure 5.4 (a) Rotational dashpot with negligible inertia. (b) Torque transmitted through a friction element.

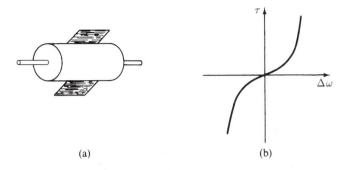

(a) (b)

Figure 5.5 (a) Rotor with damping vanes. (b) Nonlinear friction characteristic.

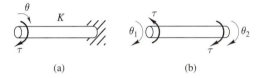

(a) (b)

Figure 5.6 (a) Rotational stiffness element with one end fixed. (b) Rotational stiffness element with $\Delta\theta = \theta_2 - \theta_1$.

relationship between τ and θ, and it is generally represented as shown in Figure 5.6(a). The angle θ is the relative angular displacement of the ends of the element from the positions corresponding to no applied torque. If both ends of the device can move as shown in Figure 5.6(b), then the element law relates τ and $\Delta\theta$, where $\Delta\theta = \theta_2 - \theta_1$. For a linear torsional spring or flexible shaft,

$$\tau = K \, \Delta\theta \tag{10}$$

where K is the stiffness constant with units of newton-meters (N·m), in contrast to newtons per meter for the parameter K in translational systems. For a thin shaft, K is directly proportional to the shear modulus of the material and to the square of the cross-sectional area, and is inversely proportional to the length of the shaft.

Because we assume that the moment of inertia of a stiffness element either is negligible or is represented by a separate element, the torques exerted on the two ends of a stiffness element must be equal in magnitude and opposite in direction, as was indicated in Figure 5.6(b). Thus for the system shown in Figure 5.7, the applied torque τ passes through the first shaft and is exerted directly on the body that has moment of inertia J.

Potential energy is stored in a twisted stiffness element and can affect the response of the system at later times. For a linear spring or shaft, the potential energy is

$$w_p = \tfrac{1}{2}K(\Delta\theta)^2 \tag{11}$$

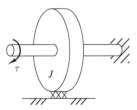

Figure 5.7 Torque transmitted through a shaft.

The relative angular displacement at time t_0 is one of the initial conditions we need in order to find the response of a system for $t \geq t_0$.

The Lever

An **ideal lever** is assumed to be a rigid bar pivoted at a point and having no mass, no friction, no momentum, and no stored energy. In all our examples the pivot point will be fixed. If the magnitude of the angle of rotation is small (say less than 0.25 rad), the motion of the ends can be considered strictly translational. In Figure 5.8, let θ denote the angular displacement of the lever from the horizontal position. For a rigid lever with a fixed pivot, the displacements of the ends are given by $x_1 \simeq d_1\theta$ and $x_2 \simeq d_2\theta$, where θ is in radians. Thus for small displacements we have

$$x_2 = \left(\frac{d_2}{d_1}\right) x_1 \tag{12}$$

and, by differentiating the foregoing equation, we find

$$v_2 = \left(\frac{d_2}{d_1}\right) v_1 \tag{13}$$

Because the sum of the moments about the pivot point vanishes, as required by the assumed absence of mass, it follows that $f_2 d_2 - f_1 d_1 = 0$, or

$$f_2 = \left(\frac{d_1}{d_2}\right) f_1 \tag{14}$$

The pivot exerts a downward force of $f_1 + f_2$ on the lever, but this does not enter into the derivation of (14), because this force exerts no moment about the pivot. The lever differs from the mass, friction, and stiffness elements in that the algebraic relationships given by (12) through (14) involve pairs of the same types of variables: two displacements, two velocities, or two forces.

If the lever's mass cannot be neglected, then we must include its moment of inertia when summing moments about the pivot point, and (14) is no longer valid. If the bar in Figure 5.8 has mass M, then its moment of inertia is given by (5). The examples in Section 5.4 include a system with a massless lever and one in which the lever's mass is not negligible.

Gears

Consider next the pair of gears shown in Figure 5.9. In order to develop the basic geometric and torque relationships, we shall assume **ideal gears**, which have no moment of inertia,

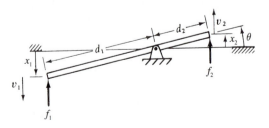

Figure 5.8 The lever.

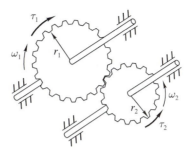

Figure 5.9 A pair of gears.

no stored energy, no friction, and a perfect meshing of the teeth. Any inertia or bearing friction in an actual pair of gears can be represented by separate lumped elements in the free-body diagrams.

The relative sizes of the two gears result in a proportionality constant for the angular displacements, angular velocities, and transmitted torques of the respective shafts. For purposes of analysis, it is convenient to visualize the pair of ideal gears as two circles, shown in Figure 5.10(a), that are tangent at the contact point and rotate without slipping. The spacing between teeth must be equal for each gear in a pair, so the radii of the gears are proportional to the number of teeth. Thus if r and n denote the radius and number of teeth, respectively, then

$$\frac{r_2}{r_1} = \frac{n_2}{n_1} = N \tag{15}$$

where N is called the **gear ratio**.

Let points A and B in Figure 5.10(a) denote points on the circles that are in contact with each other at some reference time t_0. At some later time, points A and B will have moved to the positions shown in Figure 5.10(b), where θ_1 and θ_2 denote the respective angular displacements from their original positions. Because the arc lengths PA and PB must be equal,

$$r_1\theta_1 = r_2\theta_2 \tag{16}$$

which we can rewrite as

$$\frac{\theta_1}{\theta_2} = \frac{r_2}{r_1} = N \tag{17}$$

By differentiating (16) with respect to time, we see that the angular velocities are also related by the gear ratio:

$$\frac{\omega_1}{\omega_2} = \frac{r_2}{r_1} = N \tag{18}$$

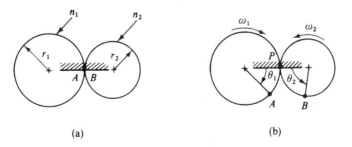

(a)   (b)

Figure 5.10 Ideal gears. (a) Reference position. (b) After rotation.

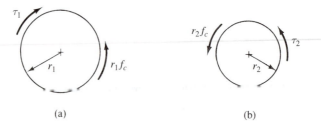

Figure 5.11 Free-body diagrams for a pair of ideal gears.

Note that the positive directions of θ_1 and θ_2, and likewise of ω_1 and ω_2, are taken in opposite directions in the figure. Otherwise a negative sign would be introduced into (16), (17), and (18).

We can derive the torque relationship for a pair of gears by drawing a free-body diagram for each gear, as shown in Figure 5.11. The external torques applied to the gear shafts are denoted by τ_1 and τ_2. The force exerted by each gear at the point of contact by its mate is f_c. By the law of reaction forces, the arrows must be in opposite directions for the two gears. The corresponding torques $r_1 f_c$ and $r_2 f_c$ are shown on the diagram. In addition to the contact force f_c, each gear must be supported by a bearing force of equal magnitude and opposite direction, because the gears have no translational motion. However, because the bearing support forces act through the center of the gear, they do not contribute to the torque and hence have been omitted from the figure. Because the gears have no inertia, the sum of the torques on each of the gears must be zero. Thus, from Figure 5.11,

$$\begin{aligned} f_c r_1 - \tau_1 &= 0 \\ f_c r_2 + \tau_2 &= 0 \end{aligned} \tag{19}$$

Eliminating the contact force f_c, the value of which is seldom of interest, we obtain

$$\frac{\tau_2}{\tau_1} = -\frac{r_2}{r_1} = -N \tag{20}$$

The minus sign in (20) should be expected, because both τ_1 and τ_2 in Figure 5.11 are shown as driving torques; that is, the reference arrows indicate that positive values of τ_1 and τ_2 both tend to make the gears move in the positive direction. The gears are assumed to have no inertia, so the two torques must actually be in opposite directions, and at any instant either τ_1 or τ_2 must be negative.

An alternative derivation of (20) makes use of the fact that the power supplied to the first gear is $p_1 = \tau_1 \omega_1$ and the power supplied to the second gear is $p_2 = \tau_2 \omega_2$. Since no energy can be stored in the ideal gears and since no power can be dissipated as heat due to the assumed absence of friction, conservation of energy requires that $p_1 + p_2 = 0$. Thus

$$\tau_1 \omega_1 + \tau_2 \omega_2 = 0$$

or

$$\frac{\tau_2}{\tau_1} = -\frac{\omega_1}{\omega_2} = -N$$

which agrees with (20).

▶ 5.3 INTERCONNECTION LAWS

The interconnection laws for rotational systems involve laws for torques and angular displacements that are analogous to Equations (2.13) and (2.14) for translational mechanical systems. The law governing reaction torques has an important modification when compared to the law governing reaction forces.

D'Alembert's Law

For a body with constant moment of inertia rotating about a fixed axis, we can write (6) as

$$\sum_i (\tau_{ext})_i - J\dot{\omega} = 0 \tag{21}$$

where the summation over i includes all the torques acting on the body. Like the translational version of D'Alembert's law, the term $-J\dot{\omega}$ can be considered an **inertial torque**. When it is included along with all the other torques acting on the body, (21) reduces to the form appropriate for a body in equilibrium:

$$\sum_i \tau_i = 0 \tag{22}$$

In the application of (22), the torque $J\dot{\omega}$ is directed opposite to the positive sense of θ, ω, and α. D'Alembert's law can also be applied to a junction point that has no moment of inertia, in which case the $J\dot{\omega}$ term vanishes.

The Law of Reaction Torques

For bodies that are rotating about the same axis, any torque exerted by one element on another is accompanied by a **reaction torque** of equal magnitude and opposite direction on the first element. In Figure 5.12, for example, a counterclockwise torque $K_1\theta_1$ exerted by the shaft K_1 on the right disk is accompanied by a clockwise torque $K_1\theta_1$ exerted by the disk on the shaft. However, for bodies not rotating about the same axis, the magnitudes of the two *torques* are not necessarily equal. For the pair of gears shown in Figure 5.9, Figure 5.10, and Figure 5.11, the contact *forces* where the gears mesh are equal and opposite, but because the gears have different radii, the *torque* exerted by the first gear on the second has a different magnitude from the *torque* exerted by the second on the first.

The Law for Angular Displacements

When examining the interconnection of elements in a system, we automatically use the fact that we can express the motions of some of the elements in terms of the motions of other elements. Although we normally do this intuitively, we can state in a more formal way a law for angular displacements analogous to the law found in Section 2.3 for linear displacements.

In Figure 5.12, suppose that the reference marks on the rims are at the top of the two disks when no torque is applied, but make the angles θ_1 and θ_2 with respect to this fixed

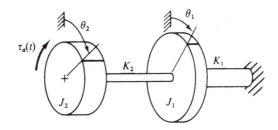

Figure 5.12 Rotational system to illustrate the laws for reaction torques and angular displacements.

reference when the torque $\tau_a(t)$ is applied to the left disk. Then the net angular displacement for the shaft K_2 with respect to its unstressed condition is $\theta_2 - \theta_1$.

This can be stated in the form of a general interconnection law by saying that at any instant the algebraic sum of the differences in angular displacement around any closed path is zero. The equation

$$\sum_i (\Delta\theta)_i = 0 \quad \text{around any closed path} \tag{23}$$

is analogous to Equation (2.14) for translational systems. It is understood that the signs in the summation take into account the direction in which the path is being traversed and that all measurements are with respect to reference positions that correspond to a single equilibrium condition for the system. If in Figure 5.12, for example, the relative differences in angular displacement are summed going from the fixed vertical reference to disk 1, then to disk 2, and finally back to the fixed vertical reference, (23) gives

$$(\theta_1 - 0) + (\theta_2 - \theta_1) + (0 - \theta_2) = 0$$

Although this is a rather trivial result, it does provide a formal basis for stating that the relative angular displacements for the shafts K_1 and K_2 are θ_1 and $\theta_2 - \theta_1$, respectively. In formulating the system equations, we normally use (23) automatically to reduce the number of displacement variables that must be shown on the diagram.

▶ ## 5.4 OBTAINING THE SYSTEM MODEL

The methods for using the element and interconnection laws to develop an appropriate mathematical model for a rotational system are the same as those discussed in Chapter 2 and Chapter 3 for translational systems. We indicate the assumed positive directions for the variables, and the assumed senses for the angular displacement, velocity, and acceleration of a body are chosen to be the same. We automatically use (23) to avoid introducing more symbols than are necessary to describe the motion. For each mass or junction point whose motion is unknown beforehand, we normally draw a free-body diagram showing all torques, including the inertial torque. We express all the torques except inputs in terms of angular displacements, velocities, or accelerations by means of the element laws. Then we apply D'Alembert's law, as given in (22), to each free-body diagram.

If we seek a set of state-variable equations, we identify the state variables and write those equations (such as $\dot{\theta} = \omega$) that do not require a free-body diagram. We write the equations obtained from the free-body diagrams by D'Alembert's law in terms of the state variables and inputs and manipulate them into the standard form of Equations (3.31) or (3.36). Each of the state-variable equations should express the derivative of a different state variable as an *algebraic* function of the state variables, inputs, and possibly time. For any output of the system that is not one of the state variables, we need a separate algebraic equation.

If, on the other hand, we seek an input-output differential equation, we normally write the equations from the free-body diagrams in terms of only angular displacements or only angular velocities. Then we combine the equations to eliminate all variables except the input and output, which can be done by the p operator described in Section 3.2. For single-input, single-output systems, the equation has the form of Equation (3.25).

We shall now consider a number of examples involving free-body diagrams, state-variable equations, and input-output equations for rotational systems. Although these

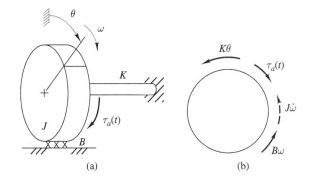

Figure 5.13 (a) Rotational system for Example 5.1. (b) Free-body diagram.

examples involve linear friction and stiffness elements, it is a simple matter to replace the appropriate terms by the nonlinear element laws when we encounter a nonlinear system. For example, a linear frictional torque $B\omega$ would become $\tau_B(\omega)$, which is a single-valued algebraic function that describes the nonlinear relationship.

▷ **EXAMPLE 5.1**

Derive the state-variable model for the rotational system shown in Figure 5.13(a) when the outputs are the angular acceleration of the disk and the counterclockwise torque exerted on the disk by the flexible shaft. Also find the input-output equation relating θ and $\tau_a(t)$.

SOLUTION The first step is to draw a free-body diagram of the disk. This is done in Figure 5.13(b), where the left face of the disk is shown. The torques acting on the disk are the stiffness torque $K\theta$, the viscous-frictional torque $B\omega$, and the applied torque $\tau_a(t)$. In addition, the inertial torque $J\dot{\omega}$ is indicated by the dashed arrow. By D'Alembert's law,

$$J\dot{\omega} + B\omega + K\theta = \tau_a(t) \tag{24}$$

The normal choices for the state variables are θ and ω, which are related to the energy stored in the shaft and in the disk, respectively. One state-variable equation is $\dot{\theta} = \omega$, and the other can be found by solving (24) for $\dot{\omega}$. Thus

$$\dot{\theta} = \omega$$
$$\dot{\omega} = \frac{1}{J}[-K\theta - B\omega + \tau_a(t)] \tag{25}$$

The output equations are

$$\alpha_M = \frac{1}{J}[-K\theta - B\omega + \tau_a(t)]$$
$$\tau_K = K\theta$$

For the input-output equation with θ designated as the output, we merely rewrite (24) with all terms on the left side expressed in terms of θ and its derivatives. Then the desired result is

$$J\ddot{\theta} + B\dot{\theta} + K\theta = \tau_a(t)$$

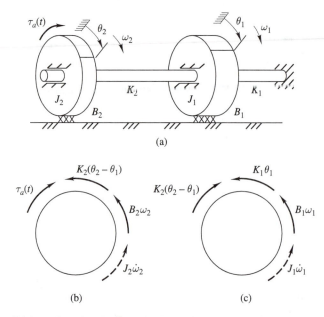

(a)

(b)   (c)

Figure 5.14 (a) Rotational system for Example 5.2. (b), (c) Free-body diagrams.

▶ EXAMPLE 5.2

In the system shown in Figure 5.14(a), the two shafts are assumed to be flexible, with stiffness constants K_1 and K_2. The two disks, with moments of inertia J_1 and J_2, are supported by bearings whose friction is negligible compared with the viscous-friction elements denoted by the coefficients B_1 and B_2. The reference positions for θ_1 and θ_2 are the positions of the reference marks on the rims of the disks when the system contains no stored energy. Find a state-variable model when the outputs are the counterclockwise torque exerted on J_2 by the shaft K_2 and the total angular momentum of the disks. Also find an input-output equation relating θ_2 and $\tau_a(t)$.

SOLUTION   The system has two inertia elements with independent angular velocities and two shafts with independent angular displacements, so four state variables are required. We choose θ_1, θ_2, ω_1, and ω_2, because they reflect the potential energy stored in each of the shafts and the kinetic energy stored in each disk.

The resulting free-body diagrams are shown in Figure 5.14(b) and Figure 5.14(c), where only the torque $K_2(\theta_2 - \theta_1)$, which is the reaction torque on the left disk by the shaft connecting the two disks, should require an explanation. The corresponding reference arrow has arbitrarily been drawn counterclockwise in Figure 5.14(b), implying that the torque is being treated as a retarding torque on disk 2. The actual torque will act in the counterclockwise direction only when $\theta_2 > \theta_1$, so the torque must be labeled $K_2(\theta_2 - \theta_1)$ and not $K_2(\theta_1 - \theta_2)$. By the law of reaction torques, the effect of the connecting shaft on disk 1 is a torque $K_2(\theta_2 - \theta_1)$ with its positive sense in the clockwise direction. We can reach the same conclusion by first selecting a clockwise sense for the arrow in Figure 5.14(c), thereby treating that torque as a driving torque on disk 1, and then noting that disk 2 will tend to drive disk 1 in the positive direction only if $\theta_2 > \theta_1$. Thus the correct expression is $K_2(\theta_2 - \theta_1)$. Of course, if we had selected a counterclockwise arrow in Figure 5.14(c), we would have labeled the arrow either $-K_2(\theta_2 - \theta_1)$ or $K_2(\theta_1 - \theta_2)$.

For each of the free-body diagrams, the algebraic sum of the torques may be set equal to zero by D'Alembert's law, giving the pair of equations

$$J_1\dot{\omega}_1 + B_1\omega_1 + K_1\theta_1 - K_2(\theta_2 - \theta_1) = 0$$
$$J_2\dot{\omega}_2 + B_2\omega_2 + K_2(\theta_2 - \theta_1) - \tau_a(t) = 0 \tag{26}$$

Two of the state-variable equations are $\dot{\theta}_1 = \omega_1$ and $\dot{\theta}_2 = \omega_2$, and we can find the other two by solving the two equations in (26) for $\dot{\omega}_1$ and $\dot{\omega}_2$, respectively. Thus

$$\dot{\theta}_1 = \omega_1$$
$$\dot{\omega}_1 = \frac{1}{J_1}[-(K_1 + K_2)\theta_1 - B_1\omega_1 + K_2\theta_2]$$
$$\dot{\theta}_2 = \omega_2 \tag{27}$$
$$\dot{\omega}_2 = \frac{1}{J_2}[K_2\theta_1 - K_2\theta_2 - B_2\omega_2 + \tau_a(t)]$$

The output equations are

$$\tau_{K_2} = K_2(\theta_2 - \theta_1)$$
$$m_T = J_1\omega_1 + J_2\omega_2$$

where m_T is the total angular momentum.

To obtain an input-output equation, we rewrite (26) in terms of the angular displacements θ_1 and θ_2. A slight rearrangement of terms yields

$$J_1\ddot{\theta}_1 + B_1\dot{\theta}_1 + (K_1 + K_2)\theta_1 - K_2\theta_2 = 0 \tag{28a}$$
$$-K_2\theta_1 + J_2\ddot{\theta}_2 + B_2\dot{\theta}_2 + K_2\theta_2 = \tau_a(t) \tag{28b}$$

Neither of the equations in (28) can be solved separately, but we want to combine them into a single differential equation that does not contain θ_1. Because θ_1 appears in (28b) but none of its derivatives do, we rearrange that equation to solve for θ_1 as

$$\theta_1 = \frac{1}{K_2}[J_2\ddot{\theta}_2 + B_2\dot{\theta}_2 + K_2\theta_2 - \tau_a(t)]$$

Substituting this result into (28a) gives

$$J_1J_2\theta_2^{(iv)} + (J_1B_2 + J_2B_1)\theta_2^{(iii)} + (J_1K_2 + J_2K_1 + J_2K_2 + B_1B_2)\ddot{\theta}_2$$
$$+ (B_1K_2 + B_2K_1 + B_2K_2)\dot{\theta}_2 + K_1K_2\theta_2 = J_1\ddot{\tau}_a + B_1\dot{\tau}_a + (K_1 + K_2)\tau_a(t) \tag{29}$$

which is the desired result. Equation (29) is a fourth-order differential equation relating θ_2 and $\tau_a(t)$, in agreement with the fact that four state variables appear in (27).

In the last example, note the signs when like terms are gathered together in the torque equations corresponding to the free-body diagrams. In (28a) for J_1, all the terms involving θ_1 and its derivatives have the same sign. Similarly in (28b) for J_2, the signs of all the terms with θ_2 and its derivatives are the same. This is consistent with the comments made after Example 2.2, and can be used as a check on the work. Some insight into the reason for this can be obtained from the discussion of stability in Section 8.2.

▶ **EXAMPLE 5.3**

Find state-variable and input-output models for the system shown in Figure 5.14(a) and studied in Example 5.2, but with the shaft connecting disk 1 to the wall removed.

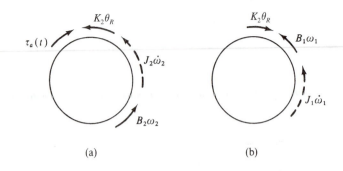

Figure 5.15 Free-body diagrams for Example 5.3.

SOLUTION   This example is analogous to Example 3.5. Because there are now only three energy-storing elements, we expect to need only three state variables. Two of these are chosen to be ω_1 and ω_2, which are related to the kinetic energy stored in the disks. The relative displacement of the ends of the connecting shaft is $\theta_2 - \theta_1$, which is related to the potential energy in that element. Hence we select as the third state variable

$$\theta_R = \theta_2 - \theta_1$$

although $\theta_1 - \theta_2$ would have been an equally good choice.

The free-body diagrams for each of the disks, with torques labeled in terms of the state variables and input, are shown in Figure 5.15. By D'Alembert's law,

$$\begin{aligned} J_1\dot{\omega}_1 + B_1\omega_1 - K_2\theta_R &= 0 \\ J_2\dot{\omega}_2 + B_2\omega_2 + K_2\theta_R &= \tau_a(t) \end{aligned} \tag{30}$$

We obtain one of the state-variable equations by noting that $\dot{\theta}_R = \dot{\theta}_2 - \dot{\theta}_1 = \omega_2 - \omega_1$, and we find the other two by rearranging the last two equations. Thus the third-order state-variable model is

$$\dot{\theta}_R = \omega_2 - \omega_1 \tag{31a}$$

$$\dot{\omega}_1 = \frac{1}{J_1}(-B_1\omega_1 + K_2\theta_R) \tag{31b}$$

$$\dot{\omega}_2 = \frac{1}{J_2}[-K_2\theta_R - B_2\omega_2 + \tau_a(t)] \tag{31c}$$

The output equations for τ_{K_2} and m_T are

$$\tau_{K_2} = K_2\theta_R$$

$$m_T = J_1\omega_1 + J_2\omega_2$$

If one of the outputs were the angular displacement θ_2, it would not be possible to write an output equation for θ_2 as an algebraic function of θ_R, ω_1, ω_2, and $\tau_a(t)$. In that case we would need four state variables. For the state-variable equations, we could either use (27) with $K_1 = 0$ or add to (31) the equation $\dot{\theta}_2 = \omega_2$.

The input-output differential equation relating θ_2 and $\tau_a(t)$ can be obtained by letting $K_1 = 0$ in (29):

$$J_1J_2\theta_2^{(\mathrm{iv})} + (J_1B_2 + J_2B_1)\theta_2^{(\mathrm{iii})} + (J_1K_2 + J_2K_2 + B_1B_2)\ddot{\theta}_2$$
$$+ (B_1 + B_2)K_2\dot{\theta}_2 = J_1\ddot{\tau}_a + B_1\dot{\tau}_a + K_2\tau_a(t)$$

Although this could be viewed as a third-order differential equation in the variable $\dot{\theta}_2$, which is the angular velocity ω_2, we would need four initial conditions if we wanted to

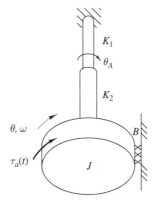

Figure 5.16 Rotational system for Example 5.4.

determine θ_2 rather than ω_2. This is consistent with the fact that four state variables are needed when the output is θ_2.

We would expect the input-output equation relating θ_R and $\tau_a(t)$ to be strictly third order, because only three state variables are needed for the output θ_R. We see from (31a) that $\omega_1 = \omega_2 - \dot{\theta}_R$. Substituting this expression into (30), we obtain

$$J_1\dot{\omega}_2 + B_1\omega_2 - (J_1\ddot{\theta}_R + B_1\dot{\theta}_R + K_2\theta_R) = 0$$
$$J_2\dot{\omega}_2 + B_2\omega_2 + K_2\theta_R = \tau_a(t)$$

We can eliminate the variable ω_2 from this pair of equations by using the p-operator method described in Section 3.2. When this is done, we find that

$$J_1J_2\dddot{\theta}_R + (J_1B_2 + J_2B_1)\ddot{\theta}_R + (J_1K_2 + J_2K_2 + B_1B_2)\dot{\theta}_R$$
$$+ (B_1 + B_2)K_2\theta_R = J_1\dot{\tau}_a + B_1\tau_a(t)$$

▶ **EXAMPLE 5.4**

The shaft supporting the disk in the system shown in Figure 5.16 is composed of two sections that have spring constants K_1 and K_2. Show how to replace the two sections by an equivalent stiffness element, and derive the state-variable model. The outputs of interest are the angular displacements θ and θ_A.

SOLUTION Free-body diagrams for each of the sections of the support shaft and for the inertia element J are shown in Figure 5.17. No inertial torques are included on the shafts, because their moments of inertia are assumed to be negligible. The quantity τ_r is the reaction torque applied by the support on the top of the shaft. Summing the torques on each of the free-body diagrams gives

$$K_1\theta_A - \tau_r = 0 \tag{32a}$$
$$K_2(\theta - \theta_A) - K_1\theta_A = 0 \tag{32b}$$
$$J\dot{\omega} + B\omega + K_2(\theta - \theta_A) - \tau_a(t) = 0 \tag{32c}$$

We need the first of these equations only if we wish to find τ_r, which is not normally the case. Equations (32b) and (32c) can also be obtained by considering the free-body

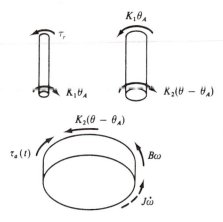

Figure 5.17 Free-body diagrams for Example 5.4.

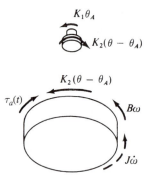

Figure 5.18 Alternative free-body diagrams for Example 5.4.

diagrams for just the disk and the massless junction of the two shafts. The corresponding free-body diagrams are shown in Figure 5.18. Applying D'Alembert's law to them yields (32b) and (32c).

We see from (32b) that θ_A and θ are proportional to each other. Specifically,

$$\theta_A = \left(\frac{K_2}{K_1 + K_2} \right) \theta \tag{33}$$

Substituting (33) into (32c) yields

$$J\dot{\omega} + B\omega + K_{eq}\theta = \tau_a(t) \tag{34}$$

where

$$K_{eq} = \frac{K_1 K_2}{K_1 + K_2}$$

The parameter K_{eq} can be regarded as an equivalent stiffness constant for the series combination of the two shafts. Selecting θ and ω as the state variables and using (34), we can write the state-variable model as

$$\dot{\theta} = \omega$$

$$\dot{\omega} = \frac{1}{J}[-K_{eq}\theta - B\omega + \tau_a(t)] \tag{35}$$

The output θ is one of the state variables. The output equation for θ_A is given by (33).

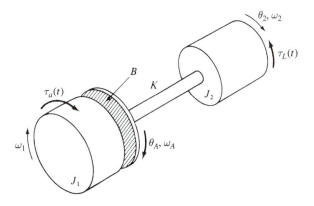

Figure 5.19 Rotational system for Example 5.5.

▶ **EXAMPLE 5.5**

The system shown in Figure 5.19 consists of a moment of inertia J_1 corresponding to the rotor of a motor or a turbine, which is coupled to the moment of inertia J_2 representing a propeller. Power is transmitted through a fluid coupling with viscous-friction coefficient B and a shaft with stiffness constant K. A driving torque $\tau_a(t)$ is exerted on J_1, and a load torque $\tau_L(t)$ is exerted on J_2. If the output is the angular velocity ω_2, find the state-variable model and also the input-output differential equation.

SOLUTION   There are three independent energy-storing elements, so we select as state variables ω_1, ω_2, and the relative displacement θ_R of the two ends of the shaft, where

$$\theta_R = \theta_A - \theta_2 \tag{36}$$

Note that the equation

$$\dot{\theta}_R = \omega_A - \omega_2 \tag{37}$$

is not yet a state-variable equation because of the symbol ω_A on the right side.

Next we draw the free-body diagrams for the two inertia elements and for the shaft, as shown in Figure 5.20. Note that the moment of inertia of the right side of the fluid coupling element is assumed to be negligible. The directions of the arrows associated with the torque $B(\omega_1 - \omega_A)$ are consistent with the law of reaction torques and also indicate that the frictional torque tends to retard the relative motion.

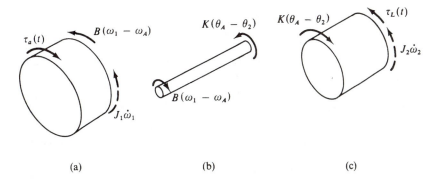

(a)   (b)   (c)

Figure 5.20 Free-body diagrams for Example 5.5.

Setting the algebraic sum of the torques on each diagram equal to zero yields the three equations

$$J_1\dot{\omega}_1 + B(\omega_1 - \omega_A) - \tau_a(t) = 0 \tag{38a}$$

$$B(\omega_1 - \omega_A) - K(\theta_A - \theta_2) = 0 \tag{38b}$$

$$J_2\dot{\omega}_2 - K(\theta_A - \theta_2) + \tau_L(t) = 0 \tag{38c}$$

Using (36), we can rewrite (38) as

$$J_1\dot{\omega}_1 + B(\omega_1 - \omega_A) - \tau_a(t) = 0 \tag{39a}$$

$$B(\omega_1 - \omega_A) = K\theta_R \tag{39b}$$

$$J_2\dot{\omega}_2 - K\theta_R + \tau_L(t) = 0 \tag{39c}$$

Substituting (39b) into (39a) and repeating (39c) give

$$J_1\dot{\omega}_1 + K\theta_R - \tau_a(t) = 0$$
$$J_2\dot{\omega}_2 - K\theta_R + \tau_L(t) = 0 \tag{40}$$

Also from (39b),

$$\omega_A = \omega_1 - \frac{K}{B}\theta_R \tag{41}$$

We substitute (41) into (37) and rearrange (40) to obtain the three state-variable equations

$$\dot{\theta}_R = -\frac{K}{B}\theta_R + \omega_1 - \omega_2 \tag{42a}$$

$$\dot{\omega}_1 = \frac{1}{J_1}[-K\theta_R + \tau_a(t)] \tag{42b}$$

$$\dot{\omega}_2 = \frac{1}{J_2}[K\theta_R - \tau_L(t)] \tag{42c}$$

Because the specified output is one of the state variables, a separate output equation is not needed as part of the state-variable model.

To obtain the input-output equation, we first rewrite (38) in terms of the angular velocities ω_1, ω_2, and ω_A and the torques $\tau_A(t)$ and $\tau_L(t)$. Differentiating (38b) and (38c) and noting that $\dot{\theta}_2 = \omega_2$ and $\dot{\theta}_A = \omega_A$, we have

$$J_1\dot{\omega}_1 + B(\omega_1 - \omega_A) = \tau_a(t)$$

$$B(\dot{\omega}_1 - \dot{\omega}_A) - K(\omega_A - \omega_2) = 0$$

$$J_2\ddot{\omega}_2 - K(\omega_A - \omega_2) + \dot{\tau}_L = 0$$

By using the p-operator technique to eliminate ω_A and ω_1 from these equations, we can obtain the input-output equation

$$\dddot{\omega}_2 + \frac{K}{B}\ddot{\omega}_2 + K\left(\frac{1}{J_1} + \frac{1}{J_2}\right)\dot{\omega}_2$$

$$= \frac{K}{J_1J_2}\tau_a(t) - \frac{1}{J_2}\ddot{\tau}_L - \frac{K}{BJ_2}\dot{\tau}_L - \frac{K}{J_1J_2}\tau_L(t) \tag{43}$$

Although this result can be viewed as a second-order differential equation in $\dot{\omega}_2$, we will need three initial conditions if we are to determine ω_2 rather than the acceleration $\dot{\omega}_2$. Note that if the load torque $\tau_L(t)$ were given as an algebraic function of ω_2, as it would be in practice, ω_2 would appear in (43). Then the input-output equation would be strictly third order.

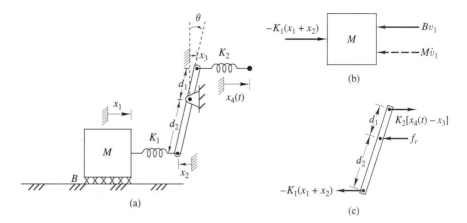

Figure 5.21 (a) Translational system containing a lever. (b), (c) Free-body diagrams.

The next three examples contain either a lever or a pendulum that does not rotate about its midpoint. In Example 5.6 the mass of the lever is assumed to be negligible. The pendulum in the next example is approximated by a point mass at the end of a rigid bar. The final example includes a lever whose mass is uniformly distributed along the bar.

▷ **EXAMPLE 5.6**

Find the state-variable equations for the system shown in Figure 5.21(a). Also find the output equation when the output is defined to be the force exerted on the pivot by the lever. The input is the displacement $x_4(t)$ of the right end of the spring K_2; it affects the mass M through the lever. The lever has a fixed pivot and is assumed to be massless yet rigid. Its angular rotation θ is small so that only horizontal motion need be considered. In a practical situation, the springs K_1 and K_2 might represent the stiffness of the lever and of associated linkages that have a certain degree of flexibility.

SOLUTION   The displacements x_2 and x_3 are directly proportional to the angle θ and hence to one another. Furthermore, the two springs appear to form a series combination somewhat similar to the one shown in Figure 2.24(a), because the lever has no mass. Hence we can express x_2, x_3, and θ as algebraic functions of x_1 and $x_4(t)$. Thus we will select only x_1 and v_1 as state variables, with $x_4(t)$ being the input. By inspection, we determine that one of the required state-variable equations is $\dot{x}_1 = v_1$.

The next step is to draw free-body diagrams for the mass M and the lever, as shown in Figure 5.21(b) and Figure 5.21(c). We must pay particular attention to the signs of the force arrows and to the expressions for the elongations of the springs. For example, the elongation of spring K_1 is $-(x_1 + x_2)$ because of the manner in which the displacements have been defined. Summing the forces on the mass M yields

$$M\dot{v}_1 + Bv_1 + K_1(x_1 + x_2) = 0 \qquad (44)$$

The forces on the lever are those exerted by the springs and the reaction force f_r of the pivot. Because the lever's angle of rotation θ is small, the motion of the lever ends can be considered to be translational, obeying the relationships $\theta = x_2/d_2 = x_3/d_1$ and

$$x_3 = \left(\frac{d_1}{d_2}\right) x_2 \qquad (45)$$

To obtain a second lever equation that involves x_1 and x_2 but not f_r, we sum moments about the pivot point, getting

$$K_2[x_4(t) - x_3]d_1 - K_1(x_1 + x_2)d_2 = 0 \qquad (46)$$

Equations (45) and (46) can also be obtained directly from (12) and (14). In order to solve (44) for $\dot{v}_1$ as a function of only the state variables and the input, we must first express x_2 in terms of x_1 and $x_4(t)$. Substituting (45) into (46) and solving for x_2, we obtain the algebraic expression

$$x_2 = \frac{(d_1/d_2)K_2 x_4(t) - K_1 x_1}{K_1 + (d_1/d_2)^2 K_2} \qquad (47)$$

We find the second of the two state-variable equations by substituting (47) into (44) and rearranging terms so as to solve for $\dot{v}_1$. Doing this, we find that the state-variable equations are

$$\dot{x}_1 = v_1$$

$$\dot{v}_1 = -\frac{1}{M}\left[Bv_1 + \alpha K_1 x_1 + \alpha\left(\frac{d_2}{d_1}\right)K_1 x_4(t)\right] \qquad (48)$$

where

$$\alpha = \frac{1}{1 + \dfrac{K_1}{K_2}\left(\dfrac{d_2}{d_1}\right)^2} \qquad (49)$$

To develop the output equation expressing f_r as an algebraic function of the state variables and the input, we first sum the *forces* on the lever in Figure 5.21(c) to obtain

$$f_r = K_2[x_4(t) - x_3] + K_1(x_1 + x_2) \qquad (50)$$

Substituting (45) and (47) into (50) gives

$$f_r = \left[K_1 + \left(K_2 - \frac{d_2}{d_1}K_1\right)\left(\frac{\alpha d_2 K_1}{d_1 K_2}\right)\right]x_1 + \left[K_2 + \alpha\left(\frac{d_2}{d_1}K_1 - K_2\right)\right]x_4(t)$$

which has the desired form.

In the last example it is instructive to consider the special case where $d_1 = d_2$, which corresponds to having the lever pivoted at its midpoint. From (49), $\alpha K_1 = K_1 K_2/(K_1 + K_2)$, which is the equivalent spring constant for a series connection of the two springs, as in Equation (2.37). Furthermore, the reader can readily verify that (48) reduces to the equations describing the system shown in Figure 5.22(a), in which the massless junction A replaces the lever. In turn, this system is equivalent to that shown in Figure 5.22(b), where a single spring with the coefficient $K_{eq} = K_1 K_2/(K_1 + K_2)$ replaces the series spring connection. Note, in determining the motion of M, that the lever changes the effective direction of the displacement input $x_4(t)$.

In each of the next two examples, an object with mass M rotates about an axis that does not pass through the center of mass. In order to obtain a complete description of all the forces acting on such an object, we would have to consider the motion of its center of mass. The acceleration of this point has one component tangential to the direction of motion and another component perpendicular to it, directed toward the pivot point. The force corresponding to the tangential acceleration is accounted for by the D'Alembert torque $J\dot{\omega}$. Corresponding to the perpendicular component of acceleration, there is a centrifugal force directed away from the pivot point. However, this centrifugal force does not result in a

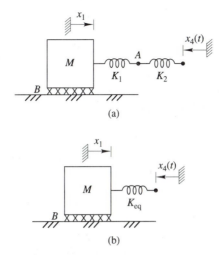

Figure 5.22 Systems equivalent to Figure 5.21(a) when $d_1 = d_2$.

torque about the pivot. Similarly, the forces exerted by the fixed pivot do not contribute to such a torque.

Obtaining expressions for the forces corresponding to the acceleration of the center of mass can become cumbersome, and the interested reader should consult a book on mechanics. When drawing free-body diagrams for rotating bodies, we will generally consider only those forces that yield a torque about the axis of rotation and will not include the centrifugal and pivot forces. We will not seek expressions for the pivot forces except when the mass of the rotating body is negligible or when the acceleration of the center of mass is zero. The first of these special cases was illustrated by Example 5.6 and Figure 5.21(c). The second occurs when the axis of rotation passes through the center of mass.

▶ **EXAMPLE 5.7**

The pendulum sketched in Figure 5.23(a) can be considered a point mass M attached to a rigid massless bar of length L, which rotates about a pivot at the other end. An external torque $\tau_a(t)$ is applied to the bar by a mechanism that is not shown. Let B denote the rotational viscous friction at the pivot point. Derive the input-output differential equation relating the angular displacement θ to the applied torque $\tau_a(t)$. Also write a set of

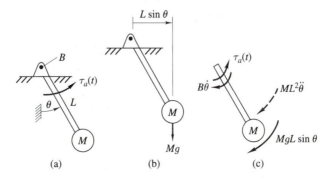

Figure 5.23 (a) Pendulum for Example 5.7. (b) Partial diagram to determine the torque produced by the weight. (c) Free-body diagram showing all torques about the pivot.

state-variable equations. Simplify the expressions for the case where $|\theta|$ is restricted to small values.

SOLUTION As can be seen from Figure 5.23(b), the weight of the mass results in a clockwise torque $MgL\sin\theta$ on the bar. Because of the assumption of a point mass, its moment of inertia about the pivot is $J = ML^2$. Applying D'Alembert's law to the free-body diagram in Figure 5.23(c) yields the input-output equation

$$ML^2\ddot{\theta} + B\dot{\theta} + MgL\,\sin\theta = \tau_a(t) \qquad (51)$$

In order to write a set of state-variable equations, we note that $\dot{\theta} = \omega$ and $\ddot{\theta} = \dot{\omega}$, where ω is the angular velocity. Then from (51) we have

$$\dot{\theta} = \omega$$
$$\dot{\omega} = \frac{1}{ML^2}[-MgL\,\sin\theta - B\omega + \tau_a(t)] \qquad (52)$$

Equations (51) and (52) are nonlinear because of the factor $\sin\theta$. In Chapter 9 we shall discuss methods for approximating nonlinear systems by linear models. For the present, we note that $\sin\theta \simeq \theta$ for small values of θ. The quality of this approximation is reasonably good for $|\theta| \le 0.5$ rad. This is indicated by the fact that when $\theta = 0.5$ rad, the deviation of θ from $\sin\theta$ is only 4.2%. Thus for small values of θ, we can approximate (51) and (52) by the linear models

$$ML^2\ddot{\theta} + B\dot{\theta} + MgL\theta = \tau_a(t)$$

and

$$\dot{\theta} = \omega$$
$$\dot{\omega} = \frac{1}{ML^2}[-MgL\theta - B\omega + \tau_a(t)]$$

▶ **EXAMPLE 5.8**

Find the state-variable equations for the system shown in Figure 5.21(a) when the lever's moment of inertia cannot be neglected. Continue to assume that the friction at the pivot point can be neglected and also that the angular rotation θ is small, so that the ends of the lever move essentially horizontally.

SOLUTION Let J denote the lever's moment of inertia about the pivot point. There are now four independent energy-storing elements: M, K_1, J, and K_2. We choose the state variables to be x_1, v_1, θ, and ω, where ω denotes the clockwise angular velocity of the lever. Two of the four state-variable equations are $\dot{x}_1 = v_1$ and $\dot{\theta} = \omega$. The other two equations can be obtained from the free-body diagrams in Figure 5.24. The diagram for the mass is the same as in the previous example, but the lever is labeled with torques rather than forces.

Summing forces on the mass M, we again have

$$M\dot{v}_1 + Bv_1 + K_1(x_1 + x_2) = 0$$

Summing torques about the lever's pivot point gives

$$J\dot{\omega} = K_2d_1[x_4(t) - x_3] - K_1d_2(x_1 + x_2)$$

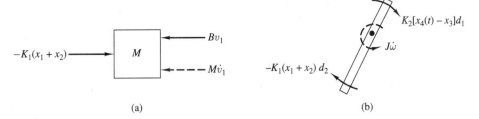

Figure 5.24 Free-body diagrams for Example 5.8. (a) Forces on M. (b) Torques on the lever.

With the substitutions $x_2 = d_2\theta$ and $x_3 = d_1\theta$, these two equations become

$$M_1\dot{v}_1 + Bv_1 + K_1x_1 + K_1d_2\theta = 0$$
$$J\dot{\omega} + K_2d_1^2\theta + K_1d_2^2\theta + K_1d_2x_1 = K_2d_1x_4(t)$$

Thus the state-variable equations are

$$\dot{x}_1 = v_1$$
$$\dot{v}_1 = \frac{1}{M_1}[-K_1x_1 - Bv_1 - K_1d_2\theta]$$
$$\dot{\theta} = \omega$$
$$\dot{\omega} = \frac{1}{J}[-K_1d_2x_1 - (K_1d_2^2 + K_2d_1^2)\theta + K_2d_1x_4(t)]$$

Each of the following two examples involves a pair of gears. The first example shows how gears can change the magnitude of the torque applied to a rotating body. When gears are used to couple two rotational subsystems, as in the second example, the parameters J, B, and K are sometimes reflected in an appropriate manner from one side of the pair of gears to the other.

▶ **EXAMPLE 5.9**

Derive the state-variable equations for the gear-driven disk shown in Figure 5.25(a). The torque $\tau_a(t)$ is applied to a gear with radius r_1. The meshing gear with radius r_2 is rigidly connected to the moment of inertia J, which in turn is restrained by the flexible shaft K and viscous damping B.

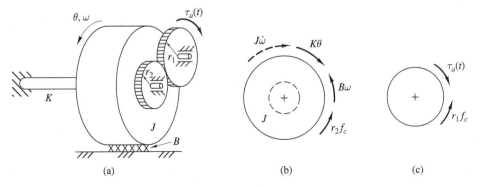

Figure 5.25 (a) System for Example 5.9. (b), (c) Free-body diagrams.

SOLUTION   The free-body diagram for the disk and the gear attached to it is shown in Figure 5.25(b), and the diagram for the other gear is shown in Figure 5.25(c). The contact force where the gears mesh is denoted by f_c, and the corresponding torques are included on the diagrams. Rather than drawing a separate free-body diagram for the shaft, we show the torque $K\theta$ that it exerts on the disk in Figure 5.25(b). By D'Alembert's law,

$$J\dot{\omega} + B\omega + K\theta - r_2 f_c = 0 \tag{53a}$$

$$r_1 f_c = \tau_a(t) \tag{53b}$$

Solving (53b) for f_c and substituting the result into (53a), we have

$$J\dot{\omega} + B\omega + K\theta = N\tau_a(t)$$

where $N = r_2/r_1$. If θ and ω are chosen as the state variables, we write

$$\dot{\theta} = \omega$$
$$\dot{\omega} = \frac{1}{J}[-K\theta - B\omega + N\tau_a(t)] \tag{54}$$

Note that (25), which describes a system identical to this one except for the gears, is identical to (54) with $N = 1$. Hence, as expected, the only effect of the gears is to multiply the applied torque $\tau_a(t)$ by the gear ratio N and to reverse its direction on the disk.

▶ **EXAMPLE 5.10**

Find the state-variable equations for the system shown in Figure 5.26(a), in which the pair of gears couples two similar subsystems.

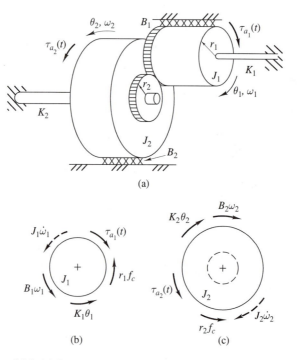

(a)

(b)

(c)

Figure 5.26   (a) System for Example 5.10. (b), (c) Free-body diagrams.

SOLUTION   Because of the two moments of inertia and the two shafts, it might appear that we could choose ω_1, ω_2, θ_1, and θ_2 as state variables. However, θ_1 and θ_2 are related by the gear ratio, as are ω_1 and ω_2. Because the state variables must be independent, either θ_1 and ω_1 or θ_2 and ω_2 constitute a suitable set.

The free-body diagrams for each of the moments of inertia are shown in Figure 5.26(b) and Figure 5.26(c). As in Example 5.9, f_c represents the contact force between the two gears. Summing the torques on each of the free-body diagrams gives

$$J_1\dot{\omega}_1 + B_1\omega_1 + K_1\theta_1 + r_1 f_c = \tau_{a_1}(t) \tag{55a}$$
$$J_2\dot{\omega}_2 + B_2\omega_2 + K_2\theta_2 - r_2 f_c = \tau_{a_2}(t) \tag{55b}$$

By the geometry of the gears,

$$\theta_1 = N\theta_2$$
$$\omega_1 = N\omega_2 \tag{56}$$

where $N = r_2/r_1$.

Selecting θ_2 and ω_2 as the state variables, we can write $\dot{\theta}_2 = \omega_2$ as the first state-variable equation and combine (55) and (56) to obtain the required equation for $\dot{\omega}_2$ in terms of θ_2, ω_2, $\tau_{a_1}(t)$, and $\tau_{a_2}(t)$. We first solve (55b) for f_c and substitute that expression into (55a). Then, substituting (56) into the result gives

$$(J_2 + N^2 J_1)\dot{\omega}_2 + (B_2 + N^2 B_1)\omega_2 + (K_2 + N^2 K_1)\theta_2 - N\tau_{a_1}(t) - \tau_{a_2}(t) = 0 \tag{57}$$

At this point, it is convenient to define the parameters

$$J_{2_{eq}} = J_2 + N^2 J_1$$
$$B_{2_{eq}} = B_2 + N^2 B_1$$
$$K_{2_{eq}} = K_2 + N^2 K_1 \tag{58}$$

which can be viewed as the combined moment of inertia, damping coefficient, and stiffness constant, respectively, when the combined system is described in terms of the variables θ_2 and ω_2. For example, it is common to say that $N^2 J_1$ is the equivalent inertia of disk 1 when that inertia is reflected to shaft 2. Similarly, $N^2 B_1$ and $N^2 K_1$ are the reflected viscous-friction coefficient and stiffness constant, respectively. Hence the parameters $J_{2_{eq}}$, $B_{2_{eq}}$, and $K_{2_{eq}}$ defined in (58) are the sums of the parameters associated with shaft 2 and the corresponding parameters reflected from shaft 1.

With the new notation, we can rewrite (57) as

$$J_{2_{eq}}\dot{\omega}_2 + B_{2_{eq}}\omega_2 + K_{2_{eq}}\theta_2 - N\tau_{a_1}(t) - \tau_{a_2}(t) = 0 \tag{59}$$

and the state-variable equations are

$$\dot{\theta}_2 = \omega_2$$
$$\dot{\omega}_2 = \frac{1}{J_{2_{eq}}}\left[-K_{2_{eq}}\theta_2 - B_{2_{eq}}\omega_2 + N\tau_{a_1}(t) + \tau_{a_2}(t)\right] \tag{60}$$

Note that the driving torque $\tau_{a_1}(t)$ applied to shaft 1 has the value $N\tau_{a_1}(t)$ when reflected to shaft 2.

If we wanted the system model in terms of θ_1 and ω_1, straightforward substitutions would lead to the equations

$$\dot{\theta}_1 = \omega_1$$
$$\dot{\omega}_1 = \frac{1}{J_{1_{eq}}}\left[-K_{1_{eq}}\theta_1 - B_{1_{eq}}\omega_1 + \tau_{a_1}(t) + \frac{1}{N}\tau_{a_2}(t)\right]$$

where the combined parameters with the elements associated with shaft 2 reflected to shaft 1 are

$$J_{1_{eq}} = J_1 + \frac{1}{N^2}J_2$$

$$D_{1_{eq}} - D_1 + \frac{1}{N^2}B_2$$

$$K_{1_{eq}} = K_1 + \frac{1}{N^2}K_2$$

With the experience we have gained in deriving the mathematical models for separate translational and rotational systems, it is a straightforward matter to treat systems that combine both types of elements. The next example uses a rack and a pinion gear to convert rotational motion to translational motion. The final example combines translational and rotational systems via a cable attached to a disk.

▷ **EXAMPLE 5.11**

Derive the state-variable model for the system shown in Figure 5.27. The moment of inertia J represents the rotor of a motor on which an applied torque $\tau_a(t)$ is exerted. The rotor is connected by a flexible shaft to a pinion gear of radius R that meshes with the linear rack. The rack is rigidly attached to the mass M, which might represent the bed of a milling machine. The outputs of interest are the displacement and velocity of the rack and the contact force between the rack and the pinion.

SOLUTION   The free-body diagrams for the moment of inertia J, the pinion gear, and the mass M are shown in Figure 5.28. The contact force between the rack and the pinion is denoted by f_c. Forces and torques that will not appear in the equations of interest (such as the vertical force on the mass and the bearing forces on the rotor and pinion gear) have been omitted. Summing the torques in Figure 5.28(a) and Figure 5.28(b) and the forces in Figure 5.28(c) yields the three equations

$$J\dot{\omega} + B_1\omega + K(\theta - \theta_A) - \tau_a(t) = 0 \tag{61a}$$

$$Rf_c - K(\theta - \theta_A) = 0 \tag{61b}$$

$$M\dot{v} + B_2 v - f_c = 0 \tag{61c}$$

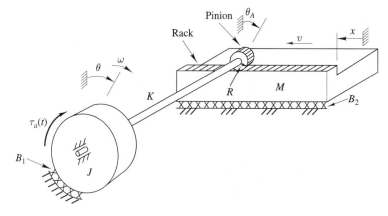

Figure 5.27 System for Example 5.11 with rack and pinion gear.

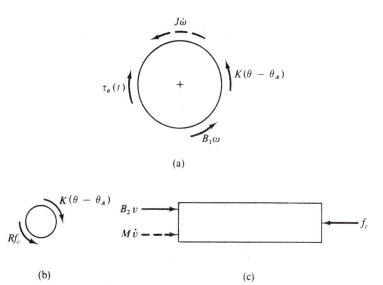

Figure 5.28 Free-body diagrams for Example 5.11. (a) Rotor. (b) Pinion gear. (c) Mass.

In addition, the geometric relationship

$$R\theta_A = x \tag{62}$$

must hold because of the contact between the rack and the pinion gear.

The fact that there are three energy-storing elements corresponding to the parameters J, M, and K suggests that the three variables ω, v, and $\theta_R = \theta - \theta_A$ might constitute a satisfactory set of state variables. However, the displacement x of the mass, which is usually of interest and which is one of the specified outputs, cannot be expressed as an algebraic function of ω, v, θ_R, and the input. Thus we need four state variables, which we choose to be θ, ω, x, and v. Using (62) to eliminate θ_A in (61a) gives

$$J\dot{\omega} + B_1\omega + K\theta - \frac{K}{R}x - \tau_a(t) = 0$$

and using (62) and (61b) to eliminate f_c in (61c) results in

$$M\dot{v} + B_2v + \frac{K}{R^2}x - \frac{K}{R}\theta = 0$$

Thus the state-variable equations are

$$\dot{\theta} = \omega$$
$$\dot{\omega} = \frac{1}{J}\left[-K\theta - B_1\omega + \frac{K}{R}x + \tau_a(t)\right]$$
$$\dot{x} = v \tag{63}$$
$$\dot{v} = \frac{1}{M}\left(\frac{K}{R}\theta - \frac{K}{R^2}x - B_2v\right)$$

The outputs x and v are also state variables. The output equation for f_c can be found by substituting (62) into (61b). It is

$$f_c = \frac{K}{R}\left(\theta - \frac{x}{R}\right)$$

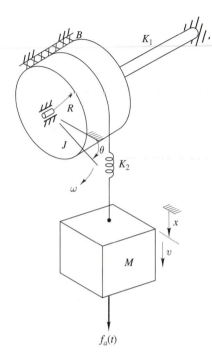

Figure 5.29 System for Example 5.12 with translational and rotational elements.

▶ **EXAMPLE 5.12**

In the system shown in Figure 5.29, the mass and spring are connected to the disk by a flexible cable. Actually, the spring might be used to represent the stretching of the cable. The mass M is subjected to the external force $f_a(t)$ in addition to the gravitational force. Let θ and x be measured from references corresponding to the position where the shaft K_1 is not twisted and the spring K_2 is not stretched. Find the state-variable model, treating $f_a(t)$ and the weight of the mass as inputs and the angular displacement θ and the tensile force in the cable as outputs.

SOLUTION   The free-body diagrams for the disk and the mass are shown in Figure 5.30, where f_2 denotes the force exerted by the spring. The downward displacement of the top end of the spring is $R\theta$, so

$$f_2 = K_2(x - R\theta) \tag{64}$$

Because of the four energy-storing elements corresponding to the parameters K_1, J, K_2, and M, we select θ, ω, x, and v as the state variables. From the free-body diagrams and with (64), we can write

$$J\dot{\omega} + B\omega + K_1\theta - RK_2(x - R\theta) = 0 \tag{65a}$$

$$M\dot{v} + K_2(x - R\theta) = f_a(t) + Mg \tag{65b}$$

Note that the reaction force $f_2 = K_2(x - R\theta)$ of the cable on the mass is not the same as the total external force $f_a(t) + Mg$ on the mass. As indicated by (65b), the difference is the inertial force $M\dot{v}$. Only if the mass were negligible would the external force be transmitted directly through the spring. From (65) and the identities $\dot{\theta} = \omega$ and $\dot{x} = v$, we can write the

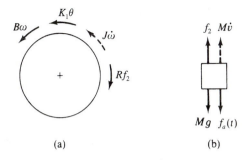

Figure 5.30 Free-body diagrams for Example 5.12. (a) Disk. (b) Mass.

state-variable equations

$$\dot{\theta} = \omega$$
$$\dot{\omega} = \frac{1}{J}\left[-(K_1 + K_2 R^2)\theta - B\omega + K_2 Rx\right]$$
$$\dot{x} = v \tag{66}$$
$$\dot{v} = \frac{1}{M}[K_2 R\theta - K_2 x + f_a(t) + Mg]$$

The only output that is not a state variable is the tensile force in the cable, for which the output equation is given by (64).

In order to emphasize the effect of the weight Mg, suppose that $f_a(t) = 0$ and that the mass and disk are not moving. Let θ_0 denote the constant angular displacement of the disk and x_0 the constant displacement of the mass under these conditions. Then (65) becomes

$$K_1\theta_0 = RK_2(x_0 - R\theta_0)$$
$$K_2(x_0 - R\theta_0) = Mg \tag{67}$$

from which

$$\theta_0 = \frac{RMg}{K_1}$$
$$x_0 = \frac{Mg}{K_2} + \frac{R^2 Mg}{K_1} \tag{68}$$

These expressions represent the constant displacements caused by the gravitational force Mg.

Now reconsider the case where $f_a(t)$ is nonzero and where the system is in motion. Let

$$\theta = \theta_0 + \phi$$
$$x = x_0 + z \tag{69}$$

so that ϕ and z represent the additional angular and vertical displacements caused by the input $f_a(t)$. Note that $\omega = \dot{\theta} = \dot{\phi}$ and $v = \dot{x} = \dot{z}$. Substituting (69) into (65) gives

$$J\dot{\omega} + B\omega + K_1(\theta_0 + \phi) - RK_2(x_0 + z - R\theta_0 - R\phi) = 0$$
$$M\dot{v} + K_2(x_0 + z - R\theta_0 - R\phi) = f_a(t) + Mg$$

Using (67) to cancel those terms involving θ_0, x_0, and Mg, we are left with

$$J\dot{\omega} + B\omega + K_1\phi - RK_2(z - R\phi) = 0$$
$$M\dot{v} + K_2(z - R\phi) = f_a(t)$$

so the corresponding state-variable equations are

$$\dot{\phi} = \omega$$
$$\dot{\omega} = \frac{1}{J}[-(K_1 + K_2 R^2)\phi - B\omega + K_2 Rz]$$
$$\dot{z} = v \tag{70}$$
$$\dot{v} = \frac{1}{M}[K_2 R\phi - K_2 z + f_a(t)]$$

We see that (66) and (70) have the same form, except that in the latter case the term Mg is missing and θ and x have been replaced by ϕ and z. As long as the stiffness elements are linear, we can ignore the gravitational force Mg if we measure all displacements from the static-equilibrium positions corresponding to no inputs except gravity. This agrees with the conclusion reached in Examples 2.5 and 2.6. Note that if one of the desired outputs is the total tensile force in the cable, we must substitute (69) into (64) to get

$$f_2 = K_2(x_0 - R\theta_0) + K_2(z - R\phi) \tag{71}$$

where x_0 and θ_0 are given by (68). The first of the two terms in (71) is the constant tensile force resulting only from the weight of the mass. The second term is the additional tensile force caused by the input $f_a(t)$.

► 5.5 COMPUTER SIMULATION

In this section, we use the techniques of Chapter 4 to simulate a rotational mechanical system. We specify a nonlinear load torque and show that this is easily added to the simulation. In Chapter 9 we will show how to treat such a nonlinearity analytically.

► EXAMPLE 5.13

Draw a Simulink diagram to solve (42) and simulate the system in Example 5.5. Assume that $\tau_L(t)$ is given by $\tau_L = \alpha|\omega_2|\omega_2$. The parameter values are $J_1 = 10$ kg·m^2, $J_2 = 150$ kg·m^2, $K = 800$ N·m, $B = 50$ N·m·s, and $\alpha = 50$ N·m·s^2. Plot ω_1 and ω_2 during the interval $0 \leq t \leq 10$ s when the input $\tau_a(t)$ is a step of 100 N·m applied at $t = 1$ s.

SOLUTION   We start the Simulink diagram with three integrators, whose inputs are the first derivatives of the three state variables. Each of these derivatives is the output of a gain block preceded by a summer. In order to avoid long crossing lines when obtaining the inputs for the summers, we use two new blocks: the Goto and From blocks, both of which are found in the Signals & Systems library. A signal that is the input to a Goto block appears at all From blocks that have the same tag. Using these two blocks can simplify the Simulink diagram by avoiding the need for long lines to carry signals created at one location to a remote site where they are needed. The complete diagram is shown in Figure 5.31.

We cannot show the variable $\tau_L(t)$ as an external input, because its value is completely determined by the variable ω_2. We create τ_L , which is $\alpha|\omega_2|\omega_2$, in the lower left corner of the diagram by using two nonlinear blocks that are found in the Math library. The block named Abs is used to create the absolute value of its input, which is ω_2. The next block is the Product function, used to produce the product of its two inputs. Here these are

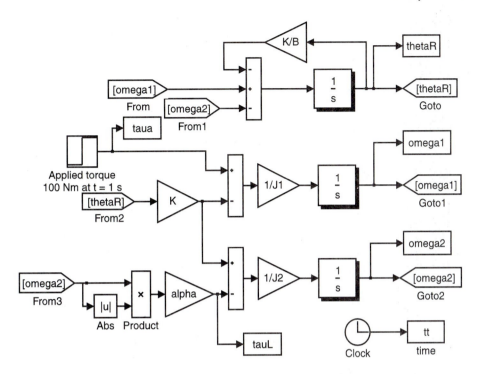

Figure 5.31 A Simulink diagram to solve (42).

ω_2 and its absolute value $|\omega_2|$. Finally, we use a gain block to multiply the output of the Product block by α.

The applied torque $\tau_a(t)$ is obtained from the Step function in the Sources library. We let the default option set the initial conditions on all the integrators equal to zero. The following M-file will produce the desired responses, which are shown in Figure 5.32.

```
J1 = 10;   J2 = 150;            % kg.m^2
K  = 800;                       % N.m
B  = 50;                        % N.m.s
alpha = 50;                     % N.m.s^2
sim('ex5_13_model')            % run the simulation
plot(tt, omega1, tt, omega2, '--'); grid
text(2, 3, '\omega_1');   text(5, 1.5, '\omega_2')
xlabel('Time (s)')
ylabel('Angular velocity (rad/s)')
```

▶ SUMMARY

In this chapter we extended the techniques of Chapters 2 and 3 to include bodies that rotate about fixed axes. Moment of inertia, friction, and stiffness elements are characterized by an algebraic relationship between the torque and the angular acceleration, velocity, or displacement, respectively. In contrast, levers and gears are described by algebraic equations that relate two variables of the same type (two displacements, for example, or two velocities). We showed how systems with levers, gears, or pulleys can have both rotational and translational motion.

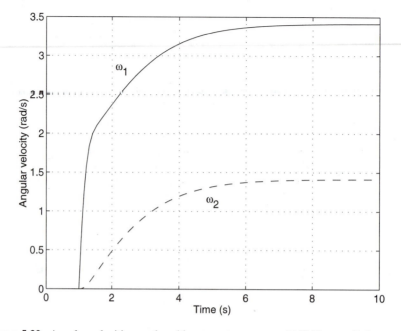

Figure 5.32 Angular velocities produced by a constant torque of 100 N·m applied at $t = 1$ s.

As in Chapter 2, we drew free-body diagrams to help us obtain the basic equations governing the motion of the system. We then combined these equations into sets of state-variable equations or input-output equations. The same general modeling procedures are used for other types of systems in later chapters.

▶ PROBLEMS

5.1. Write a differential equation for the system shown in Figure P5.1 and determine the equivalent stiffness constant.

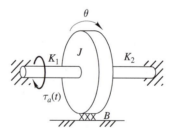

Figure P5.1

*5.2. The left side of the fluid drive element denoted by B in Figure P5.2 moves with the angular velocity $\omega_a(t)$. Find the input-output differential equation relating ω_2 and $\omega_a(t)$.

5.3.

    a.   Choose a set of state variables for the system shown in Figure P5.3, assuming that the values of the individual angular displacements $\theta_1, \theta_2, \theta_3$, and θ_4 with respect

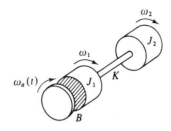

Figure P5.2

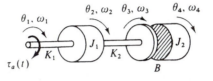

Figure P5.3

to a fixed reference are not of interest. Then write the state-variable equations describing the system.

b. Find the input-output differential equation relating ω_4 and $\tau_a(t)$.

5.4. Repeat part (a) of Problem 5.3 if the torque input $\tau_a(t)$ is replaced by the angular velocity input $\omega_1(t)$.

5.5.

a. Find a state-variable model for the system shown in Figure P5.5. The input is the applied torque $\tau_a(t)$, and the output is the viscous torque on J_2, with the positive sense counterclockwise.

b. Write the input-output differential equation relating ω_2 and $\tau_a(t)$ when $K_1 = K_2 = B = J_1 = J_2 = 1$ in a consistent set of units.

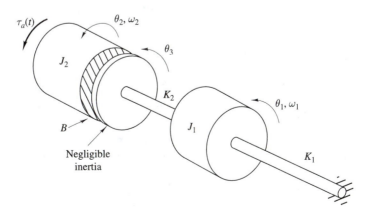

Figure P5.5

*5.6.

a. For the system shown in Figure P5.6, select state variables and write the model in state-variable form. The input is the applied torque $\tau_a(t)$, and the output is the viscous torque acting on J_2 through the fluid drive element, with the positive sense counterclockwise.

b. Write the input-output differential equation relating θ_2 and $\tau_a(t)$ when $K_1 = K_2 = B = J_1 = J_2 = 1$ in a consistent set of units.

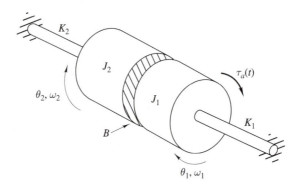

Figure P5.6

*5.7. Find the value of K_{eq} in Figure P5.7(b) such that the relationship between $\tau_a(t)$ and θ is the same as for Figure P5.7(a).

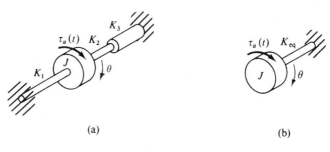

(a)   (b)

Figure P5.7

5.8. Use the p-operator technique to derive (43).

5.9. For the system shown in Figure P5.9, the angular motion of the ideal lever from the vertical position is small, so the motion of the top and midpoint can be regarded as horizontal. The input is the force $f_a(t)$ applied at the top of the lever, and the output is the support force on the lever, taking the positive sense to the right.

a. Select a suitable set of state variables and write the corresponding state-variable model.

b. Find the input-output differential equation relating x and $f_a(t)$.

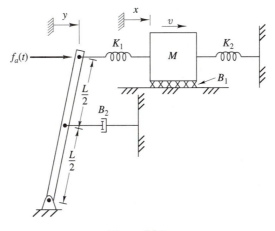

Figure P5.9

*5.10. In the mechanical system shown in Figure P5.10, the input is the applied force $f_a(t)$, and the output is the tensile force in the spring K_2. The lever is ideal and is horizontal when the system is in static equilibrium with $f_a(t) = 0$ and M supported by the spring K_1. The displacements x_1, x_2, x_3, and θ are measured with respect to this equilibrium position. The lever angle θ remains small.

    a.   Taking x_1, v_1, and θ as state variables, write the state-variable model.

    b.   Also write the algebraic output equation for the reaction force of the pivot on the lever, taking the positive sense upward.

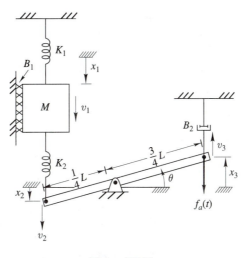

Figure P5.10

5.11. For the system shown in Figure P5.11, the input is the force $f_a(t)$, and the output is x_1. Assume that the lever is ideal and that $|\theta|$ is small.

    a.   Select a suitable set of state variables and write the corresponding state-variable model.

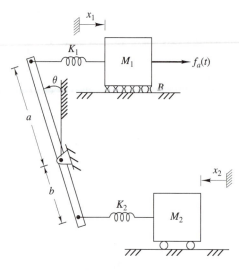

Figure P5.11

b. Find the input-output differential equation when $a = 2$ and $b = K_1 = K_2 = B = M_1 = M_2 = 1$ in a consistent set of units.

5.12. Repeat part (a) of Problem 5.9 when the moment of inertia of the lever cannot be neglected and the output is the displacement x.

*5.13. Repeat part (a) of Problem 5.10 when the moment of inertia of the lever cannot be neglected.

5.14. Repeat part (a) of Problem 5.11 when the moment of inertia of the lever cannot be neglected.

5.15. Figure P5.15 shows two pendulums suspended from frictionless pivots and connected at their midpoints by a spring. Each pendulum may be considered a point mass M at the end of a rigid, massless bar of length L. Assume that $|\theta_1|$ and $|\theta_2|$ are sufficiently small to allow use of the small-angle approximations $\sin\theta \simeq \theta$ and $\cos\theta \simeq 1$. The spring is unstretched when $\theta_1 = \theta_2$.

    a. Draw a free-body diagram for each pendulum.

    b. Define a set of state variables and write the state-variable equations.

    c. Write an algebraic output equation for the spring force. Consider the force to be positive when the spring is in tension.

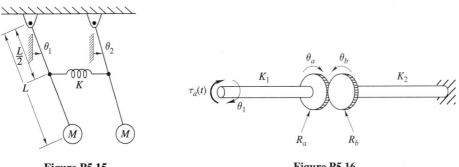

Figure P5.15   **Figure P5.16**

5.16. Find the equivalent stiffness constant K_{eq} such that the algebraic model of the gears and shafts shown in Figure P5.16 can be written as $\theta_1 = (1/K_{eq})\tau_a(t)$.

5.17. In the system shown in Figure P5.17, a torque $\tau_a(t)$ is applied to the cylinder J_1. The gears are ideal with gear ratio $N = R_2/R_1$.

    a.   Write the input-output differential equation when the inertia J_1 is reflected to J_2 and ω_2 is the output.

    b.   Write the equation when J_2 is reflected to J_1 and ω_1 is the output.

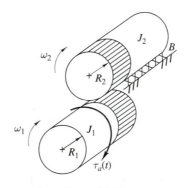

Figure P5.17

*5.18. A torque input $\tau_a(t)$ is applied to the lower gear shown in Figure P5.18. There are two outputs: the total angular momentum and the viscous torque acting on J_1, with the positive sense clockwise.

    a.   Write a pair of simultaneous differential equations describing the system, where the contact force between the gears is included.

    b.   Select state variables and write the model in state-variable form.

    c.   Derive an input-output differential equation relating θ_1 and $\tau_a(t)$.

*5.19. The input to the rotational system shown in Figure P5.19 is the applied torque $\tau_a(t)$ and the output is θ_3. The gear ratio is $N = R_2/R_3$. Using $\theta_1, \omega_1, \theta_2$, and ω_2 as the state variables, write the model in state-variable form.

5.20. Repeat Problem 5.19 when the state variables are $\theta_R = \theta_1 - \theta_2$, ω_1, θ_3, and ω_3.

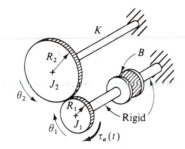

Figure P5.18

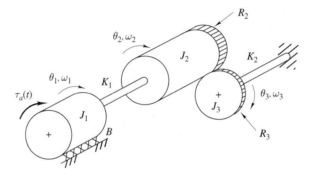

Figure P5.19

5.21. The input for the drive system shown in Figure P5.21 is the applied torque $\tau_a(t)$, and a load attached to the moment of inertia J_2 produces the load torque $\tau_L = A|\omega_2|\omega_2$.

    a.   Taking as state variables ω_1, ω_2, ω_a, $\phi_a = \theta_1 - \theta_a$ and $\phi_b = \theta_2 - \theta_b$, write the state-variable equations.

    b.   Write an algebraic output equation for the contact force on gear a, with the positive sense upward.

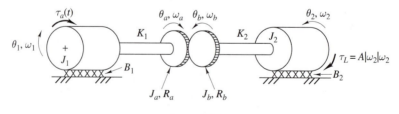

Figure P5.21

5.22. The input to the combined translational and rotational system shown in Figure P5.22 is the force $f_a(t)$ applied to the mass M. The elements K_1 and K_2 are undeflected when $x = 0$ and $\theta = 0$.

    a.   Write a single input-output differential equation for x.

    b.   Write a single input-output differential equation for θ.

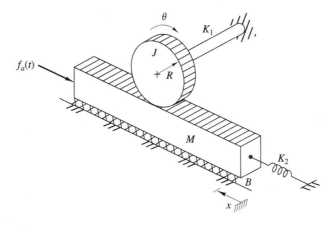

Figure P5.22

5.23. Starting with (63), find the input-output differential equation for the system shown in Figure 5.27, taking $\tau_a(t)$ as the input and x as the output.

5.24. In the mechanical system shown in Figure P5.24, the cable is wrapped around the disk and does not slip or stretch. The input is the force $f_a(t)$, and the output is the displacement x. The springs are undeflected when $\theta = x = 0$.

    a.   Write the system model as a pair of differential equations involving only the variables x, θ, $f_a(t)$, and their derivatives.

    b.   Select state variables and write the model in state-variable form.

    c.   Find an input-output differential equation relating x and $f_a(t)$ when $K_1 = K_2 = K_3 = B_1 = B_2 = M = J = R = 1$ in a consistent set of units.

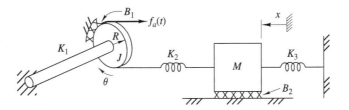

Figure P5.24

5.25. A mass and a translational spring are suspended by cables wrapped around two sections of a drum as shown in Figure P,5.25. The cables are assumed not to stretch, and the moment of inertia of the drum is J. The viscous-friction coefficient between the drum and a fixed surface is denoted by B. The spring is neither stretched nor compressed when $\theta = 0$.

    a.   Write a differential equation describing the system in terms of the variable θ.

    b.   For what value of θ will the rotational element remain motionless?

    c.   Rewrite the system's differential equation in terms of ϕ, the relative angular displacement with respect to the static-equilibrium position you found in part (b).

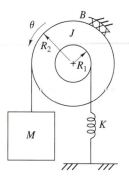

Figure P5.25

*5.26. In the system shown in Figure P5.26, the cable around the cylinder J does not stretch and does not slip. The input is the applied force $f_a(t)$ and the outputs are θ and z, where z is the additional elongation of K from its static-equilibrium position due to $f_a(t)$. The spring is undeflected when $x_1 = x_2 = \theta = 0$. Select a set of state variables and write the model in state-variable form.

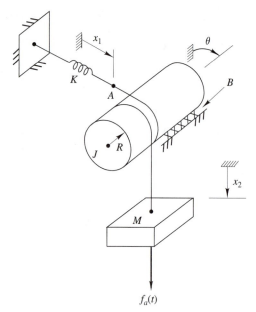

Figure P5.26

5.27. In the system shown in Figure P5.27, the two masses are equal and the cable neither stretches nor slips. The springs are undeflected when $\theta = x_1 = x_2 = 0$. The input is the force $f_a(t)$ applied to the right mass, and the output is x_1. Select a set of state variables and write the model in state-variable form.

5.28. The input to the system in Figure P5.28 is the applied torque $\tau_a(t)$. The shaft K_1 and the spring K_2 are undeflected when $\theta_1 = \theta_2 = x = 0$.

    a. Write a set of differential equations describing the system in terms of θ_1, θ_2, x, and $\tau_a(t)$.

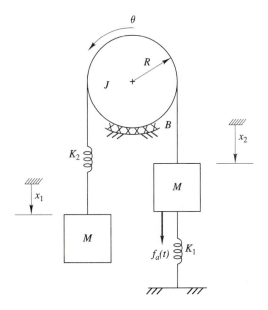

Figure P5.27

b. Select a set of state variables and write the state-variable equations.

c. Find the constant value of $\tau_a(t)$ for which the system will reach an equilibrium position with the mass remaining motionless. Determine the corresponding deflections of K_1 and K_2.

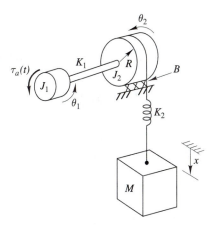

Figure P5.28

*5.29. The input to the combined translational and rotational system shown in Figure P5.29 is the displacement $x_1(t)$. The output is the torque applied to the gear by the shaft, with the positive sense clockwise. The springs are undeflected when $\theta = x_1 = x_2 = 0$. Select an appropriate set of state variables and write the model in state-variable form.

5.30. Starting with (66), find the input-output differential equation for the system shown in Figure 5.29, taking $f_a(t)$ and the gravitational force as inputs and θ as the output.

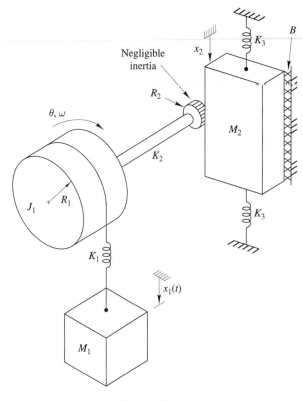

Figure P5.29

5.31.

    a. For the rotational mechanical system discussed in Example 5.3, write the state-variable equations (5.31) in matrix form. Show why we cannot obtain either θ_1 or θ_2 from this model.

    b. Add θ_1 as the fourth state variable and write the state-variable equations in matrix form.

*5.32. For the combined translational and rotational system modeled in Example 5.12, write (66) in matrix form. Note that the system has two inputs, the gravitational constant g and the applied force $f_a(t)$.

5.33. For the combined translational and rotational system modeled in Example 5.12, write (70) in matrix form. Also write a matrix output equation for the displacements θ and x defined in (69).

5.34. Draw a block diagram for the rotational mechanical system described by the following two equations, where $\theta_a(t)$ is the input.

$$2\ddot{\theta}_1 + \dot{\theta}_1 + 3(\dot{\theta}_1 - \dot{\theta}_2) + \theta_1 = \dot{\theta}_a + \theta_a(t)$$
$$2\ddot{\theta}_2 + \dot{\theta}_2 = 3(\dot{\theta}_1 - \dot{\theta}_2) + \theta_2$$

Hint: Define a new variable z such that $\dot{z} = 2\ddot{\theta}_1 - \dot{\theta}_a$ and substitute the first modeling equation into this new equation.

5.35. Draw a block diagram for the system described by Equation (5.27) that was modeled in Example 5.2.

5.36. Draw a block diagram for the system described by Equations (5.64) and (5.66) that was modeled in Example 5.12.

5.37. When the input is a constant, then we expect that all variables in the modeling equations will eventually become constants, in which case their derivatives will eventually become zero. Use this fact as a partial check on the curves in Figure 5.32. For Example 5.13, calculate what the final values of ω_1 and ω_2 should be, and compare the results with the plots obtained by MATLAB.

5.38. Draw a Simulink diagram for the system shown in Figure 5.27, and described by (63). Use the following parameter values: $M = 50$ kg, $K = 20$ N·m, $R = 0.1$ m, $J = 30$ kg·m^2, $B_1 = 50$ N·m·s, and $B_2 = 100$ N·s/m. Simulate the first 10 seconds of the response when the applied torque is a step of 1 N·m starting at $t = 1$ s. Plot the angular velocity ω and the linear velocity v.

6

ELECTRICAL SYSTEMS

Except at quite high frequencies, electrical circuits can usually be considered an interconnection of lumped elements. In such cases, which include a large and very important portion of the applications of electrical phenomena, we can model a circuit by using ordinary differential equations and we can apply the solution techniques discussed in this book.

In this chapter, we consider fixed linear circuits using the same approach we used for mechanical systems. We introduce the element and interconnection laws and then combine them to form procedures for finding the model of a circuit. After developing a general technique for finding the input-output model, we present several specialized results for the important case of resistive circuits. We then discuss systematic procedures for obtaining the model as a set of state-variable equations. After a discussion of operational amplifiers, the chapter concludes with two computer simulations.

▶ 6.1 VARIABLES

The variables most commonly used to describe the behavior of electrical circuits are

e, voltage in volts (V)

i, current in amperes (A)

The related variables

q, charge in coulombs (C)

ϕ, flux in webers (Wb)

Λ, flux linkage in weber-turns

may be used on occasion. Current is the time derivative of charge, so i and q are related by the expressions

$$i = \frac{dq}{dt} \tag{1}$$

and

$$q(t) = q(t_0) + \int_{t_0}^{t} i(\lambda)\, d\lambda \tag{2}$$

Flux and flux linkage are related by the number of turns N in a coil of wire, such that if all the turns are linked by all the flux, then $\Lambda = N\phi$.

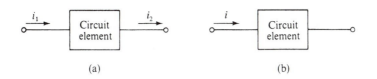

Figure 6.1 Conventions for denoting current. (a) Acceptable. (b) Preferred.

We represent the current into and out of a circuit element by arrows drawn on the circuit diagram as shown in Figure 6.1. The arrows point in the direction in which positive charge—that is, positive ions—flows when the current has a positive value. Equivalently, a positive current can also correspond to electrons (which have a negative charge) flowing in the opposite direction.

Because a net charge cannot exist within any circuit element, the current entering one end of a two-terminal element must leave the other end. Hence $i_1 = i_2$ in Figure 6.1(a) at all times, so only one current arrow need be shown, as in Figure 6.1(b).

The voltage at a point in a circuit is a measure of the difference between the electrical potential of that point and the potential of an arbitrarily established reference point called the **ground node**, or **ground** for short. The ground associated with a circuit is denoted by the symbol shown in the lower part of Figure 6.2(a). Any point in the circuit that has the same potential as the ground has a voltage of zero, by definition. The voltage e_1 shown in Figure 6.2(a) is positive if the point with which it is associated is at a higher potential than the ground; it is negative if the potential of the point with which it is associated is lower than that of the ground.

We can define the voltages of the two terminals of a circuit element individually with respect to ground by writing appropriate symbols next to the terminals, as shown in Figure 6.2(b). We define the voltage between the terminals of an element by placing a symbol next to the element and plus and minus signs on either side of the element or at the terminals, as shown in Figure 6.2(c). When the element voltage e_a is positive, the terminal marked with the plus sign is at a higher potential than the other terminal.

In Figure 6.2(d), e_1 and e_2 denote the terminal voltages with respect to ground, and e_a is the voltage across the element, with its positive sense indicated by the plus and minus

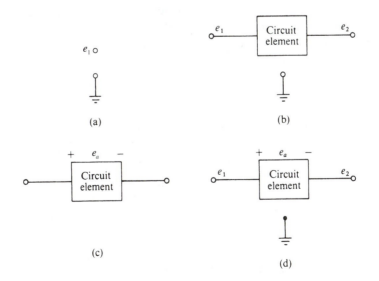

Figure 6.2 Conventions for denoting voltages.

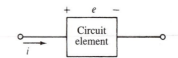

Figure 6.3 Positive senses of voltage and current for (3).

signs. These three voltages are related by the equation

$$e_a = e_1 - e_2$$

Interchanging the plus and minus signs reverses the sign of the voltage e_a in any equation in which it appears.

When we define the positive senses of the current and voltage associated with a circuit element as shown in Figure 6.3, such that a positive current is assumed to enter the element at the terminal designated by the plus sign, then the power supplied to the element is

$$p = ei \tag{3}$$

which has units of watts. If at some instant p is negative, then the circuit element is supplying power to the rest of the circuit at that instant. Because power is the time derivative of energy, the energy supplied to the element over the interval t_0 to t_1 is

$$\int_{t_0}^{t_1} p(t)\, dt$$

which has units of joules, where 1 joule = 1 volt-ampere-second.

► 6.2 ELEMENT LAWS

The elements in the electrical circuits that we shall consider are resistors, capacitors, inductors, and sources. The first three of these are referred to as **passive elements** because, although they can store or dissipate energy that is present in the circuit, they cannot introduce additional energy. They are analogous to the dashpot, mass, and spring for mechanical systems. In contrast, sources are **active elements** that can introduce energy into the circuit and that serve as the inputs. They are analogous to the force or displacement inputs for mechanical systems.

Resistor

A **resistor** is an element for which there is an algebraic relationship between the voltage across its terminals and the current through it—that is, an element that can be described by a curve of e versus i. A linear resistor is one for which the voltage and current are directly proportional to each other—that is, one described by **Ohm's law:**

$$e = Ri \tag{4}$$

or

$$i = \frac{1}{R}e \tag{5}$$

where R is the **resistance** in ohms (Ω). A resistor and its current and voltage are denoted as shown in Figure 6.4. If we reversed either the current arrow or the voltage polarity (but

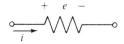

Figure 6.4 A resistor and its variables.

not both) in the figure, we would introduce a minus sign into (4) and (5). The resistance of a body of length ℓ and constant cross-sectional area A made of a material with resistivity ρ is $R = \rho\ell/A$.

A resistor dissipates any energy supplied to it by converting it into heat (in this it is analogous to the frictional element of mechanical systems). We can write the power ei dissipated by a linear resistor as

$$p = Ri^2 = \frac{1}{R}e^2$$

Capacitor

A **capacitor** is an element that obeys an algebraic relationship between the voltage and the charge, where the charge is the integral of the current. We use the symbol shown in Figure 6.5 to represent a capacitor. For a linear capacitor, the charge and voltage are related by

$$q = Ce \tag{6}$$

where C is the **capacitance** in farads (F). For a fixed linear capacitor, the capacitance is a constant. If (6) is differentiated and $\dot{q}$ replaced by i, the element law for a fixed linear capacitor becomes

$$i = C\frac{de}{dt} \tag{7}$$

To express the voltage across the terminals of the capacitor in terms of the current, we solve (7) for de/dt and then integrate, getting

$$e(t) = e(t_0) + \frac{1}{C}\int_{t_0}^{t} i(\lambda)\,d\lambda \tag{8}$$

where $e(t_0)$ is the voltage corresponding to the initial charge, and where the integral is the charge delivered to the capacitor between the times t_0 and t.

One form of a capacitor consists of two parallel metallic plates, each of area A, separated by a dielectric material of thickness d. Provided that fringing of the electric field is negligible, the capacitance of this element is $C = \varepsilon A/d$, where ε is the permittivity of the dielectric material. The values of practical capacitances are typically expressed in micro-

Figure 6.5 A capacitor and its variables.

Figure 6.6 An inductor and its variables.

farads (μF), where 1 μF $= 10^{-6}$ F. However, for numerical convenience we may use farads in our examples.

The energy supplied to a capacitor is stored in its electrical field and can affect the response of the circuit at future times. For a fixed linear capacitor, the stored energy is

$$w = \tfrac{1}{2}Ce^2$$

Because the energy stored is a function of the voltage across its terminals, the initial voltage $e(t_0)$ of a capacitor is one of the conditions we need in order to find the complete response of a circuit for $t \geq t_0$.

Inductor

An **inductor** is an element for which there is an algebraic relationship between the voltage across its terminals and the derivative of the flux linkage. The symbol for an inductor and the convention for defining its current and voltage are shown in Figure 6.6. For a linear inductor,

$$e = \frac{d}{dt}(Li)$$

where L is the **inductance** with units of henries (H). For a fixed linear inductor, L is constant and we can write the element law as

$$e = L\frac{di}{dt} \tag{9}$$

We can find an expression for the current through the inductor by using (9) to integrate di/dt, giving

$$i(t) = i(t_0) + \frac{1}{L}\int_{t_0}^{t} e(\lambda)\,d\lambda \tag{10}$$

where $i(t_0)$ is the initial current through the inductor.

For a linear inductor made by winding N turns of wire around a toroidal core of a material having a constant permeability μ, cross-sectional area A, and mean circumference ℓ, the inductance is $L = \mu N^2 A/\ell$. Typical values of inductance are usually less than 1 henry and are often expressed in millihenries (mH).

The energy supplied to an inductor is stored in its magnetic field, and for a fixed linear inductor this energy is given by

$$w = \frac{1}{2}Li^2$$

To find the complete response of a circuit for $t \geq t_0$, we need to know the initial current $i(t_0)$ for each inductor.

Sources

The inputs for electrical circuit models are provided by ideal voltage and current sources. A **voltage source** is any device that causes a specified voltage to exist between two points

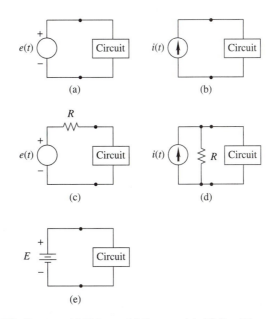

Figure 6.7 Sources. (a) Voltage. (b) Current. (c), (d) Possible representations of nonideal sources. (e) Constant voltage source.

in a circuit, regardless of the current that may flow. A **current source** causes a specified current to flow through the branch containing the source, regardless of the voltage that may be required. The symbols used to represent general voltage and current sources are shown in Figure 6.7(a) and Figure 6.7(b). We often represent physical sources by the combination of an ideal source and a resistor, as shown in parts (c) and (d) of Figure 6.7.

A voltage source that has a constant value for all time is often represented as shown in Figure 6.7(e). The symbol E denotes the value of the voltage, and the terminal connected to the longer line is the positive terminal. A battery is often represented in this fashion.

Open and Short Circuits

An **open circuit** is any element through which current cannot flow. For example, a switch in the open position provides an open circuit, as shown in Figure 6.8(a). Likewise, we can consider a current source that has a value of $i(t) = 0$ over a nonzero time interval an open circuit and can draw it as shown in Figure 6.8(b).

A **short circuit** is any element across which there is no voltage. A switch in the closed position, as shown in Figure 6.9(a), is an example of a short circuit. Another example is a voltage source with $e(t) = 0$, as indicated in Figure 6.9(b).

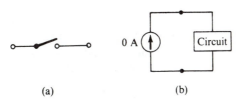

Figure 6.8 Examples of open circuits. (a) Open switch. (b) Zero current source.

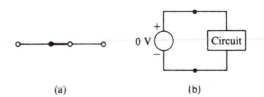

Figure 6.9 Examples of short circuits. (a) Closed switch. (b) Zero voltage source.

▷ 6.3 INTERCONNECTION LAWS

Two interconnection laws are used in conjunction with the appropriate element laws in modeling electrical circuits. These laws are known as Kirchhoff's voltage law and Kirchhoff's current law.

Kirchhoff's Voltage Law

When a closed path—that is, a loop—is traced through any part of a circuit, the algebraic sum of the voltages across the elements that make up the loop must equal zero. This property is known as **Kirchhoff's voltage law**. It may be written as

$$\sum_j e_j = 0 \qquad \text{around any loop} \tag{11}$$

where e_j denotes the voltage across the jth element in the loop.

It follows that summing the voltages across individual elements in any two different paths from one point to another will give the same result. For instance, in the portion of a circuit sketched in Figure 6.10(a), summing the voltages around the loop, going in a counterclockwise direction, and taking into account the polarities indicated on the diagram give

$$e_1 + e_2 - e_3 - e_4 = 0$$

Reversing the direction in which the loop is traversed yields

$$e_4 + e_3 - e_2 - e_1 = 0$$

Likewise, going from point A to point B by each of the two paths shown gives

$$e_1 + e_2 = e_4 + e_3$$

which is, of course, equivalent to both of the foregoing loop equations. In fact, we invoked (11) for the circuit element shown in Figure 6.2(d), which is repeated in Figure 6.10(b), when we stated that $e_a = e_1 - e_2$, because it follows from the voltage law that $e_2 + e_a - e_1 = 0$.

Kirchhoff's Current Law

When the terminals of two or more circuit elements are connected together, the common junction is referred to as a **node**. All the joined terminals are at the same voltage and can be considered part of the node. Because it is not possible to accumulate any net charge at a node, the algebraic sum of the currents at any node must be zero at all times. This property is known as **Kirchhoff's current law**. It may be written as

$$\sum_j i_j = 0 \qquad \text{at any node} \tag{12}$$

where the summation is over the currents through all the elements joined to the node.

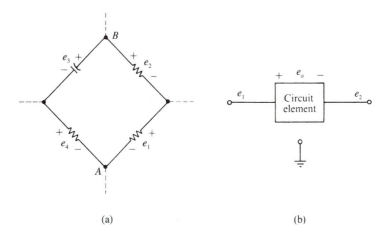

(a)   (b)

Figure 6.10 Partial circuits to illustrate Kirchhoff's voltage law.

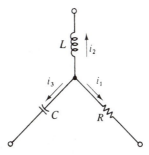

Figure 6.11 Partial circuit to illustrate Kirchhoff's current law.

In applying (12), we must take into account the directions of the current arrows. We shall use a plus sign in (12) for a current arrow directed away from the node being considered and a minus sign for a current arrow directed toward the node. This is consistent with the fact that the current i entering a node is equivalent to the current $-i$ leaving the node.[1] For the partial circuit shown in Figure 6.11, applying (12) at the node to which the three elements are connected gives $i_1 + i_2 + i_3 = 0$. If we wish, we can also use Kirchhoff's current law in (12) for any closed surface that surrounds part of the circuit.

It is a common practice to write the current-law equation directly in terms of the element values and the voltages of the nodes. Consider, for example, the circuit segment shown in Figure 6.12, where e_A, e_B, e_D, and e_F represent the voltages of the nodes with respect to ground. By Kirchhoff's voltage law, the voltage across the resistor is $e_A - e_B$; by the element law, the current through the resistor is $i_1 = (e_A - e_B)/R$. Similarly, the current through the inductor is

$$i_2 = i_2(0) + \frac{1}{L} \int_0^t (e_A - e_D) \, d\lambda$$

---

[1] Instead of interpreting the left side of (12) as the algebraic sum of the currents leaving the node, it would also be correct to use the algebraic sum of the currents entering the node.

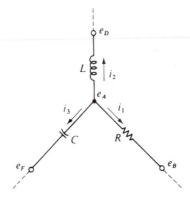

Figure 6.12 Partial circuit to illustrate Kirchhoff's current law written in terms of node voltages.

and that through the capacitor is $i_3 = C(\dot{e}_A - \dot{e}_F)$. Thus we can write the current law in terms of the node voltages and the initial inductor current as

$$\frac{1}{R}(e_A - e_B) + i_2(0) + \frac{1}{L}\int_0^t (e_A - e_D)d\lambda + C(\dot{e}_A - \dot{e}_F) = 0$$

In the following two examples, Kirchhoff's voltage law and current law are used to derive a circuit model.

▶ **EXAMPLE 6.1**

Derive the model for the circuit shown in Figure 6.13.

SOLUTION By a trivial application of Kirchhoff's current law, the same current must flow through each of the four elements in the circuit. This current is denoted by i and its positive sense is taken as clockwise, as indicated in Figure 6.13. Because they have the same current flowing through them, the four elements are said to be connected in **series**.

The voltages across the three passive elements are e_L, e_R, and e_C, and we have assigned them the polarities indicated in the diagram. Starting at the ground node and proceeding counterclockwise around the single loop, we have, by Kirchhoff's voltage law,

$$e_L + e_R + e_C - e_i(t) = 0 \tag{13}$$

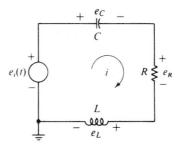

Figure 6.13 Series RLC circuit with a voltage source.

The element laws (4), (8), and (9) give expressions for e_R, e_C, and e_L:

$$e_R = Ri$$
$$e_C = e_C(0) + \frac{1}{C}\int_0^t i(\lambda)\, d\lambda \tag{14}$$
$$e_L = L\frac{di}{dt}$$

where the initial time has been taken as $t_0 = 0$ in (8). Substituting (14) into (13) and rearranging give the circuit model as the integral-differential equation

$$L\frac{di}{dt} + Ri + \frac{1}{C}\int_0^t i(\lambda)\, d\lambda = e_i(t) - e_C(0) \tag{15}$$

To eliminate the constant term and the integral, we differentiate (15) term by term, which yields

$$L\frac{d^2 i}{dt^2} + R\frac{di}{dt} + \frac{1}{C}i = \dot{e}_i$$

a second-order differential equation for the current i with the derivative of the applied voltage acting as the forcing function.

▶ **EXAMPLE 6.2**

Obtain the input-output differential equation relating the input $i_i(t)$ to the output e_o for the circuit shown in Figure 6.14.

SOLUTION   Each of the four circuit elements in Figure 6.14 has one terminal connected to the ground node and the other terminal connected to another common node. By a trivial application of Kirchhoff's voltage law, we see that the voltage across each element is e_o. Hence we say that the elements are connected in **parallel**.

Because the circuit has a single node whose voltage is unknown, we shall apply Kirchhoff's current law at that node in order to obtain the circuit model. We could also apply the current law at the ground node, but we would obtain no new information. The currents through the three passive elements are i_C, i_R, and i_L. As indicated by the arrows in Figure 6.14, each of these currents is considered positive when it flows from the upper node to the ground node.

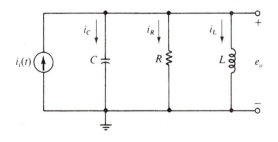

Figure 6.14   Parallel RLC circuit with a current source.

Applying Kirchhoff's current law by summing the currents leaving the upper node, we write

$$i_C + i_R + i_L - i_i(t) = 0 \tag{16}$$

From the element laws given by (5), (7), and (10), we have

$$i_R = \frac{1}{R} e_o$$

$$i_C = C \dot{e}_o \tag{17}$$

$$i_L = i_L(0) + \frac{1}{L} \int_0^t e_o(\lambda) \, d\lambda$$

where the initial time has been taken as $t_0 = 0$.

Substituting (17) into (16) and rearranging the result give the model as

$$C \dot{e}_o + \frac{1}{R} e_o + \frac{1}{L} \int_0^t e_o(\lambda) \, d\lambda = i_i(t) - i_L(0) \tag{18}$$

Differentiating (18) term by term eliminates the constant term and the integral, resulting in the input-output differential equation

$$C \ddot{e}_o + \frac{1}{R} \dot{e}_o + \frac{1}{L} e_o = \frac{di_i}{dt}$$

▶ 6.4 OBTAINING THE INPUT-OUTPUT MODEL

Two general procedures for developing input-output models of electrical circuits are the node-equation method and the loop-equation method. Example 6.1 was actually a simple illustration of the loop-equation method, and Example 6.2 used the node-equation method. In the **loop-equation method**, a rather trivial application of the current law enables us to express the current through every element in terms of one or more loop currents. We then write an appropriate set of simultaneous equations by using the voltage law and the element laws. In the **node-equation method**, we use Kirchhoff's voltage law in a trivial way to express the voltage across every element in terms of node voltages. Then we write a set of simultaneous equations by using Kirchhoff's current law and the element laws.

We shall emphasize the node-equation method, partly because in some circuits the loop-equation method requires us to use fictitious loop currents that do not correspond to measurable currents through individual elements. Furthermore, the node-equation method is well suited to handling the current sources that appear in models of transistor circuits. (References that cover both methods in detail are listed in Appendix F.)

When we use the node-equation method, we start by labeling the voltage of each node with respect to the ground node. If a voltage source is connected between a particular node and ground, the voltage of that node is the known source voltage. Where they are needed, we introduce symbols to define the voltages of the other nodes with respect to ground. Once we have done this, we can express the voltage across each passive element in terms of the node voltages by a trivial application of Kirchhoff's voltage law, as illustrated by the discussion of Figure 6.2(d) and Figure 6.12. We write a current-law equation for each of the nodes whose voltage is unknown, using the element laws to express the currents through the passive elements in terms of the node voltages. We need only combine the resulting set of equations into input-output form to complete the model.

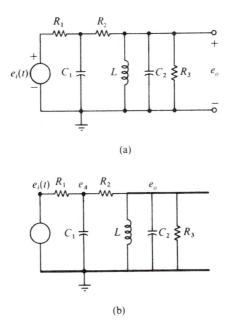

Figure 6.15 (a) Circuit for Example 6.3. (b) Circuit with node voltages shown.

▶ **EXAMPLE 6.3**

Derive the input-output equation for the circuit shown in Figure 6.15(a), using the node-equation method. The input and output voltages are $e_i(t)$ and e_o, respectively.

SOLUTION   The first step is to define all unknown node voltages and redraw the circuit diagram with all node voltages shown, as in Figure 6.15(b). We use the heavy lines to emphasize that the ground node extends across the bottom of the entire circuit and that the node whose voltage is e_o extends from L to R_3. We show the source voltage $e_i(t)$ at the upper left node and denote the voltage of the remaining node with respect to ground by e_A. Because e_A and e_o are unknown node voltages, we shall write a current-law equation at each of these nodes, using the appropriate element laws.

To assist in writing the equations, we can draw separate sketches for each node, as shown in Figure 6.16 (analogous to the free-body diagrams drawn for mechanical systems). For each element, the voltage across its terminals is shown in terms of the node voltages, with the plus sign placed at the node in question. Then we use the appropriate element law to write an expression for the current leaving the node.

We can apply Kirchhoff's current law to each of the nodes shown in Figure 6.16 by setting the algebraic sum of the currents leaving each node equal to zero. The result is the pair of equations

$$\frac{e_A - e_i(t)}{R_1} + C_1 \dot{e}_A + \frac{e_A - e_o}{R_2} = 0 \tag{19a}$$

$$\frac{e_o - e_A}{R_2} + i_L(0) + \frac{1}{L} \int_0^t e_o(\lambda)\, d\lambda + C_2 \dot{e}_o + \frac{e_o}{R_3} = 0 \tag{19b}$$

With a bit of experience, the reader should be able to write these equations directly from the circuit diagram, without drawing the sketches shown in Figure 6.16. It is worthwhile to note that the current through R_2 is labeled $(e_A - e_o)/R_2$ in the sketch for node A and is

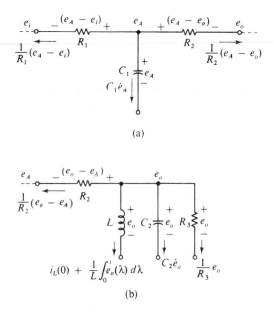

Figure 6.16 Nodes with currents expressed by element laws for Example 6.3.

labeled $(e_o - e_A)/R_2$ in Figure 6.16(b). However, the reference arrows for this current are also reversed on the two parts of Figure 6.16, so there is no inconsistency in the expressions.

We can now differentiate (19b) to eliminate the constant term and the integral. Doing this and rearranging terms give the circuit model as the following pair of coupled differential equations for the node voltages e_A and e_o:

$$C_1 \dot{e}_A + \left(\frac{1}{R_1} + \frac{1}{R_2} \right) e_A - \frac{1}{R_2} e_o = \frac{1}{R_1} e_i(t) \tag{20a}$$

$$-\frac{1}{R_2} \dot{e}_A + C_2 \ddot{e}_o + \left(\frac{1}{R_2} + \frac{1}{R_3} \right) \dot{e}_o + \frac{1}{L} e_o = 0 \tag{20b}$$

By combining these equations to eliminate e_A, we can obtain the input-output differential equation relating $e_i(t)$ and e_o. To simplify the calculations, assume that the passive elements have the numerical values $R_1 = R_2 = 2\,\Omega$, $R_3 = 4\,\Omega$, $C_1 = 1$ F, $C_2 = 4$ F, and $L = \frac{1}{2}$ H. With these parameter values, (20) becomes

$$\dot{e}_A + e_A - \tfrac{1}{2} e_o = \tfrac{1}{2} e_i(t)$$
$$-\tfrac{1}{2} \dot{e}_A + 4\ddot{e}_o + \tfrac{3}{4} \dot{e}_o + 2 e_o = 0$$

By using the p-operator method described in Section 3.2 or by using a combination of substitution and differentiation, we can show that the circuit obeys the third-order equation

$$16 \dddot{e}_o + 19 \ddot{e}_o + 10 \dot{e}_o + 8 e_o = \dot{e}_i \tag{21}$$

for the prescribed element values. In order to solve (21) for e_o for all $t \geq 0$, we would need to know the three initial conditions $e_o(0)$, $\dot{e}_o(0)$, and $\ddot{e}_o(0)$, as well as the input $e_i(t)$ for $t \geq 0$.

Several results in the last example should be specifically noted. The order of an input-output equation is generally the same as the number of energy-storing elements—that is, it is the number of capacitors plus the number of inductors. Thus we could have anticipated that (21) would be third order. In unusual cases, the order of the input-output equation might be less than the number of energy-storing elements. This can happen when the specified output does not depend on the values of some of the passive elements or when the capacitor voltages or inductor currents are not all independent. An illustration of this will be considered in Example 6.10.

In order to avoid drawing partial circuits like those in Figure 6.16, we can use the following rule for the current leaving a node through a passive element. The voltage that appears in the basic element law is replaced by the voltage of the node being considered minus the voltage at the other end of the passive element. The reader should examine (19a) for node A and (19b) for node O to see how the terms can be written down directly from Figure 6.15(b). If a current source is attached to the node, we can easily include the known source when writing Kirchhoff's current law. If, however, a voltage source is connected directly to the node, its current will not be known until after the circuit has been completely solved. To prevent introducing an additional unknown, we try to avoid summing currents at nodes to which voltage sources are attached. Thus, in the last example, we would not write a current-law equation at the junction of R_1 and $e_i(t)$.

Finally, consider the simplified node equations that remain after any integral signs have been eliminated by differentiation and after like terms have been collected together. In (20a) for node A, all the terms involving e_A and its derivatives have the same sign. Similarly in (20b) for node O, all the terms with e_o and its derivatives have the same sign. Some insight into the reason for this can be found in Section 8.2, but the reader may wish to use this general property now as a check on the work.

The next example involves a voltage source neither end of which is connected to ground. In such cases, we must take special care when labeling the node voltages and writing the current-law equations.

▶ **EXAMPLE 6.4**

Find the input-output differential equation for the circuit shown in Figure 6.17(a) where the inputs are the two voltage sources $e_1(t)$ and $e_2(t)$ and the output is the voltage e_o.

SOLUTION The voltage of each node with respect to the ground is shown in Figure 6.17(b). Two of the nodes are labeled $e_1(t)$ and e_o to correspond to the left voltage source and the output voltage, respectively. In labeling the node to the right of L and R_1, we do not introduce a new symbol (such as e_A) but take advantage of the fact that $e_2(t)$ is a source voltage. By Kirchhoff's voltage law, this remaining node voltage is $e_o + e_2(t)$, as shown on the diagram. This approach avoids the introduction of unnecessary variables whenever there is a voltage source not connected to ground.

The directions of the current reference arrows included in Figure 6.17(b) are arbitrary, but our equations must be consistent with the directions selected. Applying Kirchhoff's current law to the upper right node, we have

$$i_C + i_{R_2} + i_2 = 0 \tag{22}$$

Although we can express i_C and i_{R_2} in terms of the node voltage e_o by the element laws, the current i_2 through the voltage source cannot be directly related to $e_2(t)$. However, by applying the current law to the node labeled $e_o + e_2(t)$, we see that

$$i_2 = i_{R_3} + i_{R_1} + i_L$$

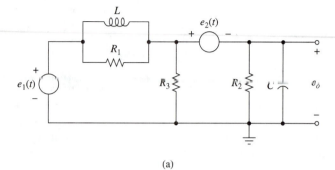

(a)

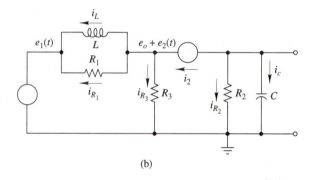

(b)

Figure 6.17 Circuit for Example 6.4. (a) As specified by the example statement. (b) With currents and node voltages defined.

which, when inserted into (22), gives

$$i_C + i_{R_2} + i_{R_3} + i_{R_1} + i_L = 0 \tag{23}$$

Using the element laws in the forms given by (5), (7), and (10), we write

$$i_{R_1} = \frac{1}{R_1}[e_o + e_2(t) - e_1(t)] \tag{24a}$$

$$i_{R_2} = \frac{1}{R_2}e_o \tag{24b}$$

$$i_{R_3} = \frac{1}{R_3}[e_o + e_2(t)] \tag{24c}$$

$$i_C = C\dot{e}_o \tag{24d}$$

$$i_L = i_L(0) + \frac{1}{L}\int_0^t (e_o + e_2 - e_1)\,d\lambda \tag{24e}$$

Substituting (24) into (23) and rearranging terms give the integral-differential equation

$$C\dot{e}_o + \left(\frac{1}{R_1} + \frac{1}{R_2} + \frac{1}{R_3}\right)e_o + \frac{1}{L}\int_0^t e_o\,d\lambda$$

$$= \frac{1}{R_1}e_1(t) - \left(\frac{1}{R_1} + \frac{1}{R_3}\right)e_2(t) + \frac{1}{L}\int_0^t (e_1 - e_2)\,d\lambda - i_L(0)$$

By differentiating this expression term by term, we obtain the desired input-output model:

$$C\ddot{e}_o + \left(\frac{1}{R_1} + \frac{1}{R_2} + \frac{1}{R_3}\right)\dot{e}_o + \frac{1}{L}e_o = \frac{1}{R_1}\dot{e}_1 - \left(\frac{1}{R_1} + \frac{1}{R_3}\right)\dot{e}_2 + \frac{1}{L}[e_1(t) - e_2(t)] \tag{25}$$

which is a second-order differential equation. To solve it, we must know the two initial conditions $e_o(0)$ and $\dot{e}_o(0)$, in addition to the source voltages $e_1(t)$ and $e_2(t)$.

The solution of a circuit model, such as the one in (25), is discussed in the next chapter. We may sometimes wish to see how the nature of the response changes when a source branch is disconnected. Even with all the sources disconnected, there will still be some output due to any energy that has been previously stored in the capacitors and inductors. The circuit in the following example contains only one energy storing element, so we should expect that it will be described by a first-order input-output equation.

▶ **EXAMPLE 6.5**

Find the differential equation for the output voltage e_o in Figure 6.18(a) when the switch is closed. Numerical values are given for the resistors but not for the capacitor. Repeat the problem when the left branch is disconnected by opening the switch.

SOLUTION   The node voltages with respect to ground, with the switch closed, are labeled in Figure 6.18(b). Summing currents at nodes A and O gives the following pair of equations:

$$C(\dot{e}_A - \dot{e}_o) + \frac{1}{3}(e_A - e_o) + \frac{1}{4}e_A + \frac{3}{4}[e_A - e_i(t)] = 0 \tag{26a}$$

$$C(\dot{e}_o - \dot{e}_A) + \frac{1}{3}(e_o - e_A) + \frac{1}{2}e_o = 0 \tag{26b}$$

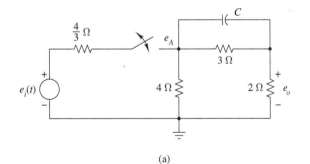

(a)

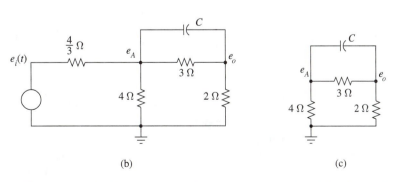

(b)   (c)

Figure 6.18   (a) Circuit for Example 6.5. (b) With the switch closed. (c) With the switch open.

We could collect like terms in each of these equations and then use the p-operator method to eliminate e_A and to obtain an equation in e_o and its derivatives. However, by adding (26a) and (26b) we see that

$$\frac{1}{2}e_o + e_A - \frac{3}{4}e_i(t) = 0$$

Replacing e_A in (26b) by $-\frac{1}{2}e_o + \frac{3}{4}e_i(t)$, we have

$$C\left[\dot{e}_o + \frac{1}{2}\dot{e}_o - \frac{3}{4}\dot{e}_i\right] + \frac{1}{3}\left[e_o + \frac{1}{2}e_o - \frac{3}{4}e_i(t)\right] + \frac{1}{2}e_o = 0$$

from which

$$\frac{3}{2}C\dot{e}_o + e_o = \frac{3}{4}C\dot{e}_i + \frac{1}{4}e_i(t) \tag{27}$$

With the switch open, the circuit reduces to the one shown in Figure 6.18(c). Once again summing currents at nodes A and O, we obtain

$$C(\dot{e}_A - \dot{e}_o) + \frac{1}{3}(e_A - e_o) + \frac{1}{4}e_A = 0$$
$$C(\dot{e}_o - \dot{e}_A) + \frac{1}{3}(e_o - e_A) + \frac{1}{2}e_o = 0 \tag{28}$$

Following the same procedure as before, we find that for part (c) of Figure 6.18,

$$2C\dot{e}_o + e_o = 0$$

Note that, as is the case in most examples, disconnecting the source has not only caused the input terms to disappear but has also changed the coefficients on the left side of the differential equation.

▶ 6.5 RESISTIVE CIRCUITS

There are many useful circuits that contain only resistors and sources, with no energy-storing elements. Such circuits are known as **resistive circuits** and are modeled by algebraic rather than differential equations. In this section we shall develop rules for finding the voltages and currents in such circuits and for replacing certain combinations of resistors by a single equivalent resistor.

The analysis of resistive circuits is important for other reasons as well. Even for circuits with energy-storing elements, we are often interested primarily in the steady-state response to a constant input, after any initial transients have died away. We shall see how any circuit reduces to a resistive one under these circumstances.

Even when we need to find the complete response of a general electrical system, a combination of two or more resistors will frequently be connected to the remainder of the circuit by a single pair of terminals, as shown in Figure 6.19(a). In such situations, it is possible to replace the entire combination of resistors by a single equivalent resistor R_{eq}, as shown in Figure 6.19(b). Provided that R_{eq} is selected such that $e = R_{eq}i$ is satisfied, the response of the remainder of the circuit is identical in both cases. Once we have found R_{eq}, it is easier to analyze the complete circuit because there are fewer nodes and thus fewer

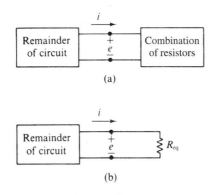

Figure 6.19 Replacement of a combination of resistors by an equivalent resistance.

equations. The two most important cases of combinations of resistors are the series and parallel connections.

Resistors in Series

Two resistors are in series when a single terminal of each resistor is connected to a single terminal of the other with no other element connected to the common node, as shown in Figure 6.20(a). Obviously, two resistors in series must have the same current flowing through them.

It follows from Ohm's law that $e_1 = R_1 i$ and $e_2 = R_2 i$ and from Kirchhoff's voltage law that $e = e_1 + e_2$. Thus

$$e = (R_1 + R_2)i \tag{29}$$

Because $e = R_{eq}i$, we know from (29) that the equivalent resistance shown in Figure 6.20(b) for the series combination shown in Figure 6.20(a) is

$$R_{eq} = R_1 + R_2 \tag{30}$$

Using (29) with the expressions for e_1 and e_2, we see that

$$e_1 = \left(\frac{R_1}{R_1 + R_2} \right) e \tag{31a}$$

$$e_2 = \left(\frac{R_2}{R_1 + R_2} \right) e \tag{31b}$$

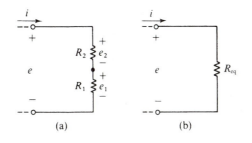

Figure 6.20 (a) Two resistors in series. (b) Equivalent resistance.

which is known as the **voltage divider rule**. From (31), the ratio of the individual resistor voltages is

$$\frac{e_1}{e_2} = \frac{R_1}{R_2} \tag{32}$$

Resistors in Parallel

Two resistors are in parallel when each terminal of one resistor is connected to a separate terminal of the other resistor, as shown in Figure 6.21(a). It is apparent that two resistors in parallel must have the same voltage across their terminals.

From Ohm's law, the individual currents are $i_1 = (1/R_1)e$ and $i_2 = (1/R_2)e$. From Kirchhoff's current law, $i = i_1 + i_2$. Thus

$$i = \left(\frac{1}{R_1} + \frac{1}{R_2}\right)e \tag{33}$$

For the equivalent resistance shown in Figure 6.21(b), we have $i = (1/R_{eq})e$. From (33), we see that for the parallel combination in Figure 6.21(a),

$$\frac{1}{R_{eq}} = \frac{1}{R_1} + \frac{1}{R_2}$$

or

$$R_{eq} = \frac{R_1 R_2}{R_1 + R_2} \tag{34}$$

To relate i_1 to the total current, we can write $i_1 = (1/R_1)e = (R_{eq}/R_1)i$. Doing this for i_2 and then using (34) to express R_{eq} in terms of R_1 and R_2, we obtain

$$i_1 = \left(\frac{R_2}{R_1 + R_2}\right)i \tag{35a}$$

$$i_2 = \left(\frac{R_1}{R_1 + R_2}\right)i \tag{35b}$$

which is known as the **current divider rule**. From (35), the ratio of the individual resistor currents is $i_1/i_2 = R_2/R_1$.

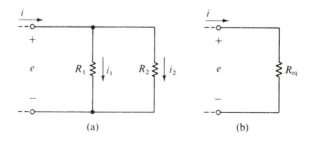

(a)   (b)

Figure 6.21 (a) Two resistors in parallel. (b) Equivalent resistance.

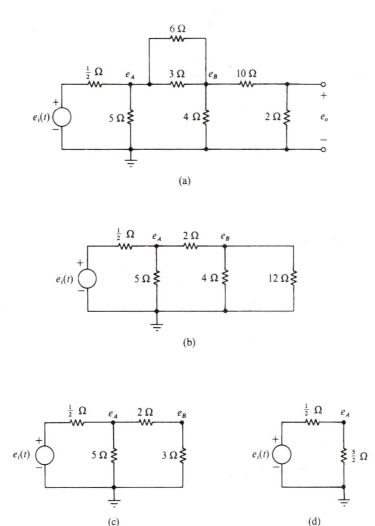

Figure 6.22 Circuits for Example 6.6. (a) Original circuit. (b), (c), (d) Equivalent circuits.

Calculating equivalent resistances and solving for the currents and voltages in most types of resistive networks can be simplified by using these rules for series and parallel combinations, as demonstrated in the following example.

▶ **EXAMPLE 6.6**

The resistive circuit shown in Figure 6.22(a) consists of a voltage source connected to a combination of seven resistors. The output is the voltage e_o. Find the equivalent resistance R_{eq} of the seven-resistor combination and evaluate e_o.

SOLUTION   To obtain R_{eq}, we use (30) and (34) repeatedly to combine series or parallel combinations of resistors into single equivalent resistors. Starting with the original circuit in Figure 6.22(a), we replace the 6-Ω and 3-Ω resistors that are in parallel with a single 2-Ω resistor. We also combine the 10-Ω and 2-Ω resistors in series into a 12-Ω resistor, which yields the intermediate circuit diagram shown in Figure 6.22(b). Note that the output voltage e_o does not appear on this diagram. Next we replace the parallel combination of the 4-Ω and 12-Ω resistors with a 3-Ω resistor, and obtain Figure 6.22(c). The series combination of 2 Ω and 3 Ω gives a 5-Ω resistor, which is in parallel with another 5-Ω branch, yielding the resistor of $\frac{5}{2}$ Ω that is shown in Figure 6.22(d). Thus the equivalent resistance connected across the voltage source is

$$R_{eq} = \tfrac{1}{2} + \tfrac{5}{2} = 3\,\Omega$$

To find the output voltage e_o, we make repeated use of the voltage-divider rule given by (31a) to obtain, in turn, e_A, e_B, and finally e_o. From Figure 6.22(d),

$$e_A = \left(\frac{\frac{5}{2}}{\frac{1}{2} + \frac{5}{2}} \right) e_i(t) = \frac{5}{6} e_i(t)$$

and from Figure 6.22(c),

$$e_B = \left(\frac{3}{2+3} \right) e_A = \frac{1}{2} e_i(t)$$

Then, from the original circuit diagram,

$$e_o = \left(\frac{2}{2+10} \right) e_B = \frac{1}{12} e_i(t)$$

Although the rules for combining series and parallel resistors often simplify the process of modeling a circuit, there are circuits in which the resistances do not occur in series or parallel combinations. In such situations, we can find an equivalent resistance by writing and solving the appropriate node equations, which will be strictly algebraic when only resistors and sources are involved.

The Steady-State Response

In the next two chapters, we present in detail general methods for obtaining the complete response of a linear system that contains energy-storing elements. The response normally consists of a transient part that dies out with time and a steady-state component that has a form similar to that of the input. If the input is constant over a long period of time, then the steady-state response will also be a constant.

If we just want to find the steady-state response to a constant input, then we can assume that all variables, including the input and output, are constants. Because the derivatives of constants are zero, we can, under these special circumstances, set all the derivatives in the modeling equations equal to zero. The equations then reduce to algebraic equations. Reconsider, for example, the circuit in Figure 6.18(a). Assume that after the switch closes, the voltage source has a constant value of 24 V, and that we want only the steady-state value of e_o, which we denote by $(e_o)_{ss}$. The input-output equation was found in (27). When we set all the derivatives equal to zero, the equation reduces to

$$(e_o)_{ss} = \frac{1}{4}(24) = 6 \text{ V}$$

For an electrical system, it is not necessary to find the general modeling equations if we only want the steady-state response to a constant input. We know that all currents and

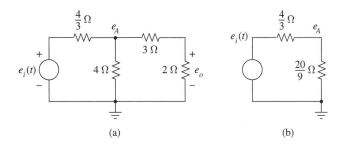

Figure 6.23 Steady-state equivalent circuit. (a) Original circuit
in steady state. (b) With equivalent resistance.

voltages then become constants. Under these circumstances, $di_L/dt = 0$ and $de_C/dt = 0$.
Then from the respective element laws, we see that the voltage across any inductor and the
current through any capacitor must be zero. Hence in the steady-state, an inductor acts like
a short circuit and a capacitor becomes an open circuit.

When finding the steady-state response to a constant input, we may redraw the circuit
with any inductor replaced by a short circuit and any capacitor replaced by an open circuit.
The original circuit then reduces to a purely resistive one. In the case of Figure 6.18(a),
with the switch closed and with $e_i(t)$ having a constant value of 24 V, we can draw the
steady-state equivalent circuit shown in Figure 6.23(a). By the rules for series and parallel
resistors, the part of the circuit to the right of the $\frac{4}{3}$-Ω resistor can be replaced by an equivalent resistance of $(4)(5)/(4+5) = \frac{20}{9}$ Ω, as shown in part (b) of Figure 6.23. Then, using the
voltage divider rule twice, we have in the steady state

$$(e_A)_{ss} = \frac{\frac{20}{9}}{\frac{4}{3} + \frac{20}{9}}(24) = 15 \text{ V}$$

and

$$(e_o)_{ss} = \frac{2}{3+2}(15) = 6 \text{ V}$$

which agrees with the answer found from the input-output equation.

6.6 OBTAINING THE STATE-VARIABLE MODEL

To obtain the model of a circuit in state-variable form, we define an appropriate set of
state variables and then derive an equation for the derivative of each state variable in terms
of only the state variables and inputs. The choice of state variables is not unique, but
they are normally related to the energy in each of the circuit's energy-storing elements.
Recalling that the energy stored in a capacitor is $\frac{1}{2}Ce^2$ and for an inductor is $\frac{1}{2}Li^2$, we
generally select the capacitor voltages and inductor currents as the state variables. For
fixed linear circuits, exceptions occur only when there are capacitor voltages or inductor
currents that are not independent of one another. This unusual situation will be illustrated in
Example 6.10.

For each capacitor or inductor, we want to express $\dot{e}_c$ or di_L/dt as an algebraic function
of state variables and inputs. We do this by writing the capacitor and inductor element laws
in their derivative forms as

$$\dot{e}_C = \frac{1}{C}i_C$$

$$\frac{di_L}{dt} = \frac{1}{L}e_L$$

(36)

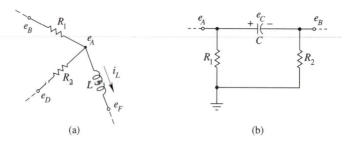

Figure 6.24 Partial circuits to illustrate writing Kirchhoff's current law in terms of state variables.

and then obtaining algebraic expressions for i_C and e_L in terms of the state variables and inputs. To find these expressions, we use the resistor element laws and Kirchhoff's voltage and current laws.

All the techniques we have discussed can still be used. The only basic difference is that we want to retain the variables e_C and i_L wherever they appear in our equations and to express other variables in terms of them. For example, in applying Kirchhoff's current law to node A in the partial circuit shown in Figure 6.24(a), we would write

$$\frac{1}{R_1}(e_A - e_B) + \frac{1}{R_2}(e_A - e_D) + i_L = 0$$

instead of

$$\frac{1}{R_1}(e_A - e_B) + \frac{1}{R_2}(e_A - e_D) + i_L(0) + \frac{1}{L}\int_0^t (e_A - e_F)\,d\lambda = 0$$

In the partial circuit of Figure 6.24(b), we do not use both e_A and e_B when summing the currents at nodes A and B. If the symbol e_B is used, the voltage at node A is written as $e_B + e_C$. If e_A is used, the voltage at node B is expressed as $e_A - e_C$.

▶ **EXAMPLE 6.7**

Write the state-variable equations for the circuit shown in Figure 6.25, for which we found the input-output equation in Example 6.2.

SOLUTION Because the circuit contains an inductor and a capacitor, we select i_L and e_C as the state variables, with the positive senses indicated on the circuit diagram. Starting with the inductor element law in the form $di_L/dt = (1/L)e_L$, we note that e_L is the same as the capacitor voltage e_C because the two elements are in parallel. Thus we obtain the first state-variable equation by replacing e_L by e_C, getting

$$\frac{di_L}{dt} = \frac{1}{L}e_C \tag{37}$$

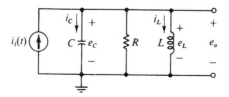

Figure 6.25 Parallel RLC circuit for Example 6.7.

For the second equation, we write the capacitor element law as $\dot{e}_C = (1/C)i_C$. To express the capacitor current i_C in terms of the state variables and input, we apply Kirchhoff's current law at the upper node, getting

$$i_C + \frac{1}{R}e_C + i_L - i_i(t) = 0 \tag{38}$$

where we have written the resistor current in terms of the state variable e_C. Solving (38) for i_C gives

$$i_C = -i_L - \frac{1}{R}e_C + i_i(t) \tag{39}$$

the right side of which is written entirely in terms of the state variables and input. Substituting (39) into the capacitor element law, we find the second state-variable equation to be

$$\dot{e}_C = \frac{1}{C}\left[-i_L - \frac{1}{R}e_C + i_i(t)\right] \tag{40}$$

Finally, we find the voltage e_o from the algebraic output equation

$$e_o = e_C \tag{41}$$

To solve the state-variable equations in (37) and (40), we must know the input and the initial values of the state variables. Note also that once we have the state-variable and output equations, we can always combine them into an input-output differential equation. In this example, we could differentiate (40), substitute (37) into it, and finally use (41) to replace e_C by e_o. The result would be the answer to Example 6.2.

There are several ways of summarizing a general procedure for constructing a state-variable model. We assume here that each capacitor voltage and inductor current is chosen to be a state variable. The unusual case where the number of state variables is less than the number of energy-storing elements will be treated later in this section.

1. We show the positive senses for each e_C and i_L on the circuit diagram, and then label i_C and e_L so that each current reference arrow enters the capacitor or inductor at the positive end of the voltage reference. Insofar as possible, we label the voltages of the nodes with respect to ground in terms of the state variables and inputs. We can then use additional symbols for any remaining node voltages, but we try to minimize the use of new variables.

2. We need to find algebraic expressions for each capacitor current i_C and each inductor voltage e_L. We make use of Kirchhoff's laws and Ohm's law, but at this time do not use the element laws for the capacitors and inductors. We require algebraic equations in this step, and the laws for the energy-storing elements will be used in the next step. In general, we may have to solve a set of simultaneous algebraic equations in order to get individual equations for each i_C and e_L in terms of the state variables and inputs.

3. For the state-variable equations, we substitute the expressions for i_C and e_L into the capacitor and inductor element laws, as given by (36). Finally, for each output that is not a state variable, we write an algebraic expression in terms of state variables and inputs.

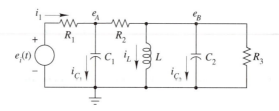

Figure 6.26 Circuit for Example 6.8.

The next two examples illustrate this general procedure for circuits of moderate complexity.

▶ **EXAMPLE 6.8**

Derive the state-variable model for the circuit shown in Figure 6.26. The outputs of interest are e_B, i_{C_2}, and i_1.

SOLUTION We choose as state variables the inductor current i_L and the capacitor voltages e_A and e_B. We need algebraic equations for the voltage across the inductor and the current through each capacitor. The inductor voltage is identical to the state variable e_B. Thus, by the element law for the inductor, one of the state-variable equations is

$$\frac{di_L}{dt} = \frac{1}{L}e_B$$

The current i_{C_1} will appear in a Kirchhoff current-law equation for node A, namely

$$i_{C_1} = \frac{1}{R_1}[e_i(t) - e_A] - \frac{1}{R_2}(e_A - e_B) \tag{42}$$

For i_{C_2} we consider node B, getting

$$i_{C_2} = \frac{1}{R_2}(e_A - e_B) - i_L - \frac{1}{R_3}e_B \tag{43}$$

Substituting (42) and (43) into the respective element-law equations gives the final two state-variable equations. The complete set of three equations is

$$\frac{di_L}{dt} = \frac{1}{L}e_B$$
$$\dot{e}_A = \frac{1}{C_1}\left[-\left(\frac{1}{R_1} + \frac{1}{R_2}\right)e_A + \frac{1}{R_2}e_B + \frac{1}{R_1}e_i(t)\right] \tag{44}$$
$$\dot{e}_B = \frac{1}{C_2}\left[-i_L + \frac{1}{R_2}e_A - \left(\frac{1}{R_2} + \frac{1}{R_3}\right)e_B\right]$$

As required, we have expressed the derivative of each of the state variables as an algebraic function of the state variables and the input $e_i(t)$. The output voltage e_B is the same as one of the state variables, and the output current i_{C_2} is given by (43). The output equation for i_1 is

$$i_1 = \frac{1}{R_1}[e_i(t) - e_A] \tag{45}$$

For a MATLAB exercise using this example, see Problem 6.42 at the end of this chapter.

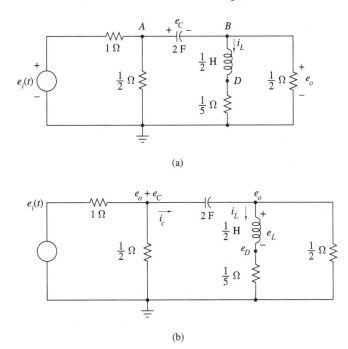

Figure 6.27 (a) Circuit for Example 6.9. (b) With additional variables added.

The next example has only two energy-storing elements and hence only two state variables. However, not all of the node voltages can be immediately expressed in terms of state variables and inputs.

▶ **EXAMPLE 6.9**

Find the state-variable model for the circuit shown in Figure 6.27(a), when e_o is the output.

SOLUTION We make the usual choice of e_C and i_L as state variables, with the positive senses shown on the diagram. The voltage at node D is $e_D = \frac{1}{5}i_L$, and node B corresponds to the output voltage e_o. Because we want to retain e_C in our equations, we use Kirchhoff's voltage law to express the voltage at node A as $e_o + e_C$. These symbols, as well as the reference directions for i_C and e_L, are added to the diagram in Figure 6.27(b). Applying Kirchhoff's current law to nodes B and A, we have

$$2e_o + i_L - i_C = 0$$
$$[e_o + e_C - e_i(t)] + 2(e_o + e_C) + i_C = 0 \tag{46}$$

The quantity inside the brackets is the current through the 1-Ω resistor—that is, it is the voltage at node A minus the source voltage all divided by 1 Ω. We now solve (46) simultaneously for e_o and i_C in terms of the state variables and the input. Doing this gives the algebraic equations

$$e_o = \frac{1}{5}[-i_L - 3e_C + e_i(t)] \tag{47a}$$

$$i_C = \frac{1}{5}[3i_L - 6e_C + 2e_i(t)] \tag{47b}$$

We also note that

$$e_L = e_o - e_D = \frac{1}{5}[-2i_L - 3e_C + e_i(t)] \tag{48}$$

To obtain the state-variable equations, we substitute (47b) and (48) into the element laws, as given by (36) and with $C = 2$ F and $L = \frac{1}{2}$ H. We also repeat the output equation (47a) to obtain the complete state-variable model:

$$\dot{e}_C = \frac{1}{10}[3i_L - 6e_C + 2e_i(t)]$$

$$\frac{di_L}{dt} = \frac{2}{5}[-2i_L - 3e_C + e_i(t)] \tag{49}$$

$$e_o = \frac{1}{5}[-i_L - 3e_C + e_i(t)]$$

In the previous three examples, we took as state variables the voltage across each capacitor and the current through each inductor. Figure 6.28 illustrates two exceptions to this procedure. For part (a) of the figure, we might first try to choose both e_A and e_B as state variables. However, by applying Kirchhoff's voltage law to the left loop, we see that

$$e_A + e_B - e_i(t) = 0 \tag{50}$$

Equation (50) is an algebraic relationship between the proposed state variables and the input, which is not allowed. In other words, the capacitor voltages e_A and e_B are not independent and thus cannot both be chosen as state variables. The basic problem is that the circuit contains a loop composed of only capacitors and voltage sources.

An analogous situation occurs when a circuit has a node to which only inductors and current sources are connected. For Figure 6.28(b), suppose that we try to choose both i_A and i_B as state variables. By Kirchhoff's current law,

$$-i_i(t) + i_A + i_B = 0$$

Again we have an algebraic relationship between the proposed state variables and the input. Because i_A and i_B are not independent, only one of them should be chosen as a state variable.

The situations illustrated in Figure 6.28 are relatively rare and may even be caused by starting with a circuit diagram that does not correspond closely enough to the physical devices. In part (a), for example, a better model of a physical source might be an ideal voltage source in series with a resistor, which could represent the "internal resistance" of the device. If the diagram were changed in this way, then there would no longer be a loop composed of only capacitors and voltage sources, and both e_A and e_B would be suitable state variables.

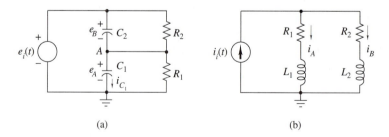

(a)   (b)

Figure 6.28 Circuits having fewer state variables than energy-storing elements.

A state-variable model for Figure 6.28(a) is developed in Example 6.10. The treatment of Figure 6.28(b) is left for a problem at the end of the chapter. In such cases, we may need to redefine a state variable in order to avoid having the derivative of the input appear on the right-hand side of the state-variable equation. For certain outputs, however, it may not always be possible to avoid an input derivative on the right-hand side of the output equation.

▶ **EXAMPLE 6.10**

Find the state-variable model for the circuit shown in Figure 6.28(a). Take the outputs to be $e_A, e_B,$ and i_{C_1}.

SOLUTION Because both e_A and e_B cannot be chosen as state variables, suppose we select e_A as the single state variable. Then we need an equation for $\dot{e}_A$ in terms of e_A and $e_i(t)$. First we write the element law for C_1 as

$$\dot{e}_A = \frac{1}{C_1} i_{C_1} \tag{51}$$

where the positive sense of i_{C_1} is downward. Then we apply Kirchhoff's current law at node A, obtaining

$$C_2(\dot{e}_A - \dot{e}_i) + \frac{1}{R_2}[e_A - e_i(t)] + \frac{1}{R_1}e_A + i_{C_1} = 0 \tag{52}$$

Solving (52) for i_{C_1} and substituting the result into (51), we find

$$\dot{e}_A = \frac{1}{C_1}\left[-C_2\dot{e}_A - \left(\frac{1}{R_1} + \frac{1}{R_2}\right)e_A + C_2\dot{e}_i + \frac{1}{R_2}e_i(t)\right]$$

which can be rearranged to yield

$$\dot{e}_A = \left(\frac{1}{C_1+C_2}\right)\left[-\left(\frac{1}{R_1} + \frac{1}{R_2}\right)e_A + C_2\dot{e}_i + \frac{1}{R_2}e_i(t)\right] \tag{53}$$

Equation (53) would be in state-variable form were it not for the term involving $\dot{e}_i$ on the right side. The derivative of the input should not appear in the final equation, so we define a new state variable, denoted by x, using the same procedure as for the mechanical system in Example 3.8. Transferring the term involving $\dot{e}_i$ to the left side of (53), we have

$$\frac{d}{dt}\left[e_A - \left(\frac{C_2}{C_1+C_2}\right)e_i(t)\right] = \left(\frac{1}{C_1+C_2}\right)\left[-\left(\frac{1}{R_1} + \frac{1}{R_2}\right)e_A + \frac{1}{R_2}e_i(t)\right] \tag{54}$$

We define the bracketed term on the left to be the new state variable

$$x = e_A - \left(\frac{C_2}{C_1+C_2}\right)e_i(t) \tag{55}$$

Then e_A is given by the output equation

$$e_A = x + \left(\frac{C_2}{C_1+C_2}\right)e_i(t) \tag{56}$$

and, when we substitute (55) and (56) into (54), the state-variable equation becomes

$$\dot{x} = \left(\frac{1}{C_1+C_2}\right)\left\{-\left(\frac{1}{R_1} + \frac{1}{R_2}\right)x + \left[\frac{1}{R_2} - \left(\frac{1}{R_1} + \frac{1}{R_2}\right)\left(\frac{C_2}{C_1+C_2}\right)\right]e_i(t)\right\} \tag{57}$$

Note that the circuit is first-order and can be modeled by a single state-variable equation. However, we did find that the state variable had to be a linear combination of the voltage across C_1 and the input. We can obtain the capacitor voltage e_A from the algebraic output equation (56) after solving the state-variable equation. For the output eqution for e_B, we combine (50) and (56) to obtain

$$e_B = \left(\frac{C_1}{C_1 + C_2}\right) e_i(t) - x$$

For the final output, we substitute (56) into the element law for C_1:

$$i_{C_1} = C_1 \dot{e}_A = C_1 \left[\dot{x} + \left(\frac{C_2}{C_1 + C_2}\right) \dot{e}_i\right]$$

and then, by (57), we write

$$i_{C_1} = \left(\frac{C_1}{C_1 + C_2}\right) \left\{-\left(\frac{1}{R_1} + \frac{1}{R_2}\right) x + \left[\frac{1}{R_2} - \left(\frac{1}{R_1} + \frac{1}{R_2}\right)\left(\frac{C_2}{C_1 + C_2}\right)\right] e_i(t) + C_2 \dot{e}_i\right\}$$

The output equations in a state-variable model should be purely algebraic whenever possible. This is an example of the unusual case where we cannot avoid a derivative of the input on the right-hand side.

6.7 OPERATIONAL AMPLIFIERS

Some important types of electrical elements, unlike those in earlier sections, have more than two terminals to which external connections can be made. Controlled sources arise in the models of transistors and other electronic devices. Rather than being independently specified, the values of such sources are proportional to the voltage or current somewhere else in the circuit. One purpose for which such devices are used is to amplify electrical signals, giving them sufficient power, for example, to drive loudspeakers, instrumentation, or various electromechanical systems. Ideal voltage and current amplifiers are shown in parts (a) and (b), respectively, of Figure 6.29.

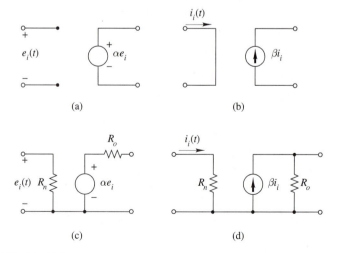

Figure 6.29 (a), (b) Ideal voltage and current amplifiers. (c), (d) Amplifiers with internal resistances causing nonideal behavior.

The models for many common devices have the two bottom terminals connected together. They may also include the added resistors shown in parts (c) and (d) of the figure in order to represent some of the imperfections of the device. For part (c) of the figure to approach part (a), R_n must be very large and R_o very small. In order for part (d) of the figure to approach part (b), R_n must be very small and R_o very large.

As is the case with any source in a circuit diagram, we assume that a controlled source can supply as much power as is required by the passive elements connected to it. Although we do not discuss in detail the internal mechanism responsible for the behavior of any element, a brief comment here should be helpful for those who encounter devices that are represented by controlled sources. The drawings in Figure 6.29 do not show all of the external connections to the physical device. In addition to the time-varying input $e_i(t)$ or $i_i(t)$, there are constant voltage sources that are normally not explicitly shown. These additional sources, which are called supply voltages, provide needed power for the time-varying output signal. The supply voltages that are needed are usually given in the specifications for the device to be used. Also specified are the maximum values of input voltage or current for which the operation of the device can be expected to remain in its linear region.

We shall assume that the values of our controlled sources are directly proportional to the signals controlling them. In addition to the sources represented in Figure 6.29, we can also have a voltage source controlled by a *current* somewhere else in the circuit, as well as a current source controlled by a *voltage*. However, we shall emphasize the voltage-controlled voltage source, because that will lead to the concept of the operational amplifier.

Electronic devices can do much more than simply amplify an input signal. In order to accomplish other objectives, additional passive elements are connected around the controlled source. In the following example, two external resistors are added to a controlled source. The controlled source is modeled by the simple circuit in Figure 6.29(a), with the constant α replaced by $-A$.

▶ **EXAMPLE 6.11**

Find e_o for the circuit shown in Figure 6.30.

SOLUTION　Summing the currents at node A gives

$$\frac{1}{R_1}[e_A - e_i(t)] + \frac{1}{R_2}(e_A - e_o) = 0$$

Because $e_o = -Ae_A$,

$$\left(\frac{1}{R_1} + \frac{1}{R_2} + \frac{A}{R_2}\right)e_A = \frac{1}{R_1}e_i(t)$$

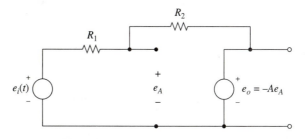

Figure 6.30　Circuit for Example 6.11.

Multiplying both sides of this equation by R_1R_2, solving for e_A, and then setting $e_o = -Ae_A$, we obtain

$$e_o = -Ae_A = \left[\frac{-AR_2}{R_2 + (1+A)R_1} \right] e_i(t) = \left[\frac{-R_2}{R_1 + \frac{1}{A}(R_1 + R_2)} \right] e_i(t)$$

Note that for very large values of A, $e_o = -(R_2/R_1)e_i(t)$. Under these conditions, the size of the voltage gain is determined solely by the ratio of the two resistors.

The **operational amplifier** (often called an **op-amp**) is a particularly important building block in the electrical part of many modern systems. The device typically contains many transistors plus a number of resistors and capacitors, and it may have ten or more external terminals. However, its basic behavior is reasonably simple. There are two input terminals for time-varying signals and one output terminal. The symbol for the device is shown in Figure 6.31(a), and a basic circuit model is given in part (b) of the figure.

Complete physical descriptions can be found in undergraduate electronics books. However, before considering any applications, we shall list without detailed explanations some practical features that users of op-amps should know about.

The input terminals marked with the minus and plus signs are called the inverting and noninverting terminals, respectively. We have denoted their voltages with respect to the ground point (which is the zero-volt reference) by e_A and e_B, although e_1 and e_2 are frequently used. One of the input terminals is often connected to the ground point, but this is not necessary.

Typical values for r_n exceed 10^6 Ω, and r_o is normally less than 100 Ω. In most applications, the resistance r_n can be replaced by an open circuit, and r_o by a short circuit, leading to the simplified model in Figure 6.31(c). Then no current can flow into the device from the left, and the output voltage is $e_o = A(e_B - e_A)$. The voltage amplification A is extremely large, typically exceeding 10^5.

Note that the symbol in Figure 6.31(a) does not show the ground point. In fact, the device itself does not have an external terminal that can be connected directly to ground.

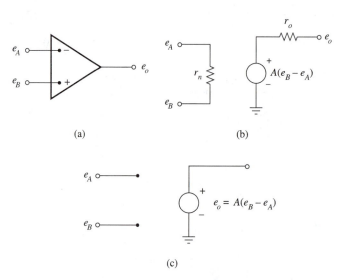

(a)

(b)

(c)

Figure 6.31 Operational amplifier. (a) Schematic representation.
(b) Equivalent circuit. (c) Idealized equivalent circuit.

There are, however, terminals for the attachment of positive and negative supply voltages. The other ends of these constant supply voltages are connected to a common junction, which is the ground point that appears in parts (b) and (c) of Figure 6.31. Circuit diagrams involving op-amps must always show which of the other elements are connected to this external ground. Sometimes ground symbols appear in several different places on the diagram, in which case they can all be connected together. However, the diagrams for our examples will already have had this done.

Because our interest is in the time-varying signals, the constant supply voltages are not normally shown on the circuit diagram. In addition to establishing an external ground point for the op-amp, they provide whatever power is needed for the output signal. Although there may also be other terminals on a practical op-amp, we need be concerned only with those shown in Figure 6.31.

For the examples in this section, we shall use Figure 6.31(c) as the equivalent circuit for the op-amp. When doing so, we shall avoid summing the currents at the output of the op-amp, because the current coming out of that terminal from the controlled voltage source would be an additional unknown. After a little practice, many people prefer not to bother to formally replace the original op-amp symbol in Figure 6.31(a) by the equivalent circuit. In this case, the reader must remember that there is no current into terminals A and B but that an unknown current leaves the output terminal. When we are dealing directly with the op-amp symbol, which does not show all the terminals, it might seem at first glance that the sum of the currents leaving the device is not zero, an apparent violation of Kirchhoff's current law. However, when the symbol in part (a) of Figure 6.31 is replaced by an equivalent circuit that includes the ground point, as in part (c), this apparent contradiction disappears.

▶ **EXAMPLE 6.12**

Find the input-output equation for the circuit shown in Figure 6.32.

SOLUTION When the op-amp is replaced by the model in Figure 6.31(c), and when terminal B is connected to the ground point, the circuit in Figure 6.32 becomes identical to the one in Figure 6.30. The results of Example 6.11 apply, so

$$e_o = \left(\frac{-R_2}{R_1 + \dfrac{1}{A}(R_1 + R_2)} \right) e_i(t)$$

which for very large values of A becomes $e_o = -(R_2/R_1)e_i(t)$.

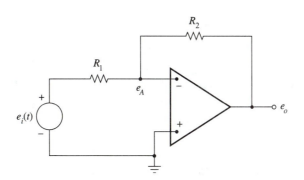

Figure 6.32 Circuit for Example 6.12.

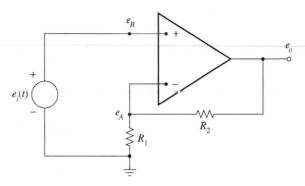

Figure 6.33 Circuit for Example 6.13.

▶ **EXAMPLE 6.13**

Find an expression for the output voltage e_o for the circuit in Figure 6.33.

SOLUTION   Because no current can flow into the input terminals of the op-amp, we can use the voltage divider rule to write

$$e_A = \frac{R_1}{R_1 + R_2} e_o$$

Then

$$e_o = A[e_i(t) - e_A] = Ae_i(t) - \left(\frac{AR_1}{R_1 + R_2}\right) e_o$$

from which

$$e_o = \left(\frac{R_1 + R_2}{R_1 + \frac{1}{A}(R_1 + R_2)}\right) e_i(t)$$

which for very large values of A becomes

$$e_o = \left(1 + \frac{R_2}{R_1}\right) e_i(t)$$

In the foregoing examples, note how the elements connected around the op-amp completely determine the behavior when we use the simple model with $A \to \infty$. We first found a general input-output equation in terms of A and then let $A \to \infty$ in order to get a simpler expression. Because A is so very large in practice, people often use an easier method to get the simpler expression directly. The output voltage of the device, given by $e_o = A(e_B - e_A)$, must be finite, so the voltage $e_B - e_A$ between the input terminals must approach zero when A is very large. In practice, this voltage really is only a tiny fraction of one volt, such as 0.1 mV.

Assuming that the voltage difference $e_B - e_A$ is virtually zero is sometimes called the **virtual-short** concept, because the voltage across a short circuit is zero. However, unlike a physical short circuit represented by an ideal wire (through which current could flow), we must still assume that no current flows into either of the input terminals.

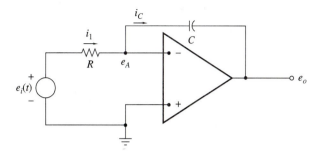

Figure 6.34 Circuit for Example 6.14.

▶ **EXAMPLE 6.14**

Use the virtual-short concept to determine the input-output equation for the circuit shown in Figure 6.34 when A is very large.

SOLUTION   Because $e_A = 0$, $i_1 = e_i(t)/R$ and $i_C = -C\dot{e}_o$. No current can flow into the input terminals of the op-amp, so $i_1 = i_C$ and

$$\dot{e}_o = -\left(\frac{1}{RC}\right) e_i(t)$$

If there is no initial stored energy, then $e_o(0) = 0$, and

$$e_o = \frac{-1}{RC} \int_0^t e_i(\lambda)\, d\lambda$$

The circuit is then called an integrator, because its output is proportional to the integral of the input.

▶ **EXAMPLE 6.15**

Find the input-output differential equation describing the circuit shown in Figure 6.35.

SOLUTION   Summing the currents at node B gives

$$C(\dot{e}_B - \dot{e}_i) + \frac{1}{R_3} e_B = 0$$

By the virtual-short concept, $e_B = e_A$, which according to the voltage-divider rule can be written as $e_B = [R_1/(R_1 + R_2)]e_o$. Substituting this into the previous equation gives

$$\frac{R_1 C}{R_1 + R_2}\dot{e}_o - C\dot{e}_i + \frac{R_1}{(R_1 + R_2)R_3} e_o = 0$$

from which

$$R_1 R_3 C\dot{e}_o + R_1 e_o = (R_1 + R_2)R_3 C\dot{e}_i$$

or

$$\dot{e}_o + \frac{1}{R_3 C} e_o = \left(1 + \frac{R_2}{R_1}\right)\dot{e}_i$$

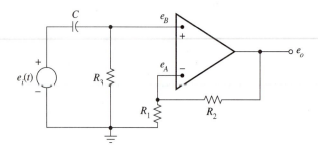

Figure 6.35 Circuit for Example 6.15.

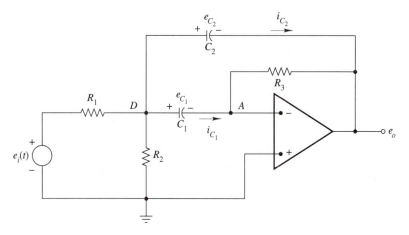

Figure 6.36 Circuit for Example 6.16.

The final example illustrates finding the state-variable model for circuits containing an op-amp. The choice of state variables and the procedure for deriving the model are essentially the same as for the examples in Section 6.6.

▶ EXAMPLE 6.16

Find a state-variable model for the circuit shown in Figure 6.36. Assume that the op-amp is ideal, with a gain large enough to allow use of the virtual-short concept. Let the output be e_o, and take as state variables the capacitor voltages e_{C_1} and e_{C_2}.

SOLUTION By the virtual-short concept, $e_A = 0$. Then the voltage at point D is e_{C_1}, and

$$e_o = e_{C_1} - e_{C_2} \tag{58}$$

which is the output equation. Because there is no current flowing into the input terminals of the op-amp, and because the voltage at node A is zero, $i_{C_1} = -(1/R_3)e_o$. Using (58), we see that

$$i_{C_1} = -\frac{1}{R_3}(e_{C_1} - e_{C_2}) \tag{59}$$

Summing currents at node D gives

$$i_{C_1} + i_{C_2} + \frac{1}{R_2}e_{C_1} + \frac{1}{R_1}[e_{C_1} - e_i(t)] = 0$$

Inserting (59) into this equation yields

$$i_{C_2} = \left(\frac{1}{R_3} - \frac{1}{R_1} - \frac{1}{R_2} \right) e_{C_1} - \frac{1}{R_3} e_{C_2} + \frac{1}{R_1} e_i(t) \qquad (60)$$

Finally, substituting (59) and (60) into the element law for a capacitor, we have the state-variable equations

$$\dot{e}_{C_1} = -\frac{1}{R_3 C_1} (e_{C_1} - e_{C_2})$$

$$\dot{e}_{C_2} = \frac{1}{C_2} \left[\left(\frac{1}{R_3} - \frac{1}{R_1} - \frac{1}{R_2} \right) e_{C_1} - \frac{1}{R_3} e_{C_2} + \frac{1}{R_1} e_i(t) \right]$$

Other op-amp circuits appear in the problems at the end of this chapter and also in Chapters 8 and 15. They include applications involving isolation, summing, inverting, integrating, and filtering of signals.

▶ 6.8 COMPUTER SIMULATION

Models of electrical circuits can be represented by block diagrams, simulated using Simulink and MATLAB, and solved numerically for their responses, as we did for mechanical systems. In this section, we consider two circuits that we have already modeled. We simulate the response of one to a sinusoidal input, and the other to a train of square pulses.

▶ EXAMPLE 6.17

Construct a Simulink diagram for (20a) and (20b), which describes the circuit in Figure 6.15 that was used for Example 6.3. Let $R_1 = R_2 = 2\ \Omega$, $R_3 = 4\ \Omega$, $C_1 = 1$ F, $C_2 = 4$ F, and $L = 0.5$ H. Find e_o and e_A when the input voltage is

$$e_i(t) = 5 \sin \omega t$$

where $\omega = 2\pi(0.1)$ rad/s. Thus, $e_i(t)$ is a sinusoid with a frequency of 0.1 Hz, a period of 10 s, and an amplitude of 5 V. Let all the initial conditions be zero. Plot $e_i(t)$, e_o, and e_A versus time for $0 \le t \le 50$ s.

SOLUTION We begin by solving (20a) and (20b) for the highest derivatives present, which are $\dot{e}_A$ and $\ddot{e}_o$, respectively. Each of these is then integrated to produce e_A and $\dot{e}_o$, and $\dot{e}_o$ is integrated again to produce e_o. Each of the quantities $\dot{e}_A$ and $\ddot{e}_o$ is formed as the output of a gain block preceded by a summer, whose inputs are chosen to satisfy the original equations.

The resulting Simulink diagram is shown in Figure 6.37. Note that it uses the `From` and `Goto` blocks introduced in Section 5.5. The simulation time has been set to 50 s, and the `Signal Generator` is set to produce sine waves with a frequency of 0.1 Hz. The following M-file will establish the required parameter values, run the simulation, and plot the desired results. When the default setting of `auto` for the `Max step size` parameter is used, the step size quickly reaches 1.0 s, which results in a rapid solution but in plots that have noticeable straight-line segments. To avoid this, we use the `Simulation > Parameters` menu to bring up the window for modifying the parameters of the differential equation solver. By experimenting, we find that a value of 0.05 s represents a good

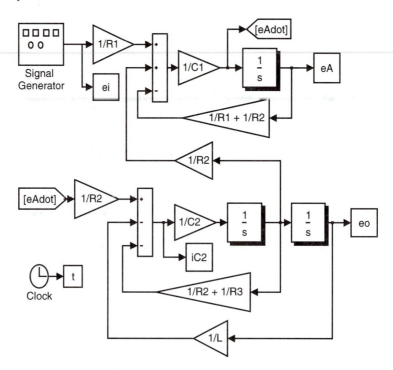

Figure 6.37 Simulink diagram to solve (20).

compromise between solution time and smoothness of the plots. The reader is encouraged to try several different values and evaluate their effects. The three lines starting with the word text cause labels to be printed on the individual curves. The first two arguments are the coordinates of the point where the text is to begin, expressed in terms of the variables being plotted. The third argument is the label that is to be printed, where the quotes tell MATLAB to treat it as a string.

In Figure 6.38 we see that the plots of e_A and e_o start at zero and then increase to eventually become constant-amplitude oscillations at the same frequency as the input. By $t = 20$ s, the transient part of e_A has essentially disappeared, but the transient part of e_o continues

```
R1 = 2; R2 = 2; R3 = 4;
L = 0.5; C1 = 1; C2 = 4;
sim('ex6_17_model')
subplot(2,1,1)              % upper plot
plot(t, ei); grid
xlabel('Time (s)')
ylabel('Voltage (V)')
text(14.0,3.9,'e_{i}')
axis([0 50 -5.3 5.3])
subplot(2,1,2)             % lower plot
plot(t, eA,  '--', t, eo); grid
xlabel('Time (s)')
ylabel('Voltage (V)')
text(15.0,2.5,'e_A')
text(18.0,-0.4,'e_o')
```

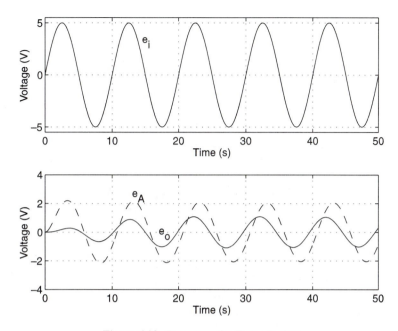

Figure 6.38 Responses for Example 6.17.

until about $t = 40$ s. The steady-state amplitudes of e_o and e_A, which are less than that of $e_i(t)$, can be checked by methods to be developed in Chapter 8. Also in the steady state, the peaks of the waveforms for e_o, e_A, and $e_i(t)$ do not occur at the same time, that is, there is a phase shift between the input and the output. Some of these characteristics are checked in Problem 8.53 at the end of Chapter 8.

▷ **EXAMPLE 6.18**

Draw a Simulink diagram for the circuit in Example 6.9 by representing Equations (49). Find the response of e_o, i_L, and e_C to an input voltage consisting of a square wave of frequency 0.05 Hz and amplitude 1 V. Plot the desired responses for $0 \leq t \leq 50$ s.

SOLUTION We begin by introducing an integrator for each of the two state variables, and then forming the inputs for those integrators according to the first two equations in (49). The output equation for e_o in (49) can then be formed from the input and the two state variables. The resulting Simulink diagram is shown in Figure 6.39. Since the parameter values are given in the original circuit, they have been included explicitly in the Simulink diagram, and do not need to be included in an M-file. The plot commands can conveniently be placed in the M-file, which produces the responses in Figure 6.40.

The response of e_o to each half of the input cycle changes abruptly, but all the variables reach constant values before the next change in the input. As a partial check, we can examine the circuit in Figure 6.27(a) under steady-state conditions, when the capacitor becomes an open circuit. No current flows to the right-hand components, so both i_L and e_o are zero. Then $e_C = e_A$, which can be found by the voltage-divider rule. Because the 1-Ω and $\frac{1}{2}$-Ω resistors are in series, $e_C = e_A = [0.5/(1 + 0.5)] \times 1 = 0.33$ V. The techniques in Chapter 8 can be used to provide further checks and insight for the output plots, as shown in Problem 8.54 at the end of that chapter.

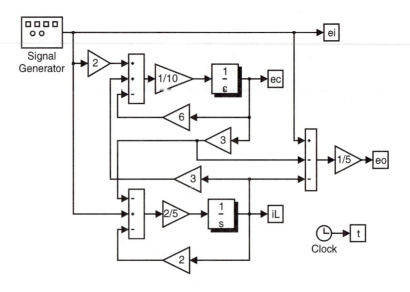

Figure 6.39 Simulink diagram representing (49).

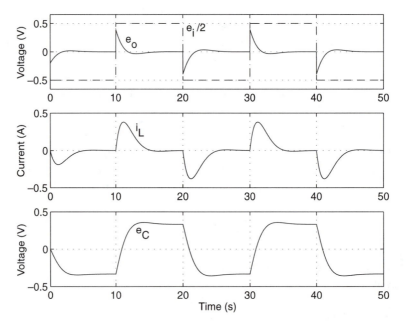

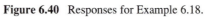

Figure 6.40 Responses for Example 6.18.

SUMMARY

After introducing the element and interconnection laws for electrical circuits, we developed systematic procedures for obtaining both the input-output differential equation and the set of state-variable equations. For the general node-equation method, we select a ground node and label the voltages of the other nodes with respect to ground. We then write current-law equations at the nodes whose voltages are unknown, using the element laws to express the currents through the passive elements in terms of the node voltages. If it becomes necessary to sum currents at a node to which a voltage source is connected, keep in mind that the current through such a source is another unknown variable.

For a state-variable model, we normally choose as state variables the voltage across each capacitor and the current through each inductor. Two types of exceptions to this choice were illustrated in Figure 6.28. As far as possible, unknown variables are labeled on the diagram in terms of state variables and inputs. By using Ohm's law and Kirchhoff's laws, we then express the capacitor currents, the inductor voltages, and any other outputs as algebraic functions of the state variables and inputs. Inserting these expressions into the element laws $\dot{e}_C = (1/C)i_C$ and $di_L/dt = (1/L)e_L$ yields the state-variable model.

The important special case of resistive circuits, including the rules for series-parallel combinations, was treated in Section 6.5. Operational amplifiers were explained in Section 6.7. For an ideal op-amp, no current flows into the input terminals, but the output current is unknown. The op-amp gain is usually large enough so that the voltage between the two input terminals can be assumed to be zero.

PROBLEMS

6.1. Find the input-output differential equation relating e_o and $i_i(t)$ for the circuit shown in Figure P6.1.

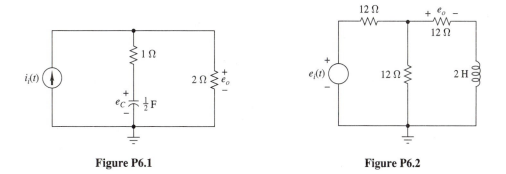

Figure P6.1   **Figure P6.2**

6.2. Find the input-output differential equation relating e_o and $e_i(t)$ for the circuit shown in Figure P6.2.

*6.3. Repeat Problem 6.2 for the circuit shown in Figure P6.3.

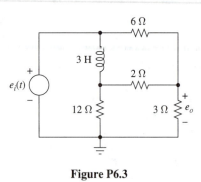

Figure P6.3

6.4. Find the input-output differential equation for the circuit shown in Figure P6.4.

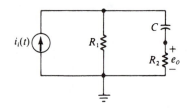

Figure P6.4

6.5. For the circuit shown in Figure P6.5, use the node-equation method to find the input-output differential equation.

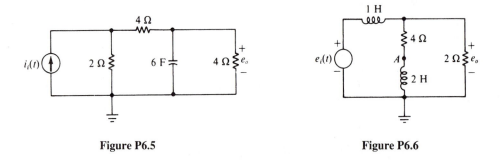

Figure P6.5   **Figure P6.6**

6.6. Repeat Problem 6.5 for the circuit shown in Figure P6.6.

*6.7. Repeat Problem 6.5 for the circuit shown in Figure P6.7.

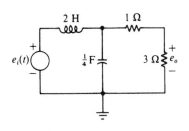

Figure P6.7

6.8. Repeat Problem 6.5 for the circuit shown in Figure P6.8.

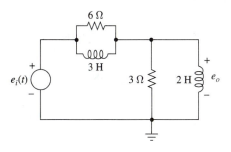

Figure P6.8

*6.9. Repeat Problem 6.5 for the circuit shown in Figure P6.9.

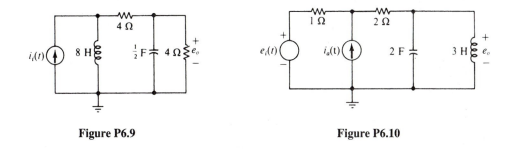

Figure P6.9   **Figure P6.10**

6.10. Find the input-output differential equation relating e_o to $e_i(t)$ and $i_a(t)$ for the circuit shown in Figure P6.10.

6.11.

    a.   For the circuit shown in Figure P6.11, write current-law equations at nodes A and B to obtain a pair of coupled differential equations in the variables e_A, e_o, and $e_i(t)$.

    b.   Find the input-output differential equation relating e_o and $e_i(t)$.

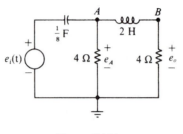

Figure P6.11

*6.12.   For the circuit shown in Figure P6.12, use the node-equation method to find the input-output differential equation relating e_o and $e_i(t)$.

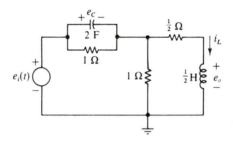

Figure P6.12

6.13. Repeat Example 6.5 with the capacitor C replaced by an inductor L.

6.14. For the circuit shown in Figure P6.14, use the rules for series and parallel combinations of resistors to find i_o and the equivalent resistance connected across the source.

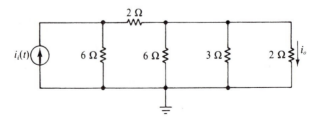

Figure P6.14

6.15. For the circuit shown in Figure P6.15, use the rules for series and parallel resistors to find e_o and the equivalent resistance connected across the source.

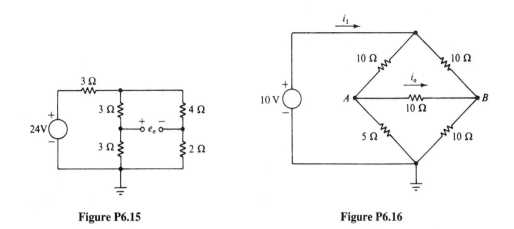

Figure P6.15   **Figure P6.16**

6.16.

a. Explain why the rules for series and parallel resistors cannot be used for the circuit shown in Figure P6.16.

b. Use the node-equation method to find the voltages of nodes A and B with respect to the ground node.

c. Find the currents i_o and i_1 and the equivalent resistance connected across the source.

*6.17. Find e_o for the circuit shown in Figure P6.17.

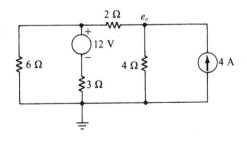

Figure P6.17

*6.18.

    a. Find a set of state-variable equations describing the circuit shown in Figure P6.18. Define the variables and show their positive senses on the diagram.

    b. Write an algebraic output equation for i_o, which is the current through the 6-V source.

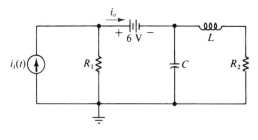

Figure P6.18

6.19. For the circuit shown in Figure P6.6, find a set of state-variable equations and an algebraic output equation for e_o.

6.20. Repeat Problem 6.19 for the circuit shown in Figure P6.7.

*6.21. Repeat Problem 6.19 for the circuit shown in Figure P6.9.

6.22. Repeat Problem 6.19 for the circuit shown in Figure P6.22.

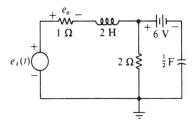

Figure P6.22

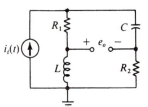

Figure P6.23

6.23. Repeat Problem 6.19 for the circuit shown in Figure P6.23.

*6.24. Repeat Problem 6.19 for the circuit shown in Figure P6.24.

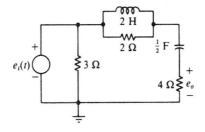

Figure P6.24

6.25. For the circuit shown in Figure P6.25, find a set of state-variable equations and write an algebraic output equation for i_o. Define the variables and show their positive senses on the diagram.

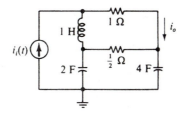

Figure P6.25

6.26. Repeat Problem 6.25 for the circuit shown in Figure P6.26.

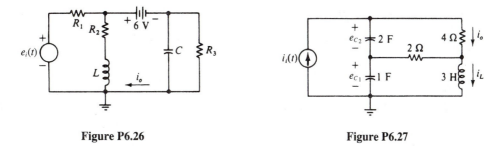

Figure P6.26   **Figure P6.27**

*6.27. Find a set of state-variable equations for the circuit shown in Figure P6.27. Write the algebraic output equation for i_o.

6.28. Find the state-variable equation for the circuit shown in Figure 6.28(a) when the initial choice of the state variable is e_B rather than e_A. Write an algebraic output equation for e_A.

6.29. Find a set of state-variable equations for the circuit shown in Figure 6.28(a) when the voltage source is replaced by the current source $i_i(t)$ with its reference arrow directed upward. Write an algebraic output equation for i_{C_1}.

6.30.

a. Find the state-variable model for the circuit shown in Figure 6.28(b).

b. Write an algebraic output equation for the voltage across the current source.

*6.31. For the op-amp circuit shown in Figure P6.31, derive the algebraic expression for the output voltage e_o in terms of the two input voltages, $e_1(t)$ and $e_2(t)$. Indicate what mathematical operation the circuit performs.

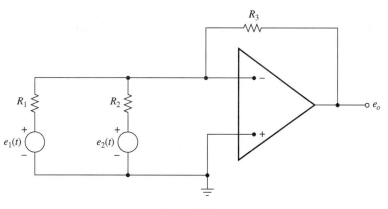

Figure P6.31

6.32. Repeat Problem 6.31 for the circuit shown in Figure P6.32.

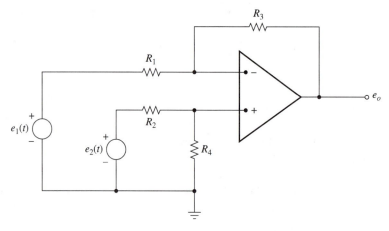

Figure P6.32

6.33. For the op-amp circuit shown in Figure P6.33, derive the input-output differential equation relating the output voltage e_o and the input voltage $e_i(t)$.

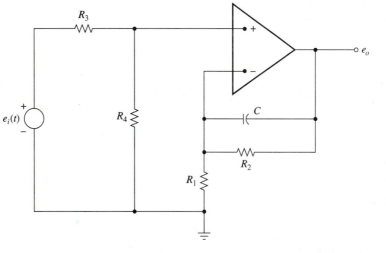

Figure P6.33

*6.34. For the op-amp circuit shown in Figure P6.34, derive the input-output differential equation relating the output voltage e_o and the input voltage $e_i(t)$.

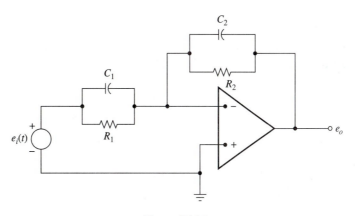

Figure P6.34

6.35.

   a. For the op-amp circuit shown in Figure P6.35, derive the input-output differential equation relating the output voltage e_o and the input voltage $e_i(t)$.

   b. Derive the state-variable model, taking the output to be the current through R_3, with the positive sense to the right.

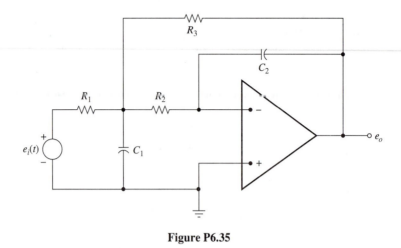

Figure P6.35

6.36. Write in matrix form the circuit model developed in Example 6.7. Identify the matrices $\mathbf{A}$, $\mathbf{B}$, $\mathbf{C}$, and $\mathbf{D}$.

6.37. Repeat Problem 6.36 for the state-variable model derived in Example 6.8 for the circuit shown in Figure 6.26.

*6.38. Repeat Problem 6.36 for the state-variable model derived in Example 6.9 for the circuit shown in Figure 6.27(a).

6.39. Draw a block diagram for the system described by (43), (44), and (45) that was modeled in Example 6.8.

6.40. Draw a block diagram for the system described by (49) that was modeled in Example 6.9.

6.41. Draw a Simulink diagram to solve (25) in Example 6.4. Use MATLAB to plot the first 10 seconds of the response to a step input of $e_1 = 1$ V at $t = 1$ s, followed by a 0.5 V step in e_2 at $t = 5$ s. Use the parameter values $R_1 = 10\,\Omega$, $R_2 = 100\,\Omega$, $R_3 = 20\,\Omega$, $C = 0.04$ F, and $L = 2.5$ H. Plot e_o versus time for 10 seconds.

6.42. Draw a Simulink diagram to solve (44) and (45) in Example 6.8. Use MATLAB to plot the first 0.2 seconds of the response when $e_i(t)$ is a square wave with amplitude 1 V and with a frequency of 10 Hz. Use the following parameter values: $R_1 = 50\,\Omega$, $R_2 = 628\,\Omega$, $R_3 = 6.28\,\Omega$, $C_1 = 100\,\mu\text{F}$, $C_2 = 1000\,\mu\text{F}$ and $L = 2.5$ mH. Plot e_A, e_B, i_1, and i_{C_2} versus time for 0.2 seconds.

7

TRANSFORM SOLUTIONS OF LINEAR MODELS

We have illustrated in previous chapters how to find the response of a system by using MATLAB and Simulink. We must now develop tools for finding the response with a pencil and paper solution. This is an essential step in the early stages of both analysis and design. Systematic analytical methods are available only for fixed linear systems, so we assume that any nonlinear model has been replaced by a linearized approximation, as will be discussed in Chapter 9.

We first discuss briefly the information that is needed for different methods of solution. Because the Laplace transform is used throughout the rest of this book, we develop its properties in some detail. We then apply it to the relatively simple case of first-order systems. In the process of doing this, we introduce two important inputs and several system concepts that will be important in Chapter 8. Finally, we present in the last two sections some procedures that will allow us to handle the more complex situations that will be encountered later.

▶ 7.1 BASIC CONSIDERATIONS

In order to be able to determine a system's response for all $t > 0$, we need the following three pieces of information: the modeling equations for $t > 0$, the inputs for all $t > 0$, and a set of initial conditions that reflect the system's history for $t < 0$. The modeling equations may be in the form of a state-variable model, an input-output differential equation, or they may be in a more primitive form (such as the free-body diagram equations for a mechanical system or the node equations for an electrical system).

Because the state variables describe the effect of the past history of a system, a natural choice for the initial conditions would be the values of each of the state variables at $t = 0$. Recall that the state variables are usually related to the energy stored in each of the system's energy-storing elements. For a translational mechanical system, we would need to know at $t = 0$ the elongation of every spring and the velocity of every mass. For an electrical system, we would need the initial voltage across each capacitor and the initial current through each inductor. The unusual cases where the number of state variables differs from the number of energy-storing elements have been illustrated in Chapters 3 and 6.

State-Variable Solutions

The form of a state-variable model for a system with three state variables, two inputs, and two outputs is given by Equations (3.1) and (3.2). The general form for an nth-order system is in Equations (3.31) and (3.32). Although Equations (3.1) and (3.31) are sets of first-order differential equations, the individual equations are generally coupled together, that is, more than one state variable appears in each of the equations. Thus the individual equations cannot be solved independently of one another. One approach is to try to find a change of variables that will uncouple the individual equations, so that each of the new equations contains only one unknown variable, although it is not always possible to do this. Alternatively, the original equations can be put into matrix form and solved as a group, by using a number of properties from linear algebra. We shall not present this approach, but the procedures can be found in the references in Appendix F.

Solution of the Input-Output Differential Equation

An nth-order system having one input and one output can be described by an input-output equation having the form of Equation (3.25). All variables except the input and output and their derivatives have been eliminated. For fixed linear systems, all the coefficients are constants. The generalization to systems with several inputs and several outputs was described in Section 3.2.

The classical solution of Equation (3.25), which is commonly taught in courses on differential equations, is outlined and illustrated in Appendix D. Briefly, the output is somewhat arbitrarily divided into a complementary solution $y_H(t)$ and a particular solution $y_P(t)$, which are then added together to give a general solution for $y(t)$ that satisfies both the differential equation and the input for all $t > 0$. This general solution will, however, contain n arbitrary constants that must be chosen to reflect the effects of the system's initial stored energy resulting from the inputs for $t < 0$.

The n initial conditions that are used in a classical solution to evaluate these constants are the initial values of the output and its first $n - 1$ derivatives, i.e., $y(0), \dot{y}(0),...,y^{(n-1)}(0)$. However, these initial conditions may be difficult to find. First, finding the initial values of derivatives is harder than finding the initial value of the output itself.

Second, the output might not be one of the state variables and hence not immediately linked to the energy in one of the energy-storing elements. The proper evaluation of $y(0)$ for a classical solution could be complicated when there is a discontinuity in the output at $t = 0$. As long as all variables remain finite, the stored energy cannot suddenly jump to a new value (because that would require that the instantaneous power flow into the element must become infinite). When the output is not one of the state variables, however, it can instantly jump to a new value.

Laplace Transform Solutions

An entirely different approach for finding the response $y(t)$ for all $t > 0$ is to transform all the variables (including the input, the output, and the modeling equations themselves) into functions of a new variable, s. When the Laplace transform is applied to fixed linear systems, the original modeling equations, which generally contain derivatives and perhaps integrals, are changed into purely algebraic equations. After these algebraic equations have been solved, some additional work is needed to transform the answer from a function of the variable s back into the desired function of time t.

In addition to avoiding the necessity of directly solving equations containing derivative and integral signs, the method has two other advantages. First, it is not necessary to find either the state-variable model or the input-output differential equation before starting the solution. In fact, it's usually best to transform immediately the original free-body-diagram or node equations.

Second, when the original modeling equations are transformed, the necessary initial conditions come into the solution automatically, and we need not wait until the very end of the solution to incorporate them. Furthermore, these initial conditions are normally directly related to the initial energy storage, and are easier to evaluate than those needed for a classical solution of the input-output differential equation.

To be able to use the Laplace transform, we need to know how to find the function of s or the function of time t whenever the other one is known, and also how to transform integral-differential equations into algebraic ones. We begin by defining the transform and then developing its properties.

7.2 TRANSFORMS OF FUNCTIONS

The **Laplace transform** converts a function of a real variable, which will always be time t in our applications, into a function of a complex variable that is denoted by s. The transform of $f(t)$ is represented symbolically by either $\mathscr{L}[f(t)]$ or $F(s)$, where the symbol $\mathscr{L}$ stands for "the Laplace transform of." One can think of the Laplace transform as providing a means of transforming a given problem from the **time domain**, where all variables are functions of t, to the **complex-frequency domain**, where all variables are functions of s.

The defining equation for the Laplace transform[1] is

$$F(s) = \int_0^\infty f(t)\epsilon^{-st} \, dt \tag{1}$$

Some authors take the lower limit of integration at $t = 0-$ rather than at $t = 0$. Either convention is acceptable provided that the transform properties are developed in a manner consistent with the defining integral.

The integration in (1) with respect to t is carried out between the limits of zero and infinity, so the resulting transform is not a function of t. In the factor ϵ^{-st} appearing in the integrand, we treat s as a constant in carrying out the integration.

The variable s is a complex quantity, which we can write as

$$s = \sigma + j\omega$$

where σ and ω are the real and imaginary parts of s, respectively. We can write the factor ϵ^{-st} in (1) as

$$\epsilon^{-st} = \epsilon^{-\sigma t}\epsilon^{-j\omega t}$$

Because the magnitude of $\epsilon^{-j\omega t}$ is always unity, $|\epsilon^{-st}| = \epsilon^{-\sigma t}$. By placing appropriate restrictions on σ, we can ensure that the the integral converges in most cases of practical interest, even if the function $f(t)$ becomes infinite as t approaches infinity. For all time functions considered in this book, convergence of the transform integral can be achieved.

To gain familiarity with the use of the transform definition and to begin the development of a table of transforms, we shall derive the Laplace transforms of several common functions. The results, along with others, are included in Appendix E.

Step Function

Suppose $f(t)$ has a constant value A for all $t > 0$. Because the range of integration in (1) is over only positive values of t, the transform of $f(t)$ is not affected by its values before

---

[1] Equation (1) defines the one-sided Laplace transform. There is a more general, two-sided Laplace transform, which is useful for theoretical work but is seldom used for solving for system responses.

$t = 0$. Then it follows from (1) that

$$\mathcal{L}[A] = \int_0^\infty A\epsilon^{-st} dt$$

$$= \left. \frac{A\epsilon^{-st}}{-s} \right|_{t=0}^{t\to\infty} \tag{2}$$

In order for the integral to converge, $|\epsilon^{-st}|$ must approach zero as t approaches infinity. Because $|\epsilon^{-st}| = \epsilon^{-\sigma t}$, the integral will converge, provided that $\sigma > 0$. Hence the expression for the transform of this function converges for all values of s in the right half of the complex plane, and (2) becomes

$$\mathcal{L}[A] = \frac{A}{s} \tag{3}$$

Note that the variable t has disappeared in the integration process and that the result is strictly a function of s.

For all functions of time with which we shall be concerned, the transform definition will converge for all values of s to the right of some vertical line in the complex s-plane. Although knowledge of the region of convergence is required in some advanced applications, we do not need such knowledge for the applications in this book. When evaluating Laplace transforms, we shall assume that σ, the real part of s, is sufficiently large to ensure convergence. If you are interested in regions of convergence, you should consult one of the more advanced references on system theory cited in Appendix F.

The unit step, denoted by $U(t)$, is an important function that will be defined and discussed in detail in Section 7.5. Note that a capital letter U is used, in contrast to the lowercase letter in the symbol $u(t)$ for a general input. The unit step function is defined to be zero for $t < 0$ and unity for $t > 0$. Its Laplace transform is given by (3) with A replaced by 1. Thus,

$$\mathcal{L}[U(t)] = \frac{1}{s} \tag{4}$$

Exponential Function

Depending on the value of the parameter a, the function $f(t) = \epsilon^{-at}$ represents an exponentially decaying function, a constant, or an exponentially growing function for $t > 0$, as shown in Figure 7.1. In any case, the Laplace transform of the exponential function is

$$\mathcal{L}[\epsilon^{-at}] = \int_0^\infty \epsilon^{-at} \epsilon^{-st} dt$$

$$= \int_0^\infty \epsilon^{-(s+a)t} dt$$

$$= \left. \frac{\epsilon^{-(s+a)t}}{-(s+a)} \right|_{t=0}^{t\to\infty}$$

The upper limit vanishes when σ, the real part of s, is greater than $-a$, so

$$\mathcal{L}[\epsilon^{-at}] = \frac{1}{s+a} \tag{5}$$

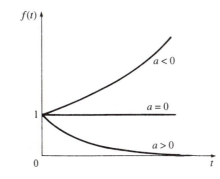

Figure 7.1 The exponential function ϵ^{-at} for various values of a.

Note that if $a = 0$, the exponential function ϵ^{-at} reduces to the constant value of 1. Likewise, (5) reduces to (4), as it must for this value of a.

Ramp Function

The unit ramp function is defined to be $f(t) = t$ for $t > 0$. Substituting for $f(t)$ in (1), we have

$$\mathcal{L}[t] = \int_0^\infty t\epsilon^{-st}\, dt \tag{6}$$

To evaluate the integral in (6), we use the formula for integration by parts:

$$\int_a^b u\, dv = uv\Big|_a^b - \int_a^b v\, du \tag{7}$$

where the limits a and b apply to the variable t. Making the identifications $u = t$ and $v = \epsilon^{-st}/(-s)$, $a = 0$, and $b \to \infty$, we can rewrite (6) as

$$\mathcal{L}[t] = \frac{t\epsilon^{-st}}{(-s)}\Big|_0^\infty - \frac{1}{(-s)}\int_0^\infty \epsilon^{-st}\, dt \tag{8}$$

It can be shown that

$$\lim_{t \to \infty} t\epsilon^{-st} = 0 \tag{9}$$

provided that σ, the real part of s, is positive. Thus

$$\mathcal{L}[t] = 0 - 0 + \frac{1}{s}\int_0^\infty \epsilon^{-st}\, dt \tag{10}$$

Because the integral in (10) is $\mathcal{L}[U(t)]$, which from (4) has the value $1/s$, (10) simplifies to

$$\mathcal{L}[t] = \frac{1}{s^2} \tag{11}$$

Trigonometric Functions

When $f(t)$ is replaced by $\sin \omega t$ or $\cos \omega t$ in the definition of the Laplace transform, we can evaluate the resulting integral by using the identities given in Table 1 in Appendix D, by performing integration by parts, or by consulting a table of integrals. Using an integral table, we find that

$$\mathcal{L}[\sin\omega t] = \int_0^\infty \sin\omega t \, \epsilon^{-st} dt$$

$$= \frac{\epsilon^{st}}{(-s)^2 + \omega^2}(-s\sin\omega t - \omega\cos\omega t)\Big|_0^\infty$$

$$= \frac{\omega}{s^2 + \omega^2} \tag{12}$$

By a similar procedure, we can show that

$$\mathcal{L}[\cos\omega t] = \frac{s}{s^2 + \omega^2} \tag{13}$$

Rectangular Pulse

The rectangular pulse shown in Figure 7.2 has a height of A and a duration of L. Its Laplace transform is

$$F(s) = \int_0^L A\epsilon^{-st} \, dt + \int_L^\infty 0\epsilon^{-st} \, dt$$

$$= \frac{A}{(-s)}\epsilon^{-st}\Big|_0^L + 0$$

$$= \frac{A}{s}(1 - \epsilon^{-sL}) \tag{14}$$

When $A = 1$ and when the pulse duration L becomes infinite, the pulse approaches the unit step function $U(t)$. Also, the term ϵ^{-sL} in (14) approaches zero for all values of s having a positive real part. Under these conditions, $F(s)$ given by (14) approaches $1/s$, which is $\mathcal{L}[U(t)]$ as expected.

▶ 7.3 TRANSFORM PROPERTIES

The Laplace transform has a number of properties that are useful in finding the transforms of functions in terms of known transforms and in solving for the responses of dynamic models. We shall state, illustrate, and in most cases derive those properties that will be useful in later work. They are tabulated in Appendix E.

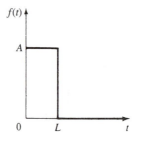

Figure 7.2 The rectangular pulse of height A and duration L.

Throughout this section, the symbols $F(s)$ and $G(s)$ denote the Laplace transforms of the arbitrary time functions $f(t)$ and $g(t)$, and a and b denote arbitrary constants. Our object is to express the transforms of various functions of $f(t)$ and $g(t)$ in terms of $F(s) = \int_0^\infty f(t)\epsilon^{-st}\,dt$ and $G(s) = \int_0^\infty g(t)\epsilon^{-st}\,dt$.

Multiplication by a Constant

To express $\mathcal{L}[af(t)]$ in terms of $F(s)$, where a is a constant and where $F(s) = \mathcal{L}[f(t)]$, we use (1) to write

$$\mathcal{L}[af(t)] = \int_0^\infty af(t)\epsilon^{-st}\,dt$$

$$= a\int_0^\infty f(t)\epsilon^{-st}\,dt$$

$$= aF(s) \tag{15}$$

Thus multiplying a function of time by a constant multiplies its transform by the same constant.

Superposition

The transform of the sum of the two functions $f(t)$ and $g(t)$ is

$$\mathcal{L}[f(t) + g(t)] = \int_0^\infty [f(t) + g(t)]\epsilon^{-st}\,dt$$

$$= \int_0^\infty f(t)\epsilon^{-st}\,dt + \int_0^\infty g(t)\epsilon^{-st}\,dt$$

$$= F(s) + G(s) \tag{16}$$

Using (15) and (16), we have the general superposition property

$$\mathcal{L}[af(t) + bg(t)] = aF(s) + bG(s) \tag{17}$$

for any constants a and b and any transformable functions $f(t)$ and $g(t)$. As an illustration of the superposition property, we can evaluate $\mathcal{L}[2 + 3\sin 4t]$ by using (17) with (3) and (12) to write

$$\mathcal{L}[2 + 3\sin 4t] = \frac{2}{s} + 3\left(\frac{4}{s^2 + 4^2}\right)$$

$$= \frac{2s^2 + 12s + 32}{s^3 + 16s}$$

Multiplication of Functions

The transform of the product of two functions $f(t)$ and $g(t)$ is generally not equal to $F(s)G(s)$, the product of the individual transforms. For example, although we know from (11) and (5) that the transforms of t and ϵ^{-at} are $1/s^2$ and $1/(s+a)$, respectively, $\mathcal{L}[t\epsilon^{-at}]$ is not $[1/s^2][1/(s+a)]$. To find the correct expression, we could insert the function of time into the defining equation in (1). However, it is easier to use either one of the two properties that are discussed next.

Multiplication by an Exponential

If we replace $f(t)$ in (1) by the function $f(t)\epsilon^{-at}$, we have

$$\mathcal{L}[f(t)\epsilon^{-at}] = \int_0^{\infty} f(t)\epsilon^{-at}\epsilon^{-st}\,dt$$

$$= \int_0^{\infty} f(t)\epsilon^{-(s+a)t}\,dt$$

$$= F(s+a) \tag{18}$$

In words, (18) states that multiplying a function $f(t)$ by ϵ^{-at} is equivalent to replacing the variable s by the quantity $s+a$ wherever it occurs in $F(s)$.

With this property, we can derive several of the transforms in Appendix E rather easily from other entries in the table. Specifically, because $\mathcal{L}[\cos \omega t] = s/(s^2 + \omega^2)$ and $\mathcal{L}[\sin \omega t] = \omega/(s^2 + \omega^2)$, we can write

$$\mathcal{L}[\epsilon^{-at}\cos \omega t] = \frac{s+a}{(s+a)^2 + \omega^2}$$

$$\mathcal{L}[\epsilon^{-at}\sin \omega t] = \frac{\omega}{(s+a)^2 + \omega^2}$$

Also, because $\mathcal{L}[t] = 1/s^2$, it follows that

$$\mathcal{L}[t\epsilon^{-at}] = \frac{1}{(s+a)^2} \tag{19}$$

Multiplication by Time

We obtain the transform of the product of $f(t)$ and the variable t by differentiating the transform $F(s)$ with respect to the complex variable s and then multiplying by -1:

$$\mathcal{L}[tf(t)] = -\frac{d}{ds}F(s) \tag{20}$$

To prove (20), we note that

$$\frac{d}{ds}F(s) = \frac{d}{ds}\left[\int_0^{\infty} f(t)\epsilon^{-st}\,dt\right]$$

$$= -\int_0^{\infty} tf(t)\epsilon^{-st}\,dt$$

$$= -\mathcal{L}[tf(t)] \tag{21}$$

Multiplying both sides of (21) by -1 results in (20).

We can illustrate the use of this property by deriving the entry in Appendix E for $\mathcal{L}[t^n]$, where n is any positive integer. Because $\mathcal{L}[1] = \mathcal{L}[U(t)] = 1/s$, it follows that

$$\mathcal{L}[t] = -\frac{d}{ds}\left(\frac{1}{s}\right) = \frac{1}{s^2}$$

$$\mathcal{L}[t^2] = -\frac{d}{ds}\left(\frac{1}{s^2}\right) = \frac{2}{s^3}$$

$$\mathcal{L}[t^3] = -\frac{d}{ds}\left(\frac{2}{s^3}\right) = \frac{2\cdot 3}{s^4}$$

and, for the general case,

$$\mathcal{L}[t^n] = \frac{n!}{s^{n+1}} \tag{22}$$

We can also use (20) to check the transform of te^{-at}. We know from (5) that $\mathcal{L}[e^{-at}] = 1/(s+a)$, so

$$\mathcal{L}[te^{-at}] = -\frac{d}{ds}\left(\frac{1}{s+a}\right) = \frac{1}{(s+a)^2}$$

which agrees with (19).

Differentiation

Because we shall need to take the Laplace transform of each term in a differential equation when solving system models for their responses, we must derive expressions for the transforms of derivatives of arbitrary order. We shall first develop and illustrate the formula for obtaining the transform of df/dt in terms of $F(s) = \mathcal{L}[f(t)]$. Then we shall use this result to derive expressions for the transforms of higher derivatives.

First Derivative
From the transform definition (1), we can write the transform of df/dt as

$$\mathcal{L}[\dot{f}] = \int_0^\infty \left(\frac{df}{dt}\right) \epsilon^{-st}\, dt \tag{23}$$

We can rewrite (23) by using the formula for integration by parts given by (7), with $a = 0$ and $b \to \infty$. The result is

$$\int_0^\infty u\, dv = uv\Big|_0^\infty - \int_0^\infty v\, du$$

If we let $u = \epsilon^{-st}$ and $dv = (df/dt)\, dt$, then $du = -s\epsilon^{-st}\, dt, v = f(t)$, and (23) becomes

$$\mathcal{L}[\dot{f}] = \epsilon^{-st} f(t)\Big|_0^\infty - \int_0^\infty f(t)(-s\epsilon^{-st})\, dt$$

For all the functions we shall encounter, there will be values of s for which $\epsilon^{-st} f(t)$ approaches zero as t approaches infinity, so

$$\mathcal{L}[\dot{f}] = [0 - f(0)] + s\int_0^\infty f(t)\epsilon^{-st}\, dt$$

$$= sF(s) - f(0) \tag{24}$$

To illustrate the application of (24), let $f(t) = \sin\omega t$. Then $F(s) = \omega/(s^2 + \omega^2)$ from (12) and $f(0) = \sin 0 = 0$, so

$$\mathcal{L}[\dot{f}] = s\left(\frac{\omega}{s^2 + \omega^2}\right) - 0 = \frac{s\omega}{s^2 + \omega^2} \tag{25}$$

We can verify this result by noting that $\dot{f} = \omega\cos\omega t$ and that

$$\mathcal{L}[\omega\cos\omega t] = \omega\mathcal{L}[\cos\omega t] = \omega\left(\frac{s}{s^2 + \omega^2}\right)$$

which agrees with (25).

Deriving and using (24) are straightforward when $f(t)$ is continuous at $t = 0$. However, when $f(t)$ is discontinuous at $t = 0$, we must take care in applying (24) because of the potential ambiguity in evaluating $f(0)$. To avoid such problems, we shall adopt the convention that $f(0) = f(0-)$, which is equivalent to saying that any discontinuity at the time origin is considered to occur just after $t = 0$. This will be considered in more detail in Section 7.5.

Second and Higher Derivatives

If we write (24) in terms of the function $g(t)$, it becomes

$$\mathcal{L}[\dot{g}] = sG(s) - g(0) \tag{26}$$

Now let $g(t) = \dot{f}(t)$. Then $\dot{g}(t) = \ddot{f}(t)$, and it follows from (26) and (24) that

$$\begin{aligned}
\mathcal{L}[\ddot{f}] &= s\mathcal{L}[\dot{f}(t)] - \dot{f}(0) \\
&= s[sF(s) - f(0)] - \dot{f}(0) \\
&= s^2 F(s) - sf(0) - \dot{f}(0) \tag{27}
\end{aligned}$$

If $\dot{f}$ is discontinuous at the time origin, we use for $\dot{f}(0)$ in (27) $\dot{f}(0-)$, the limit of $\dot{f}$ as t approaches zero from the left.

Equation (27) can be generalized to give the formula for the transform of the nth derivative of $f(t)$. The result is

$$\mathcal{L}\left[\frac{d^n f}{dt^n}\right] = s^n F(s) - s^{n-1} f(0) - \cdots - f^{(n-1)}(0) \tag{28}$$

where $f^{(n-1)}(0)$ denotes $d^{n-1}f/dt^{n-1}$ evaluated at $t = 0$. If any of the derivatives of $f(t)$ has a discontinuity at the time origin, we use its value at $t = 0-$ in the corresponding initial-condition term in (28).

Integration

The definite integral $\int_0^t f(\lambda)d\lambda$ will be a function of t because of the upper limit. From the transform definition given by (1), we can write

$$\mathcal{L}\left[\int_0^t f(\lambda)\, d\lambda\right] = \int_0^\infty \left[\int_0^t f(\lambda)\, d\lambda\right] \epsilon^{-st}\, dt$$

To evaluate the double integral on the right side of this expression, we use integration by parts with $u = \int_0^t f(\lambda)\, d\lambda$ and $dv = \epsilon^{-st}\, dt$. Then $du = f(t)\, dt, v = \epsilon^{-st}/(-s)$, and

$$\begin{aligned}
\mathcal{L}\left[\int_0^t f(\lambda)\, d\lambda\right] &= \left[\left(\frac{\epsilon^{-st}}{-s}\right)\int_0^t f(\lambda)d\lambda\right]\Bigg|_{t=0}^{t\to\infty} - \int_0^\infty \left(\frac{\epsilon^{-st}}{-s}\right) f(t)\, dt \\
&= [0 - 0] + \frac{1}{s}\int_0^\infty f(t)\epsilon^{-st}\, dt \\
&= \frac{1}{s}F(s) \tag{29}
\end{aligned}$$

For example, if $f(t) = \cos \omega t$, then $F(s) = s/(s^2 + \omega^2)$, so

$$\begin{aligned}
\mathcal{L}\left[\int_0^t \cos \omega\lambda\, d\lambda\right] &= \frac{1}{s}\left(\frac{s}{s^2 + \omega^2}\right) \\
&= \frac{1}{s^2 + \omega^2}
\end{aligned}$$

To check this result, we note that

$$\int_0^t \cos \omega \lambda \, d\lambda = \frac{1}{\omega} \sin \omega t$$

and

$$\mathcal{L}\left[\frac{1}{\omega} \sin \omega t\right] = \frac{1}{\omega}\left(\frac{\omega}{s^2 + \omega^2}\right) = \frac{1}{s^2 + \omega^2}$$

which agrees with the result we obtained by using (29).

In the next section, we will need to know the function of time that corresponds to the transformed quantity $A/[s(s+a)]$. One way to do this is to use (29) with $F(s) = A/(s+a)$ and $f(t) = A\epsilon^{-at}$. Then

$$\mathcal{L}^{-1}\left[\frac{A}{s(s+a)}\right] = \int_0^t A\epsilon^{-a\lambda} d\lambda = \frac{A\epsilon^{-a\lambda}}{-a}\Big|_0^t = \frac{A}{a}(1 - \epsilon^{-at}) \qquad (30)$$

When solving dynamic models, we may need to transform a term such as $g(t) = g(0) + \int_0^t f(\lambda)d\lambda$. To do this, we note that $g(0)$ is a constant and thus has the transform $g(0)/s$, whereas we can transform the integral part of the term by using (29). The result is

$$G(s) = \frac{g(0)}{s} + \frac{F(s)}{s} \qquad (31)$$

► 7.4 FIRST-ORDER SYSTEMS

In this section, we consider the solution of first-order systems, which usually consist of one energy-storing element and any number of dissipative elements. Although it is not difficult to handle such systems without using the Laplace transform, we take this opportunity to illustrate the basic steps needed for any transform solution, without being distracted by the algebraic details involved with higher-order systems. We also introduce some definitions that will be used throughout the rest of the book. Although they are illustrated in the context of first-order systems, we shall define the terms in a general way to make their extension to systems of higher order obvious.

For the examples in this section, we only need a few results from Sections 7.2 and 7.3. Because of Equation (17), we can transform any equation term by term and can take any multiplying constant outside of the Laplace transform sign. By (3), the transform of any constant A is A/s. For derivatives and integrals, we use (24) and (29):

$$\mathcal{L}[\dot{f}] = sF(s) - f(0)$$

$$\mathcal{L}\left[\int_0^t f(\lambda) \, d\lambda\right] = \frac{F(s)}{s}$$

We must also be able to determine the function of time for $t > 0$ when the function of s is known. In our examples, we shall use (3), (5), and (30), which are repeated here for convenience:

$$\mathcal{L}^{-1}\left[\frac{A}{s}\right] = A \qquad (32)$$

$$\mathcal{L}^{-1}\left[\frac{A}{s+a}\right] = A\epsilon^{-at} \qquad (33)$$

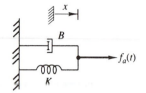

Figure 7.3 First-order mechanical system for Example 7.1.

$$\mathscr{L}^{-1}\left[\frac{A}{s(s+a)}\right] = \frac{A}{a}(1-\epsilon^{-at}) \tag{34}$$

▶ **EXAMPLE 7.1**

In the mechanical system shown in Figure 7.3, the spring is unstreched for all $t < 0$. The applied force $f_a(t)$ is zero for $t < 0$, but has a constant value of A for all $t > 0$. Find an expression for the displacement x for all $t > 0$.

SOLUTION   Summing forces gives the first-order differential equation

$$B\dot{x} + Kx = f_a(t)$$

Transforming this equation term by term, we have

$$B\mathscr{L}[\dot{x}(t)] + K\mathscr{L}[x(t)] = \mathscr{L}[f_a(t)]$$

Because $f_a(t)$ has a constant value of A for all $t > 0$, its Laplace transform is A/s. The displacement $x(t)$ is unknown, so we cannot write down its transform explicitly. Thus we use the symbol $X(s)$ to denote its transform.[2] For the first term in this equation, we use the rule for the transform of a derivative, in order to write $\mathscr{L}[\dot{x}(t)] = sX(s) - x(0)$. Then the transformed equation becomes

$$B[sX(s) - x(0)] + KX(s) = \frac{A}{s}$$

Note that the original differential equation has been converted into an algebraic equation. Because the initial spring elongation is zero, $x(0) = 0$. Solving the transformed equation algebraically for $X(s)$ gives

$$X(s) = \frac{A/B}{s(s+K/B)}$$

Except for a different multiplying constant A/B, this expression has the form of (34), with $a = K/B$. Thus, for all $t > 0$,

$$x(t) = \left(\frac{A}{K}\right)[1 - \epsilon^{-(K/B)t}]$$

---

[2]As a rule, functions of time are denoted by lowercase letters and their Laplace transforms by the corresponding capital letters followed by the transform variable s in parentheses.

As a partial check on the answer to Example 7.1, we note that $x(0) = 0$, as expected. Furthermore, in the steady state when t becomes indefinitely large, the expression for $x(t)$ reduces to A/K. When the system is motionless, no force is exerted by the friction element, so the constant force $f_a(t) = A$ is applied entirely to the spring K. If we denote by x_{ss} the displacement in this steady-state condition, then $Kx_{ss} = A$ and $x_{ss} = A/K$. The same result can also be obtained by setting the derivative in the original modeling equation equal to zero, because when the input is constant, all variables become constants in the steady state.

General Procedure

For first-order systems, the modeling equations for $t > 0$ will usually contain only one derivative or integral sign, and only one initial condition is needed to describe the energy stored within the system at $t = 0$. For higher-order systems, there may be any number of integral-differential equations, and any number of energy-storing elements and initial conditions. The examples in this section will be first-order systems, but we shall summarize the procedure in general terms. Application of the Laplace transform to solve for the response of dynamic systems of any order consists of the following three steps:

1. Write and immediately transform the differential or integral-differential equations describing the system for $t > 0$, evaluating all the initial-condition terms that appear in the transformed equations.

2. Solve these algebraic equations for the transform of the output.

3. Evaluate the inverse transform to obtain the output as a function of time.

The first step transforms the original equations into a set of algebraic equations in the variable s. Then we may solve the algebraic equations by any convenient method to obtain the transform of the response. The inverse transform gives us the complete response for all $t > 0$ and, unlike the classical approach, contains no unknown constants that must still be evaluated. Any necessary initial-condition terms will automatically appear in step 1.

General First-Order Model

The above procedure applies to a system of any order, and for any set of inputs and initial conditions. But in order to develop some definitions and concepts, we shall now restrict ourselves to systems described by a single first-order input-output differential equation having the form

$$\dot{y} + \frac{1}{\tau}y = f(t)$$

where τ is a real, nonzero constant. The function $f(t)$ is either the input itself or is a function that is completely determined by the input. To find the response for $t > 0$, we must know $f(t)$ for all positive values of time and also the initial condition $y(0)$. For the rest of this section we shall assume that $f(t)$ has a constant value A for all $t > 0$. Transforming the differential equation term by term gives

$$sY(s) - y(0) + \frac{1}{\tau}Y(s) = \frac{A}{s}$$

Collecting the terms involving $Y(s)$ on the left side of the equation and putting the other terms on the right side, we have

$$\left(s + \frac{1}{\tau}\right)Y(s) = y(0) + \frac{A}{s}$$

Solving this algebraic equation for $Y(s)$ yields

$$Y(s) = \frac{y(0)}{s + \dfrac{1}{\tau}} + \frac{A}{s\left(s + \dfrac{1}{\tau}\right)}$$

We can find the inverse transforms of the two terms on the right by using (33) and (34). Thus we can write

$$y(t) = y(0)\epsilon^{-t/\tau} + \tau A - \tau A \epsilon^{-t/\tau}$$

Rearranging terms, we obtain

$$y(t) = \tau A + [y(0) - \tau A]\epsilon^{-t/\tau} \tag{35}$$

Parts of the Solution

It is sometimes useful to divide the complete response of a system into two parts: the zero-input response and the zero-state response. The **zero-input response** $y_{zi}(t)$ is the output for the special case when $f(t)$ is zero for all $t > 0$, but when some initial stored energy results in nonzero initial values for some of the state variables. The **zero-state response** is the complete response to the input when the initial values of the state variables are zero.

Consider the zero-input response, when $A = 0$ and when the system is merely responding to some inital energy stored within it, as reflected by the value of $y(0)$. Then from (35), $y_{zi}(t) = y(0)\epsilon^{-t/\tau}$. Typical curves for $y_{zi}(t)$ are shown in Figure 7.4. If $y_{zi}(t)$ decays to zero as t approaches infinity, the system is said to be **stable**. If, on the other hand, $y_{zi}(t)$ increases without limit as t becomes large, the system is **unstable**. A first-order system is

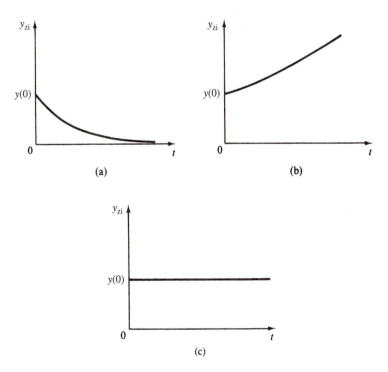

Figure 7.4 The zero-input response for a first-order system. (a) $\tau > 0$. (b) $\tau < 0$. (c) τ infinite.

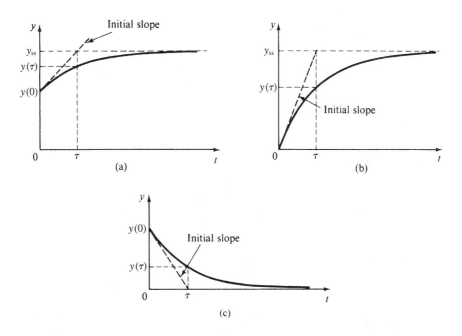

Figure 7.5 Complete response of a stable first-order system to a constant input. (a) y_{ss} and $y(0)$ both nonzero. (b) $y(0) = 0$. (c) $y_{ss} = 0$.

stable if $\tau > 0$ and unstable if $\tau < 0$. If the magnitude of τ approaches infinity, $y_{zi}(t)$ becomes constant as shown in Figure 7.4(c). Such a system is said to be **marginally stable**.

The complete response can also be viewed as the sum of transient and steady-state components. The **transient response** $y_{tr}(t)$ consists of those terms that decay to zero as t approaches infinity. The **steady-state response** $y_{ss}(t)$ is the part of the solution that remains after the transient terms have disappeared. For a first-order system whose response to a constant input is given by (35), and for which τ is positive, the steady-state part of the response is $y_{ss}(t) = \tau A$. Then we may rewrite (35) as

$$y(t) = y_{ss} + [y(0) - y_{ss}]\epsilon^{-t/\tau} \tag{36}$$

which is shown in Figure 7.5(a). Special cases of this result are shown in parts (b) and (c) of the figure.

To give significance to the value of $y(\tau)$ shown in Figure 7.5(a), first note from (36) that, for $t = \tau$, we have

$$y(\tau) = y_{ss} + [y(0) - y_{ss}]\epsilon^{-1}$$

Because $\epsilon^{-1} = 0.3679$,

$$y(\tau) = y(0) + 0.6321[y_{ss} - y(0)]$$

Thus after τ seconds, the response to a constant input is approximately 63 percent of the way from the initial value to the steady-state value. Because $\epsilon^{-4} = 0.0183$, the response after 4τ seconds is approximately 98 percent of the way from the initial value to the steady-state value. We see that the system parameter τ, which is called the **time constant**, is a measure of the system's speed of response. Another interpretation of the time constant results from considering the initial slope of the response. Differentiation of (36) gives

$$\dot{y}(t) = \frac{1}{\tau}[y_{ss} - y(0)]\epsilon^{-t/\tau}$$

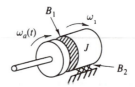

Figure 7.6 First-order rotational mechanical system for Example 7.2.

and setting $t = 0$ yields

$$y_{ss} = y(0) + \tau \dot{y}(0)$$

Thus if the slope of the response curve were maintained at its initial value of $\dot{y}(0)$, it would take τ seconds, instead of an infinite time, for the response to reach its steady-state value.

▶ **EXAMPLE 7.2**

Find the response of the first-order rotational system shown in Figure 7.6. The system is at rest with no stored energy at $t = 0$, and for $t > 0$ the input is $\omega_a(t) = A$. The output is the angular velocity ω_1.

SOLUTION The system is described by the differential equation

$$J\dot{\omega}_1 + (B_1 + B_2)\omega_1 = B_1\omega_a(t)$$

or

$$\dot{\omega}_1 + \left(\frac{B_1 + B_2}{J}\right)\omega_1 = \frac{B_1}{J}\omega_a(t)$$

The time constant is

$$\tau = \frac{J}{B_1 + B_2} \tag{37}$$

With $\omega_a(t) = A$ for all positive time, $(\omega_1)_{ss} = AB_1/(B_1 + B_2)$. Because $\omega_1(0) = 0$,

$$\omega_1 = \left(\frac{AB_1}{B_1 + B_2}\right)\left(1 - \epsilon^{-t/\tau}\right) \tag{38}$$

As should be expected, the time needed to approach steady-state conditions is directly proportional to the inertia of the disk. However, once the disk reaches a constant speed, the only torques on it are those from the friction elements. Thus the steady-state angular velocity is determined by the relative sizes of B_1 and B_2.

▶ **EXAMPLE 7.3**

Find an expression for the output voltage $e_o(t)$ for all $t > 0$ for the system shown in Figure 7.7, when the applied voltage $e_i(t)$ is zero for all $t < 0$ but has a constant value A for all $t > 0$.

SOLUTION If we try to use (36) directly, we encounter a problem trying to evaluate the initial condition $e_o(0)$. Therefore, we shall go back and apply the general three-step

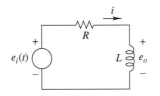

Figure 7.7 *RL* circuit for Example 7.3.

transform procedure to the modeling equations. Because all the elements are in series, we could sum voltages around the closed path to get a differential equation in which the only unknown is the current $i(t)$. However, the output is specified as e_o, so we choose instead to write a node equation at the upper right corner:

$$\frac{1}{R}[e_o - e_i(t)] + \left[i(0) + \frac{1}{L}\int_0^t e_o \, d\lambda\right] = 0 \tag{39}$$

where the quantity in brackets is the element law for the current $i(t)$ through the inductor. The initial current $i(0)$ must be zero since there was no excitation for $t < 0$. The input $e_i(t)$ is a constant A for all $t > 0$, so its transform is A/s. Transforming the node equation gives

$$\frac{1}{R}\left[E_o(s) - \frac{A}{s}\right] + \left(\frac{1}{L}\right)\left(\frac{1}{s}\right)E_o(s) = 0$$

Solving algebraically for the transform of the output gives

$$E_o(s) = \frac{A}{s + \dfrac{R}{L}}$$

Taking the inverse transform, we have

$$e_o(t) = A\epsilon^{-Rt/L} \quad \text{for} \quad t > 0 \tag{40}$$

When checking this result, we note that $e_o(t)$ goes to zero as t approaches infinity. In Section 6.5, we saw that when the input is a constant, a steady-state circuit can be drawn by replacing any capacitor by an open circuit, and any inductor by a short circuit. When this is done for our example, the output is the voltage across a short circuit, so it should indeed be zero in the steady state.

Also notice that the expression for $e_o(t)$ becomes equal to A when t is replaced by zero, even though $e_o(t) = 0$ for $t < 0$ because there was no excitation until $t = 0$. The voltage across an inductor can suddenly jump to a new value at $t = 0$, because it is not directly related to the energy stored within the device. In our Laplace transform solution, however, the only inital condition encountered was $i(0)$, which cannot change instantaneously. We shall discuss the initial condition problem in general terms next.

Initial Conditions

As in the last example, it is possible for the output to undergo, at $t = 0$, an instantaneous change, caused by the sudden application of an input or by some other event. When we must consider functions that have a discontinuity at $t = 0$, it is customary to use the notations $y(0-)$ and $y(0+)$ for the limiting values of $y(t)$ as t approaches zero through negative

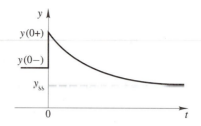

Figure 7.8 Typical response of a first-order circuit.

and positive values, respectively. The function $y(t)$ shown in Figure 7.8 has a discontinuity at $t = 0$, where it jumps instantaneously from $y(0-)$ to $y(0+)$. This is typical in the response of many electrical systems. To account for the possibility of discontinuous behavior, we modify (36) slightly to read

$$y(t) = y_{ss} + [y(0+) - y_{ss}]\epsilon^{-t/\tau} \tag{41}$$

Using $y(0+)$, the value of $y(t)$ immediately following any discontinuity, is consistent with the fact that the equation should be valid for all $t > 0$.

To understand which variables might suddenly jump to new values, recall that the energy stored in any passive element cannot change instantaneously unless there is an infinitely large power flow. A translational mechanical system may have potential energy stored in a stretched spring or kinetic energy stored in a moving mass. The amount of this initial stored energy is determined by x_K or v_M, respectively. Thus, as long as all variables remain finite, x_K and v_M cannot change instantly. Similarly, in a rotational system, θ_K and ω_J cannot change instantaneously.

An electrical system may have energy stored within a capacitor or an inductor. Such energy is determined by e_C or i_L, respectively. Thus, as long as all variables remain finite, e_C and i_L cannot instantly jump to new values.

In any mechanical or electrical system, the initial stored energy can be accounted for by knowing the values of $x_K, v_M, \theta_K, \omega_J, e_C$, and i_L at $t = 0$. When the Laplace transform is applied to the original modeling equations, these are the only initial conditions that should be expected, even in the case of higher-order systems. Because these quantities cannot normally change instantaneously, they do not suddenly jump to new values at $t = 0$ even if the input suddenly changes.

The output $e_o(t)$ in Example 7.3 had the form shown in Figure 7.8 with $y(0-) = 0$, $y(0+) = A$, and $y_{ss} = 0$. In that example, recall that the only initial condition that was encountered was the current through the inductor, which could not change instantly. Thus there was no confusion about the proper value to use for $i(0)$. The Laplace transform solution then automatically gave the correct value of A for $e_o(0+)$. Avoiding problems with difficult initial conditions is one of the advantages of the transform method as compared with a classical solution. If we choose not to transform the original modeling equations immediately, then we may have to deal directly with a variable that has a discontinuity at $t = 0$. That situation will be discussed in Section 7.5 and again in Section 8.1.

▶ **EXAMPLE 7.4**

The circuit shown in Figure 7.9(a) contains one energy-storing element and three dissipative elements. The voltage across the capacitor is the node voltage e_A. For $t < 0$, the switch is open and no energy is stored in the capacitor. The switch then closes at $t = 0$. Find an expression for $e_o(t)$ for all $t > 0$.

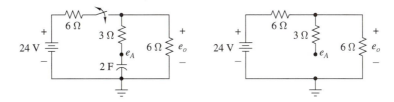

Figure 7.9 (a) Circuit for Examples 7.4 and 7.5. (b) Steady-state circuit with the switch closed.

SOLUTION   Summing currents at nodes A and O for $t > 0$ with the switch closed, we have

$$2\dot{e}_A + \frac{1}{3}(e_A - e_o) = 0$$

$$\frac{1}{6}e_o + \frac{1}{3}(e_o - e_A) + \frac{1}{6}(e_o - 24) = 0$$

Transforming these equations term by term gives

$$2[sE_A(s) - e_A(0)] + \frac{1}{3}[E_A(s) - E_o(s)] = 0$$

$$\frac{1}{6}E_o(s) + \frac{1}{3}[E_o(s) - E_A(s)] + \frac{1}{6}\left[E_o(s) - \frac{24}{s}\right] = 0$$

We set $e_A(0)$ equal to zero and then solve algebraically for $E_o(s)$, obtaining

$$E_o(s) = \frac{6s + 1}{s\left(s + \dfrac{1}{12}\right)} = \frac{6}{\left(s + \dfrac{1}{12}\right)} + \frac{1}{s\left(s + \dfrac{1}{12}\right)}$$

Using (33) and (34) to take the inverse transform, we get

$$e_o(t) = 6\epsilon^{-t/12} + 12(1 - \epsilon^{-t/12}) = 12 - 6\epsilon^{-t/12} \quad \text{for} \quad t > 0$$

Note from this expression that $e_o(0+) = 6$ V, even though e_o was zero for $t < 0$. Although the voltage across the capacitor must remain zero at $t = 0+$, there is nothing to stop the voltage across a resistor from jumping instantly to a new value.

For very large values of t, after the transient term has died out, our answer reduces to $(e_o)_{ss} = 12$ V. In order to check this result, we can redraw the circuit with the switch closed and with the capacitor replaced by an open circuit, as shown in Figure 7.9(b). The two 6-Ω resistors are in series, because no current can flow down through the 3-Ω resistor. Then by the voltage-divider rule, $(e_o)_{ss} = [6/(6 + 6)](24) = 12$ V. Also note that $(e_A)_{ss} = (e_o)_{ss} = 12$ V, since there is no voltage across the 3-Ω resistor in the steady state. Finally, note that the transform of the output has a pole at $s = -1/\tau$ and that the transient term in $e_o(t)$ contains the factor $\epsilon^{-t/\tau}$, where $\tau = 12$ s. Plots of both e_A and e_o are shown in Figure 7.10(a). The time constant τ is 12 s. When $t = \tau$, the response is approximately 63 percent of the way from the initial to the steady-state value. Thus $e_o(\tau) \approx 6 + (0.63)(12 - 6) = 9.8$ V, while $e_A(\tau) \approx (0.63)(12) = 7.6$ V.

▶ **EXAMPLE 7.5**

Repeat Example 7.4 when the switch in Figure 7.9(a) has been closed for a long time for $t < 0$, so that steady-state conditions have been reached. The switch then opens at $t = 0$. Again find $e_o(t)$ for $t > 0$.

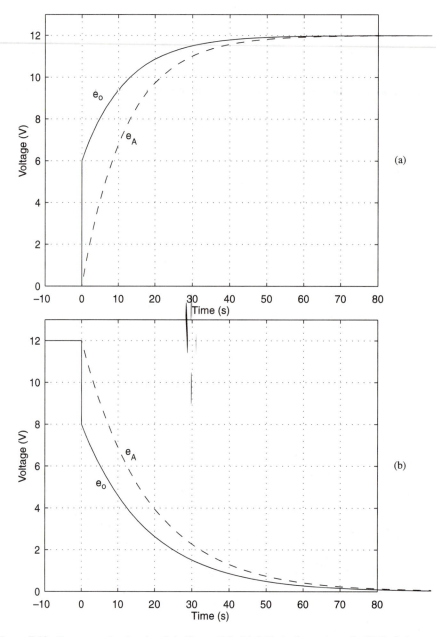

Figure 7.10 Responses for the circuit in Figure 7.9. (a) Switch closes at $t = 0$. (b) Switch opens at $t = 0$.

SOLUTION We need to know the voltage across the capacitor at $t = 0-$ in order to include the effect of the circuit's history for $t < 0$. The steady-state circuit for $t < 0$ is shown in Figure 7.9(b), which gives 12 V for e_A in the steady state. This voltage cannot change instantaneously, so $e_A(0+) = 12$ V. The two left-hand elements in Figure 7.9(a) are disconnected for all $t > 0$, so

$$2\dot{e}_A + \frac{1}{3}(e_A - e_o) = 0$$

$$\frac{1}{6}e_o + \frac{1}{3}(e_o - e_A) = 0$$

Transforming these equations, we have

$$2[sE_A(s) - e_A(0)] + \frac{1}{3}[E_A(s) - E_o(s)] = 0$$

$$\frac{1}{6}E_o(s) + \frac{1}{3}[E_o(s) - E_A(s)] = 0$$

Replacing $e_A(0)$ by 12 V and combining these two algebraic equations, we find that

$$E_o(s) = \frac{8}{s + \dfrac{1}{18}}$$

Hence, we can write

$$e_o(t) = 8\epsilon^{-t/18} \quad \text{for all} \quad t > 0$$

We see that $e_o(0+) = 8$ V, so once again the output voltage jumps to a new value at $t = 0$. Notice that $e_o(t)$ goes to zero for large values of t. This is expected since there is no external source of energy for $t > 0$. The factor $\epsilon^{-t/18}$ indicates that the time constant τ is 18 s. Plots of e_A and e_o for this case are shown in part (b) of Figure 7.10.

The last two examples show that the time constant τ for the circuit in Figure 7.9(a) is 12 s when the switch is closed and 18 s when the switch is open. The time constant is a measure of how quickly any initial stored energy would be dissipated when the external input is zero for $t > 0$. With the switch closed, current from the discharging capacitor can flow through the 3-Ω resistor and then through two different 6-Ω paths. With the switch open, one of the two 6-Ω paths has been removed for $t > 0$, so we would expect it would take longer to dissipate any initial energy stored in the capacitor.

▶ 7.5 THE STEP FUNCTION AND IMPULSE

Two of the most important inputs we encounter in the analysis of dynamic systems are the unit step function and the unit impulse. We need to understand their properties and be able to find the response of a system to each of them.

The Unit Step Function

One frequently encounters inputs that are zero before some reference time and that have a nonzero constant value thereafter. To treat such inputs mathematically, we define the **unit step function**, which is denoted by $U(t)$. This function is defined to be zero for $t \leq 0$ and unity for $t > 0$; it is shown in Figure 7.11(a).[3] If the step discontinuity occurs at some later time t_1, as shown in Figure 7.11(b), the function is defined by

$$U(t - t_1) = \begin{cases} 0 & \text{for } t \leq t_1 \\ 1 & \text{for } t > t_1 \end{cases} \tag{42}$$

---

[3]Note that a capital U is used for the unit step function $U(t)$, in contrast to the lowercase letter in the symbol $u(t)$ for a general input. The value of the unit step function at time zero could be defined to be unity, or its value could be left undefined at this instant. Defining $U(0) = 0$ will be convenient for our applications.

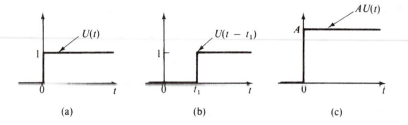

Figure 7.11 Step functions. (a) $U(t)$. (b) $U(t - t_1)$. (c) $AU(t)$.

This notation is consistent with the fact that when any function $f(t)$ is plotted versus t, replacing every t in $f(t)$ by $t - t_1$ shifts the curve t_1 units to the right. If the height of the step is A rather than unity, we simply have A times the unit step function, as shown in Figure 7.11(c).

From the definition of $U(t - t_1)$, we note that

$$f(t)U(t - t_1) = \begin{cases} 0 & \text{for } t \le t_1 \\ f(t) & \text{for } t > t_1 \end{cases} \tag{43}$$

Thus the output of a system that was at rest for $t \le 0$ is often written in the form $f(t)U(t)$, where the multiplying factor $U(t)$ is used in place of the phrase "for $t > 0$."

We define the **unit step response** of a system, denoted by $y_U(t)$, as the output that occurs when the input is the unit step function $U(t)$ and when the system contains no initial stored energy—that is, $y_U(t)$ is the zero-state response to the input $U(t)$. In Section 7.4, we found the response for several examples when the input was zero for $t \le 0$ and A for $t > 0$. To obtain the unit step response for those systems, we merely replace A by 1. For the translational system shown in Figure 7.3,

$$y_U(t) = \frac{1}{K}[1 - \epsilon^{-Kt/B}] \quad \text{for } t \ge 0$$

whereas for the rotational system shown in Figure 7.6,

$$y_U(t) = \frac{B_1}{B_1 + B_2}[1 - \epsilon^{-(B_1 + B_2)t/J}] \quad \text{for } t \ge 0$$

We can represent any function that consists of horizontal and vertical lines as the sum of step functions. Consider, for example, the pulse shown in Figure 7.12 (a). This function is the sum of the two functions shown in parts (b) and (c) of the figure, so

$$f_1(t) = AU(t) - AU(t - t_1) \tag{44}$$

Suppose that the pulse in (44) is the input to a linear first-order system described by the differential equation

$$\dot{y} + \frac{1}{\tau}y = u(t) \tag{45}$$

and for which $y(0) = 0$. From Section 7.4, we know that the response to $AU(t)$ is $A\tau(1 - \epsilon^{-t/\tau})$ for $t \ge 0$, and this is also the response for $0 \le t \le t_1$ to the input in (44). For $t > t_1$, we may use superposition and sum the responses to the components $AU(t)$ and $-AU(t - t_1)$ to obtain

$$y(t) = A\tau[1 - \epsilon^{-t/\tau}] - A\tau[1 - \epsilon^{-(t-t_1)/\tau}]$$
$$= A\tau(-1 + \epsilon^{t_1/\tau})\epsilon^{-t/\tau}$$

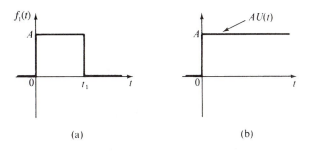

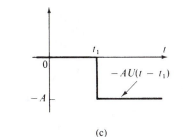

Figure 7.12 (a) Rectangular pulse. (b), (c) Formation of the rectangular pulse by the sum of two step functions.

Hence we may write

$$y(t) = \begin{cases} A\tau(1 - \epsilon^{-t/\tau}) & \text{for } 0 \leq t \leq t_1 \\ A\tau(\epsilon^{t_1/\tau} - 1)\epsilon^{-t/\tau} & \text{for } t > t_1 \end{cases} \tag{46}$$

which is shown in Figure 7.13(a). As expected, the response for the first t_1 seconds is the same as in Figure 7.5(b), whereas for $t > t_1$ the output decays exponentially to zero with a time constant τ. The output is the superposition of the two functions shown in Figure 7.13(b) and Figure 7.13(c). It can also be written in the alternative form

$$y(t) = A\tau(1 - \epsilon^{-t/\tau})U(t) - A\tau(1 - \epsilon^{-(t-t_1)/\tau})U(t - t_1)$$

It is instructive to rewrite the pulse response for $t > t_1$ in (46) with $\epsilon^{t_1/\tau}$ replaced by its Taylor-series expansion

$$\epsilon^{t_1/\tau} = 1 + \frac{t_1}{\tau} + \frac{1}{2!}\left(\frac{t_1}{\tau}\right)^2 + \cdots$$

Then, for $t > t_1$,

$$y(t) = A\tau\left[\frac{t_1}{\tau} + \frac{1}{2!}\left(\frac{t_1}{\tau}\right)^2 + \cdots\right]\epsilon^{-t/\tau}$$

Suppose that the pulse width is small compared to the time constant of the system, so that $t_1 \ll \tau$. Then we can neglect all the terms inside the brackets except the first and write

$$y(t) \simeq At_1\epsilon^{-t/\tau} \quad \text{for } t > t_1 \tag{47}$$

where At_1 is the area underneath the input pulse. We could also consider input pulses that have somewhat different shapes but have the same area underneath the curve. We would find that as long as the width t_1 of any pulse that is the input to a first-order system is small compared to the system's time constant, the response for $t > t_1$ depends on the area underneath the pulse but not on its shape.

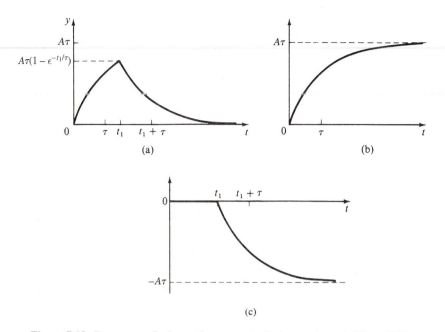

Figure 7.13 Responses of a first-order system to the inputs shown in Figure 7.12.

As background for another important property, note that the response of the system modeled by (45) to a step function of height At_1, with $y(0) = 0$, is At_1 times the unit step response:

$$At_1 y_U(t) = At_1 \tau (1 - \epsilon^{-t/\tau}) \quad \text{for } t \geq 0$$

Note that

$$\frac{d}{dt}[At_1 y_U(t)] = At_1 \epsilon^{-t/\tau} \tag{48}$$

for all positive values of t. Because the right side of (48) is identical to (47), we see that if $t_1 \ll \tau$, the response for $t > t_1$ to a pulse of area At_1 is the derivative of the response to a step function of height At_1.

The Unit Impulse

The response of a first-order system to a pulse of given area appears to be independent of the pulse shape as long as the pulse width t_1 is small compared to the time constant τ, so it is reasonable to try to define an idealized pulse function whose width is small compared to the time constant of all first-order systems. However, in order to have $t_1 \ll \tau$ for all nonzero values of τ, t_1 must be infinitesimally small, and to have a nonzero pulse area, the height of the pulse must become infinitely large. Although such an idealized pulse creates conceptual and mathematical difficulties, let us reconsider the rectangular input pulse that is shown in Figure 7.14(a). Note that $f_1(t)$ in part (a) of the figure is the derivative of the function $f_2(t)$ shown in part (b).

Let us specify $A = 1/t_1$ in Figure 7.14 such that the area underneath $f_1(t)$ is unity and such that the value of $f_2(t)$ for $t > t_1$ is also unity. If we continually decrease the value of t_1, then the height of the pulse $f_1(t)$ increases in order to maintain unit area, while $f_2(t)$ rises to its final value more rapidly. The dashed lines in Figure 7.15 show the changes in $f_1(t)$ and $f_2(t)$ when t_1 is halved.

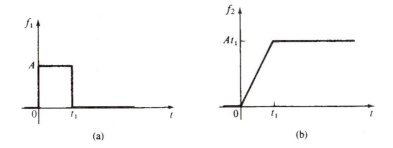

Figure 7.14 (a) Rectangular pulse with area At_1. (b) $f_2(t) = \int_0^t f_1(\lambda)\, d\lambda$.

As t_1 approaches zero, $f_1(t)$ approaches a pulse of infinitesimal width, infinite height, and unit area. The limit of this process is called the **unit impulse**, denoted by the symbol $\delta(t)$. It is represented graphically as shown in Figure 7.16(a). The number 1 next to the head of the arrow indicates the area underneath the function that approached the impulse. In the limit, as t_1 approaches zero, $f_2(t)$ becomes the unit step function $U(t)$, shown in Figure 7.16(b). Because $f_1(t)$ in Figure 7.14 is the time derivative of $f_2(t)$, it appears that in the limit,

$$\delta(t) = \frac{d}{dt}U(t) \tag{49}$$

The arguments we used in the preceding paragraph are heuristic and not mathematically rigorous. Questions can be raised about (49) because differentiation is a limiting process and we introduced $\delta(t)$ as the result of another limiting process. Then, without mathematical justification, we interchanged the order of the two limiting processes. Furthermore, if we were to put our description of the unit impulse into equations, we might write

$$\delta(t) = 0 \quad \text{for } t \neq 0 \tag{50a}$$

$$\int_{-\varepsilon}^{\varepsilon} \delta(t)\, dt = 1 \quad \text{for } \varepsilon > 0 \tag{50b}$$

where (50b) suggests that the unit impulse has unit area. Although we shall not discuss the matter in detail here, it happens that (50) violates the axioms of real-function theory and is an impermissible way of defining a function. In fact, $\delta(t)$ is not a function in the usual

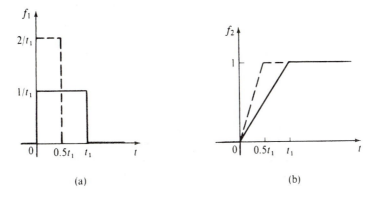

Figure 7.15 The functions $f_1(t)$ and $f_2(t)$ shown in Figure 7.14 when $A = 1/t_1$.

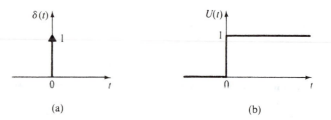

Figure 7.16 (a) The unit impulse. (b) The unit step function.

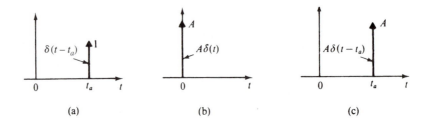

Figure 7.17 Impulses. (a) $\delta(t - t_a)$. (b) $A\delta(t)$. (c) $A\delta(t - t_a)$.

sense, and we have purposely avoided calling it the unit impulse "function." However, results obtained in this section can be justified by rigorous mathematical arguments.[4]

We can define the unit impulse formally in terms of an integral and in a way that is consistent with distribution theory. For any function $f(t)$ that is continuous at $t = 0$, the unit impulse $\delta(t)$ must satisfy the integral expression

$$\int_{-a}^{b} f(t)\delta(t)\,dt = f(0) \quad \text{for } a > 0, b > 0 \tag{51}$$

Note that (50b) is a special case of this equation, with $f(t) = 1$ for all values of t.

A unit impulse that occurs at time t_a rather than at $t = 0$ is denoted by $\delta(t - t_a)$ and is shown in Figure 7.17(a). Furthermore, for any function $f(t)$ that is continuous at $t = t_a$, we can replace (51) by

$$\int_{b}^{c} f(t)\delta(t - t_a)\,dt = f(t_a) \quad \text{for } b < t_a < c \tag{52}$$

which is referred to as the **sampling property** of the impulse. The product $A\delta(t)$, which is shown in Figure 7.17(b), is called an impulse of weight A. We can visualize it as the limit of a high, narrow pulse of area A. Equation (49) may be replaced by the more general relationship

$$A\delta(t - t_a) = \frac{d}{dt}[AU(t - t_a)] \tag{53}$$

In other words, differentiating a step function of height A occurring at t_a gives rise to an impulse of weight A at $t = t_a$, which is indicated in Figure 7.17(c).

The **unit impulse response** $h(t)$ is defined as the output that occurs when the input is $\delta(t)$ and when the system contains no stored energy before the impulse is applied. Thus

---

[4]The unit impulse is also called the **Dirac delta function** and is used in many areas of science and engineering. Both the unit impulse and ordinary functions can be regarded as special cases of generalized functions or distributions.

$h(t)$ is the zero-state response to $\delta(t)$. The zero-state response for $t > t_1$ to a pulse of any shape is approximately equal to $h(t)$ times the area underneath the original pulse, as long as the pulse width t_1 is small compared to the time constant of the system.

A unit impulse applied to a system can cause an infinite power flow that instantaneously changes the energy stored in the system. One way to find the response to the unit impulse is to use the property described in the following paragraph.

Assume that we know the response to a certain input for a linear system that contains no initial stored energy. Then, if we substitute a new input that is the derivative of the original input, the new response will be the derivative of the old response. Because $\delta(t)$ is the derivative of $U(t)$, the unit impulse response and the unit step response for a linear system are related by

$$h(t) = \frac{d}{dt} y_U(t) \tag{54}$$

Thus once the unit step response $y_U(t)$ is known, we can obtain $h(t)$ from it by using (54). A formal proof of this property will be found in Section 8.3.

▶ **EXAMPLE 7.6**

For the translational system shown in Figure 7.18, find $y_U(t)$ and $h(t)$ when the output is the velocity $v(t)$.

SOLUTION   The system is described by the differential equation

$$M\dot{v} + Bv = f_a(t)$$

or

$$\dot{v} + \frac{B}{M} v = \frac{1}{M} f_a(t)$$

Because the system's time constant is $\tau = M/B$, and $v_{ss} = 1/B$ if $f_a(t) = 1$ for all $t > 0$, the unit step response is

$$y_U(t) = \frac{1}{B}(1 - \epsilon^{-Bt/M}) \quad \text{for } t > 0 \tag{55}$$

From (54), the unit impulse response can be found by differentiating $y_U(t)$. It is

$$h(t) = \frac{1}{M} \epsilon^{-Bt/M} \quad \text{for } t > 0 \tag{56}$$

These responses are shown in Figure 7.19.

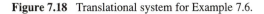

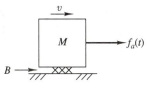

Figure 7.18 Translational system for Example 7.6.

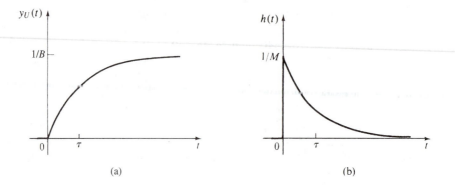

Figure 7.19 Responses for the system shown in Figure 7.18 when the output is the velocity.
(a) Unit step response. (b) Unit impulse response.

Note in Figure 7.19(a) that when $f_a(t)$ is a step function, the velocity of the mass does not change instantaneously at $t = 0$. When $f_a(t)$ is an impulse, however, the velocity of the mass does undergo an instantaneous change from zero to $1/M$. In the first case, there is no change in the kinetic energy of the mass at $t = 0$. In the latter case, the impulse causes an instantaneous increase in the energy of the mass.

Consider further the velocity expression in (56) that is caused by a unit impulse for $f_a(t)$, and recall some of the equations in Chapter 2. According to Equation (2.6), the kinetic energy that the impulse inserts into the mass is

$$\frac{1}{2}Mv^2(0+) = \frac{1}{2M}$$

The applied force $f_a(t)$ returns to zero for all time after the impulse, but there is still the friction force $f_B = Bv$. From Equation (2.1), the power dissipated in the form of heat is

$$p = f_B v = Bv^2 = \frac{B}{M^2}\epsilon^{-2Bt/M}$$

which gradually decays to zero as the mass slows down. As explained in Chapter 2, the energy supplied between times t_0 and t_1 is $\int_{t_0}^{t_1} p(t)\,dt$. Thus the total energy dissipated by friction for all $t > 0$ is

$$\int_0^\infty p(t)\,dt = \frac{B}{M^2}\int_0^\infty \epsilon^{-2Bt/M}\,dt = \frac{B}{M^2}\cdot\frac{M}{2B}\left[-\epsilon^{-2Bt/M}\right]_0^\infty$$

which reduces to $\frac{1}{2M}$. As expected, all the energy that was inserted into the system by the impulse is eventually lost in the form of heat.

▶ **EXAMPLE 7.7**

Let $M = 0.1$ kg and $B = 0.1$ N·s/m for the system discussed in Example 7.6. Let the input $f_a(t)$ be a rectangular pulse having a width L and a height $1/L$. Construct a Simulink diagram in which $f_a(t)$ is represented by the superposition of two step functions: a positive step starting at $t = 0$ and a negative step starting at $t = L$. Plot the output velocity when $L = 3$ s, $L = 0.5$ s, and when $L = 0.1$ s.

SOLUTION The Simulink diagram is shown in Figure 7.20, and the M-file is given below. The resulting output plots are shown in Figure 7.21. Although the input step occurs

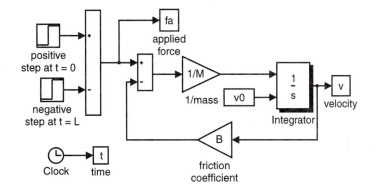

Figure 7.20 Simulink model for Example 7.7

at $t = 0$, for clarity we have plotted the velocity starting at $t = -1$ s by setting the Start time to -1 in the Simulation > Parameters dialog box. The Stop time was set to 8 s. Notice that the time constant τ for the system is 1 s. The area underneath each of the input pulses is unity since they begin at $t = 0$, end at $t = L$, and have an amplitude of $1/L$, all set by the two Step blocks at the left of the figure.

```
v0 = 0.;    % initial velocity, m/s
M = 0.1;    % kg
B = 0.1;    % Ns/m
%----- narrow pulse -------
L = 0.1;
sim('ex7_7_model')
plot(t,v, t,fa,'--'); grid; hold on
%----- medium-width pulse ------
L = 0.5;
sim('ex7_7_model')
plot(t,v, t,fa,'--')
%----- wide pulse ------
L = 3.0;
sim('ex7_7_model')
plot(t,v, t,fa,'--'); hold off
%----- make plot more informative --------
ylabel('Velocity (m/s); Force (N)')
xlabel(' Time (s)')
```

The peak value of the velocity is reached at the end of the input pulse, after which it decays exponentially to zero. The form of the response for $L = 3$ s, which is three times the time constant, is similar to the one shown in Figure 7.13(a).

The response when $L = 0.1$ s, which is $\frac{1}{10}$ of the time constant, is very close to the unit impulse response found in Example 7.6. However, instead of the velocity rising instantly to its peak value, it takes 0.1 s to do so. Furthermore, this peak value is only about 9.5 m/s, instead of the 10 m/s reached for the impulse response. The reader may wish to use the M-file to plot the response for other values of L. As L is made extremely small compared with the system's time constant, the response becomes virtually identical with the impulse response, which itself can be plotted from the Simulink diagram by removing the input and letting the initial condition be $v(0) = 10$, the value obtained from (56) at $t = 0$.

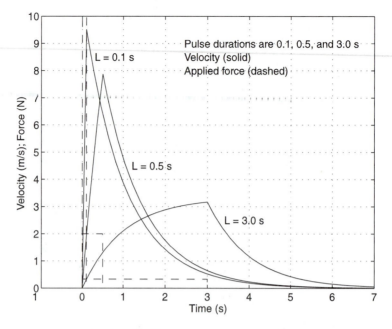

Figure 7.21 Pulse responses for Example 7.7.

The Transform of the Unit Impulse

From (1), we can write the transform of the unit impulse as

$$\mathscr{L}[\delta(t)] = \int_0^\infty \delta(t)\epsilon^{-st}\, dt$$

This expression presents a dilemma because the impulse $\delta(t)$ occurs at $t = 0$, which is the lower limit of integration. To avoid this problem, we shall agree that when we are working with Laplace transforms, the unit impulse $\delta(t)$ occurs just after $t = 0$, namely at $t = 0+$. Then the impulse will be contained within the limits of integration in the definition of the Laplace transform. With this stipulation, (52) is applicable with $t_a = 0+, b = 0$, and $c \to \infty$, and we can say that

$$\int_0^\infty \delta(t)\epsilon^{-st}\, dt = \epsilon^{-st}\Big|_{t=0+} = 1$$

so

$$\mathscr{L}[\delta(t)] = 1 \qquad (57)$$

In books where the lower limit of the transform integral in (1) is taken to be $t = 0-$, the impulse can be considered to occur at $t = 0$. This convention also yields (57).

It is instructive to revisit the rectangular pulse shown in Figure 7.2. If we choose $A = 1/L$, the pulse has unit area, occurs immediately after $t = 0$, and by (14) has the Laplace transform $(1/sL)(1 - \epsilon^{-sL})$. If we now take the limit as L approaches zero, the pulse becomes a unit impulse occuring immediately after $t = 0$. In the limit, the expression for its transform becomes an indeterminate form, because both the numerator and the denominator approach zero. To evaluate this indeterminate form, we can use L'Hôpital's rule, differentiate separately the numerator and denominator with respect to L, and then finally take the new limit. Thus

$$\mathcal{L}[\delta(t)] = \lim_{L \to 0} \frac{1 - \epsilon^{-sL}}{sL} = \lim_{L \to 0} \frac{s\epsilon^{-sL}}{s} = 1$$

which agrees with (57).

If the input to a system is the unit impulse, then the resulting infinite power flow can instantly change the energy stored within the system. Taking $\mathcal{L}[\delta(t)] = 1$, however, means that we are regarding the impulse as occuring at $t = 0+$, and that we must therefore use for the initial conditions the values of the variables just before the impulse takes place.

The transform of any function that is always zero except for $t < 0$ is of course zero. Suppose that when the input is $\delta(t)$ we were to regard the impulse as occurring just before $t = 0$, in which case its transform would be zero. Then the initial-condition terms that appear in the solution would have to include the effect of the additional energy that had been instantly inserted into the system by the impulse. Although it is possible to obtain a solution this way, the evaluation of the necessary initial conditions would be difficult. We shall always assume that all impulses occur after $t = 0$ and that $\mathcal{L}[\delta(t)] = 1$.

In a transform solution, there can be other cases where we must be careful to distinguish between $t = 0-$ and $t = 0+$. For example, the formulas for the derivatives of functions contain terms of the form $f(0)$ and $\dot{f}(0)$. In the event of a discontinuity in $f(t)$ or $\dot{f}$ at $t = 0$, we shall use for the initial condition the limit of $f(t)$ or $\dot{f}$ as t approaches zero from the left, i.e., the value just before the discontinuity occurs. Consider, for example, the formula for the transform of a first derivative, which was derived in (24):

$$\mathcal{L}[\dot{f}] = sF(s) - f(0) \tag{58}$$

To illustrate the application of (58) when $f(t)$ has a discontinuity at $t = 0+$, consider the product of $\cos \omega t$ and the unit step function $U(t)$:

$$f(t) = [\cos \omega t]U(t)$$

$$= \begin{cases} 0 & \text{for } t \le 0 \\ \cos \omega t & \text{for } t > 0 \end{cases} \tag{59}$$

which is shown in Figure 7.22(a). Because $f(t)$ differs from $\cos \omega t$ only at the point $t = 0$ within the interval $0 \le t < \infty$, its transform is the same[5] as that of $\cos \omega t$:

$$F(s) = \frac{s}{s^2 + \omega^2} \tag{60}$$

We see from (59), however, that $f(0) = 0$ because of the presence of the unit step function. Thus (58) gives

$$\mathcal{L}[\dot{f}] = s\left(\frac{s}{s^2 + \omega^2}\right) - 0$$

$$= \frac{s^2}{s^2 + \omega^2} \tag{61}$$

To check this result, consider $\dot{f}$, the derivative of (59), which is shown in Figure 7.22(b). Because of the discontinuity in $f(t)$ at $t = 0+$, the function $\dot{f}$ contains a unit impulse at $t = 0+$. Consistent with the facts that $\dot{f}(t) = 0$ for all $t < 0$ and that the impulse occurs at

---

[5]The value of a definite integral, and hence the result of using (1), is not affected if the value of the integrand at a single point is changed to another finite value.

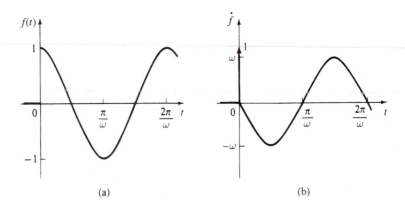

Figure 7.22 The function $f(t) = [\cos \omega t]U(t)$ and its derivative.

$t = 0+$, we define $\dot{f}(0) = 0$. Thus

$$\dot{f} = \begin{cases} 0 & \text{for } t \le 0 \\ \delta(t) - \omega \sin \omega t & \text{for } t > 0 \end{cases} \tag{62}$$

Transforming (62) gives

$$\mathcal{L}[\dot{f}] = 1 - \omega \left(\frac{\omega}{s^2 + \omega^2} \right)$$

$$= \frac{s^2}{s^2 + \omega^2}$$

which is in agreement with (61).

For comparison, consider the related function

$$g(t) = \cos \omega t \qquad \text{for all } t$$

which is continuous for all t and for which $g(0) = 1$. From Equation (24), the transform of its derivative is

$$\mathcal{L}[\dot{g}] = s \left(\frac{s}{s^2 + \omega^2} \right) - 1$$

$$= -\frac{\omega^2}{s^2 + \omega^2}$$

The derivative of $g(t)$ is $\dot{g} = -\omega \sin \omega t$ for all t. Taking the transform of this expression gives $-\omega^2/(s^2 + \omega^2)$, which agrees with the foregoing result. In summary, whenever we use (58) we shall consider any discontinuities in $f(t)$ at the time origin to occur at $t = 0+$.

There are two ways to find a system's response to the unit impulse. One is to first find the response to a unit step and then differentiate that answer, according to (54). We can also find the response directly from the modeling equations by transforming them, using 1 for the transform of the input, and setting the initial-condition terms equal to zero.

▶ **EXAMPLE 7.8**

Find the unit impulse response of the *RL* circuit shown in Figure 7.7 by transforming the modeling equation directly. Then check the answer by differentiating the unit step response.

SOLUTION   The general modeling equation, as given in (39), is

$$\frac{1}{R}[e_o - e_i(t)] + \left[i(0) + \frac{1}{L}\int_0^t e_o\, d\lambda\right] = 0$$

We transform this equation, with $i(0) = 0$ and with $\mathcal{L}[e_i(t)] = \mathcal{L}[\delta(t)] = 1$, to get

$$\frac{1}{R}[E_o(s) - 1] + \frac{1}{L}\cdot\frac{1}{s}E_o(s) = 0$$

which gives

$$E_o(s) = \frac{s}{s + R/L}$$

Until now, we have not encountered the inverse transform of an expression like this, and it is not found in Appendix E. As part of the more detailed treatment of inverse transforms in Section 7.6, we shall explain how a step of long division can be used in this case to write

$$E_o(s) = 1 - \frac{R/L}{s + R/L}$$

For now, this result can be verified easily by recombining the two terms on the right over a common denominator. Because we have dealt with both of these two individual terms before, we can write for $t \geq 0$

$$e_o(t) = \delta(t) - \frac{R}{L}\epsilon^{-(R/L)t} \tag{63}$$

To check this result we next find the unit step response, which is given by (40) with $A = 1$. Thus

$$y_U(t) = \begin{cases} 0 & \text{for } t \leq 0 \\ \epsilon^{-(R/L)t} & \text{for } t > 0 \end{cases}$$

as shown in Figure 7.23(a). To get the unit impulse response for $t > 0$, we use (54) and write

$$\frac{d}{dt}\epsilon^{-(R/L)t} = -\frac{R}{L}\epsilon^{-(R/L)t}$$

This is not the entire expression for $h(t)$, however, because we must also differentiate the discontinuity of unity height that occurs at $t = 0$ in $y_U(t)$. This gives us an impulse of weight 1 as part of the answer for the unit impulse response. Thus we again get the expression in (63), which is shown in Figure 7.23(b).

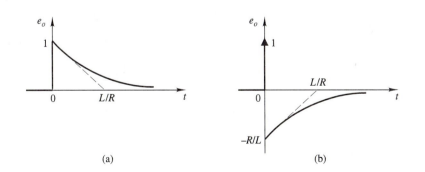

(a)   (b)

Figure 7.23   Responses for Example 7.8. (a) Unit step response. (b) Unit impulse response.

In the last example, notice that the current $i(t)$ through the inductor is

$$i(t) = \frac{1}{R}[e_i(t) - e_o] = \frac{1}{L}\epsilon^{-(R/L)t} \quad \text{for } t > 0$$

The current jumps from zero to $1/L$ at $t = 0$, thereby instantly inserting energy into the inductor. For $t > 0$, when $e_i(t)$ is again zero, this energy is dissipated in the resistor as the current gradually returns to zero. As a general rule, the current through an inductor cannot change instantaneously unless there is an impulse of voltage. Similarly, the voltage across a capacitor cannot change instantaneously unless there is an impulse of current.

▶ 7.6 TRANSFORM INVERSION

When we use Laplace transforms to solve for the response of a system, we find the transform $F(s)$ of a particular variable, such as the output, first. The final step in the process, known as **transform inversion**, is to determine the corresponding time function $f(t)$ where $f(t) = \mathcal{L}^{-1}[F(s)]$, which is read "$f(t)$ is the inverse transform of $F(s)$."

For the types of problems encountered in this book, we do not need a completely general method of transform inversion. In this section, we present and illustrate a method that uses a partial-fraction expansion of $F(s)$. An extension of this method that permits handling an additional category of transformed functions appears in Section 7.7.

Assume that we can write the transform $F(s)$ as the ratio of two polynomials $N(s)$ and $D(s)$, such that

$$F(s) = \frac{N(s)}{D(s)} = \frac{b_m s^m + \cdots + b_0}{s^n + a_{n-1}s^{n-1} + \cdots + a_0} \tag{64}$$

Functions that can be written in the form of (64) are called **rational functions**. A **proper rational function** is one for which $m \leq n$; while a **strictly proper rational function** is one for which $m < n$.

The equation $D(s) = 0$ will have n roots denoted by $s_1, s_2, \ldots, s_n$, and $D(s)$ can be written in factored form as

$$D(s) = (s - s_1)(s - s_2) \cdots (s - s_n) \tag{65}$$

Note that the coefficient of the highest power of s in the denominator polynomial $D(s)$ has been assumed to be unity. If this is not the case for a given $F(s)$, we can always make the coefficient unity by dividing both $N(s)$ and $D(s)$ by a constant. The quantities $s_1, s_2, \ldots, s_n$ are called the **poles** of $F(s)$ and are those values of s for which $F(s)$ becomes infinite.[6]

The **method of partial-fraction expansion** is applicable to any strictly proper rational function. Briefly, it allows us to express a known transform $F(s)$ as the sum of less complicated transforms. Using the table in Appendix E, we can identify the time functions that correspond to the individual transforms in the expansion and then use the superposition theorem to write $f(t)$.

We consider first the case where all the poles of $F(s)$ are distinct. We next modify the procedure to include repeated poles, where two or more of the quantities $s_1, s_2, \ldots, s_n$ are equal. Then we examine the case where the poles are complex numbers and discuss what to do if the degree of $N(s)$ is not less than that of $D(s)$. Throughout the section, we assume that the polynomial $D(s)$ has been factored, so that the values of the n poles are known.

---

[6]It is assumed that $N(s) \neq 0$ at any of the poles.

Distinct Poles

The partial-fraction expansion theorem states that if $F(s)$ is a strictly proper rational function with distinct poles, it can be written as

$$F(s) = \frac{A_1}{s - s_1} + \frac{A_2}{s - s_2} + \cdots + \frac{A_n}{s - s_n} \tag{66}$$

where $A_1, A_2, \ldots, A_n$ are constants. We can write (66) with a summation sign as follows:

$$F(s) = \sum_{i=1}^{n} A_i \left(\frac{1}{s - s_i} \right) \tag{67}$$

From (5), the term $1/(s - s_i)$ is the transform of the time function $\epsilon^{s_i t}$. Then, from the superposition formula given in (17), it follows that for $t > 0$,

$$f(t) = A_1 \epsilon^{s_1 t} + A_2 \epsilon^{s_2 t} + \cdots + A_n \epsilon^{s_n t}$$

$$= \sum_{i=1}^{n} A_i \epsilon^{s_i t} \tag{68}$$

We can find the n poles s_i by factoring the denominator of $F(s)$ or, more generally, by finding the roots of $D(s) = 0$. We shall now develop a procedure for evaluating the n coefficients A_i so that we can write $f(t)$ as a sum of exponential time functions by using (68).

Multiplying both sides of (66) by the term $(s - s_1)$ yields

$$(s - s_1)F(s) = A_1 + A_2 \frac{(s - s_1)}{(s - s_2)} + \cdots + A_n \frac{(s - s_1)}{(s - s_n)}$$

Because this equation must be a mathematical identity for all values of the variable s, we can set s equal to s_1 throughout the equation. The poles are distinct, so $s_1 \neq s_j$ for $j = 2, 3, \ldots, n$. Thus each term on the right side will vanish except the term A_1, and we can write

$$A_1 = (s - s_1)F(s)|_{s=s_1} \tag{69}$$

To make it clear why (69) does not give a value of zero, we write $F(s)$ as $N(s)/D(s)$ with $D(s)$ in factored form. Then

$$A_1 = \left. \frac{(s - s_1)N(s)}{(s - s_1)(s - s_2) \cdots (s - s_n)} \right|_{s=s_1}$$

The term $(s - s_1)$ in the numerator will be canceled by the corresponding term in $D(s)$ before s is replaced by s_1.

Repeating the above process with s_1 replaced by s_2, we have

$$A_2 = (s - s_2)F(s)|_{s=s_2}$$

The general expression for the coefficients is

$$A_i = (s - s_i)F(s)|_{s=s_i} \qquad i = 1, 2, \ldots, n \tag{70}$$

In a numerical problem we can check the calculation of $A_1, A_2, \ldots, A_n$ by combining the terms on the right side of (66) over a common denominator, which should give the original function $F(s)$.

▶ **EXAMPLE 7.9**

Find the inverse transform of

$$F(s) = \frac{-s+5}{(s+1)(s+4)}$$

SOLUTION   Comparing $F(s)$ to (64), we see that $N(s) = -s+5$ so $m = 1$, and $D(s) = (s+1)(s+4)$ so $n = 2$. Because $D(s)$ is already in factored form, we see by inspection that the poles are $s_1 = -1$ and $s_2 = -4$. Thus we can rewrite the transform $F(s)$ in the form of (66) as

$$F(s) = \frac{A_1}{s+1} + \frac{A_2}{s+4}$$

Using (70) with $i = 1$ and 2, we find the coefficients of the partial-fraction expansion to be

$$A_1 = \frac{(s+1)(-s+5)}{(s+1)(s+4)}\bigg|_{s=-1} = \frac{6}{3} = 2$$

$$A_2 = \frac{(s+4)(-s+5)}{(s+1)(s+4)}\bigg|_{s=-4} = \frac{9}{-3} = -3$$

Hence the partial-fraction expansion of the transform is

$$F(s) = \frac{2}{s+1} - \frac{3}{s+4}$$

and, from (68), the time function $f(t)$ is

$$f(t) = 2\epsilon^{-t} - 3\epsilon^{-4t} \qquad \text{for } t > 0$$

Repeated Poles

If two or more of the n roots of $D(s) = 0$ are identical, these roots, which are poles of $F(s)$, are said to be **repeated**. When $F(s)$ contains repeated poles, (66) and (68) no longer hold. If $s_1 = s_2$, for example, the first two terms in (68) become $A_1 e^{s_1 t}$ and $A_2 e^{s_1 t}$, which are identical except for the multiplying constant, and (68) is not valid. If $s_1 = s_2$ and if the remaining poles are distinct, then (66) must be modified to be

$$F(s) = \frac{A_{11}}{(s-s_1)^2} + \frac{A_{12}}{s-s_1} + \frac{A_3}{s-s_3} + \cdots + \frac{A_n}{s-s_n} \qquad (71)$$

Referring to Appendix E, we see that the first term on the right-hand side of (71) is the transform of $A_{11} t e^{s_1 t}$, whereas the second term has the same form as the remaining ones. Thus the inverse transform is

$$f(t) = A_{11} t e^{s_1 t} + A_{12} e^{s_1 t} + A_3 e^{s_3 t} + \cdots + A_n e^{s_n t} \qquad (72)$$

Note that the repeated pole at $s = s_1$ introduces into the time function a term of the form $t e^{s_1 t}$.

In order to evaluate any of the $n - 2$ coefficients $A_3, \ldots, A_n$ we can use the procedure given by (70). This formula cannot be used to find A_{11} and A_{12} because of the term $(s - s_1)^2$ in the denominator of $F(s)$. However, multiplying both sides of (71) by $(s - s_1)^2$ gives

$$(s-s_1)^2 F(s) = A_{11} + A_{12}(s-s_1) + A_3 \frac{(s-s_1)^2}{(s-s_3)} + \cdots + A_n \frac{(s-s_1)^2}{(s-s_n)} \qquad (73)$$

Setting s equal to s_1 throughout (73) results in

$$A_{11} = (s - s_1)^2 F(s)|_{s=s_1} \tag{74}$$

To find A_{12}, we note that the right-hand side of (73) has the terms $A_{11} + A_{12}(s - s_1)$ and that all the remaining terms contain $(s - s_1)^2$ in their numerators. Thus if we differentiate both sides of (73) with respect to s, we have

$$\frac{d}{ds}[(s - s_1)^2 F(s)] = A_{12} + (s - s_1)G(s) \tag{75}$$

where $G(s)$ is a rational function without a pole at $s = s_1$. Note that in (75) the coefficient A_{11} is not present, A_{12} stands alone, and the function $G(s)$ that contains all the other coefficients is multiplied by the quantity $s - s_1$. Hence setting s equal to s_1 in (75) gives

$$A_{12} = \left\{ \frac{d}{ds}[(s - s_1)^2 F(s)] \right\} \bigg|_{s=s_1} \tag{76}$$

where the differentiation must be performed before s is set equal to s_1.

If $F(s)$ has two or more pairs of identical poles, then each pair of poles contributes terms of the form $\epsilon^{s_i t}$ and $t \epsilon^{s_i t}$, and we can evaluate the coefficients of these terms by using (74) and (76) with the appropriate indices. If $F(s)$ has three or more identical poles, (71) must be modified further. For example, if $s_1 = s_2 = s_3$ and if the remaining poles are distinct, the partial-fraction expansion has the form

$$F(s) = \frac{A_{11}}{(s - s_1)^3} + \frac{A_{12}}{(s - s_1)^2} + \frac{A_{13}}{s - s_1} + \frac{A_4}{s - s_4} + \cdots + \frac{A_n}{s - s_n} \tag{77}$$

By a procedure similar to that used in the derivation of (74) and (76), we find that

$$A_{11} = (s - s_1)^3 F(s)|_{s=s_1}$$

$$A_{12} = \left\{ \frac{d}{ds}[(s - s_1)^3 F(s)] \right\} \bigg|_{s=s_1}$$

$$A_{13} = \frac{1}{2!} \left\{ \frac{d^2}{ds^2}[(s - s_1)^3 F(s)] \right\} \bigg|_{s=s_1} \tag{78}$$

and

$$A_i = (s - s_i)F(s)|_{s=s_i} \quad \text{for } i = 4, 5, \ldots, n.$$

The time function corresponding to (77) and (78) is

$$f(t) = \left(\frac{1}{2!} A_{11} t^2 + A_{12} t + A_{13} \right) \epsilon^{s_1 t} + A_4 \epsilon^{s_4 t} + \cdots + A_n \epsilon^{s_n t} \tag{79}$$

▶ **EXAMPLE 7.10**

Find the inverse Laplace transform of

$$F(s) = \frac{5s + 16}{(s + 2)^2 (s + 5)}$$

SOLUTION   The denominator of $F(s)$ is given in factored form, so we note that the poles are $s_1 = s_2 = -2$ and $s_3 = -5$. Because of the repeated pole, the partial-fraction expansion of $F(s)$ has the form

$$F(s) = \frac{A_{11}}{(s+2)^2} + \frac{A_{12}}{s+2} + \frac{A_3}{s+5}$$

Using (74), (76), and (70) in order, we find that the coefficients are

$$A_{11} = (s+2)^2 F(s)\big|_{s=-2} = \frac{5s+16}{s+5}\bigg|_{s=-2} = 2$$

$$A_{12} = \left\{\frac{d}{ds}\left[\frac{5s+16}{s+5}\right]\right\}\bigg|_{s=-2} = \frac{9}{(s+5)^2}\bigg|_{s=-2} = 1$$

$$A_3 = (s+5)F(s)\big|_{s=-5} = \frac{5s+16}{(s+2)^2}\bigg|_{s=-5} = -1$$

Using numerical values for the coefficients gives the partial-fraction expansion of the transform as

$$F(s) = \frac{2}{(s+2)^2} + \frac{1}{s+2} - \frac{1}{s+5}$$

The corresponding time function is

$$f(t) = 2t\epsilon^{-2t} + \epsilon^{-2t} - \epsilon^{-5t} \quad \text{for } t > 0$$

Complex Poles

The form of $f(t)$ given by (68) is valid for complex poles as well as for real poles. When (68) is used directly with complex poles, however, it has the disadvantage that the functions $\epsilon^{s_i t}$ and the coefficients A_i are complex. Hence, further developments are necessary in order to write $f(t)$ directly in terms of real functions and real coefficients. We shall present two methods for finding $f(t)$ that lead to slightly different but equivalent forms. First we show how mathematical identities are used to combine any complex terms in the partial-fraction expansion of $F(s)$ into real functions of time. Then we present the alternative method known as completing the square. Basic operations with complex numbers are reviewed in Appendix C.

For simplicity, we shall first assume that $F(s)$ has only two complex poles and that the order of its numerator is less than the order of its denominator. Any other poles, whether real or complex, will lead to additional terms in the partial-fraction expansion. With these restrictions, we can write the transform as

$$F(s) = \frac{Bs+C}{(s+a-j\omega)(s+a+j\omega)} \tag{80}$$

which has poles at $s_1 = -a + j\omega$ and $s_2 = -a - j\omega$.

Partial-Fraction Expansion

As indicated in (80), the complex poles of $F(s)$ always occur in complex conjugate pairs. Using the partial-fraction expansion of (66) for distinct poles, we can write $F(s)$ as

$$F(s) = \frac{K_1}{s+a-j\omega} + \frac{K_2}{s+a+j\omega} \tag{81}$$

where

$$K_1 = (s+a-j\omega)F(s)|_{s=-a+j\omega} \qquad (82a)$$

$$K_2 = (s+a+j\omega)F(s)|_{s=-a-j\omega} \qquad (82b)$$

Because (82a) is identical to (82b) except for the sign of the imaginary term $j\omega$ wherever it appears, the coefficient K_2 is the complex conjugate of K_1; that is, $K_2 = K_1^*$. Hence we can write K_1 and K_2 in polar form as $K_1 = K\epsilon^{j\phi}$ and $K_2 = K\epsilon^{-j\phi}$, where K and ϕ are the magnitude and angle, respectively, of the complex number K_1. Thus both K and ϕ are real quantities, and $K \geq 0$.

Rewriting (81) with K_1 and K_2 in polar form, we have

$$F(s) = \frac{K\epsilon^{j\phi}}{s+a-j\omega} + \frac{K\epsilon^{-j\phi}}{s+a+j\omega} \qquad (83)$$

which has the form of (67), although the coefficients and poles are complex. From (68) with $n = 2$ and with the appropriate values substituted for A_i and s_i, the complex form of the time function is

$$f(t) = K\epsilon^{j\phi}\epsilon^{(-a+j\omega)t} + K\epsilon^{-j\phi}\epsilon^{(-a-j\omega)t}$$

To obtain $f(t)$ as a real function of time, we factor the term $2K\epsilon^{-at}$ out of each term on the right-hand side of this equation and combine the remaining complex exponentials. This yields

$$f(t) = 2K\epsilon^{-at}\left[\frac{\epsilon^{j(\omega t+\phi)} + \epsilon^{-j(\omega t+\phi)}}{2}\right]$$

Recognizing from Table 1 in Appendix D that the term in the brackets is $\cos(\omega t + \phi)$, we can write

$$f(t) = 2K\epsilon^{-at}\cos(\omega t + \phi) \qquad \text{for } t > 0 \qquad (84)$$

where the parameters a, ω, K, and ϕ are all real. Note that (83) and (84) constitute one of the entries in Appendix E.

Completing the Square
To develop the second of the two forms of $f(t)$ when $F(s)$ has complex poles, we multiply the denominator factors in (80) to get

$$F(s) = \frac{Bs+C}{s^2 + 2as + a^2 + \omega^2}$$

Next we write the denominator as the sum of the perfect square $(s+a)^2$ and the constant ω^2, so that

$$F(s) = \frac{Bs+C}{(s+a)^2+\omega^2} \qquad (85)$$

Now we rearrange the numerator of $F(s)$ to have the term $(s+a)$ appear:

$$F(s) = \frac{B(s+a) + (C-aB)}{(s+a)^2+\omega^2}$$

$$= B\left[\frac{s+a}{(s+a)^2+\omega^2}\right] + \left(\frac{C-aB}{\omega}\right)\left[\frac{\omega}{(s+a)^2+\omega^2}\right]$$

Referring to Appendix E, we see that the quantities within the brackets are the transforms of $\epsilon^{-at}\cos\omega t$ and $\epsilon^{-at}\sin\omega t$, respectively. Thus

$$f(t) = B\epsilon^{-at}\cos\omega t + \left(\frac{C-aB}{\omega}\right)\epsilon^{-at}\sin\omega t \tag{86}$$

Equations (85) and (86) form another one of the entries in Appendix E. Although (84) and (86) look somewhat different, we can always use Table 1 in Appendix D to show that they are equivalent functions of time.

▶ **EXAMPLE 7.11**

Use each of the two methods that we have described to find $f(t)$ when

$$F(s) = \frac{4s+8}{s^2+2s+5}$$

SOLUTION   The poles of $F(s)$ are the roots of $s^2+2s+5=0$, namely $s_1 = -1+j2$ and $s_2 = -1-j2$. Hence $a=1$, $\omega=2$, and the partial-fraction expansion of $F(s)$ is

$$F(s) = \frac{K_1}{s+1-j2} + \frac{K_2}{s+1+j2}$$

Solving for K_1 according to (82a), we find

$$K_1 = \frac{(s+1-j2)(4s+8)}{(s+1-j2)(s+1+j2)}\Bigg|_{s=-1+j2}$$

$$= 2-j1$$

$$= \sqrt{5}\epsilon^{-j0.4636}$$

Thus $K=\sqrt{5}$ and $\phi = -0.4636$ rad. Substituting the known values of a, ω, K, and ϕ into (84) yields

$$f(t) = 2\sqrt{5}\epsilon^{-t}\cos(2t - 0.4636) \qquad \text{for } t>0 \tag{87}$$

Having found $f(t)$ from the partial-fraction expansion of $F(s)$, we repeat the problem by the method of completing the square. Because the denominator of $F(s)$ is s^2+2s+5, we can obtain a perfect-square term by adding and subtracting 1 to give

$$(s^2+2s+1)+(5-1) = (s+1)^2+4$$

so

$$F(s) = \frac{4s+8}{(s+1)^2+2^2}$$

By comparison with (85), we see that $B=4, C=8, a=1$, and $\omega=2$. With these values, (86) gives

$$f(t) = 4\epsilon^{-t}\cos 2t + 2\epsilon^{-t}\sin 2t \quad \text{for } t>0$$

which we can convert to the form given by (87) by using the appropriate entry in Table 1 in Appendix D.

Comparison of the two methods for finding $f(t)$ indicates that the partial-fraction expansion requires the manipulation of complex numbers. It is frequently employed, however, and we shall need it for a derivation in the next chapter. The method of completing the square has the advantage of avoiding the use of complex numbers. The following example illustrates the two methods when $F(s)$ has more than the two poles specified by (80) and (85).

▶ **EXAMPLE 7.12**

Find the inverse transform of

$$F(s) = \frac{5s^2 + 8s - 5}{s^2(s^2 + 2s + 5)}$$

$$= \frac{5s^2 + 8s - 5}{s^2(s + 1 - j2)(s + 1 + j2)}$$

SOLUTION In order to use (84), we write the partial-fraction expansion as

$$F(s) = \frac{A_{11}}{s^2} + \frac{A_{12}}{s} + \frac{K_1}{s + 1 - j2} + \frac{K_1^*}{s + 1 + j2} \tag{88}$$

where

$$A_{11} = \left. \frac{5s^2 + 8s - 5}{s^2 + 2s + 5} \right|_{s=0} = -1$$

$$A_{12} = \left\{ \frac{d}{ds} \left[\frac{5s^2 + 8s - 5}{s^2 + 2s + 5} \right] \right\} \Bigg|_{s=0} = \left. \frac{2s^2 + 60s + 50}{(s^2 + 2s + 5)^2} \right|_{s=0} = 2$$

$$K_1 = \left. \frac{5s^2 + 8s - 5}{s^2(s + 1 + j2)} \right|_{s=-1+j2} = \frac{-28 - j4}{16 - j12} = \frac{-7 - j1}{4 - j3}$$

Although it is possible to evaluate the quotient of two complex numbers directly on many hand calculators, an alternative procedure is to rationalize the fraction by multiplying both halves by the complex conjugate of the denominator. Then

$$K_1 = \frac{-7 - j1}{4 - j3} \cdot \frac{4 + j3}{4 + j3} = \frac{1}{25}(-25 - j25)$$

$$= -1 - j1 = \sqrt{2}\epsilon^{-j2.356}$$

Inserting the values of A_{11}, A_{12}, and K_1 into (88) and then using (11), (3), and (84), we obtain

$$f(t) = -t + 2 + 2\sqrt{2}\epsilon^{-t} \cos(2t - 2.356) \quad \text{for } t > 0 \tag{89}$$

In order to avoid complex numbers, we may choose not to factor the quadratic $s^2 + 2s + 5$ that appears in the denominator of $F(s)$. For the corresponding term in a partial-fraction expansion, however, we must then assume a numerator of the form $Bs + C$. Thus, for an alternative solution to this example, we write

$$F(s) = \frac{5s^2 + 8s - 5}{s^2(s^2 + 2s + 5)}$$

$$= \frac{A_{11}}{s^2} + \frac{A_{12}}{s} + \frac{Bs + C}{s^2 + 2s + 5}$$

Recombining the terms on the right side of this equation over a common denominator gives

$$F(s) = \frac{(A_{12}+B)s^3 + (A_{11}+2A_{12}+C)s^2 + (2A_{11}+5A_{12})s + 5A_{11}}{s^2(s^2+2s+5)}$$

By equating corresponding coefficients in the numerators, we have

$$A_{12}+B = 0$$
$$A_{11}+2A_{12}+C = 5$$
$$2A_{11}+5A_{12} = 8$$
$$5A_{11} = -5$$

from which $A_{11} = -1$, $A_{12} = 2$, $B = -2$, and $C = 2$. If we also complete the square in the factor s^2+2s+5, we may write the partial-fraction expansion as

$$F(s) = \frac{-1}{s^2} + \frac{2}{s} + \frac{-2s+2}{(s+1)^2+(2)^2}$$

From (11), (3), and (86), the corresponding function of time is

$$f(t) = -t + 2 - 2\epsilon^{-t}\cos 2t + 2\epsilon^{-t}\sin 2t \quad \text{for } t > 0 \qquad (90)$$

It is easy to show by Table 1 in Appendix D that (89) and (90) are equivalent.

Preliminary Step of Long Division

Remember that the techniques discussed so far are subject to the restriction that $F(s)$ is a strictly proper rational function—in other words, that m, the degree of the numerator polynomial $N(s)$, is less than n, the degree of the denominator polynomial $D(s)$. Otherwise, the partial-fraction expansion given for distinct poles by (67) or for a pair of repeated poles by (71) is not valid. In order to find the inverse transform of $F(s)$ when $m = n$, we must first write $F(s)$ as the sum of a constant[7] and a fraction whose numerator is of degree $n - 1$ or less. We can accomplish this by dividing the numerator by the denominator so that the remainder is of degree $n - 1$ or less. Then we may write

$$F(s) = A + F'(s) \qquad (91)$$

where A is a constant and is the transform of $A\delta(t)$. The function $F'(s)$ is a ratio of polynomials having the same denominator as $F(s)$ but a numerator of degree less than n. We can find the inverse transform of $F'(s)$ by using the techniques described previously. Thus

$$f(t) = A\delta(t) + \mathcal{L}^{-1}[F'(s)] \qquad (92)$$

▶ **EXAMPLE 7.13**

Find $f(t)$ when

$$F(s) = \frac{2s^2+7s+8}{s^2+3s+2}$$

---

[7] When $m > n$, we can write $F(s)$ as the sum of a polynomial in s and a strictly proper rational function, but we shall not encounter such cases here.

SOLUTION   Both the numerator and the denominator of $F(s)$ are quadratic in s, so we have $m = n = 2$. Before carrying out the partial-fraction expansion, we must rewrite $F(s)$ in the form of (91) by dividing its numerator by its denominator, as follows:

$$
\begin{array}{r}
2 \\
s^2 + 3s + 13 \overline{\smash{)}2s^2 + 7s + 8} \\
\underline{2s^2 + 6s + 4} \\
s + 4
\end{array}
$$

Because the poles of $F(s)$ are $s_1 = -1$ and $s_2 = -2$, we can write

$$
F(s) = 2 + \frac{s+4}{(s+1)(s+2)}
$$

$$
= 2 + \frac{A_1}{s+1} + \frac{A_2}{s+2}
$$

where

$$
A_1 = \left. \frac{(s+1)(s+4)}{(s+1)(s+2)} \right|_{s=-1} = 3
$$

$$
A_2 = \left. \frac{(s+2)(s+4)}{(s+1)(s+2)} \right|_{s=-2} = -2
$$

Thus

$$
F(s) = 2 + \frac{3}{s+1} - \frac{2}{s+2}
$$

and

$$
f(t) = 2\delta(t) + 3\epsilon^{-t} - 2\epsilon^{-2t} \quad \text{for } t > 0
$$

which is the sum of an impulse at $t = 0+$ and two decaying exponentials.

7.7 ADDITIONAL TRANSFORM PROPERTIES

We now introduce several useful transform properties that were not needed for the examples in Section 7.4. The first two of these concern functions that are shifted in time. The others provide certain information about the time function directly from its transform, without our having to carry out the transform inversion.

Time Delay

If the function $f(t)$ is delayed by a units of time, we denote the delayed function as $f(t-a)$, where $a > 0$. In order to develop a general expression for the transform of the delayed function $f(t-a)$ in terms of the transform of $f(t)$, we must ensure that any part of $f(t)$ that is nonzero for $t < 0$ does not fall within the range $0 < t < \infty$ for the delayed function. Otherwise, a portion of the original time function will contribute to the transform of $f(t-a)$ but not to that of $f(t)$. To illustrate this point, consider the functions $f(t)$ and $f(t-a)$ shown in Figure 7.24. The shaded portion of $f(t)$ that is nonzero for $t < 0$ does not affect $\mathcal{L}[f(t)]$, because it is outside the limits of integration in (1). It does affect $\mathcal{L}[f(t-a)]$, however, because for $a > 0$, at least part of it falls within the interval $0 < t < \infty$, as shown in Figure 7.24(b).

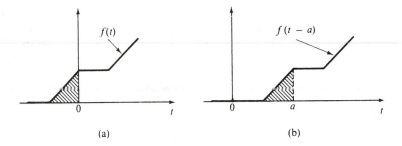

Figure 7.24 A function for which the time-delay theorem is not applicable.

We shall consider functions of the form
$$f_1(t) = f(t)U(t)$$
which are the product of any transformable function $f(t)$ and the unit step function $U(t)$. Because $U(t) = 0$ for all $t \leq 0$, the function $f_1(t)$ will be zero for all $t \leq 0$, and the delayed function
$$f_1(t-a) = f(t-a)U(t-a)$$
will be zero for all $t \leq a$. The functions $f_1(t)$ and $f_1(t-a)$ corresponding to the $f(t)$ defined in Figure 7.24(a) are shown in Figure 7.25.

To express the transform of $f(t-a)U(t-a)$, where $a > 0$, in terms of $F(s) = \mathcal{L}[f(t)]$, we start with the transform definition in (1) and write
$$\mathcal{L}[f(t-a)U(t-a)] = \int_0^\infty f(t-a)U(t-a)\epsilon^{-st}\,dt$$

Because
$$U(t-a) = \begin{cases} 0 & \text{for } t \leq a \\ 1 & \text{for } t > a \end{cases}$$

we can rewrite the transform as
$$\mathcal{L}[f(t-a)U(t-a)] = \int_a^\infty f(t-a)\epsilon^{-st}\,dt$$

$$= \epsilon^{-sa}\int_a^\infty f(t-a)\epsilon^{-s(t-a)}dt$$

$$= \epsilon^{-sa}\int_0^\infty f(\lambda)\epsilon^{-s\lambda}d\lambda$$

$$= \epsilon^{-sa}F(s) \tag{93}$$

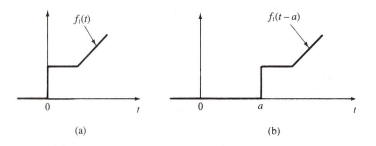

Figure 7.25 A function for which the time-delay theorem is applicable.

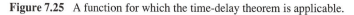

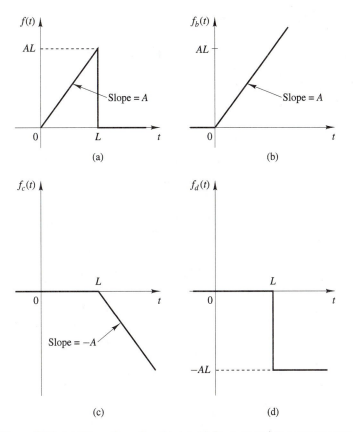

Figure 7.26 (a) Triangular pulse. (b), (c), (d) Its ramp and step components.

where $F(s)$ is the transform of $f(t)$ and where $a > 0$. This theorem is one of the entries in Appendix E.

▶ **EXAMPLE 7.14**

Use the time-delay theorem to derive the transform of the triangular pulse shown in Figure 7.26(a) and defined by the equation

$$f(t) = \begin{cases} 0 & \text{for } t \leq 0 \\ At & \text{for } 0 < t \leq L \\ 0 & \text{for } t > L \end{cases}$$

SOLUTION   Any pulse that consists of straight lines can be decomposed into a sum of step functions and ramp functions. The triangular pulse shown in Figure 7.26(a) can be regarded as the superposition of the three functions shown in Figure 7.26(b), Figure 7.26(c), and Figure 7.26(d). These are a ramp starting at $t = 0$, a delayed ramp starting at $t = L$, and a delayed step function starting at $t = L$. Thus

$$f(t) = AtU(t) - A(t-L)U(t-L) - ALU(t-L) \tag{94}$$

From Appendix E, we note that

$$\mathscr{L}[AtU(t)] = \frac{A}{s^2} \tag{95}$$

Using (93) with (95), we have

$$\mathcal{L}[-A(t-L)U(t-L)] = -\frac{A\epsilon^{-sL}}{s^2}$$

From (93) and the fact that $\mathcal{L}[U(t)] = 1/s$,

$$\mathcal{L}[-ALU(t-L)] = -\frac{AL\epsilon^{-sL}}{s}$$

Using the superposition theorem, we obtain the transform of the triangular pulse as

$$F(s) = \frac{A}{s^2}(1 - \epsilon^{-sL}) - \frac{AL}{s}\epsilon^{-sL} \qquad (96)$$

Inversion of Some Irrational Transforms

The use of a partial-fraction expansion to find an inverse transform is restricted to transforms that are rational functions of s. However, the transform given by (96) is not a rational function because of the factor ϵ^{-sL}. For transforms that would be rational functions except for multiplicative factors in the numerator such as ϵ^{-sL}, we may use the time-delay theorem in (93).

Assume that an irrational transform can be written as

$$F(s) = F_1(s) + F_2(s)\epsilon^{-sa} \qquad (97)$$

where $F_1(s)$ and $F_2(s)$ are rational functions and where a is a positive constant. Then we can find the inverse transforms of $F_1(s)$ and $F_2(s)$ by using partial-fraction expansions and, if the order of the numerator is not less than that of the denominator, a preliminary step of long division. Denote the inverse transforms of $F_1(s)$ and $F_2(s)$ by $f_1(t)$ and $f_2(t)$, respectively. Then, by (93),

$$f(t) = f_1(t) + f_2(t-a)U(t-a) \qquad (98)$$

▶ **EXAMPLE 7.15**

Find the inverse transform of

$$F(s) = \frac{A}{s^2}(1 - \epsilon^{-sL}) - \frac{AL}{s}\epsilon^{-sL}$$

SOLUTION We can rewrite the transform $F(s)$ in the form of (97) as

$$F(s) = \frac{A}{s^2} - \left(\frac{A}{s^2} + \frac{AL}{s}\right)\epsilon^{-sL}$$

Hence $F_1(s) = A/s^2$ and $f_1(t) = At$ for $t > 0$. The rational portion of the remaining term is

$$F_2(s) = -\left(\frac{A}{s^2} + \frac{AL}{s}\right)$$

which has as its inverse transform

$$f_2(t) = -At - AL \qquad \text{for } t > 0$$

Using (98) with $a = L$, we can write the complete inverse transform as

$$f(t) = At - A(t - L)U(t - L) - ALU(t - L) \qquad \text{for } t > 0$$

which agrees with (94) and Figure 7.26(a).

Initial-Value and Final-Value Theorems

It is possible to determine the limits of $f(t)$ as time approaches zero and infinity directly from its transform $F(s)$ without having to find $f(t)$ for all $t > 0$. First we consider the limit of $f(t)$ as time approaches zero through positive values (that is, from the right). This limit of $f(t)$ is denoted by $f(0+)$. To evaluate this limit directly from $F(s)$, we use the **initial-value theorem**, which states that

$$f(0+) = \lim_{s \to \infty} sF(s) \tag{99}$$

where the limit exists.

If $F(s)$ is a rational function, (99) will yield a finite value provided that the degree of the numerator polynomial is less than that of the denominator—in other words, provided that $m < n$. If we attempt to use (99) when $m = n$, the result will be infinite. Recall from Section 7.6 that to find $f(t)$ when $m = n$, we must use a preliminary step of long division to write $F(s)$ as the sum of a constant and a transform for which $m < n$. Because the inverse transform of the constant is an impulse at $t = 0+$, the value of $f(0+)$ is undefined when $m = n$.

The **final-value theorem** states that

$$f(\infty) = \lim_{s \to 0} sF(s) \tag{100}$$

provided that $F(s)$ has no poles in the right half of the complex plane and, with the possible exception of a single pole at the origin, has no poles on the imaginary axis. The symbol $f(\infty)$ denotes the limit of $f(t)$ as t approaches infinity.

To gain some insight into the effect of this restriction on the use of the final-value theorem, we recall that the forms of the terms in a partial-fraction expansion are dictated by the locations of the poles of $F(s)$. Suppose, for example, that

$$F(s) = \frac{A_1}{s} + \frac{A_2(s + \alpha)}{(s + \alpha)^2 + \beta^2} + \frac{A_3}{s - b} + \frac{A_4 \omega}{s^2 + \omega^2}$$

where $\alpha, \beta, b,$ and ω are positive real constants. The expansion implies that the poles of $F(s)$ are $s_1 = 0$, $s_2 = -\alpha + j\beta$, $s_3 = -\alpha - j\beta$, $s_4 = b$, $s_5 = j\omega$, and $s_6 = -j\omega$. The corresponding time function for $t > 0$ is

$$f(t) = A_1 + A_2 \epsilon^{-\alpha t} \cos \beta t + A_3 \epsilon^{bt} + A_4 \sin \omega t$$

The limits of the first two terms as t approaches infinity are A_1 and zero, respectively. However, $A_3 \epsilon^{bt}$ increases without limit, whereas $A_4 \sin \omega t$ oscillates continually without approaching a constant value. Thus, because of the poles of $F(s)$ at $s_4 = b$, $s_5 = j\omega$, and $s_6 = -j\omega$, the function $f(t)$ does not approach a limit as t approaches infinity. As another example, double poles of $F(s)$ at the origin will cause the partial-fraction expansion to have terms of the form

$$F(s) = \frac{A_{11}}{s^2} + \frac{A_{12}}{s} + \cdots$$

The corresponding time function for $t > 0$ is

$$f(t) = A_{11}t + A_{12} + \cdots$$

which again does not approach a limit.

The use of the initial-value and final-value theorems is illustrated in the following example and in the next chapter. In Example 7.17 we consider a transform for which neither theorem is applicable.

▶ **EXAMPLE 7.16**

Use the initial-value and final-value theorems to find $f(0+)$ and $f(\infty)$ when

$$F(s) = \frac{s^2 + 2s + 4}{s^3 + 3s^2 + 2s} \tag{101}$$

SOLUTION   From (99), the initial value of $f(t)$ is

$$f(0+) = \lim_{s \to \infty} \frac{s(s^2 + 2s + 4)}{s^3 + 3s^2 + 2s}$$

$$= \lim_{s \to \infty} \frac{s^3 + 2s^2 + 4s}{s^3 + 3s^2 + 2s} \tag{102}$$

Because $f(0+)$ is the limit of a ratio of polynomials in s as s approaches infinity, we need to consider only the highest powers in s in both the numerator and denominator. Hence (102) reduces to

$$f(0+) = \lim_{s \to \infty} \frac{s^3}{s^3} = 1$$

Before applying the final-value theorem, we must verify that the conditions necessary for it to be valid are satisfied. In this case, we can rewrite (101) with its denominator in factored form as

$$F(s) = \frac{s^2 + 2s + 4}{s(s+1)(s+2)} \tag{103}$$

which has distinct poles at $s = 0$, -1, and -2. The function $F(s)$ has a single pole at the origin with the remaining poles inside the left half of the s-plane, so we can apply the final-value theorem. Using (100), we find that

$$f(\infty) = \lim_{s \to 0} \frac{s(s^2 + 2s + 4)}{s(s^2 + 3s + 2)}$$

$$= \lim_{s \to 0} \frac{s^2 + 2s + 4}{s^2 + 3s + 2} = 2$$

In this example, it is a simple task to evaluate $f(t)$ for all $t > 0$ by writing $F(s)$ as given by (103) in its partial-fraction expansion. The result is

$$f(t) = 2 - 3\epsilon^{-t} + 2\epsilon^{-2t} \qquad \text{for } t > 0$$

and it is apparent that the values we found for $f(0+)$ and $f(\infty)$ are correct.

▶ **EXAMPLE 7.17**

Explain why the initial-value and final-value theorems are not applicable to the transform

$$F(s) = \frac{s^3 + 2s^2 + 6s + 8}{s^3 + 4s} \tag{104}$$

SOLUTION   Attempting to apply the initial-value theorem, we would write

$$f(0+) = \lim_{s \to \infty} \frac{s(s^3 + 2s^2 + 6s + 8)}{s^3 + 4s}$$

$$= \lim_{s \to \infty} \frac{s^4 + 2s^3 + 6s^2 + 8s}{s^3 + 4s}$$

which is infinite. This is expected, because $m = n = 3$ in (104).

As for the final-value theorem, we see from (104) that we can write the denominator of $F(s)$ as $s(s^2 + 4) = s(s - j2)(s + j2)$. Hence $F(s)$ has a pair of imaginary poles at $s_2 = j2$ and $s_3 = -j2$ which violates the requirements for the final-value theorem. If we carry out the partial-fraction expansion of $F(s)$, we find that

$$F(s) = 1 + \frac{2}{s} + \frac{2}{s^2 + 4}$$

which is the transform of the time function

$$f(t) = \delta(t) + 2 + \sin 2t \quad \text{for } t > 0$$

The initial-value theorem is invalid because of the impulse at $t = 0+$, and the final-value theorem is invalid because of the constant-amplitude sinusoidal term.

▶ SUMMARY

The Laplace transform can be used to convert the integral-differential equations that describe fixed, linear dynamic systems into algebraic equations. The general procedure consists of three steps: transforming the system equations, solving the resulting algebraic equations for the transform of the output, and taking the inverse transform. Any initial conditions that need to be evaluated automatically appear in the first step. These initial conditions are often easier to find than those that are needed for a time-domain solution of the input-output equation.

Although the basic definition can be used to find the Laplace transform of a given function of time, transforms of the most common functions are tabulated in Appendix E. The inverse transform, which should be expressed as a real function of time, can usually be found by writing a partial-fraction expansion and by again using the table in Appendix E. The appendix also summarizes the properties and theorems developed throughout this chapter.

In addition to being an important tool for finding the response of a system to specified inputs, the Laplace transform can be used to develop a number of important general concepts. This further development, which builds on the notions of stability, time constant, step response, and impulse response from this chapter, will be carried out in the next chapter.

▷ **PROBLEMS**

7.1. Using the definition of the Laplace transform in Equation (1), evaluate the transforms of the following functions.

    a. $f_1(t) = t^2$

    b. $f_2(t) = \epsilon^{-at} \cos \omega t$

    c. $f_3(t) = t\epsilon^{-at}$

    d. $f_4(t) = \sin 2t$ for $0 < t < \pi$ and zero elsewhere

7.2.

    a. Derive the expressions given in Equations (12) and (13) for the transforms of $\sin \omega t$ and $\cos \omega t$ by using the identities in Table 1 in Appendix D and then applying Equation (5).

    b. Using Equations (12) and (13) with Table 1 in Appendix D, derive expressions for the transforms of $\sin(\omega t + \phi)$ and $\cos(\omega t + \phi)$.

*7.3. Use the properties tabulated in Appendix E to find the Laplace transform of each of the following functions of time.

    a. $f_1(t) = t\epsilon^{-2t} \cos 3t$

    b. $f_2(t) = t^2 \sin \omega t$

    c. $f_3(t) = \dfrac{d}{dt}(t^2 \epsilon^{-t})$

    d. $f_4(t) = \displaystyle\int_0^t \lambda^2 \epsilon^{-\lambda}\, d\lambda$

7.4.

    a. Prove that $\mathcal{L}[f(t/a)] = aF(as)$.

    b. Apply this property with $f(t) = \cos \omega t$ to find $\mathcal{L}[\cos 2\omega t]$.

7.5.

    a. For the mechanical system shown in Figure P7.5, find the input-output equation relating x_o to the displacement input $x_a(t)$.

    b. What is the time constant associated with this first-order system?

    c. Let $K_1 = 1$ N/m, $K_2 = 2$ N/m, and $B = 1$ N·s/m. Find x_o as a function of time for all $t > 0$ when there is no energy stored in the springs for $t < 0$ and when $x_a(t) = 1 - \epsilon^{-2t}$ for $t > 0$.

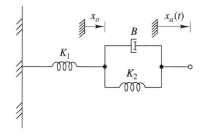

Figure P7.5

7.6. A velocity input $v_a(t)$ is applied to point A in the mechanical system shown in Figure P7.6.

    a.   Write the system's differential equation in terms of the velocity v_1.

    b.   What is the time constant τ for the system?

    c.   Sketch the response when $v_a(t) = 0$ for $t \geq 0$ and $v_1(0) = 10$.

    d.   Repeat parts (a), (b), and (c) when the velocity input is replaced by a force $f_a(t)$ applied at point A, with the positive sense to the right. Explain why the expression for τ differs from the answer to part (b).

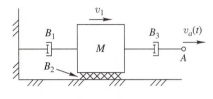

Figure P7.6

*7.7. Assume that the circuit shown in Figure P6.1 contains no stored energy for $t < 0$ and that the input is the unit step function. The input-output differential equation, found in Problem 6.1, is

$$\frac{3}{4}\dot{e}_o + \frac{1}{2}e_o = \frac{1}{2}\frac{di_i}{dt} + i_i(t)$$

    a.   Verify the differential equation and determine the time constant.

    b.   Find and sketch e_o versus t for all $t > 0$ when $i_i(t)$ is the unit step function. Notice that $e_o(t)$ has a discontinuity at $t = 0$.

    c.   Check the steady-state response by replacing the capacitor by an open circuit.

7.8. Repeat Problem 7.7 for the circuit shown in Figure P6.4, for which the input-output differential equation, found in Problem 6.4, is

$$C(R_1 + R_2)\dot{e}_o + e_o = CR_1R_2\frac{di_i}{dt}$$

7.9. Assume that the circuit shown in Figure P6.2 contains no stored energy for $t < 0$ and that the input is the unit step function. The input-output differential equation, found in Problem 6.2, is

$$\dot{e}_o + 9e_o = 3e_i(t)$$

    a.   Verify the differential equation and determine the time constant.

    b.   Find and sketch e_o versus t for all $t > 0$.

    c.   Check the steady-state response by replacing the inductor by a short circuit.

7.10. The switch shown in Figure P7.10 has been closed for a long time, so the circuit is in the steady state at $t = 0-$. If the switch opens at $t = 0$, find and sketch e_o versus t. Show on the same scale the curves for $R_o = 5\,\Omega$ and for $R_o = 10\,\Omega$. What would be the circuit's response if R_o were to approach infinity?

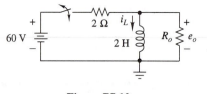

Figure P7.10

*7.11. Consider a first-order system with input $u(t)$ and output y that is described by $\dot{y} + 0.5y = u(t)$.

 a. Find and sketch the unit step response $y_U(t)$.

 b. Find and sketch the response to $u(t) = 2$ for $t > 0$ and with $y(0) = -1$.

 c. Find and sketch the response to $u(t) = U(t) - U(t-2)$ with $y(0) = 0$.

 d. Find and sketch the unit impulse response $h(t)$.

7.12. The voltage source in Figure P7.12 is zero for $t < 0$, but has a constant value of A for all $t > 0$.

 a. Find and sketch the output voltage e_o for all $t > 0$.

 b. Check the steady-state response by replacing the capacitor with an open circuit.

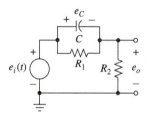

Figure P7.12

7.13. Let the voltage source in Figure 6.18(a) have a constant value of 24 V and let $C = 2$ F.

 a. The switch has been open for all $t < 0$, so there is no initial stored energy in the capacitor. The switch then closes at $t = 0$. Transform the modeling equations for $t > 0$, which are given by Equations (6.26), and find e_o for all $t > 0$.

 b. The switch has been closed for a long time, so that the circuit is in the steady state at $t = 0-$. The switch then opens at $t = 0$. Transform the modeling equations for $t > 0$, which are given by Equations (6.28), and find e_o for all $t > 0$. As part of your solution show that the initial voltage across the capacitor is $e_A(0) - e_o(0) = 9$ V.

 c. Find the time constants for parts (a) and (b) and explain why they are not the same.

7.14. The unit impulse response for the circuit in Figure 7.7 was found in Example 7.8. From Section 6.2, the energy stored in an inductor is $w = \frac{1}{2}Li^2$, and the power dissipated in a resistor is $p = Ri^2$.

a. Find the energy that the unit impulse instantaneously inserts into the inductor.

b. Find the total energy dissipated in the resistor for $t > 0$.

c. Compare the two results.

For Problems 7.15 through 7.17, find $f(t)$ for the given $F(s)$.

7.15.

a. $F(s) = \dfrac{2s^3 + 3s^2 + s + 4}{s^3}$

b. $F(s) = \dfrac{3s^2 + 9s + 24}{(s-1)(s+2)(s+5)}$

c. $F(s) = \dfrac{4}{s^2(s+1)}$

d. $F(s) = \dfrac{3s}{s^2 + 2s + 26}$

7.16.

a. $F(s) = \dfrac{s}{s^2 + 8s + 16}$

b. $F(s) = \dfrac{1}{s(s^2 + \omega^2)}$

c. $F(s) = \dfrac{8s^2 + 20s + 74}{s(s^2 + s + 9.25)}$

d. $F(s) = \dfrac{2s^2 + 11s + 16}{(s+2)^2}$

*7.17.

a. $F(s) = \dfrac{s^3 + 2s + 4}{s(s+1)^2(s+2)}$

b. $F(s) = \dfrac{4s^2 + 10s + 10}{s^3 + 2s^2 + 5s}$

c. $F(s) = \dfrac{3(s^3 + 2s^2 + 4s + 1)}{s(s+3)^2}$

d. $F(s) = \dfrac{s^3 - 4s}{(s+1)(s^2 + 4s + 4)}$

*7.18.

a. Find the inverse transform of

$$F(s) = \dfrac{3s^2 + 2s + 2}{(s+2)(s^2 + 2s + 5)}$$

by writing a partial-fraction expansion and using Equation (84).

b. Repeat part (a) by completing the square and using Equation (86).

7.19. Find the inverse transform of the following function of s by first finding the constants $A, B,$ and C and then completing the square in the denominator of the term having complex poles.

$$F(s) = \frac{4(s^2 + 8s + 72)}{s^3 + 8s^2 + 32s} = \frac{A}{s} + \frac{Bs + C}{s^2 + 8s + 32}$$

*7.20. Use (29) and the results of Example 7.9 to find the inverse transform of

$$F(s) = \frac{-s + 5}{s(s+1)(s+4)}$$

Check your answer by writing a partial-fraction expansion for $F(s)$.

7.21. Use (24) and the results of Example 7.9 to find the inverse transform of

$$F(s) = \frac{s(-s+5)}{(s+1)(s+4)}$$

Check your answer by writing a partial-fraction expansion for $F(s)$.

7.22.

    a.   Use the time-delay theorem in (93) to derive the Laplace transform of the rectangular pulse shown in Figure 7.2.

    b.   Find the Laplace transform of the triangular pulse shown in Figure 7.26(a) by using Equation (1) directly, rather than by decomposing the pulse into ramp and step components. Compare your results to (96).

*7.23. Using the definition of the Laplace transform in Equation (1), evaluate the transforms of the following functions.

    a.   The function $f_1(t)$ shown in Figure P7.23(a).

    b.   The function $f_2(t)$ shown in Figure P7.23(b).

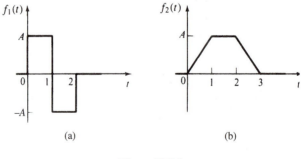

            (a)   (b)

Figure P7.23

7.24. Repeat Problem 7.23 by decomposing the functions into step functions and ramp functions, as appropriate.

7.25. Sketch the time functions corresponding to each of the following Laplace transforms.

a. $F(s) = \dfrac{1}{s^2 + 1}\left(1 + \epsilon^{-\pi s}\right)$

b. $F(s) = \dfrac{1}{s^2}\left(1 - 2\epsilon^{-s} + \epsilon^{-2s}\right)$

*7.26. Apply the initial-value and final-value theorems to find $f(0+)$ and $f(\infty)$ for each of the four transforms in Problem 7.17. If either theorem is not applicable to a particular transform, explain why this is so.

7.27. Repeat Problem 7.26 for the following transforms.

a. $F(s)$ in Problem 7.19.

b. The functions in part (c) and part (d) of Problem 7.16.

c. $F(s) = \dfrac{4s^3}{(5s^2 + 3)^2}$

7.28.

a. Using Equation (24) and the initial-value theorem, show that
$$\dot{f}(0+) = \lim_{s \to \infty}\left[s^2 F(s) - sf(0)\right]$$
provided that the limit exists.

b. Use the property derived in part (a) to find $\dot{f}(0+)$ for the transform $F(s)$ given in Equation (101). Assume that $f(t) = 1$ for $t \le 0$. If this property is not applicable, explain why this is so.

c. Repeat part (b), assuming that $f(t) = 0$ for $t \le 0$.

d. Check your answers to part (b) and part (c) by differentiating the expression given for $f(t)$ in Example 7.16.

8

TRANSFER FUNCTION ANALYSIS

We presented in Section 7.4 the general procedure for finding the response of any fixed linear system by use of the Laplace transform. We transformed the modeling equations for $t > 0$, thereby converting them into algebraic equations. We solved these transformed equations for the transform of the output, and then took the inverse transform in order to obtain the output as a function of time. However, we applied the procedure only to systems with one energy-storing element.

In this chapter, we first consider the complete response of more complex systems. We then examine in some detail two important special cases: the zero-input response, where the system's excitation consists only of some initial stored energy, and then the response to a given input when the initial stored energy is zero. From the second case, we develop and illustrate the important concept of the transfer function.

We give special attention to the responses to the unit impulse, the unit step function, and sinusoidal functions. We also show that finding the transfer function for electrical systems can be simplified by introducing the concept of impedances.

▶ 8.1 THE COMPLETE SOLUTION

The basic procedure for higher-order systems is the same as in Section 7.4, even though the algebra may be more involved. If the system is described by several modeling equations, we transform them individually to obtain a set of algebraic equations. We must then solve these equations simultaneously to get an expression for the transformed output, with all other unknowns eliminated. The resulting expression may be fairly complicated, but we can normally use the partial-fraction techniques in Section 7.6 to find the output as a real function of time.

The first examples in this section consider a mechanical system that has three energy-storing elements, but which is at rest for $t < 0$. We then examine a second-order electrical system that contains some initial stored energy as a result of an input that was present for $t < 0$.

▶ EXAMPLE 8.1

The translational system shown in Figure 8.1(a) was modeled in Example 3.7 and in Example 3.11. Let $M_1 = 1$ kg, $B_1 = B_2 = 1$ N·s/m, and $K_1 = K_2 = 1$ N/m. The springs are undeflected when $x_1 = x_2 = 0$. Take the output to be the displacement x_1. Find the unit step response by transforming the equations obtained from the free-body diagrams.

SOLUTION   The free-body diagrams for the mass M and for the massless junction A are shown in parts (b) and (c) of the figure, respectively. Summing the forces on these diagrams gives

$$M\ddot{x}_1 + B_1\dot{x}_1 + K_1(x_1 - x_2) = f_a(t)$$
$$B_2\dot{x}_2 + K_2x_2 + K_1(x_2 - x_1) = 0$$

Transforming these equations, with all the element values equal to unity and with $\mathcal{L}[f_a(t)] = F_a(s)$, we have

$$[s^2X_1(s) - sx_1(0) - \dot{x}_1(0)] + [sX_1(s) - x_1(0)] + X_1(s) - X_2(s) = F_a(s) \tag{1}$$
$$[sX_2(s) - x_2(0)] + 2X_2(s) - X_1(s) = 0$$

Because we assume that the initial stored energy is zero when finding the unit step response, the initial elongations of the springs and the initial velocity of the mass must be

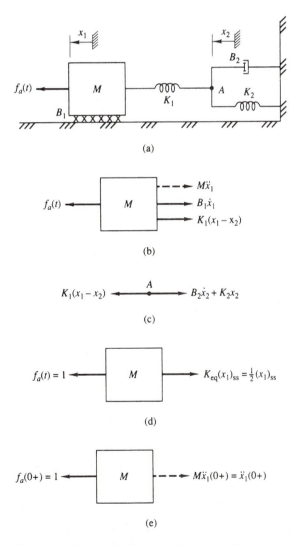

(a)

(b)

(c)

(d)

(e)

Figure 8.1   (a) Translational system for Example 8.1. (b),(c) General free-body diagrams. (d) Free-body diagram for the steady state. (e) Free-body diagram for $t = 0+$.

zero. Thus $x_1(0) = x_2(0) = 0$ and $\dot{x}_1(0) = 0$, so (1) becomes

$$(s^2 + s + 1)X_1(s) - X_2(s) = F_a(s)$$
$$-X_1(s) + (s + 2)X_2(s) = 0$$

Eliminating $X_2(s)$ from these equations, we obtain

$$X_1(s) = \left[\frac{s+2}{s^3 + 3s^2 + 3s + 1} \right] F_a(s) \tag{2}$$

Factoring the denominator, noting that $F_a(s) = \mathcal{L}[U(t)] = 1/s$, and then writing a partial-fraction expansion, we have

$$X_1(s) = \frac{s+2}{s(s+1)^3} = \frac{A_0}{s} + \frac{A_{11}}{(s+1)^3} + \frac{A_{12}}{(s+1)^2} + \frac{A_{13}}{s+1} \tag{3}$$

where

$$A_0 = \frac{s+2}{(s+1)^3}\bigg|_{s=0} = 2$$

$$A_{11} = \frac{s+2}{s}\bigg|_{s=-1} = -1$$

$$A_{12} = \left[\frac{d}{ds}\left(\frac{s+2}{s} \right) \right]_{s=-1} = -2$$

$$A_{13} = \frac{1}{2}\left[\frac{d^2}{ds^2}\left(\frac{s+2}{s} \right) \right]_{s=-1} = -2$$

To find the output as a function of time, we insert these numbers into (3) and take the inverse transform of each term. This gives us

$$x_1(t) = 2 - (\tfrac{1}{2}t^2 + 2t + 2)\epsilon^{-t} \quad \text{for } t > 0 \tag{4}$$

This equation reveals that the steady-state response is $(x_1)_{ss} = 2$ m, which we can check by referring to Figure 8.1(a). When $f_a(t)$ has a constant value of 1, the mass will eventually become motionless and there will be no inertial or friction forces. Because the friction elements can be disregarded, the two springs are then in series and can be replaced by a single equivalent spring. Using Equation (2.37), we see that $K_{eq} = \tfrac{1}{2}$ N/m. The only forces acting on the mass are those shown in Figure 8.1(d), so we can again conclude that $(x_1)_{ss} = 2$ m.

By differentiating (4), we can show that the velocity and acceleration of the mass are

$$\begin{aligned} \dot{x}_1 &= (\tfrac{1}{2}t^2 + t)\epsilon^{-t} \quad \text{for } t > 0 \\ \ddot{x}_1 &= (-\tfrac{1}{2}t^2 + 1)\epsilon^{-t} \quad \text{for } t > 0 \end{aligned} \tag{5}$$

Replacing t by zero in (4) and (5), we see that

$$x_1(0+) = 0 \tag{6a}$$
$$\dot{x}_1(0+) = 0 \tag{6b}$$
$$\ddot{x}_1(0+) = 1 \tag{6c}$$

The fact that x_1 and $\dot{x}_1$ remain zero at $t = 0+$ is expected; the elongations of the springs and the velocity of the mass cannot change instantaneously. Because of this, we also know that at $t = 0+$ there is no force on the mass from K_1 or B_1. Thus the only forces on the mass at $t = 0+$ are the applied and inertial forces, as shown in Figure 8.1(e). From that figure, we see that the initial acceleration must be $\ddot{x}_1(0+) = 1$ m/s^2, which serves as a check on (6c).

▷ **EXAMPLE 8.2**

Repeat Example 8.1 by transforming the state-variable equations.

SOLUTION The state-variable model for Figure 8.1(a) was found in Example 3.7. From Equations (3.14), with all the element values set equal to unity, we have

$$\dot{x}_1 = v_1$$
$$\dot{v}_1 = -x_1 - v_1 + x_2 + f_a(t)$$
$$\dot{x}_2 = x_1 - 2x_2$$

Transforming these equations gives

$$sX_1(s) - x_1(0) = V_1(s) \tag{7a}$$
$$sV_1(s) - v_1(0) = -X_1(s) - V_1(s) + X_2(s) + F_a(s) \tag{7b}$$
$$sX_2(s) - x_2(0) = X_1(s) - 2X_2(s) \tag{7c}$$

where we can again set the initial-condition terms equal to zero. Then, by substituting (7a) into (7b) and rearranging the last two equations, we obtain

$$(s^2 + s + 1)X_1(s) - X_2(s) = F_a(s)$$
$$-X_1(s) + (s+2)X_2(s) = 0$$

from which

$$X_1(s) = \left[\frac{s+2}{s^3 + 3s^2 + 3s + 1} \right] F_a(s)$$

which agrees with (2).

▷ **EXAMPLE 8.3**

Repeat Example 8.1 by transforming the input-output differential equation.

SOLUTION The input-output equation for Figure 8.1(a) was derived in Example 3.11. From Equations (3.27) with the element values set equal to unity, we have

$$\dddot{x}_1 + 3\ddot{x}_1 + 3\dot{x}_1 + x_1 = \dot{f}_a + 2f_a(t)$$

We transform this equation, getting

$$[s^3 X_1(s) - s^2 x_1(0) - s\dot{x}_1(0) - \ddot{x}_1(0)] + 3[s^2 X_1(s) - sx_1(0) - \dot{x}_1(0)]$$
$$+ 3[sX_1(s) - x_1(0)] + X_1(s) = sF_a(s) - f_a(0) + 2F_a(s) \tag{8}$$

We note that (8) involves four different initial conditions: $x_1(0), \dot{x}_1(0), \ddot{x}_1(0)$, and $f_a(0)$. Because there is zero initial stored energy and because the step-function input is assumed to occur just after $t = 0$ (that is, at $t = 0+$), each of these initial conditions is zero. Then

(8) reduces to

$$(s^3 + 3s^2 + 3s + 1)X_1(s) = (s+2)F_a(s)$$

which is equivalent to (2).

Although somewhat different, the methods used in the last three examples gave the same equations for the transformed output $X_1(s)$ and for $x_1(t)$ for all $t > 0$. When the equations from the free-body diagrams were transformed immediately or when the state-variable equations were transformed, the only initial conditions needed were for functions that did not change instantaneously.

This was not the case, however, when we obtained the input-output equation before applying the Laplace transform. In this method, it is not unusual for one or more of the initial conditions to involve functions that have discontinuities. For the acceleration $\ddot{x}_1$ in the last example, $\ddot{x}_1(0-) = 0$, but it was found that $\ddot{x}_1(0+) = 1$ m/s². Whenever there is a discontinuity at the time origin, we use the value at $t = 0-$ for the initial-condition term, for the reasons explained in Section 7.5. This is in contrast to the classical solution of an input-output differential equation which requires the values of the initial conditions at $t = 0+$. One advantage of the transform method is that evaluating the initial-condition terms is often easier.

In the last three examples, we chose to denote the transform of the unit step input $f_a(t)$ by $F_a(s)$ rather than to immediately replace it by $\mathcal{L}[U(t)] = 1/s$. Thus the expression for $X_1(s)$ in (2) was a somewhat more general result than necessary. Equation (2) holds for any input, provided only that the input has a Laplace transform and that the initial stored energy is zero. The quantity inside the brackets is called the transfer function and will become increasingly important. When finding the response to a particular input, however, we more often replace its transform immediately by the appropriate expression from Appendix E.

▶ EXAMPLE 8.4

After steady-state conditions have been reached, the switch in Figure 8.2(a) opens at $t = 0$. Find the voltage e_o across the capacitor for all $t > 0$.

SOLUTION   The circuit for $t > 0$ is shown in Figure 8.2(b), with the switch open and with the node voltages e_o and e_A shown. The current-law equations at these two nodes are

$$e_o - 12 + \frac{1}{4}\dot{e}_o + i_L(0) + 4\int_0^t (e_o - e_A)\, d\lambda = 0$$

$$\frac{1}{2}(e_A - 12) - i_L(0) + 4\int_0^t (e_A - e_o)d\lambda + \frac{1}{2}e_A = 0$$

Transforming these equations, we have

$$E_o(s) - \frac{12}{s} + \frac{1}{4}[sE_o(s) - e_o(0)] + \frac{i_L(0)}{s} + \frac{4}{s}[E_o(s) - E_A(s)] = 0$$

$$\frac{1}{2}\left[E_A(s) - \frac{12}{s}\right] - \frac{i_L(0)}{s} + \frac{4}{s}[E_A(s) - E_o(s)] + \frac{1}{2}E_A(s) = 0$$

We must find the numerical values of the initial conditions $e_o(0)$ and $i_L(0)$ that appear in these transformed equations. Because e_o and i_L are measures of the energy stored in the capacitor and inductor, respectively, they cannot change instantaneously and do not have discontinuities at the time origin.

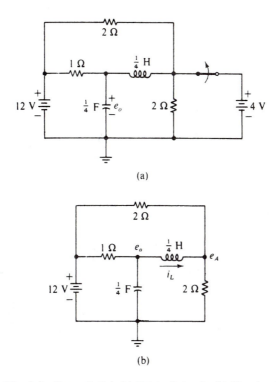

Figure 8.2 Circuit for Example 8.4. (a) Original circuit. (b) Circuit valid for $t > 0$.

When the circuit is in the steady state with the switch closed, the capacitor and inductor may be replaced by open and short circuits, respectively, as explained at the end of Section 6.5. This is done in Figure 8.3(a), from which we find that $e_o(0) = 4$ V and $i_L(0) = 8$ A. Substituting these initial conditions into the transformed equations and collecting like terms, we get

$$\left[\frac{1}{4}s + 1 + \frac{4}{s}\right] E_o(s) - \frac{4}{s} E_A(s) = 1 + \frac{4}{s} \tag{9a}$$

$$-\frac{4}{s} E_o(s) + \left[1 + \frac{4}{s}\right] E_A(s) = \frac{14}{s} \tag{9b}$$

We want to find the capacitor voltage $e_o(t)$, so the next step is to solve (9) for $E_o(s)$ by eliminating $E_A(s)$. Noting from (9b) that

$$E_A(s) = \frac{4E_o(s) + 14}{s + 4}$$

and substituting this expression into (9a), we find that

$$E_o(s) = \frac{4(s^2 + 8s + 72)}{s^3 + 8s^2 + 32s}$$

One pole of $E_o(s)$ is $s = 0$, and the remaining two are the roots of $s^2 + 8s + 32 = 0$. Hence the poles are $s_1 = 0$, $s_2 = -4 + j4$, and $s_3 = -4 - j4$, and we can expand the transformed output into the form

$$E_o(s) = \frac{A_1}{s} + \frac{A_2}{s + 4 - j4} + \frac{A_3}{s + 4 + j4} \tag{10}$$

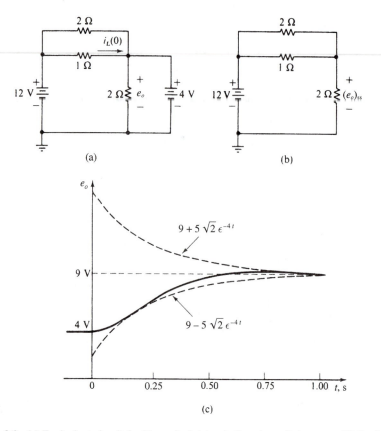

Figure 8.3 (a) Equivalent circuit for Example 8.4 just before the switch opens. (b) Equivalent circuit valid as t approaches infinity. (c) Complete response.

where the coefficients are

$$A_1 = sE_o(s)|_{s=0} = \frac{(4)(72)}{32} = 9$$

$$A_2 = (s+4-j4)E_o(s)|_{s=-4+j4}$$

$$= \frac{4[(-4+j4)^2 + 8(-4+j4) + 72]}{(-4+j4)(-4+j4+4+j4)}$$

$$= \frac{4(40)}{(-4+j4)(j8)} = \frac{5}{-(1+j1)} = \frac{5}{\sqrt{2}}\epsilon^{j3\pi/4}$$

and

$$A_3 = A_2^* = \frac{5}{\sqrt{2}}\epsilon^{-j3\pi/4}$$

Comparing the second and third terms on the right side of (10) to Equation (7.83), we see that $a = 4$, $\omega = 4$, $K = 5/\sqrt{2}$, and $\phi = \frac{3}{4}\pi$ rad. Using Equation (7.84) to write the part of the response that corresponds to the pair of complex poles, we have

$$e_o(t) = 9 + 5\sqrt{2}\epsilon^{-4t}\cos\left(4t + \frac{3}{4}\pi\right) \qquad \text{for } t > 0$$

As a check on the work, note that as t approaches infinity, this expression gives a constant value of 9. We can also find the steady-state behavior with the switch open from

Figure 8.3(b), where the capacitor and inductor have again been replaced by open and short circuits, respectively. The parallel combination of the 2-Ω and 1-Ω resistors is equivalent to $(2)(1)/(2+1) = \frac{2}{3}\Omega$. By the voltage-divider rule,

$$e_o(\infty) = \frac{2}{2+\frac{2}{3}}(12) = 9 \text{ V}$$

which agrees with the general equation for $t > 0$.

The complete response is shown in Figure 8.3(c). The transient component has an envelope that decays with a time constant of 0.25 s and has a period of $\frac{1}{2}\pi$ s.

Parts of the Solution

We discussed in Section 7.4 how the response of first-order systems could be divided into the sum of zero-input and zero-state components, and also into the sum of transient and steady-state components. To illustrate the extension of these concepts to higher-order systems and to examine further the role of any initial stored energy, we consider in the next example a simple second-order mechanical system. Part of the solution will be carried out in literal form. In order to emphasize the concepts, the algebraic details of finding the coefficients in the partial-fraction expansions will be left for the reader. A problem at the end of the chapter presents the electrical analog of the mechanical system.

▷ **EXAMPLE 8.5**

In the system shown in Figure 8.4(a), the input is the applied force $f_a(t)$, and the output is v, the velocity of the mass. The effects of the input for $t < 0$ are summarized by the initial conditions $x(0)$ and $v(0)$, which determine the initial energy stored in the spring and in the mass, and which we assume are known. First find a general expression for the transform of the output in terms of the transformed input and the initial conditions. Then assume that $M = 1$ kg, $B = 3$ N·s/m, and $K = 2$ N/m and identify the transforms of the zero-state and zero-input components of the output.

For $t > 0$, let the input be the sum of constant and sinusoidal components, specifically let $f_a(t) = 1 + \sin t$. Find $v(t)$ for all $t > 0$ and identify the steady-state and transient components.

SOLUTION   The free-body diagram is in Figure 8.4(b). However, because the output is v rather than x, we shall write the displacement in terms of the definite integral of the velocity: $x(t) = x(0) + \int_0^t v \, d\lambda$. Then

$$M\dot{v} + Bv + K \left[x(0) + \int_0^t v \, d\lambda \right] = f_a(t)$$

Denoting the transform of the input by $F_a(s)$ and noting that the transform of the constant $x(0)$ is $x(0)/s$, we write

$$M[sV(s) - v(0)] + BV(s) + \frac{K}{s}x(0) + \frac{K}{s}V(s) = F_a(s)$$

We now multiply both sides of this equation by s and collect the terms involving $V(s)$ on the left side. Solving for $V(s)$ yields

$$V(s) = \frac{sF(s)}{Ms^2 + Bs + K} + \frac{sMv(0) - Kx(0)}{Ms^2 + Bs + K}$$

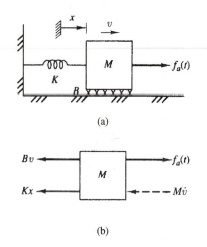

(a)

(b)

Figure 8.4 A mass-spring-friction system for Example 8.5. (a) The system. (b) Free-body diagram.

Inserting numerical values gives

$$V(s) = \left[\frac{s}{s^2 + 3s + 2}\right] F_a(s) + \frac{sv(0) - 2x(0)}{s^2 + 3s + 2} \tag{11}$$

The zero-state response corresponds to the case where the initial values of the state variables are zero, so the transform of the zero-state response is the first of the two terms in (11). With $s^2 + 3s + 2 = (s+1)(s+2)$,

$$V_{zs}(s) = \left[\frac{s}{(s+1)(s+2)}\right] F_a(s) \tag{12}$$

Note that

$$F_a(s) = \mathcal{L}[1 + \sin t] = \frac{1}{s} + \frac{1}{s^2 + 1} = \frac{s^2 + s + 1}{s(s^2 + 1)}$$

so

$$V_{zs}(s) = \left[\frac{s}{(s+1)(s+2)}\right]\left[\frac{s^2 + s + 1}{s(s^2 + 1)}\right] \tag{13}$$

Because $s^2 + 1$ has the complex factors $(s - j1)(s + j1)$, we write the partial-fraction expansion in the following form:

$$V_{zs}(s) = \frac{A_1}{s+1} + \frac{A_2}{s+2} + \frac{A_3}{s} + \frac{Bs + C}{s^2 + 1}$$

Using the techniques in Section 7.6, we find in a straightforward way that $A_1 = 1/2, A_2 = -3/5, A_3 = 0, B = 1/10$, and $C = 3/10$. Then for $t > 0$

$$v_{zs}(t) = \frac{1}{2}\epsilon^{-t} - \frac{3}{5}\epsilon^{-2t} + \frac{1}{10}\cos t + \frac{3}{10}\sin t$$

Two sinusoidal terms having the same angular frequency can be combined into a single term with a phase angle. Using Table 1 in Appendix D, we can write

$$v_{zs}(t) = \frac{1}{2}\epsilon^{-t} - \frac{3}{5}\epsilon^{-2t} + 0.3162 \sin(t + 0.3218) \tag{14}$$

The transform of the zero-input response is the second of the two terms in (11). Thus

$$V_{zi}(s) = \frac{sv(0) - 2x(0)}{(s+1)(s+2)} = \frac{A_4}{s+1} + \frac{A_5}{s+2}$$

It is straightforward to show that $A_4 = -v(0) - 2x(0)$ and that $A_5 = 2v(0) + 2x(0)$, so

$$v_{zi}(t) = -[v(0) + 2x(0)]\epsilon^{-t} + 2[v(0) + x(0)]\epsilon^{-2t} \tag{15}$$

The complete response of the system is merely the sum of the expressions in (14) and (15). The steady-state response, which is the part that remains as t approaches infinity, is

$$v_{ss}(t) = 0.3162 \sin(t + 0.3218) \tag{16}$$

The transient response, which dies out as t approaches infinity, is

$$v_{tr}(t) = \frac{1}{2}\epsilon^{-t} - \frac{3}{5}\epsilon^{-2t} - [v(0) + 2x(0)]\epsilon^{-t} + 2[v(0) + x(0)]\epsilon^{-2t}$$

$$= \left[\frac{1}{2} - v(0) - 2x(0)\right]\epsilon^{-t} + \left[-\frac{3}{5} + 2v(0) + 2x(0)\right]\epsilon^{-2t} \tag{17}$$

We conclude this section by examining further the results of the last example. For the case when there is no initial stored energy, we see from (12) that the transform of the zero-state response is the product of the function of s within the brackets and the transformed input, no matter what the input function might be. The quantity inside the brackets, which is determined by the system configuration and the element values, is called the **transfer function** and is often represented by the symbol $H(s)$. In this example,

$$H(s) = \frac{s}{s^2 + 3s + 2} = \frac{s}{(s+1)(s+2)}$$

which has poles, as defined in Section 7.6, at $s = -1$ and $s = -2$. The transformed input is

$$F_a(s) = \frac{s^2 + s + 1}{s(s^2 + 1)}$$

which has poles at $s = 0$, $j1$, and $-j1$. The poles of $V_{zs}(s)$, some of which come from the denominator of $H(s)$ and some of which come from the denominator of the transformed input, determined the form of the partial-fraction expansion. Reviewing the process of taking the inverse transform and finding $v_{zs}(t)$, we see that the poles of $H(s)$ gave rise to the transient terms, while the poles of the transformed input gave rise to the steady-state terms.

The input in the last example consisted of a constant and a sinusoidal component. We would normally expect that the steady-state output would contain terms of the same kind. We see from (16) that $v_{ss}(t)$ does contain a sinusoidal term having the same angular frequency as the one in $f_a(t)$. There is not, however, a constant term in the steady-state output.

In the partial-fraction expansion for $V_{zs}(s)$, note that the value of A_3 in the term A_3/s was zero. This could have been anticipated, because a numerator factor in $H(s)$ canceled the s in the denominator of $F_a(s)$. In effect, the system refused to respond in the steady state to the constant term in the input. This can also be seen from Figure 8.4(a). When the applied force is a constant, the spring will eventually hold the mass in a fixed position so that the velocity becomes zero.

To appreciate the role of the initial stored energy, consider the zero-input response in (15). This consists only of transient terms that have the same form as the transient terms in $v_{zs}(t)$. Any initial stored energy does not contribute to the steady-state response at all, nor does it affect the form of the transient terms. Its only effect is to change the size of the transient terms in the complete response.

▶ 8.2 THE ZERO-INPUT RESPONSE

The zero-input response is defined to be the output $y(t)$ when the input $u(t)$ is zero for all $t > 0$ and when the initial conditions are nonzero. Because it is understood that the input is zero, we shall generally omit the subscript zi throughout this entire section. The general form of the input-output differential equation for a fixed linear nth-order system was given in Equation (3.25). When the input terms are zero, this becomes

$$a_n y^{(n)} + a_{n-1} y^{(n-1)} + \cdots + a_1 \dot{y} + a_0 y = 0 \tag{18}$$

where $y^{(n)}$ denotes $d^n y / dt^n$. Using the expressions for the transforms of derivatives given in Appendix E, we can transform (18) term by term to obtain

$$a_n[s^n Y(s) - s^{n-1} y(0) - \cdots - y^{(n-1)}(0)]$$
$$+ a_{n-1}[s^{n-1} Y(s) - s^{n-2} y(0) - \cdots - y^{(n-2)}(0)]$$
$$+ \cdots + a_1[sY(s) - y(0)] + a_0 Y(s) = 0 \tag{19}$$

where $Y(s) = \mathcal{L}[y(t)]$. If we retain the terms involving $Y(s)$ on the left-hand side and collect those involving the initial conditions on the right-hand side, (19) becomes

$$(a_n s^n + a_{n-1} s^{n-1} + \cdots + a_1 s + a_0)Y(s)$$
$$= a_n y(0) s^{n-1} + [a_n \dot{y}(0) + a_{n-1} y(0)] s^{n-2} + \ldots$$
$$+ [a_n y^{(n-1)}(0) + a_{n-1} y^{(n-2)}(0) + \cdots + a_1 y(0)]$$

Thus the transform of the zero-input response is

$$Y(s) = \frac{F(s)}{P(s)} \tag{20}$$

where

$$F(s) = a_n y(0) s^{n-1} + [a_n \dot{y}(0) + a_{n-1} y(0)] s^{n-2} + \ldots$$
$$+ [a_n y^{(n-1)}(0) + a_{n-1} y^{(n-2)}(0) + \cdots + a_1 y(0)] \tag{21}$$

and

$$P(s) = a_n s^n + a_{n-1} s^{n-1} + \cdots + a_1 s + a_0 \tag{22}$$

Because $P(s)$ is of degree n and $F(s)$ is at most of degree $n-1$, $Y(s)$ is a strictly proper rational function. The numerator polynomial $F(s)$ depends on the initial conditions. We shall show in the next section that the denominator polynomial $P(s)$ is identical to the denominator of the transfer function. It is a characteristic of the system and is sometimes called the characteristic polynomial. To factor $P(s)$, we can find the roots of the characteristic equation $P(s) = 0$. There are explicit formulas for doing this for first- and second-order polynomials, but we may have to use a computer or hand calculator for higher-order cases. When $P(s)$ is factored, it will have the form

$$P(s) = a_n(s - s_1)(s - s_2) \ldots (s - s_n)$$

The quantities $s_1, s_2, \ldots, s_n$ are the poles of the transformed output (as well as the poles of the transfer function); hence they determine the form of the zero-input response. For a first-order system, where $n = 1$,

$$P(s) = a_1 s + a_0$$
$$= a_1(s + 1/\tau)$$

where τ is the time constant. The zero-input response will have one of the forms shown in Figure 7.4.

Second-Order Systems

For second-order systems, where $n = 2$, $P(s) = a_2 s^2 + a_1 s + a_0$. The transform of the zero-input response will have two poles s_1 and s_2, which are the roots of the characteristic equation

$$a_2 s^2 + a_1 s + a_0 = 0 \tag{23}$$

These roots may be real and distinct, repeated, or complex. Finding the function of time for each of these cases was discussed and illustrated in Section 7.6. If the roots are real and distinct, then the zero-input response has the form

$$y(t) = K_1 \epsilon^{s_1 t} + K_2 \epsilon^{s_2 t} \tag{24}$$

For $s_1 = s_2$, the two roots are not distinct, and we must replace (24) by

$$y(t) = K_1 \epsilon^{s_1 t} + K_2 t \epsilon^{s_1 t} \tag{25}$$

If the roots are complex, they must have the form $s_1 = \alpha + j\beta$ and $s_2 = \alpha - j\beta$. As in Appendix E, we can write

$$y(t) = \epsilon^{\alpha t}[K_1 \cos \beta t + K_2 \sin \beta t]$$

or, equivalently,

$$y(t) = K \epsilon^{\alpha t} \cos(\beta t + \phi) \tag{26}$$

Examples of $y(t)$ are shown in Figure 8.5 for the three cases represented by (24) through (26) when s_1, s_2, and α are negative numbers. The light lines in Figure 8.5(a) and Figure 8.5(b) indicate the two individual functions that are added to give $y(t)$, which is the heavy curve. In Figure 8.5(c), dashed lines labeled $K\epsilon^{\alpha t}$ and $-K\epsilon^{\alpha t}$ form the envelope of the damped oscillations. The values of the arbitrary constants K_1, K_2, K, and ϕ in (24) through (26) depend on the initial conditions.

The Complex Plane

When the roots of the characteristic equation are plotted in a complex plane, inspection of the plot reveals the nature of the system's zero-input response. The root locations corresponding to the responses shown in Figure 8.5 are indicated by the crosses in the respective parts of Figure 8.6.

Recall from Section 7.4 that a stable system is one for which the zero-input response decays to zero as t approaches infinity. All the examples in Figure 8.5 fall into this category. Examples of unstable systems, where $y(t)$ increases without bound, are shown in Figure 8.7. The characteristic-root locations appear directly under the corresponding sketches of the response. Figure 8.7(a) and Figure 8.7(b) correspond to a positive value of s_1 in (24), Figure 8.7(c) and Figure 8.7(d) correspond to a positive value of α in (26), and Figure 8.7(e) and Figure 8.7(f) correspond to $s_1 = 0$ in (25).

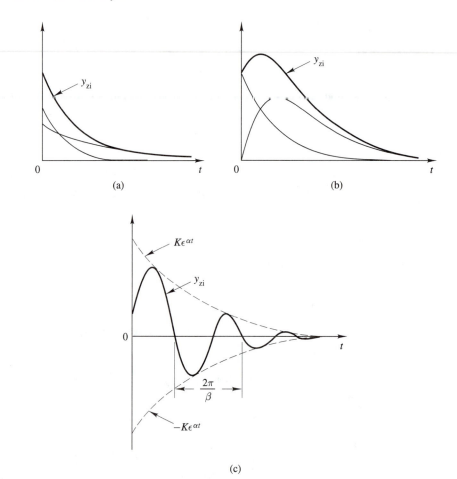

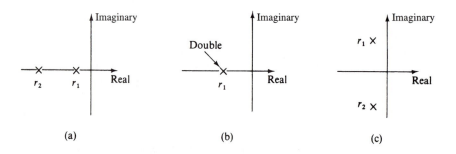

Figure 8.5 Typical curves for the zero-input response of a second-order system. (a) Real distinct negative roots of the characteristic equation. (b) Identical negative roots. (c) Complex roots with $\alpha < 0$.

Figure 8.6 Roots of the characteristic equation corresponding to Figure 8.5. (a) Real distinct negative roots. (b) Identical negative roots. (c) Complex roots with $\alpha < 0$.

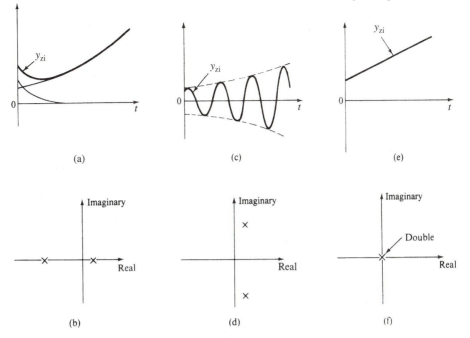

Figure 8.7 Examples of unstable second-order systems. (a), (b) One real root in the right half-plane. (c), (d) Complex roots in the right half-plane. (e), (f) Double root at the origin of the complex plane.

In addition to these stable and unstable classes of systems, it is possible (at least in an idealized case) for a linear system to be marginally stable and have a zero-input response that neither decays to zero nor grows without bound. For second-order systems, such a response occurs when the characteristic equation has either a single root at $s_1 = 0$ with the remaining root in the left half-plane or a pair of imaginary roots at $s_1 = j\beta$ and $s_2 = -j\beta$. The first case corresponds to the characteristic equation $s^2 + a_1 s = 0$ and to the zero-input response

$$y(t) = K_1 + K_2 \epsilon^{-a_1 t}$$

where $a_1 > 0$. The second case corresponds to the characteristic equation $s^2 + \beta^2 = 0$ and to the response

$$y(t) = K \cos(\beta t + \phi)$$

A system whose zero-input response is a constant-amplitude sine or cosine function is often called a **simple harmonic oscillator**. Sample root locations and typical zero-input responses for the two types of marginally stable second-order systems are shown in Figure 8.8.

We can summarize the discussion about stability in a way that is applicable to fixed linear systems of any order as follows: If all the roots of the characteristic equation are inside the left half-plane, the system is stable. If there are any roots inside the right half-plane or repeated roots on the imaginary axis, the system is **unstable**. If all the roots are inside the left half-plane except for one or more distinct roots on the imaginary axis, the system is **marginally stable**.

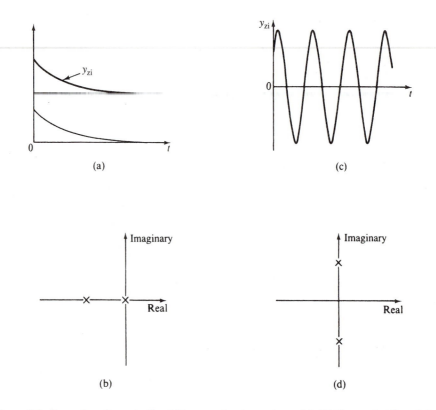

Figure 8.8 Examples of marginally stable second-order systems. (a), (b) One root at the origin of the complex plane. (c) (d) A pair of complex roots on the imaginary axis.

A Property of Stable Systems

The following property of polynominals can be useful when we are considering stability. If the signs of the coefficients in the characteristic equation are not all the same, then there is at least one root inside the right half-plane and the system is unstable. However, if all the signs are the same, we can be sure that there are no right half-plane roots only in the case of first- and second-order systems; for higher-order systems, all the signs being the same is not sufficient to guarantee the absence of right half-plane roots.

In Chapters 2, 5, and 6 we noted a property that can be used as a check on the equations obtained by examining the free-body diagrams for a mechanical system or the nodes of an electrical circuit. Suppose we sum the forces at a mass M_i and collect like terms. Then all the terms involving the displacement of M_i and its derivatives must have the same sign. We can make a similar statement when summing torques for a rotational system. For an electrical circuit, suppose we sum the currents that leave a node whose voltage is e_A and then collect like terms. All the terms involving e_A and its derivatives must have the same sign.

Although we will not prove this property, we do now provide an explanation for it. All the systems that we have considered, except for those in Section 6.7, were excited by external independent inputs and consisted only of passive elements (such as $M, J, B, K, C, R,$ and L). Systems composed of passive elements cannot be unstable, because there are no permanent sources of energy inside them. Thus all the coefficients in the characteristic equation, which are identical to those on the left-hand side of the input-output differential equation, must have the same sign. Let us first write and combine the system equations in literal form, with the element values represented by letters rather than numbers. Then the coefficients on the left-hand side of the input-output equation must have the same sign for all combinations of positive numbers that might be substituted for the element values. If, however, the property described in the previous paragraph were not satisfied, then we would find that we can make the signs of the coefficients different merely by choosing some sufficiently large element values.

▶ **EXAMPLE 8.6**

Consider the following pair of coupled first-order differential equations:

$$a_1 \dot{v}_1 + a_0 v_1 - b v_2 = f_1(t) \tag{27a}$$
$$-c v_1 + d_1 \dot{v}_2 + d_0 v_2 = f_2(t) \tag{27b}$$

which describe several simple systems. Two such systems will be presented immediately after this example. Assume, without loss of generality, that a_1 and d_1 are positive. If this were not the case for either of the original equations, we could make it so by multiplying that equation by -1. The magnitudes of a_0 and d_0 can be made arbitrarily large by an appropriate choice of element values.

Find the input-output equation relating v_2 to the inputs $f_1(t)$ and $f_2(t)$. Show that a_0 and d_0 must be positive in order for the system described by (27) to be stable.

SOLUTION From (27b),

$$v_1 = \frac{1}{c}[d_1 \dot{v}_2 + d_0 v_2 - f_2(t)]$$

Substituting this expression into (27a) and collecting terms, we obtain

$$a_1 d_1 \ddot{v}_2 + (a_1 d_0 + a_0 d_1)\dot{v}_2 + (a_0 d_0 - bc)v_2 = c f_1(t) + a_1 \dot{f}_2 + a_0 f_2(t)$$

For a stable system, the coefficient of $\dot{v}_2$ must be positive. Because this must be true regardless of the magnitudes of a_0 and d_0, we require that a_0 and d_0 be positive. We also require that $a_0 d_0 > bc$, which will be satisfied for passive systems, as illustrated by the two cases that are considered next.

For the mechanical system shown in Figure 8.9(a), it is easy to show that (27a) and (27b) are the D'Alembert-law equations written for M_1 and M_2, respectively, where

$$a_1 = M_1, \quad a_0 = B_1 + B_3, \quad b = c = B_3$$
$$d_1 = M_2, \quad d_0 = B_2 + B_3$$

Thus we should be able to anticipate that a_1 and a_0 must have the same sign, as must d_1 and d_0. Similarly, for the electrical circuit in Figure 8.9(b), we can show that (27a) and (27b) are the current-law equations at nodes 1 and 2, with the following change of symbols:

$$v_1 = e_1, \quad v_2 = e_2, \quad f_1(t) = i_1(t), \quad f_2(t) = i_2(t)$$
$$a_1 = C_1, \quad a_0 = \frac{1}{R_1} + \frac{1}{R_3}, \quad b = c = \frac{1}{R_3}$$
$$d_1 = C_2, \quad d_0 = \frac{1}{R_2} + \frac{1}{R_3}$$

Even though we have not justified the sign rule for the general case, it is helpful to use it as a partial check on the equations for any passive system.

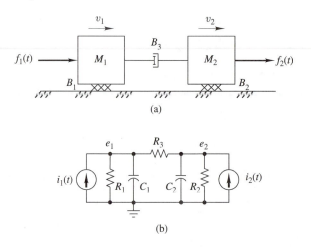

Figure 8.9 Two systems that can be described by (27).

Damping Ratio and Undamped Natural Frequency

Reconsider the characteristic equation $P(s) = 0$ for a second-order system, as given in (23). We can always make the coefficient a_2 equal to unity, if necessary by dividing every term by a constant. Then when the roots of the characteristic equation are complex, the parameter a_0 is positive. With $a_2 = 1$ and $a_0 > 0$, it is useful to rewrite the characteristic equation in the standard form

$$s^2 + 2\zeta\omega_n s + \omega_n^2 = 0 \tag{28}$$

The parameter ω_n is called the **undamped natural frequency** and has units of radians per second. The parameter ζ is dimensionless and is known as the **damping ratio**. For $\zeta > 1$, the roots are distinct negative numbers, and $y(t)$ consists of two decaying exponentials. For $\zeta = 1$, there is a repeated root at $s = -\omega_n$, and $y(t)$ consists of terms having the form $\epsilon^{-\omega_n t}$ and $t\epsilon^{-\omega_n t}$. For $0 \leq \zeta < 1$, the roots are complex and are

$$s_1 = -\zeta\omega_n + j\omega_n\sqrt{1-\zeta^2}$$

$$s_2 = -\zeta\omega_n - j\omega_n\sqrt{1-\zeta^2}$$

Then by (26), the zero-input response is

$$y(t) = K\epsilon^{-\zeta\omega_n t}\cos(\omega_n\sqrt{1-\zeta^2}\,t + \phi) \tag{29}$$

For $\zeta < 0$, the system is unstable.

The advantage of introducing the parameters ζ and ω_n becomes apparent when the characteristic roots are complex and are plotted in the complex plane, as indicated in Figure 8.10(a). Their distances from the origin are the same and are denoted by d. The distance d is the square root of the sum of the squares of the real and imaginary parts of the root. For the upper root,

$$d = \sqrt{(-\zeta\omega_n)^2 + \omega_n^2(1-\zeta^2)} = \omega_n$$

Hence, when $0 \leq \zeta < 1$, the complex characteristic roots lie on a circle of radius ω_n centered at the origin. It is easy to show that their locations on the circle depend only on the damping ratio ζ. Specifically, the angle θ between the negative real axis and the line from the origin

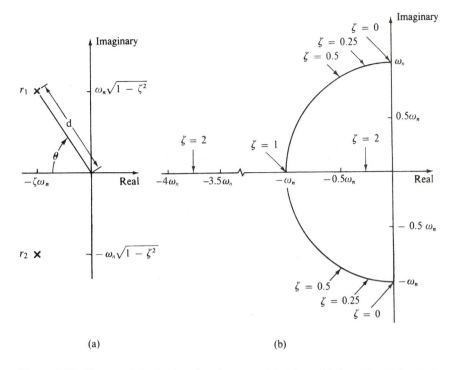

Figure 8.10 Characteristic-root locations in terms of ζ and ω_n. (a) General complex roots. (b) Locations for constant ω_n and varying ζ.

to s_1 in Figure 8.10(a) satisfies the relationship $\cos\theta = \zeta\omega_n/\omega_n = \zeta$. Thus

$$\theta = \cos^{-1}\zeta \tag{30}$$

The geometric relationships between ζ, ω_n, and the roots of (28) are summarized in Figure 8.10(b) for several different values of ζ.

It is instructive to observe the effect of the damping ratio ζ on the responses of a system described by the equation

$$\ddot{y} + 2\zeta\omega_n\dot{y} + \omega_n^2 y = \omega_n^2 u(t) \tag{31}$$

When $u(t) = 0$ for $t > 0$, and when $y(0) = 0$ and $\dot{y}(0) = \omega_n^2$, we can use MATLAB to obtain the output curves shown in Figure 8.11. The reason for this particular choice of initial conditions will be explained in the next section. Curves are shown for several values of ζ, with ω_n held constant.

The value of ζ determines the nature of the zero-input response (which for stable systems also shows the form of the general transient response) and is often a key design parameter. For $\zeta > 1$, the transient response decays to zero monotonically. For $0 < \zeta < 1$, it consists of decaying oscillations. The case $\zeta = 1$ is the boundary between responses that oscillate and those that do not, and the system is said to have **critical damping**. Second-order systems for which $\zeta > 1$ have more than critical damping; those for which $0 < \zeta < 1$ are less than critically damped. When $\zeta = 0$ the system is **undamped**. The choice of ζ depends on the particular application. However, a value of ζ much greater than one implies a slow transient decay and a relatively sluggish system. A value much smaller than one would result in transient oscillations that take a long time to die out.

We consider next an example in numerical form. Then we discuss briefly the role of friction and resistance in determining the damping ratio. We conclude by examining two approximations of a nonlinear second-order system that exhibit quite different behaviors.

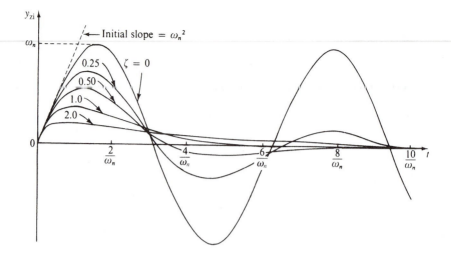

Figure 8.11 The zero-input response for a second-order system described by (31).

▶ **EXAMPLE 8.7**

The differential equation describing a certain fixed linear system is

$$2\ddot{y} + \alpha\dot{y} + 50y = f(t)$$

where $f(t)$ consists of terms involving the input for $t > 0$. Determine ω_n and ζ for $\alpha = 12$ and for $\alpha = 52$. Write the form of the zero-input response for each case. For what values of α does the zero-input response consist of decaying oscillations?

SOLUTION   Dividing by 2 to make the coefficient of $\ddot{y}$ equal to unity gives

$$\ddot{y} + \frac{\alpha}{2}\dot{y} + 25y = \frac{1}{2}f(t)$$

Thus $\omega_n^2 = 25$ and $2\zeta\omega_n = \alpha/2$, from which $\omega_n = 5$ rad/s and $\zeta = \alpha/20$. When $\alpha = 12$, $\zeta = 3/5$, which corresponds to less than critical damping. The roots of the characteristic equation $s^2 + 6s + 25 = 0$ are at $-3 \pm j4$, so $y(t) = K\epsilon^{-3t}\cos(4t + \phi)$.

When $\alpha = 52$, $\zeta = 13/5$, showing that the system is more than critically damped. The characteristic equation for this case is $s^2 + 26s + 25 = 0$, which has roots at $s_1 = -1$ and $s_2 = -25$, so $y(t) = K_1\epsilon^{-t} + K_2\epsilon^{-25t}$. The zero-input response will contain decaying oscillations when $0 < \zeta < 1$, which requires $0 < \alpha < 20$.

Although the damping ratio normally depends on the values of all the passive elements, it is instructive to consider the role of friction and resistance for the four systems shown in Figure 8.12. For the translational system in part (a) of Figure 8.12, which was examined in Example 2.1, the differential equation was shown to be

$$\ddot{x} + \frac{B}{M}\dot{x} + \frac{K}{M}x = \frac{1}{M}f_a(t)$$

and the damping ratio is

$$\zeta = \frac{B}{2\sqrt{MK}} \tag{32}$$

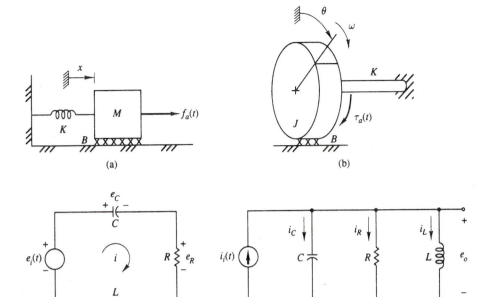

Figure 8.12 Second-order systems illustrating the role of friction and resistance in determining ζ.

For the rotational system in part (b) of Figure 8.12, which was modeled in Example 5.1, the differential equation is

$$\ddot{\theta} + \frac{B}{J}\dot{\theta} + \frac{K}{J}\theta = \frac{1}{J}\tau_a(t)$$

and the damping ratio is

$$\zeta = \frac{B}{2\sqrt{JK}} \qquad (33)$$

The series circuit in part (c) of Figure 8.12 was treated in Example 6.1 and is described by

$$\frac{d^2i}{dt^2} + \frac{R}{L}\frac{di}{dt} + \frac{1}{LC}i = \frac{1}{L}\dot{e}_i$$

The damping ratio is

$$\zeta = \frac{R}{2}\sqrt{\frac{C}{L}} \qquad (34)$$

Finally, for the parallel circuit in part (d) of Figure 8.12, we have from Example 6.2,

$$\ddot{e}_o + \frac{1}{RC}\dot{e}_o + \frac{1}{LC}e_o = \frac{1}{C}\frac{di_i}{dt}$$

and the damping ratio is

$$\zeta = \frac{1}{2R}\sqrt{\frac{L}{C}} \qquad (35)$$

If there is no friction in the two mechanical cases, then we see from (32) and (33) that $\zeta = 0$. The zero-input response will consist of a constant-amplitude oscillation. Because there is no element that can dissipate energy in the form of heat, any initial stored energy will be interchanged undiminished between potential energy in K and kinetic energy in

M or J. For positive values of B that do not exceed $2\sqrt{MK}$ or $2\sqrt{JK}$, $y(t)$ will contain decaying oscillations. If B is increased still further, $y(t)$ will have two decaying exponential terms.

When $R = 0$ in part (c) of Figure 8.12—that is, when the resistor is replaced by a short circuit—we see from (34) that $\zeta = 0$. Then any initial stored energy is continually swapped back and forth between the capacitor and inductor, and the zero-input response does not die out. When R takes on positive values, energy is lost from the circuit in the form of heat, and $y(t)$ decays to zero.

For the circuit in part (d) of Figure 8.12, the undamped case of no energy loss in the form of heat occurs when R is replaced by an open circuit—that is, when $R \to \infty$. Then $y(t)$ will again be a constant-amplitude sinusoidal oscillation. As R takes on smaller positive values, ζ increases.

▶ **EXAMPLE 8.8**

The pendulum shown in Figure 8.13(a) was considered in Example 5.7. From (5.51), its input-output differential equation is

$$\ddot{\theta} + \frac{B}{ML^2}\dot{\theta} + \frac{g}{L}\sin\theta = \frac{1}{ML^2}\tau_a(t) \tag{36}$$

Discuss the motion of the pendulum for small variations about each of the two equilibrium positions shown in Figure 8.13(b) and Figure 8.13(c).

SOLUTION   For the positions shown in parts (b) and (c) of Figure 8.13, $\theta = 0$ and $\theta = \pi$ rad, respectively. Note that (36) is satisfied if $\tau_a(t) = 0$ and if θ has a constant value of either zero or π rad. This indicates that if the pendulum is stationary in either of these two positions, it will not move unless there is an applied torque.

Equation (36) is nonlinear because of the factor $\sin\theta$. In Example 5.7, we noted that $\sin\theta \simeq \theta$ for small values of θ. Thus for small deviations about the vertical position shown in Figure 8.13(b), we can approximate (36) by the linear equation

$$\ddot{\theta} + \frac{B}{ML^2}\dot{\theta} + \frac{g}{L}\theta = \frac{1}{ML^2}\tau_a(t) \tag{37}$$

The zero-input response will then have the form of (29) with

$$\omega_n = \sqrt{\frac{g}{L}}$$

$$\zeta = \frac{B}{2Mg^{1/2}L^{3/2}}$$

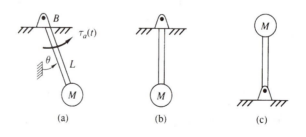

(a)   (b)   (c)

Figure 8.13   (a) Pendulum for Example 8.8. (b) Stable equilibrium position. (c) Unstable equilibrium position.

As expected, the system is stable as long as the pendulum remains close to the position shown in Figure 8.13(b). Decreasing the friction coefficient B decreases the damping ratio ζ. If $B = 0$, then (37) becomes

$$\ddot{\theta} + \frac{g}{L}\theta = \frac{1}{ML^2}\tau_a(t)$$

for which

$$\theta(t) = K\cos\left(\sqrt{\frac{g}{L}}t + \psi\right)$$

where K and ψ are arbitrary constants that depend on the initial conditions. Such a system is marginally stable. If $\tau_a(t) = 0$ for $t > 0$ but if the pendulum is given some initial angular displacement θ_0, then it will have a sinusoidal oscillation with an angular frequency of $\sqrt{g/L}$ radians per second.

Finally, we consider small displacements about the position shown in Figure 8.13(c). If $\phi(t)$ represents the deviation from that position, then $\theta(t) = \phi(t) + \pi$. We note that $\dot{\theta} = \dot{\phi}$ and $\ddot{\theta} = \ddot{\phi}$, and we also use the mathematical identity $\sin(\phi + \pi) = -\sin\phi \simeq -\phi$. Thus for motion close to the position shown in Figure 8.13(c), (36) reduces to

$$\ddot{\phi} + \frac{B}{ML^2}\dot{\phi} - \frac{g}{L}\phi = \frac{1}{ML^2}\tau_a(t) \tag{38}$$

This linearized model is unstable, because one root of the characteristic equation is in the right half-plane. For the case where $B = 0$,

$$\ddot{\phi} - \frac{g}{L}\phi = \frac{1}{ML^2}\tau_a(t)$$

and

$$\phi(t) = K_1\epsilon^{\sqrt{g/L}\,t} + K_2\epsilon^{-\sqrt{g/L}\,t}$$

This agrees with our expectation that the equilibrium position in Figure 8.13(c) would be unstable. Even a very small initial displacement away from the upright position would cause the pendulum to fall.

Mode Functions

The transformed zero-input response $Y(s)$, as given by (20), is a strictly proper rational function, so we can expand it in partial fractions by the methods of Section 7.6. If the characteristic equation $P(s) = 0$ has the distinct roots $s_1, s_2, \ldots, s_n$, then

$$Y(s) = \frac{F(s)}{a_n(s - s_1)(s - s_2)\ldots(s - s_n)}$$

$$= \frac{A_1}{s - s_1} + \frac{A_2}{s - s_2} + \cdots + \frac{A_n}{s - s_n} \tag{39}$$

where the coefficients $A_1, A_2, \ldots, A_n$ depend on the initial conditions $y(0), \dot{y}(0), \ldots, y^{(n-1)}(0)$. Because $A_i/(s - s_i) = \mathcal{L}[A_i\epsilon^{s_it}]$, the inverse transform of (39) is

$$y(t) = A_1\epsilon^{s_1t} + A_2\epsilon^{s_2t} + \cdots + A_n\epsilon^{s_nt} \tag{40}$$

The exponential functions ϵ^{s_it} that make up $y(t)$ are referred to as the **mode functions** or **modes** of the system's zero-input response. The s_i are the roots of the characteristic equation, so the mode functions are properties of the system. However, the coefficients A_i in (40) are the weightings of the individual mode functions, and they depend on the particular initial conditions. In fact, we can select specific initial conditions so as to eliminate any of the mode functions in the zero-input response by forcing the corresponding A_i to be zero. The following example illustrates the manner in which the initial conditions affect the weighting of the system's modes.

▷ **EXAMPLE 8.9**

In Example 8.5, we found that the transform of the zero-input response for the system shown in Figure 8.4(a) is

$$V(s) = \frac{sMv(0) - Kx(0)}{Ms^2 + Bs + K}$$

so the characteristic polynomial is $P(s) = Ms^2 + Bs + K$. Identify the mode functions when $M = 1$ kg, $B = 3$ N·s/m, and $K = 2$ N/m. Find the combinations of initial conditions that will suppress one of the two modes. Repeat the example when $M = 1$ kg, $B = 2$ N·s/m, and $K = 2$ N/m.

SOLUTION   For the first set of element values, the characteristic equation is $P(s) = s^2 + 3s + 2 = 0$. We showed in Example 8.5 that the roots are $s_1 = -1$ and $s_2 = -2$, and that

$$v(t) = -[v(0) + 2x(0)]\epsilon^{-t} + 2[v(0) + x(0)]\epsilon^{-2t}$$

The mode functions are ϵ^{-t} and ϵ^{-2t}. If $v(0) = -2x(0)$, then the response reduces to

$$v(t) = 2[v(0) + x(0)]\epsilon^{-2t} = -2x(0)\epsilon^{-2t}$$

which looks like the response of a first-order system having a time constant $\tau = 0.5$ s. Similarly, if $v(0) = -x(0)$, then

$$v(t) = -[v(0) + 2x(0)]\epsilon^{-t} = -x(0)\epsilon^{-t}$$

which is the response of a first-order system with $\tau = 1$ s. Hence one should not attempt to deduce the order of a system or the character of its mode functions solely on the basis of a single sample of the zero-input response. Rather, one should be certain that each of the mode functions appears in the observed responses.

For the second set of element values,

$$V(s) = \frac{sv(0) - 2x(0)}{s^2 + 2s + 2}$$

and $P(s) = s^2 + 2s + 2$. The roots of the characteristic equation are $s_1 = -1 + j1$ and $s_2 = -1 - j1$. Because s_1 and s_2 are complex, and because we want to choose modes that are real functions, we do not factor $P(s)$. Instead, we use (7.85) and (7.86) to obtain

$$v(t) = v(0)\epsilon^{-t}\cos t + [2x(0) - v(0)]\epsilon^{-t}\sin t$$

The two mode functions are $\epsilon^{-t}\cos t$ and $\epsilon^{-t}\sin t$. To suppress the first of these two modes, let $v(0) = 0$. To suppress the second mode, let $v(0) = 2x(0)$.

▷ **8.3 THE ZERO-STATE RESPONSE**

The zero-state response $y_{zs}(t)$ is the response to a nonzero input when the initial values of the state variables are zero, that is, when the initial stored energy is zero. Because we assume this throughout this entire section, unless otherwise stated, we generally omit the subscript zs.

One illustration of a zero-state response was the response found in Examples 8.1, 8.2, and 8.3 for the system shown in Figure 8.1(a). Consider now a system described by the general nth-order input-output equation

$$a_n y^{(n)} + a_{n-1} y^{(n-1)} + \cdots + a_1 \dot{y} + a_0 y = b_m u^{(m)} + \cdots + b_0 u(t) \tag{41}$$

We assume that the input starts at $t = 0+$ and that $y(0), \dot{y}(0), \ldots, y^{(n-1)}(0), u(0), \dot{u}(0), \ldots,$ $u^{(m-1)}(0)$ are zero.

The Transfer Function

We transform both sides of (41), with the initial-condition terms set equal to zero. Collecting the remaining terms, we obtain the algebraic equation

$$(a_n s^n + a_{n-1} s^{n-1} + \cdots + a_1 s + a_0) Y(s) = (b_m s^m + \cdots + b_0) U(s)$$

which can be rearranged to give the transform of the output as

$$Y(s) = \left(\frac{b_m s^m + \cdots + b_0}{a_n s^n + a_{n-1} s^{n-1} + \cdots + a_1 s + a_0} \right) U(s) \tag{42}$$

Hence the output transform $Y(s)$ is the product of the input transform $U(s)$ and a rational function of the complex variable s whose coefficients are the coefficients in the input-output differential equation. This rational function of s is known as the system's **transfer function**. It plays a key role in the analysis of linear systems. Denoting the transfer function by $H(s)$, we rewrite (42) as

$$Y(s) = H(s) U(s) \tag{43}$$

Note that when we know a system's input-output differential equation, we can write its transfer function directly as

$$H(s) = \frac{b_m s^m + \cdots + b_0}{a_n s^n + a_{n-1} s^{n-1} + \cdots + a_1 s + a_0} \tag{44}$$

Even if the system model is in a form other than the input-output differential equation, the transfer function can be found from (43) or from its equivalent,

$$H(s) = \frac{Y(s)}{U(s)} \tag{45}$$

We denote the input by a general symbol, rather than using a specific function of time. We can then transform any version of the modeling equations, replace all of the initial-condition terms by zero, and solve for the transformed output.

In Example 8.1 we transformed the equations that were written directly from the free-body diagrams in Figure 8.1. The resulting transfer function was the quantity inside the brackets in (11). In Example 8.2, we obtained the same result by transforming the state-variable model. Finally, in Example 8.3 we started with the input-output equation. Notice that the coefficients in the transfer function were the same as those in the system's input-output equation.

Poles and Zeros

When the two polynomials in s that constitute $H(s)$ are factored, the transfer function will have the form

$$H(s) = K \left[\frac{(s - z_1)(s - z_2) \ldots (s - z_m)}{(s - p_1)(s - p_2) \ldots (s - p_n)} \right] \tag{46}$$

The quantities $z_1, z_2, \ldots, z_m$ are those values of s for which the numerator of $H(s)$ is zero, and they are called the **zeros** of the transfer function.[1] The quantities $p_1, p_2, \ldots, p_n$ are those values of s for which the denominator of $H(s)$ vanishes and for which $H(s)$ becomes infinite. They are the **poles** of the transfer function. From a comparison of (44) and (46), it follows that $K = b_m/a_n$, where we have assumed that neither b_m nor a_n may be zero. In the following discussion, we shall assume that none of the zeros coincide with any of the poles—that is, that $z_i \neq p_j$ for all $1 \leq i \leq m$ and all $1 \leq j \leq n$. If a pole and a zero of $H(s)$ were coincident, some of the factors in (46) could be canceled, in which case we could not reconstruct the input-output differential equation from $H(s)$. Then it would not be possible to find the response of all the system's modes from the transfer function.

From (46), it is apparent that a transfer function can be completely specified by its poles, its zeros, and the multiplying constant K. The poles and zeros may be complex numbers and can be represented graphically by points in a complex plane. This plane is called the **s-plane** or the **complex-frequency plane**, and because $s = \sigma + j\omega$, its real and imaginary axes are labeled σ and ω, respectively. The poles of $H(s)$ are indicated by crosses and the zeros by circles placed at the appropriate points.

Referring to (22), we see that the denominator of $H(s)$ in (44) is the characteristic polynomial of the system. Thus the poles $p_1, p_2, \ldots, p_n$ of $H(s)$ are identical to the characteristic roots that determine the form of the zero-input response. As a consequence, we can write down the form of the zero-input response as soon as we know the poles of $H(s)$.

To lend further significance to the transfer-function concept, recall that the stability of a system is determined by the roots of its characteristic equation. Hence the stability can also be characterized by the locations of the poles of the transfer function. If all the poles are inside the left half of the s-plane ($\sigma < 0$), the system is stable; if at least one pole is in the right half of the s-plane ($\sigma > 0$), the system is unstable. If all the poles of $H(s)$ are in the left half of the s-plane except for distinct poles on the imaginary axis ($\sigma = 0$), the system is marginally stable. In addition, a system is unstable if its transfer function has repeated poles on the imaginary axis. Before discussing the zero-state response in more detail, we shall illustrate some of these notions in the following example.

▶ **EXAMPLE 8.10**

Find the transfer function and draw the corresponding pole-zero plot in the s-plane for the system described by the input-output equation

$$\dddot{y} + 7\ddot{y} + 15\dot{y} + 25y = 2\ddot{u} + 6\dot{u}$$

Also comment on the stability of the system and give the form of its zero-input response.

SOLUTION   Transforming the input-output equation with zero initial conditions and with $u(0) = \dot{u}(0) = 0$, we have

$$(s^3 + 7s^2 + 15s + 25)Y(s) = (2s^2 + 6s)U(s)$$

which, when we solve for the ratio $Y(s)/U(s)$, gives

$$H(s) = \frac{2s^2 + 6s}{s^3 + 7s^2 + 15s + 25}$$

Alternatively, we could have used (44) to write $H(s)$ by inspection of the input-output equation.

---

[1] If $m < n$, which is often the case, $H(s)$ will have a zero of multiplicity $n - m$ at infinity. However, one is usually interested in only the numerator zeros $z_1, z_2, \ldots, z_m$.

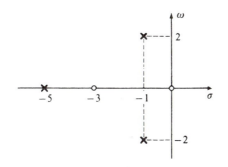

Figure 8.14 Pole-zero plot for $H(s)$ in Example 8.10.

In order to draw the pole-zero plot representing $H(s)$, we must factor its numerator and denominator. Although the numerator is readily factored, the denominator is a cubic in s. It turns out that

$$H(s) = \frac{2s(s+3)}{(s+5)(s^2+2s+5)}$$

$$= 2\left[\frac{s(s+3)}{(s+5)(s+1-j2)(s+1+j2)}\right] \tag{47}$$

Thus $H(s)$ has two real zeros (at $s = 0$ and $s = -3$) and three poles (a real one at $s = -5$ and a complex pair at $s = -1 + j2$ and $s = -1 - j2$), all of which can be represented by the pole-zero plot shown in Figure 8.14. In addition, the multiplying constant is $K = 2$.

Because the transfer function has all three of its poles in the left half of the s-plane, the system is stable. The fact that one of the zeros (at $s = 0$) is not inside the left half-plane has no bearing on the system's stability. For that matter, $H(s)$ can have zeros in the right half-plane and still correspond to a stable system. Knowing the three poles of the transfer function, we can immediately write down the form of the zero-input response in either of the following two equivalent forms:

$$\begin{aligned} y_{zi}(t) &= K_1\epsilon^{-5t} + \epsilon^{-t}(K_2\cos 2t + K_3\sin 2t) \\ y_{zi}(t) &= K_1\epsilon^{-5t} + K_4\epsilon^{-t}\cos(2t + \phi) \end{aligned} \tag{48}$$

Transient and Steady-State Components

Recall that the transient component of the response consists of those terms that decay to zero as t becomes large, whereas the remaining terms constitute the steady-state component. For a stable system and for an input that does not decay to zero, we saw in Example 8.5 how the poles of $H(s)$ give rise to the transient terms, and the poles of the transformed input give rise to the steady-state terms.

We now examine the relation $Y(s) = H(s)/U(s)$ in a somewhat more general way to see how the poles and zeros are related to the transient and steady-state components of the output. Assume that the transform of the input $u(t)$ can be written as a rational function of s:

$$U(s) = \frac{N(s)}{D(s)} \tag{49}$$

where $N(s)$ and $D(s)$ are polynomials. For cases where the input contains a term delayed by a units of time, $N(s)$ will contain a factor ϵ^{-as}, which can be treated as in Example 7.15. Other inputs, such as $u(t) = 1/t$, that have Laplace transforms that do not fit (49) are beyond the scope of our consideration.

Rewriting (43) using (46) for $H(s)$ and (49) for $U(s)$, we have

$$Y(s) = K \left[\frac{(s - z_1) \ldots (s - z_m)}{(s - p_1)(s - p_2) \ldots (s - p_n)} \right] \cdot \frac{N(s)}{D(s)} \tag{50}$$

Recognizing that $U(s)$ has its own poles and zeros, we see that the poles and zeros of $Y(s)$ are the combination of those of $H(s)$ and those of $U(s)$. If we expand $Y(s)$ in a partial fraction expansion and all of its poles are real and distinct, the expansion will have the form

$$Y(s) = \frac{A_1}{s - p_1} + \frac{A_2}{s - p_2} + \cdots + \frac{A_n}{s - p_n} + \frac{A_{n+1}}{s - p_{n+1}} + \cdots + \frac{A_q}{s - p_q} \tag{51}$$

where the poles $p_1, p_2, \ldots, p_n$ are the poles of $H(s)$ and the poles $p_{n+1}, \ldots, p_q$ are the poles of $U(s)$. Taking the inverse transform of each term, we have

$$y(t) = A_1 \epsilon^{p_1 t} + A_2 \epsilon^{p_2 t} + \cdots + A_n \epsilon^{p_n t} + A_{n+1} \epsilon^{p_{n+1} t} + \cdots + A_q \epsilon^{p_q t} \tag{52}$$

For a stable system, the first n terms will be transient terms that decay to zero as t becomes large.[2] Then, if the terms in the input do not decay to zero,

$$y_{tr}(t) = A_1 \epsilon^{p_1 t} + \cdots + A_n \epsilon^{p_n t}$$
$$y_{ss}(t) = A_{n+1} \epsilon^{p_{n+1} t} + \cdots + A_q \epsilon^{qt}$$

Keep in mind that when complex poles are present, it may be more efficient to include the corresponding second-order term (quadratic denominator and linear numerator) in the partial-fraction expansion, as in (7.85), rather than to factor it. Also keep in mind that even if there were some initial stored energy, that would affect only the size and not the form of the transient component.

▶ **EXAMPLE 8.11**

Find the zero-state response of a system for which

$$H(s) = \frac{9s + 14}{3(s + 1)(s + 3)}$$

when the input is $u(t) = 6 \cos 2t$ for $t > 0$. Identify the transient and steady-state terms.

SOLUTION   The transform of the input is

$$U(s) = \frac{6s}{s^2 + 4}$$

When $H(s)$ and $U(s)$ are substituted into (43), it follows that

$$Y(s) = \frac{9s + 14}{3(s + 1)(s + 3)} \cdot \frac{6s}{(s^2 + 4)} \tag{53}$$

Referring to the procedures in Section 7.6, we can write the partial-fraction expansion for $Y(s)$ in either of the following two equivalent forms:

$$Y(s) = \frac{A_1}{s + 1} + \frac{A_2}{s + 3} + \frac{A_3}{s - j2} + \frac{A_3^*}{s + j2} \tag{54a}$$

$$Y(s) = \frac{A_1}{s + 1} + \frac{A_2}{s + 3} + \frac{A_4 s + A_5}{s^2 + 4} \tag{54b}$$

---

[2] If a system is marginally stable, it can exhibit a steady-state response due to initial stored energy, even without an input. An unstable system can yield an unbounded response without an input. Hence the designations *transient* and *steady-state* are most useful for stable systems.

where A_3 is a complex number and A_3^* denotes its complex conjugate. Using the methods of Section 7.6 to evaluate the constants, we find that $A_1 = -1$, $A_2 = -3$, $A_3 = 2\sqrt{2}\epsilon^{-j\pi/4}$, $A_4 = 4$, and $A_5 = 8$. Then, by the entries in Appendix E, we obtain the following expressions for the inverse transform for $t > 0$.

$$y(t) = -\epsilon^{-t} - 3\epsilon^{-3t} + 4\sqrt{2}\cos(2t - \pi/4) \tag{55a}$$

$$y(t) = -\epsilon^{-t} - 3\epsilon^{-3t} + 4\cos 2t + 4\sin 2t \tag{55b}$$

By using Table 1 in Appendix D, it is easy to show that these two forms for $y(t)$ are equivalent. For either form, the first two terms constitute the transient response. These terms come from the poles of $H(s)$, which are entirely within the left half of the s-plane. The rest of the expressions represent the steady-state response, which is a constant-amplitude oscillation resulting from the poles of $U(s)$, which are on the imaginary axis of the s-plane.

The Impulse Response

Because the impulse response is a zero-state response, we can use (43), which is

$$Y(s) = H(s)U(s) \tag{56}$$

where $Y(s)$ is the transformed output, $H(s)$ is the system's transfer function, and $U(s)$ is the transform of the input. The transform of the unit impulse is unity, so $U(s) = 1$ and (56) reduces to

$$Y(s) = H(s)$$

In words, the transform of the unit impulse response is merely the system's transfer function $H(s)$. In addition, $h(t)$ must be zero for all $t \leq 0$, because it is the zero-state response and the impulse does not occur until $t = 0+$. Using the symbol $h(t)$ to denote the response to the unit impulse, we can write

$$h(t) = \begin{cases} 0 & \text{for } t \leq 0 \\ \mathcal{L}^{-1}[H(s)] & \text{for } t > 0 \end{cases} \tag{57}$$

which is a significant result in spite of its simplicity. In particular, (57) serves to tie together the system's time-domain characterization in terms of $h(t)$ and its complex-frequency-domain characterization $H(s)$. For example, one can relate the locations in the s-plane of the poles and zeros of the transfer function to the character of the system's impulse response. Furthermore, although the subject is beyond the scope of this book, the impulse response can be related to the response to arbitrary inputs by an integral expression known as the convolution integral.

▶ **EXAMPLE 8.12**

Find the unit impulse response for a second-order system described by the input-output equation

$$\ddot{y} + 2\zeta\omega_n\dot{y} + \omega_n^2 y = \omega_n^2 u(t) \tag{58}$$

Plot the response for several values of the damping ratio ζ, with ω_n held constant.

SOLUTION   Using (44) and (57), we can write for $t > 0$

$$h(t) = \mathcal{L}^{-1}[H(s)] = \mathcal{L}^{-1}\left[\frac{\omega_n^2}{s^2 + 2\zeta\omega_n s + \omega_n^2}\right]$$

For the case when $0 < \zeta < 1$, we complete the square in the denominator to obtain

$$h(t) = \mathcal{L}^{-1}\left[\frac{\omega_n^2}{(s + \zeta\omega_n)^2 + (1 - \zeta^2)\omega_n^2}\right]$$

Then for $t > 0$

$$h(t) = \omega_n^2 e^{-\zeta\omega_n t} \sin(\omega_n \sqrt{1 - \zeta^2}\, t)$$

We can differentiate this expression to get $\dot{h}(t)$ for $t > 0$. If we want the values of $h(t)$ and its first derivative at $t = 0+$, we can replace every t by 0 to find that $h(0+) = 0$ and $\dot{h}(0+) = \omega_n^2$. Alternatively, we can use the initial-value theorem to obtain the same results.

Finding expressions for $h(t)$ when $\zeta = 1$ and when $\zeta > 1$ is left to the reader. The values of $h(0+)$ and $\dot{h}(0+)$ are the same as before.

An input that is the unit impulse instantaneously inserts some energy into the system at $t = 0$, but then becomes zero for all $t > 0$. Recall that in Figure 8.11 we plotted the zero-input response for (58) when $y(0) = 0$ and $\dot{y}(0) = \omega_n^2$. Thus that figure also represents the system's unit impulse response.

For a stable system, the response to an impulse will decay to zero as t approaches infinity. The system in the next example, however, is only marginally stable, because $H(s)$ has a single pole on the imaginary axis of the complex plane.

▶ **EXAMPLE 8.13**

Find and sketch the unit impulse response of the system whose input-output equation is

$$\dddot{y} + 4\ddot{y} + 3\dot{y} = 2\dot{u} + u(t)$$

SOLUTION   By inspection of the input-output differential equation, we see that the transfer function is

$$H(s) = \frac{2s + 1}{s^3 + 4s^2 + 3s}$$

$$= \frac{2s + 1}{s(s + 1)(s + 3)}$$

Carrying out the partial-fraction expansion, we find that

$$H(s) = \frac{\frac{1}{3}}{s} + \frac{\frac{1}{2}}{s + 1} - \frac{\frac{5}{6}}{s + 3}$$

Thus the unit impulse response is

$$h(t) = \begin{cases} 0 & \text{for } t \leq 0 \\ \frac{1}{3} + \frac{1}{2}e^{-t} - \frac{5}{6}e^{-3t} & \text{for } t > 0 \end{cases}$$

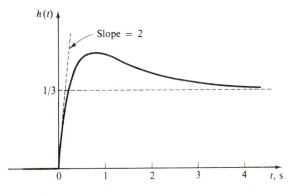

Figure 8.15 Impulse response for the system in Example 8.13.

which is shown in Figure 8.15. In this case,

$$h(0+) = \lim_{s \to \infty} \frac{s(2s+1)}{s(s^2+4s+3)} = 0$$

and

$$\lim_{t \to \infty} h(t) = \lim_{s \to 0} \frac{s(2s+1)}{s(s^2+4s+3)} = \frac{1}{3}$$

both of which agree with the sketch of $h(t)$. Here $h(t)$ approaches a nonzero constant value because of the single pole at $s = 0$ in $H(s)$.

The Step Response

Because the transform of the unit step function is $1/s$, it follows from (56) that the transform of $y_U(t)$, the zero-state response of a system to a unit step-function input, is

$$Y_U(s) = H(s) \cdot \frac{1}{s} \tag{59}$$

Thus the unit step response $y_U(t)$ is zero for $t \leq 0$, and for $t > 0$ it is the inverse transform of $H(s)/s$. The poles of $Y_U(s)$ consist of the poles of $H(s)$ and a pole at $s = 0$. Provided that $H(s)$ has no pole at $s = 0$, the time functions that make up $y_U(t)$ will be the mode functions of the system's zero-input response and a constant term resulting from the pole of $Y_U(s)$ at $s = 0$. In fact, if the system is stable, all the mode functions will decay to zero as t approaches infinity, and the steady-state portion of the step response will be due entirely to the pole at $s = 0$. The coefficient in the partial-fraction expansion of this steady-state term, and thus $(y_U)_{ss}$ itself, will be

$$(y_U)_{ss} = sY_U(s)|_{s=0} = H(0) \tag{60}$$

We can also get this result by applying the final-value theorem to $Y_U(s)$ as given by (59):

$$y_U(\infty) = \lim_{s \to 0} sY_U(s) = H(0)$$

Because the initial conditions do not affect the steady-state response of a stable system, the result applies even when the initial stored energy is not zero. If $H(s)$ has a single pole at $s = 0$, then $Y_U(s)$ as given by (59) will have a double pole at $s = 0$, and the step response will contain a ramp function in addition to a constant term. Then $y_U(t)$ will grow without bound as t approaches infinity.

▷ **EXAMPLE 8.14**

Find the unit step response for a second-order system described by the input-output equation

$$\ddot{y} + 2\zeta\omega_n\dot{y} + \omega_n^2 y = \omega_n^2 u(t) \tag{61}$$

Plot the response for several values of the damping ratio ζ, with ω_n held constant.

SOLUTION From (59),

$$y_U(t) = \mathcal{L}^{-1}\left[H(s) \cdot \frac{1}{s}\right] = \mathcal{L}^{-1}\left[\frac{\omega_n^2}{s(s^2 + 2\zeta\omega_n s + \omega_n^2)}\right]$$

We shall omit the details of taking the inverse transform. For $0 < \zeta < 1$, for example, we find that

$$y_U(t) = 1 - \left(\frac{\zeta}{1-\zeta^2}\right) \epsilon^{-\zeta\omega_n t} \sin(\omega_n\sqrt{1-\zeta^2}\, t + \phi)$$

where

$$\phi = \tan^{-1}\left(\zeta/\sqrt{1-\zeta^2}\right)$$

The unit step response is shown in Figure 8.16 for several values of ζ, with ω_n held constant. For $\zeta > 0$, the steady-state response is unity. Note that the value of ζ determines to what extent, if any, the response will overshoot its steady-state value. The overshoot is 100 percent for $\zeta = 0$ and decreases to zero when the damping ratio is unity. For $\zeta > 1$, the response approaches its steady-state value monotonically.

The case $\zeta = 1$ is the boundary between responses that oscillate and those that do not, and the system is said to have **critical damping**. Second-order systems for which $\zeta > 1$ have more than critical damping; those for which $0 < \zeta < 1$ are less than critically damped. When $\zeta = 0$ the system is **undamped**. Figure 8.11 shows the unit impulse response of the general second-order system described by (61) for the values of ζ used in Figure 8.16. Note that the steady-state response is zero for $\zeta > 0$ and that the value of the damping ratio establishes the character of the response.

In the design of systems, the choice of ζ depends upon several factors, but consider the following two simple examples. In many electrical measuring instruments, we want the

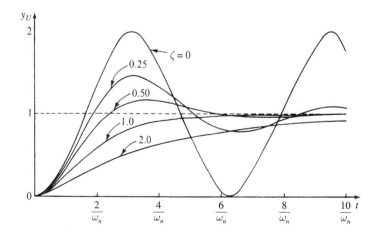

Figure 8.16 The unit step response for a second-order system described by (61).

output to settle down quickly, so that a reading may be obtained promptly. An appropriate value of ζ might be a little less than one, perhaps 0.8. On the other hand, if we want a robot arm to move out and paint a flat surface as it passes by, we would need ζ to be somewhat greater than one. Otherwise, the arm might overshoot and dent the surface that is to be painted.

▶ **EXAMPLE 8.15**

Use Laplace transforms to find the unit step response of the marginally stable system described by $\dddot{y} + 4\ddot{y} + 3\dot{y} = 2\dot{u} + u(t)$, for which we found the impulse response in Example 8.13.

SOLUTION Inspection of the system's differential equation reveals the transfer function to be

$$H(s) = \frac{2s+1}{s^3 + 4s^2 + 3s}$$

Hence the transform of the unit step response is

$$Y_U(s) = \frac{2s+1}{s^2(s^2 + 4s + 3)}$$

whose partial-fraction expansion can be shown to be

$$Y_U(s) = \frac{1/3}{s^2} + \frac{2/9}{s} - \frac{1/2}{s+1} + \frac{5/18}{s+3}$$

Thus

$$y_U(t) = \frac{1}{3}t + \frac{2}{9} - \frac{1}{2}\epsilon^{-t} + \frac{5}{18}\epsilon^{-3t} \quad \text{for } t > 0$$

As shown in Figure 8.17, the steady-state portion of the unit step response contains a ramp component with a slope of $\frac{1}{3}$, in addition to the constant of $\frac{2}{9}$. This is because $H(s)$ has a pole at $s = 0$. This pole combines with the pole at $s = 0$ that is due to the step-function input to give a double pole of $Y(s)$ at $s = 0$.

As an aid in sketching the unit step response, we note that in addition to $y_U(0+) = 0$, the initial slope is zero, because $h(t)$ is the derivative of the step response and (from Figure 8.15) $h(0+) = 0$.

Differentiating the Input

In Section 7.5, we stated that the unit impulse response is the derivative of the unit step response. To verify the relationship between $h(t)$ and $y_U(t)$, we use the theorem for the transform of an integral to write

$$\frac{1}{s}H(s) = \mathcal{L}\left[\int_0^t h(\lambda)d\lambda\right]$$

where λ is a dummy variable of integration. Because $\mathcal{L}[y_U(t)] = H(s)/s$, it follows that

$$y_U(t) = \int_0^t h(\lambda)d\lambda \tag{62}$$

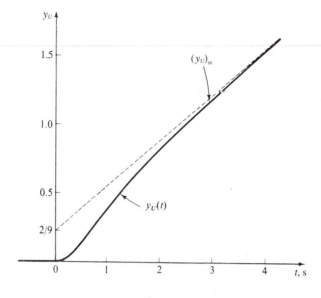

Figure 8.17 Step response for Example 8.15.

In graphical terms, (62) states that the unit step response $y_U(t)$ is the area from 0 to t underneath the curve of the unit impulse response $h(t)$.

We can also use the theorem for a derivative to write

$$\mathcal{L}[\dot{y}_U(t)] = sY_U(s) - y_U(0)$$

When we are dealing with zero-state responses, we assume that the output, the input, and all of their derivatives are zero for $t \leq 0$. Any discontinuities are assumed to occur at $t = 0+$. With $Y_U(s) = H(s)/s$ and with $y_U(0) = 0$,

$$\mathcal{L}[\dot{y}_U(t)] = H(s) = \mathcal{L}[h(t)]$$

Noting that $y_U(t)$ and $h(t)$ must be zero for $t \leq 0$, we conclude that

$$h(t) = \frac{d}{dt} y_U(t) \tag{63}$$

which agrees with Equation (7.54). Of course, either (62) or (63) follows immediately from the other equation.

For a linear system that contains no initial stored energy, the above arguments can be used to prove a more general statement. Suppose we know the response to a certain input. If we substitute a new input that is the derivative of the old input, then the new response is the derivative of the old response. This property may be useful when constructing block diagrams for certain types of system models.

▶ 8.4 FREQUENCY RESPONSE

In discussing the impulse and step responses, we were interested in both the transient and steady-state components. However, when considering the sinusoidal input

$$u(t) = \sin \omega t \tag{64}$$

we are normally interested only in the steady-state response. Throughout this section, we shall assume that the systems are stable, so that the zero-input responses would decay to zero. As we saw in Example 8.11, the steady-state response will have the same frequency as the input but, in general, its amplitude and phase will differ from those of the input. Denoting the amplitude and phase angle of the steady-state response by A and ϕ, respectively, we can write

$$y_{ss}(t) = A\sin(\omega t + \phi) \tag{65}$$

which is referred to as the **sinusoidal steady-state response**. It plays a key role in many aspects of system analysis, including electronic circuits and feedback control systems. We now show how we can express A and ϕ in terms of the transfer function $H(s)$.

The initial conditions do not affect the steady-state response of a stable system, so we start with the transform of the zero-state response as given by (56). Because $\mathcal{L}[\sin\omega t] = \omega/(s^2 + \omega^2)$, the transformed response to (64) is

$$Y(s) = H(s)\left[\frac{\omega}{s^2 + \omega^2}\right] \tag{66}$$

Hence the poles of $Y(s)$ will be the poles of $H(s)$ plus the two imaginary poles of $U(s)$ at $s = j\omega$ and $s = -j\omega$. For a stable system, all the poles of $H(s)$ will be in the left half of the complex plane, and all the terms in the response corresponding to these poles will decay to zero as time increases. Thus the steady-state response will result from the imaginary poles of $U(s)$ and will be a sinusoid at the frequency ω. To determine the steady-state component of the response, we can write (66) as

$$Y(s) = H(s)\left[\frac{\omega}{(s - j\omega)(s + j\omega)}\right]$$

$$= \frac{C_1}{s - j\omega} + \frac{C_2}{s + j\omega} + [\text{terms corresponding to poles of } H(s)] \tag{67}$$

where

$$C_1 = (s - j\omega)Y(s)|_{s=j\omega}$$

$$= H(s)\left.\frac{\omega}{(s + j\omega)}\right|_{s=j\omega}$$

$$= \frac{H(j\omega)}{2j}$$

The constant C_2 is the complex conjugate of C_1; that is,

$$C_2 = -\frac{H(-j\omega)}{2j}$$

Because all the terms in (67) corresponding to the poles of $H(s)$ will decay to zero if the system is stable, the transform of the steady-state response is

$$Y_{ss}(s) = \frac{H(j\omega)/2j}{s - j\omega} - \frac{H(-j\omega)/2j}{s + j\omega} \tag{68}$$

In general, $H(j\omega)$ is a complex quantity and may be written in polar form as

$$H(j\omega) = M(\omega)\epsilon^{j\theta(\omega)} \tag{69}$$

where $M(\omega)$ is the magnitude of $H(j\omega)$ and $\theta(\omega)$ is its angle. Both M and θ depend on the value of ω, as is emphasized by the ω within the parentheses. The quantity $H(-j\omega)$ is the complex conjugate of $H(j\omega)$, and we can write it as

$$H(-j\omega) = M(\omega)\epsilon^{-j\theta(\omega)} \tag{70}$$

Substituting (69) and (70) into (68) gives

$$Y_{ss}(s) = \frac{M}{2j}\left(\frac{\epsilon^{j\theta}}{s - j\omega} - \frac{\epsilon^{-j\theta}}{s + j\omega}\right)$$

Taking the inverse transforms of the two terms in $Y_{ss}(s)$, we find that

$$y_{ss}(t) = \frac{M}{2j}\left[\epsilon^{j\theta}\epsilon^{j\omega t} - \epsilon^{-j\theta}\epsilon^{-j\omega t}\right]$$
$$= \frac{M}{2j}\left[\epsilon^{j(\omega t + \theta)} - \epsilon^{-j(\omega t + \theta)}\right]$$

Using the exponential form of the sine function, as in Table 1 in Appendix D, we obtain

$$y_{ss}(t) = M\sin(\omega t + \theta) \tag{71}$$

which has the form predicted in (65), where $A = M(\omega)$ and $\phi = \theta(\omega)$. One can use a similar derivation to show that for a stable system, the steady-state response to $u(t) = B\sin(\omega t + \phi_1)$ is

$$y_{ss}(t) = BM\sin(\omega t + \phi_1 + \theta) \tag{72}$$

Likewise, the steady-state response to $u(t) = B\cos(\omega t + \phi_2)$ is

$$y_{ss}(t) = BM\cos(\omega t + \phi_2 + \theta) \tag{73}$$

In summary, the sinusoidal steady-state response of a stable linear system is a sinusoid having the same frequency as the input, an amplitude that is $M(\omega)$ times that of the input, and a phase angle that is $\theta(\omega)$ plus the input angle, where $M(\omega)$ and $\theta(\omega)$ are the magnitude and angle of $H(j\omega)$, respectively. As a consequence, the function $H(j\omega)$, which is the transfer function evaluated for $s = j\omega$, is known as the **frequency-response function**.

Calculating and interpreting the frequency-response function are illustrated in the following examples. In the first example, we find the steady-state response to a sinusoidal input that has a specified frequency. In the next two examples, we sketch curves of $M(\omega)$ and $\theta(\omega)$ as functions of ω. Such curves indicate how the magnitude and angle of the sinusoidal steady-state response change as the frequency of the input is changed.

▶ **EXAMPLE 8.16**

Use (69) and (73) to find the steady-state response of the system described by the transfer function

$$H(s) = \frac{9s + 14}{3(s^2 + 4s + 3)}$$

to the input $u(t) = 6\cos 2t$, and compare the result to that from Example 8.11.

SOLUTION   We replace s by $j\omega$ with $\omega = 2$ rad/s, the frequency of the input, to form

$$H(j2) = \frac{j18 + 14}{3(-4 + j8 + 3)} = \frac{14 + j18}{3(-1 + j8)}$$

The magnitude of $H(j2)$ is the magnitude of its numerator divided by the magnitude of its denominator:

$$M = \frac{|14 + j18|}{3|-1 + j8|} = \frac{22.80}{3(8.062)} = 0.9427$$

The angle of $H(j2)$ is the angle of its numerator minus the angle of its denominator:

$$\theta = \tan^{-1}(18/14) - \tan^{-1}(-8)$$
$$= 0.9098 - 1.695 = -0.7853 \text{ rad}$$

Thus, by (73),

$$y_{ss}(t) = 6(0.9427)\cos(2t - 0.7853)$$
$$= 5.656\cos(2t - 0.7853)$$

which agrees with the steady-state portion of the response found in Example 8.11.

▶ **EXAMPLE 8.17**

Evaluate and sketch the magnitude and phase angle of the frequency-response function for the system described by the first-order equation $\dot{y} + (1/\tau)y = u(t)$.

SOLUTION The system transfer function is $H(s) = 1/(s+1/\tau)$. When we replace s by $j\omega$, we find the system's frequency-response function to be

$$H(j\omega) = \frac{1}{j\omega + 1/\tau} \tag{74}$$

In order to identify the magnitude $M(\omega)$ and angle $\theta(\omega)$ for (74), we convert it to polar form. To find $M(\omega)$, we divide the magnitude of the numerator of (74) by the magnitude of the denominator, obtaining

$$M(\omega) = \frac{1}{[\omega^2 + (1/\tau)^2]^{1/2}}$$

Subtracting the angle of the denominator from the numerator angle, we find that the phase angle[3] of $H(j\omega)$ is

$$\theta(\omega) = \arg[1] - \arg[j\omega + 1/\tau]$$
$$= 0 - \tan^{-1}\omega\tau$$
$$= -\tan^{-1}\omega\tau$$

To assist us in sketching M versus ω and θ versus ω, we note that at $\omega = 0$ we have $M = \tau$ and $\theta = 0$. As ω approaches infinity, M diminishes monotonically to zero, and θ decreases to $-\frac{1}{2}\pi$ rad. At $\omega = 1/\tau$, $M = \tau/\sqrt{2}$, and $\theta = -\frac{1}{4}\pi$ rad. The actual functions are shown in Figure 8.18.

▶ **EXAMPLE 8.18**

The output for the circuit shown in Figure 8.19 is i_o. Find the frequency-response function $H(j\omega)$ and sketch its magnitude versus frequency.

SOLUTION We start by evaluating the circuit's transfer function $H(s)$, which is the ratio $I_o(s)/I_i(s)$ when the initial stored energy is zero. Applying Kirchhoff's current law at

---

[3]The notation arg[z] denotes the angle of the complex quantity z. It is expressed in radians.

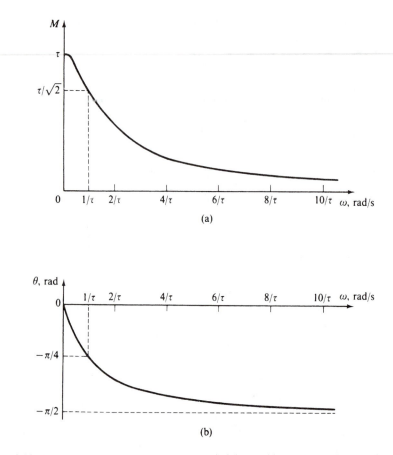

Figure 8.18 Frequency-response function for $\dot{y} + (1/\tau)y = u(t)$. (a) Magnitude $M(\omega)$. (b) Phase angle $\theta(\omega)$.

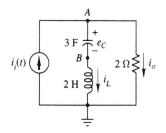

Figure 8.19 Circuit for Example 8.18.

nodes A and B gives

$$3(\dot{e}_A - \dot{e}_B) + \frac{1}{2}e_A - i_i(t) = 0$$

$$3(\dot{e}_B - \dot{e}_A) + i_L(0) + \frac{1}{2}\int_0^t e_B(\lambda)d\lambda = 0$$

The output current is given by

$$i_o = \frac{1}{2}e_A$$

Transforming these equations with zero initial voltage across the capacitor and zero initial current through the inductor gives

$$(3s + \frac{1}{2})E_A(s) - 3sE_B(s) = I_i(s)$$

$$-3sE_A(s) + \left(3s + \frac{1}{2s}\right)E_B(s) = 0$$

$$I_o(s) = \frac{1}{2}E_A(s)$$

We can solve this set of three transformed equations for $H(s) = I_o(s)/I_i(s)$. The result is

$$H(s) = \frac{6s^2 + 1}{6s^2 + 6s + 1}$$

We find the system's frequency-response function by setting s equal to $j\omega$ in $H(s)$. Noting that $(j\omega)^2 = -\omega^2$, we get

$$H(j\omega) = \frac{-6\omega^2 + 1}{-6\omega^2 + j6\omega + 1}$$

The magnitude function is

$$M(\omega) = \frac{|1 - 6\omega^2|}{|1 - 6\omega^2 + j6\omega|}$$

$$= \frac{|1 - 6\omega^2|}{[(1 - 6\omega^2)^2 + (6\omega)^2]^{1/2}}$$

which is plotted versus ω in Figure 8.20. The phase angle $\theta(\omega)$ is given by

$$\theta(\omega) = \arg[1 - 6\omega^2] - \arg[1 - 6\omega^2 + j6\omega]$$

which could also be plotted versus frequency.

Note that when $\omega = 0$, $M(\omega)$ is unity. A sinusoidal function for which $\omega = 0$ reduces to a constant, so $H(0)$ is the steady-state response to the unit step function, which is consistent with our earlier discussion. We can check the fact that $H(0) = 1$ for Figure 8.19 by observing from Section 6.5 that when finding the steady-state response to a constant input, we can replace the inductor and capacitor by short and open circuits, respectively.

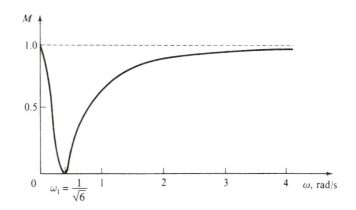

Figure 8.20 Frequency-response magnitude for the circuit in Example 8.18.

We see from Figure 8.20 that the sinusoidal steady-state response is zero at a frequency of $\omega_1 = 1/\sqrt{6}$ rad/s. Because of the shape of the plot of $M(\omega)$ versus ω, such a circuit is often called a *notch filter*. For sinusoidal inputs with frequencies $\omega \ll \omega_1$ or $\omega \gg \omega_1$, the magnitude of the frequency-response function is approximately unity, which means that the amplitude of $(i_o)_{ss}$ will be close to that of $i_i(t)$. For inputs with $\omega \simeq 1/\sqrt{6}$ rad/s, however, the amplitude of $(i_o)_{ss}$ will be much less than that of $i_i(t)$. Hence sinusoidal inputs in this frequency range are substantially attenuated by the circuit. Such a circuit can be used to filter out unwanted signals that have frequencies close to ω_1 without significantly affecting signals at other frequencies.

Curves showing how the magnitude and angle of $H(j\omega)$ change with frequency constitute one of the important tools for the analysis and design of feedback control systems and will be used extensively in Chapters 14 and 15. It will become convenient to plot the curves somewhat differently, in the form of Bode diagrams, but the basic procedure is illustrated by the previous two examples.

▶ 8.5 PROPERTIES OF THE TRANSFER FUNCTION

In the previous sections of this chapter, we defined the transfer function $H(s)$ and showed how to find it by transforming the system's modeling equations and assuming that there is no initial stored energy. The transfer function plays a central role in the description, analysis, and design of linear systems. The diagram in Figure 8.21 illustrates this central role and is the basis for the examples in this section. After going through the examples, the reader should understand how to follow the solid arrows to go from one characterization of the system to another one. For the path from the pole-zero plot to $H(s)$, the multiplying constant K that appears in (46) must be specified separately. Although the figure does not show all the possible paths, those shown are the ones that we shall illustrate.

Before presenting the examples, we first summarize some of the key relationships that we have developed earlier in this chapter. For a system described by the input-output equation

$$a_n y^{(n)} + a_{n-1} y^{(n-1)} + \cdots + a_1 \dot{y} + a_0 y = b_m u^{(m)} + \cdots + b_0 u(t) \tag{75}$$

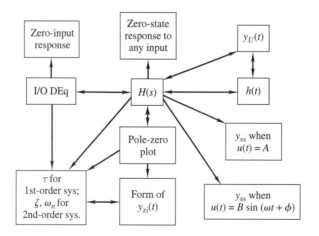

Figure 8.21 The central role of the transfer function.

the transfer function is

$$H(s) = \frac{b_m s^m + \cdots + b_0}{a_n s^n + a_{n-1} s^{n-1} + \cdots + a_1 s + a_0} \tag{76}$$

Frequently, both halves of the transfer function are divided by a constant in order to make the coefficient of s^n be unity. Then, for a first-order system, the denominator polynomial can be written in the standard form

$$P(s) = s + \frac{1}{\tau} \tag{77}$$

where τ is the time constant. For a second-order system, the standard form of the denominator is

$$P(s) = s^2 + 2\zeta\omega_n s + \omega_n^2 \tag{78}$$

where ζ is the damping ratio, and ω_n is the undamped natural frequency.

For any input $u(t)$, the transform of the zero-state response is

$$Y_{zs}(s) = H(s)U(s) \tag{79}$$

As special cases of this relationship, the transforms of the unit step and unit impulse responses are

$$\mathcal{L}[y_U(t)] = H(s) \cdot \frac{1}{s} \tag{80}$$

$$\mathcal{L}[h(t)] = H(s) \tag{81}$$

Although the last equation may seem trivial, it confirms that it is consistent to denote the unit impulse response by $h(t)$ and the transfer function by $H(s)$.

The steady-state response to a constant input A is

$$y_{ss}(t) = H(0) \cdot A \tag{82}$$

For the steady-state response to a sinusoidal input such as $u(t) = B\sin(\omega t + \phi)$, we evaluate $H(j\omega)$ in polar form. If $H(j\omega) = M\epsilon^{j\theta}$, then

$$y_{ss}(t) = BM\sin(\omega t + \phi + \theta) \tag{83}$$

In the first of the following examples, the transfer function is given. When it is not completely given as part of the problem statement, it is often best to find it before answering the questions that are asked. The details of finding the coefficients in routine partial-fraction expansions and of simplifying complex numbers are left to the reader.

▷ **EXAMPLE 8.19**

The transfer function of a certain linear system is $H(s) = 2/(s+2)^2$. Find the steady-state response to a unit step function, and then find the complete response. Next, find the steady-state response to $u(t) = 6\sin(2t + \pi/4)$. Finally, find the complete response $y(t)$ for all $t > 0$ when the input $u(t) = 0$ for $t > 0$, but when $y(0) = 1$ and $\dot{y}(0) = -2$.

SOLUTION   For the steady-state response to a unit step function, we use (82) with $H(0) = 0.5$ to write $y_{ss} = (0.5)(1) = 0.5$. For the complete response, we use (80):

$$Y_U(s) = \frac{2}{s(s+2)^2} = \frac{0.5}{s} - \frac{1}{(s+2)^2} - \frac{0.5}{s+2}$$

so for $t > 0$

$$y_U(t) = 0.5 - t\epsilon^{-2t} - 0.5\epsilon^{-2t}$$

Note that the steady-state component of this expression is 0.5, which agrees with our previous answer for y_{ss}.

To find the steady-state response to a sinusoid with an angular frequency of 2 rad/s, we need

$$H(j2) = \frac{2}{(2+j2)^2} = 0.25\epsilon^{-j\pi/2}$$

Using (83) with $M = 0.25$ and $\theta = -\pi/2$ rad gives $y_{ss}(t) = 1.5\sin(2t - \pi/4)$.

Finally, for the last part of the example, the input is zero for $t > 0$, but the initial conditions are nonzero. We cannot make use of (79), which is valid only for the zero-state response. By (76), however, we know that the system's input-output equation is $\ddot{y} + 4\dot{y} + 4y = u(t)$. Noting that the right side is zero for $t > 0$, we transform this equation to get

$$[s^2Y(s) - sy(0) - \dot{y}(0)] + 4[sY(s) - y(0)] + 4Y(s) = 0$$

Inserting the given numerical values for the initial conditions and solving algebraically for the transform of the output, we obtain $Y(s) = 1/(s+2)$, so $y(t) = \epsilon^{-2t}$ for all $t > 0$. For the given transfer function, we would normally expect the zero-input response to contain modes of the form ϵ^{-2t} and $t\epsilon^{-2t}$, but the second of these is not excited by our choice of initial conditions.

► **EXAMPLE 8.20**

The pole-zero plot for a certain transfer function is shown in Figure 8.22. It is also known that the steady-state response to a unit step is 0.5. Find the system's input-output equation.

SOLUTION   Because the transfer function has zeros at -1 and -3, and poles at -2 and -4,

$$H(s) = \frac{K(s+1)(s+3)}{(s+2)(s+4)}$$

To evaluate the multiplying constant K, we note that $H(0) = 3K/8$, which by (82) must equal 0.5. Thus $K = 4/3$. Multiplying out the numerator and denominator of $H(s)$ gives

$$H(s) = \frac{(4/3)(s^2 + 4s + 3)}{s^2 + 6s + 8}$$

so $\ddot{y} + 6\dot{y} + 8y = (4/3)[\ddot{u} + 4\dot{u} + 3u(t)]$.

► **EXAMPLE 8.21**

The unit-step response of a certain system is $y_U(t) = 4 + \epsilon^{-2t}(\cos t - 18\sin t)$ for $t > 0$. Find the damping ratio ζ. Also find the steady-state response to the input $u(t) = 4 + 3\sin(t + \pi/3) + 2\cos(2t + \pi/4)$.

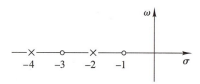

Figure 8.22 Pole-zero plot for Example 8.20.

SOLUTION   By (80), the transfer function is

$$H(s) = s\left[\frac{4}{s} + \frac{(s+2)-18}{(s+2)^2+1}\right] = \frac{5s^2+20}{s^2+4s+5}$$

To obtain the damping ratio, we can compare the denominator polynomial to the standard form given in (78). We see that $\omega_n^2 = 5$ and $2\zeta\omega_n = 4$. Thus $\omega_n = \sqrt{5}$ and $\zeta = 2/\sqrt{5} = 0.8945$. The fact that ζ is less than one is consistent with the fact that the transient part of $y_U(t)$ consists of decaying sinusoids rather than decaying exponentials.

The damping ratio could have been obtained without evaluating the transfer function. The form of the transient component in $y_U(t)$ corresponds to roots of the characteristic equation at $s_1 = -2 + j1$ and $s_2 = -2 - j1$. Thus the factored form of the characteristic polynomial must be $P(s) = (s+2-j1)(s+2+j1)$, so we see once again that $P(s) = s^2 + 4s + 5$.

To find the steady-state response to the given input, we must find the individual steady-state responses to each of the three components, and then add them together by superposition. Because $H(0) = 4$, the steady-state response to the constant component is $(4)(4) = 16$. Because

$$H(j1) = \frac{-5+20}{-1+j4+5} = \frac{15}{4+j4} = M\epsilon^{j\theta}$$

where $M = 2.652$ and $\theta = -\pi/4$ rad, the second term in the output is $7.956\sin(t - \pi/12)$. Finally,

$$H(j2) = \frac{-20+20}{-4+j8+5} = 0$$

so the steady-state response to the third component of the input is zero. In summary, $y_{ss}(t) = 16 + 7.956\sin(t - \pi/12)$.

Additional illustrations of the relationships shown in Figure 8.21 are included in the problems at the end of the chapter. The fact that many of the connecting lines in Figure 8.21 have one-way arrows should not be surprising. Specifying the sinusoidal steady-state response for just a single frequency, for example, does not give enough information to be able to determine $H(s)$ and the transient response. In Chapter 14, however, we show how complete frequency-response curves, that is plots of $M(\omega)$ and $\theta(\omega)$ for all values of ω, can indeed serve as a complete description of a linear system. Instead of showing the curves in the form illustrated in Figure 8.18, however, we shall find it convenient to use alternate forms known as Bode diagrams.

▶ 8.6 IMPEDANCES

To find the transfer function for any linear system, we can transform the system's equations while assuming that there is no initial energy in the energy-storing elements. We can then solve for the transform of the output and after that, write the transfer function down by (45). However, when this needs to be done for the electrical part of a system, an easier method can be developed.

Consider the element laws for the resistor, inductor, and capacitor:

$$e(t) = Ri(t)$$

$$e(t) = L\frac{di}{dt} \tag{84}$$

$$i(t) = C\frac{de}{dt}$$

Transforming these equations, using the symbols $E(s) = \mathcal{L}[e(t)]$ and $I(s) = \mathcal{L}[i(t)]$, gives

$$E(s) = RI(s)$$
$$E(s) = L[sI(s) - i(0)] \tag{85}$$
$$I(s) = C[sE(s) - e(0)]$$

When determining the transfer function, we always assume that there is no initial stored energy. In the case of an inductor, this means that $i(0) = 0$; for a capacitor, it means that $e(0) = 0$. Under these circumstances, the previous transformed equations for the resistor, inductor, and capacitor reduce to

$$E(s) = RI(s) \tag{86a}$$
$$E(s) = sLI(s) \tag{86b}$$
$$E(s) = \frac{1}{sC}I(s) \tag{86c}$$

Equation (86a) looks exactly like Ohm's law for a resistor, except that transformed quantities are used for the voltage and current. Note that (86b) and (86c) have this same form, except that R is replaced by sL and $1/sC$, respectively. Thus we are led to define the **impedance** $Z(s)$ as the ratio of the transformed voltage to the transformed current when there is no initial stored energy. For the three types of passive circuit elements,

$$Z_R(s) = R \tag{87a}$$
$$Z_L(s) = sL \tag{87b}$$
$$Z_C(s) = \frac{1}{sC} \tag{87c}$$

Note that the impedance depends on the element value and on s. However, (86) and (87) are *algebraic* equations, with no derivative or integral signs and with no initial-condition terms. Furthermore, transforming the algebraic equations given by the interconnection laws (Kirchhoff's voltage and current laws) yields the same equations, but with the variables replaced by their Laplace transforms. When finding the transfer function, therefore, we can use all the rules for resistive circuits, even if the elements are not of the same type. This includes the procedures in Section 6.5 for series and parallel combinations, such as equivalent resistance and the voltage-divider and current-divider rules. For this reason, the impedance $Z(s)$ is sometimes viewed as a generalized resistance. It can be treated as a resistance, even though it is a function of s.

In order to avoid writing any time-domain equations, we redraw the circuit with the passive elements characterized by their impedances and the voltages and currents characterized by their Laplace transforms. The result is called the s-domain circuit. We can obtain the transfer function directly from it, using only the techniques for resistive circuits.

▶ **EXAMPLE 8.22**

Find the transfer function $H(s) = E_o(s)/E_i(s)$ for the circuit shown in Figure 8.23(a).

SOLUTION   The s-domain circuit is drawn in Figure 8.23(b). The passive elements have been characterized by their impedances, as given in (87), and the input and output voltages by their Laplace transforms. The two right-hand elements are in parallel and can be replaced by the equivalent impedance

$$Z_2(s) = \frac{R_2/sC}{R_2 + 1/sC} = \frac{R_2}{1 + sCR_2}$$

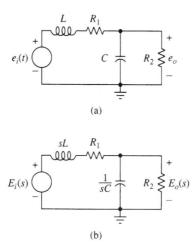

Figure 8.23 Circuit for Example 8.22. (a) Time-domain circuit. (b) *s*-domain circuit.

Then, by the voltage-divider rule,

$$E_o(s) = \frac{Z_2(s)}{sL + R_1 + Z_2(s)} E_i(s)$$

$$= \frac{R_2}{(1 + sCR_2)(sL + R_1) + R_2} E_i(s)$$

$$= \left[\frac{R_2}{LCR_2 s^2 + (L + R_1 R_2 C)s + (R_1 + R_2)} \right] E_i(s) \qquad (88)$$

The quantity inside the brackets in (88) is the transfer function $H(s)$.

We saw from (41) and (44) how the transfer function and the input-output differential equation are directly related to one another. In the last example, the input-output equation for the transfer function in (88) would be

$$LCR_2 \ddot{e}_o + (L + R_1 R_2 C)\dot{e}_o + (R_1 + R_2)e_o = R_2 e_i(t)$$

Note that for a constant input voltage $e_i(t) = A$, we can find the steady-state response by setting the derivatives of the output equal to zero, which yields

$$(e_0)_{ss} = \frac{R_2}{R_1 + R_2} A \qquad (89)$$

This same result is achieved by replacing s by zero for the transfer function in (88). Alternatively, if we look at the original circuit in Figure 8.23(a) with the inductor and capacitor replaced by short and open circuits, respectively, we see that (89) follows immediately from the voltage-divider rule.

In the design of electrical networks, it is sometimes convenient to represent individual subnetworks by their impedances. Once a general expression for the transfer function has been found, we can then decide what to put inside the subnetworks in order to obtain a specific result. We conclude this section with two such examples involving an operational amplifier.

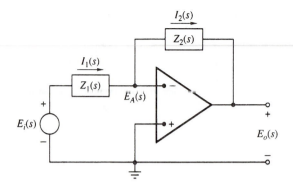

Figure 8.24 Circuit for Example 8.23.

▶ **EXAMPLE 8.23**

Find the transfer function $H(s)$ for the s-domain circuit shown in Figure 8.24.

SOLUTION From the virtual short concept discussed near the end of Section 6.7 for the ideal op-amp, we know that $E_A(s) = 0$ and that $I_1(s) = I_2(s)$. Thus

$$\frac{1}{Z_1(s)}E_i(s) = -\frac{1}{Z_2(s)}E_o(s)$$

from which the transfer function is seen to be

$$H(s) = \frac{E_o(s)}{E_i(s)} = -\frac{Z_2(s)}{Z_1(s)}$$

If, in the last example, we wanted $H(s) = K/s$, corresponding to an integrator, we could put a resistor R in the first subnetwork and a capacitor C in the second one. Then $Z_1(s) = R$, $Z_2(s) = 1/sC$, and $H(s) = -1/RCs$. The basic configuration in Figure 8.24 can be used to obtain a variety of useful transfer functions, as illustrated by some of the problems at the end of the chapter.

▶ **EXAMPLE 8.24**

Find the transfer function $H(s) = E_o(s)/E_i(s)$ for the s-domain circuit shown in Figure 8.25. Let $K = (R_1 + R_2)/R_1$.

SOLUTION This circuit can be relatively complex, with perhaps several elements inside each of the four boxes, but the general analysis is simplified by the use of impedances. Because no current flows into the input terminals of the op-amp, we can use the voltage-divider rule to write

$$E_A(s) = \frac{R_1}{R_1 + R_2}E_o(s) = \frac{1}{K}E_o(s)$$

$$E_B(s) = \frac{Z_4(s)}{Z_2(s) + Z_4(s)}E_C(s)$$

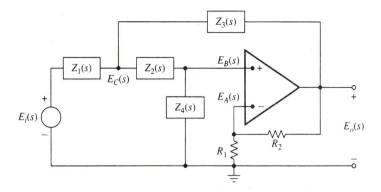

Figure 8.25 Circuit for Example 8.24.

Using the virtual-short concept for the ideal op-amp, we know that $E_A(s) = E_B(s)$, so

$$E_C(s) = \frac{Z_2(s) + Z_4(s)}{KZ_4(s)} E_o(s) \tag{90}$$

Summing the currents at node C yields

$$\frac{1}{Z_1(s)}[E_C(s) - E_i(s)] + \frac{1}{Z_3(s)}[E_C(s) - E_o(s)] + \left[\frac{1}{Z_2(s) + Z_4(s)}\right] E_C(s) = 0 \tag{91}$$

Inserting (90) into (91), solving for $E_o(s)$, and simplifying, we obtain (after a little algebra) the following result:

$$H(s) = \frac{E_o(s)}{E_i(s)} = \frac{KZ_3(s)Z_4(s)}{Z_1(s)Z_2(s) + (1-K)Z_1(s)Z_4(s) + Z_3(s)[Z_1(s) + Z_2(s) + Z_4(s)]} \tag{92}$$

Even by considering very simple subnetworks for the blocks labeled $Z_1(s)$ through $Z_4(s)$ in Figure 8.25, we can obtain a number of useful results. Suppose, for example, that we want the denominator of the transfer function to be a quadratic in s and the numerator to be just a constant. We can put a single resistor inside blocks 1 and 2 and a single capacitor inside blocks 3 and 4. For simplicity, let $Z_1(s) = Z_2(s) = R$ and $Z_3(s) = Z_4(s) = 1/s\,C$. Then the general expression for the transfer function in (92) becomes

$$H(s) = \frac{\dfrac{K}{(RC)^2}}{s^2 + \dfrac{1}{RC}(3-K)s + \left(\dfrac{1}{RC}\right)^2}$$

We can control the position of the poles of $H(s)$ by varying the parameter K. If, for example, $K = 3$ (corresponding to $R_2 = 2R_1$), the poles are on the imaginary axis of the complex plane. If $K = 1$ (corresponding to $R_2 = 0$), both the poles are on the negative real axis at $s = -1/RC$. Examples of the importance of controlling the pole and zero positions will be given in Chapter 15.

▶ SUMMARY

The transform of the zero-input response has the form $Y(s) = F(s)/P(s)$, where $F(s)$ is a polynomial that depends on the initial conditions. The mode functions, which characterize the form of the zero-input response, are determined from the roots of the characteristic equation $P(s) = 0$. The weighting of the mode functions depends on the initial energy in each of the energy-storing elements.

When the initial stored energy is zero, the transform of the output is given by $Y(s) = H(s)U(s)$, where $H(s)$ is the transfer function and $U(s)$ is the transform of the input. The transfer function plays a central role in system analysis and design, as indicated in Figure 8.21. It can be written down by inspection of the input-output differential equation, can be found by transforming the system equations with the initial-condition terms set equal to zero, or (in the case of electrical systems) can be found by using impedances. The poles and zeros are the values of s for which $H(s)$ becomes infinite or vanishes, respectively. The poles of $H(s)$ are identical to the roots of the characteristic equation $P(s) = 0$.

Two important special cases of the zero-state response are the unit impulse response $h(t)$ and the unit step response $y_U(t)$. Because they are found when the system is initially at rest, they are both zero for $t \leq 0$. For $t > 0$, $h(t) = \mathcal{L}^{-1}[H(s)]$ and $y_U(t) = \mathcal{L}^{-1}[H(s)/s]$.

For a stable system with an input that does not decay to zero, the steady-state response does not depend on the initial conditions. For the steady-state response to sinusoidal inputs, we can use the frequency-response function $H(j\omega)$. If $M(\omega)$ and $\theta(\omega)$ denote the magnitude and angle of $H(j\omega)$, then the steady-state response to $u(t) = B\sin(\omega t + \phi_1)$ is $y_{ss}(t) = BM\sin(\omega t + \phi_1 + \theta)$.

▶ PROBLEMS

8.1. A system obeys the differential equation

$$3\ddot{y} + 12\dot{y} + 9y = 9\dot{u} + 14u(t)$$

Find the complete response when the initial conditions are $y(0) = 2$, $\dot{y}(0) = 0$ and when the input is $u(t) = 3\epsilon^{-2t}$ for $t > 0$ and zero otherwise.

8.2. Repeat Problem 8.1 when $y(0) = 2$ and $\dot{y}(0) = 0$ and when $u(t) = 6\cos 2t$ for $t > 0$ and zero otherwise.

8.3. For a system described by the equation $\ddot{y} + 3\dot{y} + 2y = u(t)$, use the Laplace transform to find $y(t)$ for $t > 0$ when the input is $u(t) = 5t$ for $t > 0$ and when the initial conditions are $y(0) = 1$ and $\dot{y}(0) = -1$. Sketch $y(t)$.

8.4. The circuit in Figure 8.2(a) has reached steady-state conditions with the switch open. If the switch then closes at $t = 0$, find $e_o(t)$ for all $t > 0$. Show that your answer yields the expected values when $t = 0+$ and when t becomes indefinitely large.

8.5. For the system shown in Figure P8.5, the input is $i_i(t)$ and the output is e_o. Assume that the initial conditions $e_o(0)$ and $i_L(0)$, which summarize the effects of the input for $t < 0$, are known.

   a. Find a general expression for $E_o(s)$ in terms of the transformed input and the initial conditions.

   b. Let $C = 1$ F, $R = \frac{1}{3}$ Ω, and $L = \frac{1}{2}$ H. Identify the transforms of the zero-state and zero-input components of the output.

   c. Let $i_i(t) = 1 + \sin t$ for $t > 0$ and find the zero-state and zero-input components of $e_o(t)$.

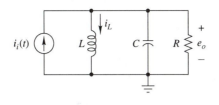

Figure P8.5

d. Identify the transfer function and explain why there is no constant term in the steady-state response.

8.6. Consider the circuit shown in Figure P8.6.

a. Verify that $E_o(s)$, the transform of the output voltage, can be written as

$$E_o(s) = \frac{-se_C(0) - 2i_L(0) + (s^2 + 2s/R)E_i(s)}{s^2 + (1 + 2/R)s + 16}$$

b. Determine the poles of $E_o(s)$ for $R = 2/9\ \Omega, 2/7\ \Omega$, and $2/3\ \Omega$.

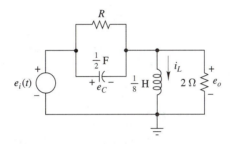

Figure P8.6

8.7. Consider the electrical circuit shown in Figure 6.15 and discussed in Example 6.3.

a. Verify that, for the parameter values given in the example, (6.19) can be written as

$$2\dot{e}_A + 2e_A - e_o = e_i(t)$$

$$-2e_A + 16\dot{e}_o + 3e_o + 8\int_0^t e_o(\lambda)d\lambda = -4i_L(0)$$

b. Find the Laplace transform of the zero-input response of the output e_o to an initial voltage $e_A(0)$ on the capacitor C_1. Give your answer as a ratio of polynomials.

8.8. The unit step responses of four second-order systems are shown in Figure P8.8, and four pairs of characteristic roots are listed below. Match each pair of roots with the corresponding step-response curve.

A	$s = -0.5 \pm j2$
B	$s = -0.5 \pm j4$
C	$s = -1 \pm j4$
D	$s = -2 \pm j4$

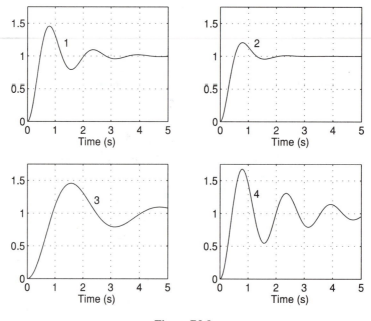

Figure P8.8

8.9. The unit impulse responses of four second-order systems are shown in Figure P8.9, and four pairs of characteristic roots are listed below. Match each pair of roots with the corresponding impulse-response curve.

A	$s = -0.2 \pm j1$
B	$s = -0.2 \pm j3$
C	$s = -0.2, -0.3$
D	$s = -0.5 \pm j1$

8.10. If the spring K_1 in Figure 2.19 is replaced by a cable that does not stretch, the system can be shown to obey the differential equation

$$(M_1 + M_2)\ddot{x} + (B_1 + B_2)\dot{x} + K_2 x = f_a(t) - M_2 g$$

where $x_1 = x_2 = x$.

    a.   Verify that the differential equation is correct.

    b.   Find expressions for the damping ratio ζ and the undamped natural frequency ω_n.

    c.   Find the steady-state response when the applied force is the unit step function.

8.11. The input-output differential equation for the mechanical system shown in Figure P5.22 is

$$(M + J/R^2)\ddot{x} + B\dot{x} + (K_2 + K_1/R^2)x = -f_a(t)$$

    a.   Verify that the differential equation is correct.

    b.   Find expressions for the damping ratio ζ and the undamped natural frequency ω_n.

    c.   Find the steady-state response when the applied force is the unit step function. Check your answer by examining Figure P5.22 directly in the steady state.

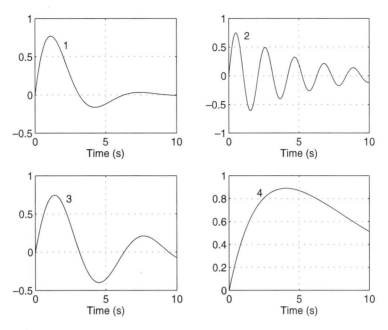

Figure P8.9

8.12. The input-output differential equation for the circuit shown in Figure P8.12 is

$$C\ddot{e}_o + \left(\frac{1}{R_1} + \frac{1}{R_2}\right)\dot{e}_o + \frac{1}{L}e_o = \frac{1}{R_1}\dot{e}_i$$

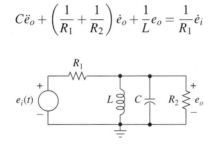

Figure P8.12

a. Verify that the differential equation is correct.

b. Find expressions for the damping ratio ζ and the undamped natural frequency ω_n.

8.13. The rotational mechanical system shown in Figure 5.13(a) has no applied torque for $t > 0$. Let $J = 1$ kg·m², $B = 5$ N·m·s, and $K = 6$ N·m.

a. Find the zero-input response in terms of $\theta(0)$ and $\dot{\theta}(0)$.

b. Identify the mode functions. For each mode function, give the restrictions on the initial conditions needed to eliminate it from the zero-input response.

8.14. Repeat Problem 8.13 for each of the following sets of parameters:

a. $J = 1$ kg·m², $B = 2$ N·m·s, and $K = 5$ N·m.

b. $J = 1$ kg·m², $B = 4$ N·m·s, and $K = 4$ N·m.

8.15. Use the expression for $E_o(s)$ in the statement of Problem 8.6 to obtain the zero-input response of the circuit shown in Figure P8.6 when $R = \frac{2}{9}\,\Omega$. Identify the mode functions. For each mode function, give the restrictions on the initial conditions needed to eliminate it from the zero-input response.

8.16. Repeat Problem 8.15 when $R = \frac{2}{7}\,\Omega$.

8.17. Repeat Problem 8.15 when $R = \frac{2}{3}\,\Omega$.

8.18. Find ω_n and ζ for the system in Example 4.5 when $B = 4, 8, 12$, and 25 N·s/m. Use the results as a check on the curves in Figure 4.15. For what range of values of B will the the displacement x never overshoot the steady-state value?

8.19.

    a. Find the transfer function $H(s)$ for the circuit shown in Figure P8.6.

    b. Find expressions for the damping ratio ζ and the undamped natural frequency ω_n in terms of the resistance R.

    c. Find the zero-state response to the input $e_i(t) = 1 + \epsilon^{-2t}$ for $t > 0$ when $R = \frac{2}{7}\,\Omega$. Explain why there is no steady-state response even though there is a constant term in the input.

*8.20. Find the transfer function $X_R(s)/F_a(s)$ for the third-order mechanical system modeled in Example 3.5.

8.21.

    a. For the translational mechanical system described in Problem 2.4, let $M_1 = M_2 = B_1 = B_2 = B_3 = K_1 = K_2 = K_3 = 1$ in a consistent set of units. Verify that the system can be modeled by the pair of differential equations

$$\ddot{x}_1 + 2\dot{x}_1 + 2x_1 - \dot{x}_2 - x_2 = f_a(t)$$
$$-\dot{x}_1 - x_1 + \ddot{x}_2 + 2\dot{x}_2 + 2x_2 = 0$$

    b. Find the transfer function when the output is x_1.

    c. Repeat part (b) when the output is x_2. Comment on the similarities and dissimilarities between the two transfer functions.

8.22.

    a. Find the transfer function for the translational system modeled in Example 2.6 when the output is z_1, the displacement of the frame M_1 relative to its equilibrium position when the applied force is zero. Take the parameter values to be $M_1 = M_2 = B = K_1 = K_2 = K_3 = 1$ in a consistent set of units.

    b. Repeat part (a) when the output is z_2.

*8.23. Find the transfer function of the rotational mechanical system modeled in Example 5.11 when the output is x.

8.24. Find the transfer function of the mechanical system modeled in Example 5.12, which has both translational and rotational elements, when the output is z.

*8.25. Find the transfer function $E_o(s)/E_i(s)$ for the circuit modeled in Example 6.9.

8.26. Use the Laplace transform to find the unit impulse response for the system in Example 4.5. Show that the curves in Figure 4.16 represent x and v when $f_a(t) = 2\delta(t)$.

8.27.

    a.   Plot the pole-zero pattern for the transfer function

$$H(s) = \frac{s+1}{s^2 + 5s + 6}$$

    b.   Write the general form of the zero-input response and find the system's input-output differential equation.

    c.   Use the initial-value and final-value theorems, if they are applicable, to determine the values of the unit step response at $t = 0+$ and when t approaches infinity.

    d.   Find the unit step response $y_U(t)$ and the unit impulse response $h(t)$ using $H(s)$.

*8.28. Repeat Problem 8.27 for

$$H(s) = \frac{s^2 + 2s + 2}{s^2 + 4s + 4}$$

8.29. Repeat Problem 8.27 for

$$H(s) = \frac{s+3}{s^3 + 7s^2 + 10s}$$

8.30. Repeat Problem 8.27 for

$$H(s) = \frac{12}{s(s^2 + 2s + 4)}$$

*8.31. Repeat Problem 8.27 for

$$H(s) = \frac{s^2}{(2s+1)(s^2 + 4)}$$

8.32. For the differential equation $\ddot{y} + 5\dot{y} + 4y = u(t)$, find and sketch the unit step response $y_U(t)$ and the unit impulse response $h(t)$.

8.33. A second-order system is described by the equation $\ddot{y} + 4\dot{y} + 25y = 50u(t)$.

    a.   Find the damping ratio ζ and the undamped natural frequency ω_n.

    b.   Find and sketch the unit step response $y_U(t)$ and the unit impulse response $h(t)$.

*8.34. Repeat Problem 8.33 for the differential equation $\ddot{y} + 2\dot{y} + 2y = u(t)$.

8.35. A second-order system is described by $\ddot{y} + 2\dot{y} + 4y = 4u(t)$.

    a.   Find the damping ratio ζ and the undamped natural frequency ω_n.

    b.   Find and sketch the response when $u(t) = 2$ for $t > 0$ and when the initial conditions are $y(0) = 0$ and $\dot{y}(0) = 1$.

*8.36. The unit step response of a certain linear system is

$$y_U(t) = [2 + 2\epsilon^{-t}\cos(2t + \pi/4)]U(t)$$

a. Find the numerical values of the damping ratio ζ and the undamped natural frequency ω_n.

b. Find the unit impulse response $h(t)$. Sketch $y_U(t)$ and $h(t)$.

*8.37.

a. Verify that the transfer function for the circuit shown in Figure P8.37 is

$$H(s) = \frac{s^2 + 2s + 1}{s^2 + 4s + 4}$$

b. Find the unit impulse response.

c. Find the unit step response.

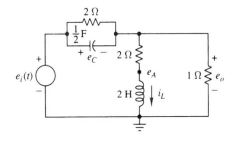

Figure P8.37

8.38. The mechanical system described in Problem 8.10 has two inputs, the gravitational force $M_2 g$ and the force $f_a(t)$ applied to M_1.

a. Solve for x_0, the constant displacement caused by gravity when the applied force $f_a(t)$ is zero.

b. Find the transformed output $X(s)$ when $x(0) = x_0$ and $f_a(t)$ is the unit step function.

c. Find $x(t)$ for the conditions in part (b) when $M_1 = M_2 = B_1 = B_2 = K_2 = 1$ in a consistent set of units. Check your answer by evaluating x_{ss} and $x(0+)$.

8.39. Use the Laplace transform to solve for the unit step response of the mechanical system described in Problem 8.11 when the parameter values are $M = J = R = K_1 = K_2 = 1$ and $B = 5$ in a consistent set of units.

8.40. For the rotational system shown in Figure 5.19 and discussed in Example 5.5, let $J_1 = J_2 = 0.5, B = 1$, and $K = 2$ in a consistent set of units.

a. Find a general expression for $\Omega_2(s) = \mathcal{L}[\omega_2(t)]$ by transforming Equation (5.42).

b. If the system is initially at rest and the inputs are $\tau_L(t) = 0$ and $\tau_a(t) = \epsilon^{-2t}$ for $t > 0$, find and sketch ω_2 for $t > 0$.

*8.41. Find and sketch e_o versus t for the circuit shown in Figure 6.17 and discussed in Example 6.4. The parameter values are $C = 0.5$ F, $R_1 = 2\ \Omega, R_2 = R_3 = 1\ \Omega$, and $L = 0.5$ H. The inputs are $e_1(t) = 2$ V for all t and $e_2(t) = U(t)$ V. Steady-state conditions exist at $t = 0-$.

8.42. Use the Laplace transform to solve for the unit step response of the circuit described in Problem 8.12 when the parameter values are $C = L = R_1 = R_2 = 1$ in a consistent set of units.

8.43.

a. For the circuit shown in Figure P8.43, verify that the application of Kirchhoff's current law to nodes A and O leads to the following pair of coupled equations.

$$\frac{1}{2}[e_A - e_i(t)] + \frac{1}{2}e_A + 2(\dot{e}_A - \dot{e}_o) = 0$$

$$2(\dot{e}_o - \dot{e}_A) + i_L(0) + 4\int_0^t e_o(\lambda)d\lambda + 2e_o = 0$$

b. Use the Laplace transform to find the unit impulse response when e_o is the output.

c. Use the Laplace transform to solve for the unit step response.

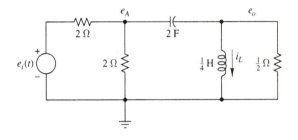

Figure P8.43

8.44. By a derivation similar to the one used to obtain Equation (71), show that the steady-state response of a stable system to the input $u(t) = B\cos\omega t$ is $y_{ss}(t) = BM\cos(\omega t + \theta)$, where M and θ are given by Equation (69).

8.45. Derive the expression given in Equation (72) for the steady-state response to $u(t) = B\sin(\omega t + \phi_1)$.

8.46. The modeling equation for Figure 2.12(a) was given in Equation (2.15) and was represented by the Simulink diagram in Figure 4.13. As in Example 4.8, let $M = 2$ kg, $B = 4$ N·s/m, $K = 16$ N/m, and $f_a(t) = 8\sin\omega t$. The response when $\omega = 2\pi(0.05)$ rad/s was plotted in the top half of Figure 4.20. Problem 4.10 asked for the steady-state response for several other values of ω.

a. Use the frequency-response function $H(j\omega)$ to check the plot for $\omega = 2\pi(0.05)$ rad/s.

b. Use $H(j\omega)$ to check the magnitude and angle of the steady-state response when $\omega = 2\pi(0.5)$ rad/s.

8.47. For each of the following transfer functions, plot the pole-zero pattern, draw curves of $M(\omega)$ versus ω and $\theta(\omega)$ versus ω, and comment briefly on your results. For the function in part (c), include the numerical values for $\omega = 9.9$, 10.0, and 10.1 rad/s.

a. $H(s) = \dfrac{2}{s^2 + 2s + 1}$

b. $H(s) = \dfrac{2s^2}{s^2 + 2s + 1}$

c. $H(s) = \dfrac{s}{s^2 + 0.2s + 100}$

*8.48. Repeat Problem 8.47 for the following transfer functions.

a. $H(s) = \dfrac{s-1}{s+1}$

b. $H(s) = \dfrac{(s+1)^2}{s(s+5)}$

c. $H(s) = \dfrac{s}{(s^2 + 0.2s + 100)^2}$

*8.49. The steady-state response of a stable system to the input

$$u(t) = \sin 5t + \sin 10t + \sin 15t$$

has the form

$$y_{ss}(t) = A\sin(5t + \theta_1) + B\sin(10t + \theta_2) + C\sin(15t + \theta_3)$$

For a system described by the transfer function in part (c) of Problem 8.47, use Equation (71) to find the values of A, B, and C, and calculate the ratios A/B and C/B.

8.50. The steady-state response of a system described by the transfer function in part (a) of Problem 8.47 to the input

$$u(t) = 1 + \sin t + \sin 10t$$

has the form

$$y_{ss}(t) = A + B\sin(t + \theta_2) + C\sin(10t + \theta_3)$$

Calculate A, B, and C, and comment on the ratios B/A and C/A.

*8.51.

a. For the series *RLC* circuit shown in Figure 6.13, find the transfer function $H(s) = I(s)/E_i(s)$.

b. By examining $H(j\omega)$, determine the value of ω such that the steady-state response to $e_i(t) = \sin\omega t$ is $i_{ss}(t) = (1/R)\sin\omega t$.

8.52.

a. Find the transfer function $H(s) = E_o(s)/I_i(s)$ for the parallel *RLC* circuit shown in Figure 6.14.

b. By examining $H(j\omega)$, determine the value of ω for which the steady-state response to $i_i(t) = \sin\omega t$ is $e_o(t) = R\sin\omega t$.

8.53. In Example 6.17 we used Simulink and MATLAB to plot the response of the circuit in Figure 6.15 to the sinusoidal input $e_i(t) = 5\sin\omega t$, where $\omega = 0.2\pi$. The input-output equation is given by Equation (6.21) for the parameter values used in the simulation.

a. Determine the transfer function $H(s)$ relating $E_o(s)$ and $E_i(s)$.

b. Evaluate the frequency-response function $H(j\omega)$ when $\omega = 0.2\pi$, and use this to calculate the magnitude and angle of $(e_o)_{ss}$. Compare the results with the plot of e_o in Figure 6.38.

c. Find the poles and zeros of $H(s)$. Determine the time constant associated with the transient terms that decay most slowly. Use this to estimate how quickly the transient terms will die out. Compare the result with the plot of e_o in Figure 6.38.

8.54. In Example 6.18 we used Simulink and MATLAB to plot the response of the second-order circuit in Figure 6.27 when $e_i(t)$ is a square wave. The state-variable model is given in Equations (6.49). The first half cycle of the input has a constant value of -1 for $0 < t < 10$ s. Except for the minus sign, $e_i(t)$ is the same as a unit step function during this time.

a. Show that the transfer function relating $E_o(s)$ and $E_i(s)$ is $H(s) = (5s^2 + 2s)/(25s^2 + 35s + 21)$.

b. Find the poles and zeros, and calculate the damping ratio ζ and the undamped natural frequency ω_n. Estimate the time constant associated with the transient terms and show that the transient will essentially disappear well before 10 seconds. Your value of ζ should indicate that the step response has a small overshoot. This overshoot may be easier to see on the plot of e_C than on the plot of e_o in Figure 6.40.

c. Calculate $(e_o)_{ss}$ and $e_o(0+)$ when the input is a unit step function. Compare these values to those on the plot of e_o in Figure 6.40.

8.55. The unit step response of a certain system is $y_U(t) = t\epsilon^{-2t}$ for $t > 0$.

a. Find the values of ζ and ω_n.

b. Find the complete response for $t > 0$ when the input is $u(t) = tU(t)$.

c. Find the steady-state response when $u(t) = 3 + 5\sin(2t + \pi/4)$.

8.56. A transfer function is known to have the form

$$H(s) = \frac{s^2 + as + b}{s^2 + cs + d}$$

Use the following facts to determine the values of the real constants a, b, c, and d.

The steady-state response to a unit step is 2.

The steady-state response to the input $u(t) = A\sin 2t$ is zero.

The transient response is critically damped ($\zeta = 1$).

8.57. Find the transfer function for the circuit shown in Figure 8.19 by characterizing the passive elements by their impedances.

8.58.

a. Draw the s-domain version of the circuit shown in Figure P8.37 using impedances.

b. Using the circuit drawn in part (a), determine the transfer function $E_o(s)/E_i(s)$.

8.59.

a. Draw the s-domain circuit for Figure P6.1, with the passive elements characterized by their impedances.

    b. Use the result of part (a) to determine the circuit's transfer function.

    c. From the answer to part (b), write the input-output differential equation.

8.60. Repeat Problem 8.59 for the circuit shown in Figure P6.6.

*8.61. Repeat Problem 8.59 for the circuit shown in Figure P6.12.

8.62.

    a. Use impedances and the results of Example 8.23 to find the transfer function for the op-amp circuit shown in Figure P8.62.

    b. Plot the pole-zero pattern.

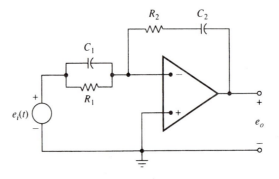

Figure P8.62

8.63. Repeat Problem 8.62 for the circuit shown in Figure P6.34.

9

DEVELOPING A LINEAR MODEL

In nearly all the examples used in earlier chapters, the elements were assumed to be linear. In practice, however, many elements are inherently nonlinear and may be considered linear over only a limited range of operating conditions.

When confronted with a mathematical model that contains nonlinearities, the analyst has essentially three choices: (1) to attempt to solve the differential equations directly, (2) to derive a fixed linear approximation that can be analyzed, or (3) to obtain computer solutions of the response for specific numerical cases. The first alternative is possible only in specialized cases and will not be pursued. We discussed computer solutions in Chapter 4, and we now present the linearization approach in this chapter.

We first develop a method for linearizing an element law where the two variables, such as force and displacement, are not directly proportional. We then show how to incorporate the linearized element law into the system model. We consider mechanical systems with nonlinear stiffness or friction elements, as well as nonlinear electrical systems. We conclude with an example that uses different models for different stages of the solution.

▶ 9.1 LINEARIZATION OF AN ELEMENT LAW

The object of linearization is to derive a linear model whose response will agree closely with that of the nonlinear model. Although the responses of the linear and nonlinear models will not agree exactly and may differ significantly under some conditions, there will generally be a set of inputs and initial conditions for which the agreement will be satisfactory. In this section, we consider the linearization of a single element law that is a nonlinear function of a single variable. We can express such an element law as an algebraic function $f(x)$. If x represents the total length of a nonlinear spring and $f(x)$ the force on the spring, the function $f(x)$ might appear as shown in Figure 9.1(a), where x_0 denotes the free or unstretched length.

We shall carry out the linearization of the element law with respect to an **operating point**, which is a specific point on the nonlinear characteristic denoted by $\bar{x}$ and $\bar{f}$. A sample operating point is shown in Figure 9.1(b). We discuss the procedure for determining the operating point in the following section; for now, we shall assume that the values of $\bar{x}$ and $\bar{f}$ are known.

We can write $x(t)$ as the sum of a constant portion, which is its value at the operating point, and a time-varying portion $\hat{x}(t)$ such that

$$x(t) = \bar{x} + \hat{x}(t) \tag{1}$$

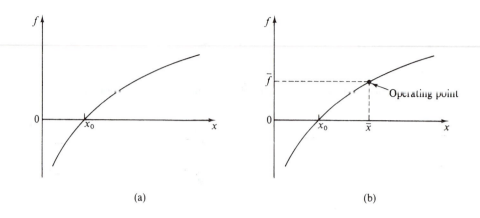

Figure 9.1 (a) A nonlinear spring characteristic. (b) Nonlinear spring characteristic with operating point.

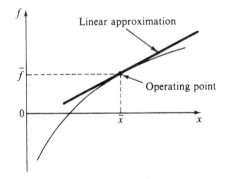

Figure 9.2 Nonlinear spring characteristic with linear approximation.

The constant term $\bar{x}$ is called the **nominal value** of x, and the time-varying term $\hat{x}(t)$ is the **incremental variable** corresponding to x. Likewise, we can write $f(t)$ as the sum of its nominal value $\bar{f}$ and the incremental variable $\hat{f}(t)$:

$$f(t) = \bar{f} + \hat{f}(t) \tag{2}$$

where the dependence of $\hat{f}$ on time is shown explicitly. Because $\bar{x}$ and $\bar{f}$ always denote a point that lies on the curve for the nonlinear element law,

$$\bar{f} = f(\bar{x}) \tag{3}$$

Having defined the necessary terms, we shall develop two equivalent methods of linearizing the element law to relate the incremental variables $\hat{x}$ and $\hat{f}$. The first uses a graphical approach; the second is based on a Taylor-series expansion.

Graphical Approach

Figure 9.2 shows the nonlinear element law $f(x)$ with the tangent to the curve at the operating point appearing as the straight line. For the moment, we note that the tangent line will be a good approximation to the nonlinear curve provided that the independent variable x does not deviate greatly from its nominal value $\bar{x}$ and that the curvature of the curve $f(x)$

is small in the vicinity of the operating point. The slope of the tangent line is

$$k = \frac{df}{dx}\Big|_{x=\bar{x}}$$

or, written more concisely,

$$k = \frac{df}{dx}\Big|_{\bar{x}} \tag{4}$$

where the subscript $\bar{x}$ after the vertical line indicates that the derivative must be evaluated at $x = \bar{x}$. The tangent passes through the operating point that has the coordinates $(\bar{x}, \bar{f})$ and is described by the equation

$$f = \bar{f} + k(x - \bar{x})$$

which can be written as

$$f - \bar{f} = k(x - \bar{x}) \tag{5}$$

Noting from (1) and (2) that the incremental variables are

$$\hat{x} = x - \bar{x} \tag{6a}$$

$$\hat{f} = f - \bar{f} \tag{6b}$$

we see that (5) reduces to

$$\hat{f} = k\hat{x} \tag{7}$$

where k is given by (4).

We can represent (7) in graphical form by redrawing the nonlinear function $f(x)$ with a coordinate system whose axes are $\hat{x}$ and $\hat{f}$ and whose origin is located at the operating point, as depicted in Figure 9.3. The incremental variables $\hat{x}$ and $\hat{f}$ are linearly related, the constant of proportionality k being the slope of the tangent line at the operating point.

Obviously, the accuracy of the linear approximation depends on the curvature of $f(x)$ in the vicinity of the operating point $(\bar{x}, \bar{f})$ and on the extent to which x deviates from $\bar{x}$ as the system responds to its excitations. We expect that the linearized model should be a good approximation for those values of x for which the straight line closely approximates the original curve. If the nonlinear spring is part of a larger system, we are unlikely to know in advance what range of values we will encounter for x. The problem is further complicated by the fact that usually we are primarily interested in how well certain responses of the overall system are approximated by calculations made on the linearized model. For example, we can expect some of the elements in a particular system to play a more critical role than others in determining the response of interest. Generally, the only way to assess

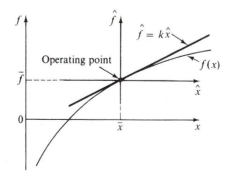

Figure 9.3 Nonlinear spring characteristic with incremental-variable coordinates.

with certainty the quality of the approximations is to compare computer solutions for the nonlinear model and the linearized model.

Series-Expansion Approach

As an alternative to the geometric arguments just presented, we can derive the linearized approximation in (7) by expressing $f(x)$ in terms of its **Taylor-series expansion** about the operating point $(\bar{x}, \bar{f})$. This expansion is

$$f(x) = f(\bar{x}) + \left.\frac{df}{dx}\right|_{\bar{x}} (x - \bar{x}) + \frac{1}{2!}\left.\frac{d^2f}{dx^2}\right|_{\bar{x}} (x - \bar{x})^2 + \cdots$$

where the subscript $\bar{x}$ after the vertical line indicates that the associated derivative is evaluated at $x = \bar{x}$. We can find the first two terms of this expansion provided that f and its first derivative exist for $x = \bar{x}$. We seek a linear approximation to the actual curve, so we shall neglect subsequent terms, which are higher-order in $x - \bar{x}$. The justification for truncating the series after the first two terms is that if x is sufficiently close to $\bar{x}$, then the higher-order terms are negligible compared to the constant and linear terms. Hence we write

$$f(x) \simeq f(\bar{x}) + \left.\frac{df}{dx}\right|_{\bar{x}} (x - \bar{x}) \tag{8}$$

Because $\bar{f} = f(\bar{x})$ and $k = [df/dx]|_{\bar{x}}$, this equation is equivalent to (5), which reduces to (7).

The accuracy of the linearized approximation in (8) depends on the extent to which the higher-order terms we have omitted from the Taylor-series expansion are truly negligible. This in turn depends on the magnitude of $x - \bar{x}$, which is the incremental independent variable, and on the values of the higher-order derivatives of $f(x)$ at the operating point. Before addressing the linearization of the complete model, we shall consider two numerical examples that illustrate the linearization of a nonlinear element law.

▶ EXAMPLE 9.1

A nonlinear translational spring obeys the force-displacement relationship $f(x) = |x|x$, where x is the elongation of the spring from its unstretched length. Determine the linearized element law in numerical form for each of the operating points corresponding to the nominal spring elongations $\bar{x}_1 = -1, \bar{x}_2 = 0, \bar{x}_3 = 1$, and $\bar{x}_4 = 2$.

SOLUTION We can rewrite $f(x)$ as

$$f(x) = \begin{cases} -x^2 & \text{for } x < 0 \\ x^2 & \text{for } x \geq 0 \end{cases} \tag{9}$$

The slope of the tangent at the operating point is

$$k = \left.\frac{df}{dx}\right|_{\bar{x}} = \begin{cases} -2\bar{x} & \text{for } \bar{x} < 0 \\ 2\bar{x} & \text{for } \bar{x} \geq 0 \end{cases}$$

$$= 2|\bar{x}| \quad \text{for all } \bar{x} \tag{10}$$

Thus the linear approximation to the spring characteristic is

$$\hat{f} = 2|\bar{x}|\hat{x}$$

where the coefficient $k = 2|\bar{x}|$ can be thought of as an effective spring constant whose numerical value depends on the nominal value of the spring's elongation $\bar{x}$. Substituting

TABLE 9.1 **Nominal Elongations, Nominal Forces, and Effective Spring Constants**

i	$\bar{x}_i$	$\bar{f}_i$	k_i
1	−1	−1	2
2	0	0	0
3	1	1	2
4	2	4	4

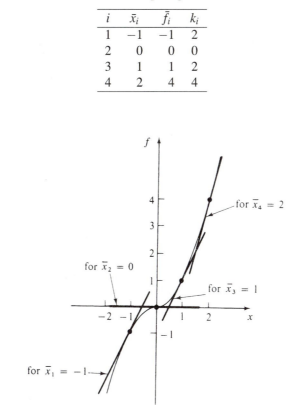

Figure 9.4 Nonlinear spring characteristic and linear approximations for four values of $\bar{x}$.

the four specified values of $\bar{x}$ into (9) and (10) gives the values of $\bar{f}$ and k that appear in Table 9.1. Figure 9.4 shows the four linear approximations superimposed on the nonlinear spring characteristic. Note that the value of the effective spring constant k is strongly dependent on the location of the operating point. In fact, k vanishes for $\bar{x} = 0$, which implies that the spring would not appear in the linearized model of a system having $\bar{x} = 0$ as its operating point. It is also interesting to note that k is the same for $\bar{x} = -1$ and $\bar{x} = +1$, although the values of $\bar{f}$ differ.

Concerning the accuracy of the approximation, one might say that for deviations in x up to 0.25 from the operating point, the approximation seems to be quite good; for deviations exceeding 1.0, it would be poor. It is difficult to make a definitive statement, however, without knowing the system in which the element is to appear.

▷ **EXAMPLE 9.2**

A torque τ_M exerted on a body that can rotate with an angular displacement θ is given by the equation $\tau_M = D\sin\theta$. Determine the linearized element law relating $\hat{\tau}_M$ and $\hat{\theta}$. Consider the five operating points corresponding to $\bar{\theta}_1 = 0, \bar{\theta}_2 = \pi/4, \bar{\theta}_3 = \pi/2, \bar{\theta}_4 = 3\pi/4$, and $\bar{\theta}_5 = \pi$.

TABLE 9.2   Operating Points and Effective Stiffness Constants for Example 9.2

i	$\bar{\theta}_i$	$\bar{\tau}_{M_i}$	k_i
1	0	0	D
2	$\pi/4$	$D/\sqrt{2}$	$D/\sqrt{2}$
3	$\pi/2$	D	0
4	$3\pi/4$	$D/\sqrt{2}$	$-D/\sqrt{2}$
5	π	0	$-D$

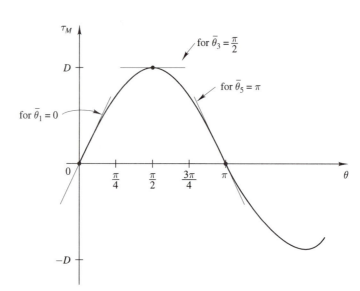

Figure 9.5   Nonlinear characteristic curve and linear approximations for Example 9.2.

SOLUTION   The Taylor-series expansion for τ_M is

$$\tau_M = D\sin\bar{\theta} + \frac{d}{d\theta}D\sin\theta\bigg|_{\bar{\theta}}(\theta - \bar{\theta}) + \cdots$$

$$= D\sin\bar{\theta} + D(\cos\bar{\theta})\hat{\theta} + \cdots$$

Using the first two terms in this series, and noting that $\bar{\tau}_M = D\sin\bar{\theta}$, we can write

$$\hat{\tau}_M = D(\cos\bar{\theta})\hat{\theta} = k\hat{\theta}$$

where $k = D\cos\bar{\theta}$ is the linearized stiffness constant. In Table 9.2, the values of $\bar{\theta}, \bar{\tau}_M$, and k are listed for each of the five specified operating points.

The values of k_i, which were found from $k_i = D\cos\bar{\theta}_i$, are also the slopes of the tangents to the characteristic curve drawn at the operating points. The tangent lines at three of these operating points are shown in Figure 9.5.

In the free-body diagram for the pendulum shown in Figure 5.23, one of the torques was $\tau_M = MgL\sin\theta$. In Examples 5.7 and 8.8, we used the small-angle approximation for θ and a trigonometric identity to obtain a linearized element law for small variations about the operating points $\bar{\theta} = 0$ and $\bar{\theta} = \pi$. We found that $\hat{\tau}_M = (MgL)\hat{\theta}$ for $\bar{\theta} = 0$ and that $\hat{\tau}_M = -(MgL)\hat{\theta}$ for $\bar{\theta} = \pi$. In Example 9.2, these same results were obtained by a more general linearization method.

▶ 9.2 LINEARIZATION OF THE MODEL

We shall now consider the process of incorporating one or more linearized element laws into a system model. Starting with a given nonlinear model, we need to do the following:

1. Determine the operating point of the model by writing and solving the appropriate nonlinear algebraic equations. Select the proper operating-point value if extraneous solutions also appear.

2. Rewrite all linear terms in the mathematical model as the sum of their nominal and incremental variables, noting that the derivatives of constant terms are zero.

3. Replace all nonlinear terms by the first two terms of their Taylor-series expansions—that is, the constant and linear terms.

4. Using the algebraic equation(s) defining the operating point, cancel the constant terms in the differential equations, leaving only linear terms involving incremental variables.

5. Determine the initial conditions of all incremental variables in terms of the initial conditions of the variables in the nonlinear model.

For all situations we shall consider, the operating point of the system will be a condition of equilibrium in which each variable will be constant and equal to its nominal value and in which all derivatives will be zero. Inputs will take on their nominal values, which are typically selected to be their average values. For example, if a system input were $u(t) = A + B\sin\omega t$, then the nominal value of the input would be taken as $\bar{u} = A$. Under these conditions, the differential equations reduce to algebraic equations that we can solve for the operating point, using a computer if necessary.

Upon completion of step 4, the terms remaining in the model should involve only incremental variables, and the terms should all be linear with constant coefficients. In general, the coefficients involved in those terms that came from the expansion of nonlinear terms depend on the equilibrium conditions. Hence we must find a specific operating point before we can express the linearized model in numerical form. The entire procedure will be illustrated by several examples.

▶ EXAMPLE 9.3

Derive a linearized model of the translational mechanical system shown in Figure 9.6(a), where the nonlinear spring characteristic $f_K(x)$ is given in Figure 9.6(b) and where the average value of the applied force $f_a(t)$ is zero.

SOLUTION First we derive the nonlinear model by drawing the free-body diagram shown in Figure 9.6(c) and summing forces. This yields

$$M\ddot{x} + B\dot{x} + f_k(x) = f_a(t) \tag{11}$$

To find the operating point, we replace $f_a(t)$ by its average value $\bar{f}_a$ and x by $\bar{x}$:

$$M\ddot{\bar{x}} + B\dot{\bar{x}} + f_K(\bar{x}) = \bar{f}_a$$

Noting that $\dot{\bar{x}} = \ddot{\bar{x}} = 0$ because $\bar{x}$ is a constant and that $\bar{f}_a$ was specified to be zero, we see that

$$\bar{f}_K = f_K(\bar{x}) = 0$$

Thus the operating point is at $\bar{x} = 0, \bar{f}_K = 0$, which corresponds to the origin of the spring characteristic in Figure 9.6(b).

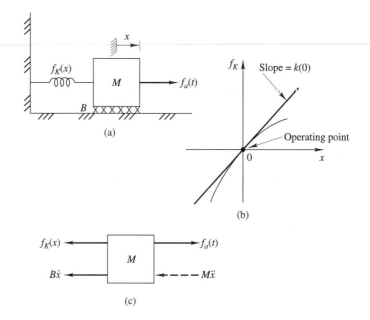

Figure 9.6 (a) Nonlinear system for Example 9.3. (b) Nonlinear spring characteristic. (c) Free-body diagram.

The next step is to rewrite the linear terms in (11) in terms of the incremental variables $\hat{x} = x - \bar{x}$ and $\hat{f}_a(t) = f_a(t) - \bar{f}_a$. This yields

$$M(\ddot{\bar{x}} + \ddot{\hat{x}}) + B(\dot{\bar{x}} + \dot{\hat{x}}) + f_K(x) = \bar{f}_a + \hat{f}_a(t)$$

Because $\dot{\bar{x}} = \ddot{\bar{x}} = 0$, we can rewrite the equation as

$$M\ddot{\hat{x}} + B\dot{\hat{x}} + f_K(x) = \bar{f}_a + \hat{f}_a(t) \tag{12}$$

Expanding the spring force $f_K(x)$ about $\bar{x} = 0$ gives

$$f_K(x) = f_K(0) + \left.\frac{df_K}{dx}\right|_{x=0} \hat{x} + \cdots$$

Substituting the first two terms into (12) yields the approximate equation

$$M\ddot{\hat{x}} + B\dot{\hat{x}} + f_K(0) + k(0)\hat{x} = \bar{f}_a + \hat{f}_a(t)$$

The constant $k(0)$ denotes the derivative df_K/dx evaluated at $x = 0$ and is the slope of the tangent to the spring characteristic at the operating point, as indicated in Figure 9.6(b). The spring force at the operating point is $f_K(0) = \bar{f}_a = 0$, so the linearized model is

$$M\ddot{\hat{x}} + B\dot{\hat{x}} + k(0)\hat{x} = \hat{f}_a(t) \tag{13}$$

which is a fixed linear differential equation in the incremental variable $\hat{x}$ with the incremental input $\hat{f}_a(t)$. The coefficients are the constants M, B, and $k(0)$. To solve (13), we must know the initial values $\hat{x}(0)$ and $\dot{\hat{x}}(0)$, which we find from the initial values $x(0)$ and $\dot{x}(0)$. Because $\hat{x}(t) = x(t) - \bar{x}$ and $\dot{\hat{x}}(t) = \dot{x}(t) - \dot{\bar{x}}$ for all values of t,

$$\hat{x}(0) = x(0) - \bar{x}$$
$$\dot{\hat{x}}(0) = \dot{x}(0) - \dot{\bar{x}}$$

In this example, $\bar{x} = \dot{\bar{x}} = 0$, so $\hat{x}(0) = x(0)$ and $\dot{\hat{x}}(0) = \dot{x}(0)$. Once we have solved the linearized model, we find the approximate solution of the nonlinear model by adding the nominal value $\bar{x}$ to the incremental solution $\hat{x}(t)$. Remember that the sum of the terms, $\bar{x} + \hat{x}(t)$, is only an approximation to the actual solution of the nonlinear model.

Should we want to put the linearized model given by (13) into state-variable form, we need only define the incremental velocity $\hat{v} = v - \bar{v}$, where $\bar{v} = 0$. Then $\hat{v} = \dot{\hat{x}}$, and we can write the pair of first-order equations

$$\dot{\hat{x}} = \hat{v}$$

$$\dot{\hat{v}} = \frac{1}{M}[-k(0)\hat{x} - B\hat{v} + \hat{f}_a(t)]$$

Because $\bar{x} = \bar{v} = 0$, the appropriate initial conditions are $\hat{x}(0) = x(0)$ and $\hat{v}(0) = v(0)$.

▷ EXAMPLE 9.4

Repeat Example 9.3 with the applied force $f_a(t)$ having a nonzero average value of $\bar{f}_a$ for positive time, as shown in Figure 9.7(a).

SOLUTION   The form of the nonlinear model given by (11) is unaffected by the value of $\bar{f}_a$. However, a new operating point will exist that is defined by the equation

$$f_K(\bar{x}) = \bar{f}_K = \bar{f}_a \tag{14}$$

For the spring characteristic shown in Figure 9.7(b) and the value of $\bar{f}_a$ shown in part (a) of the figure, the operating point is point A. In graphical terms, the straight line $\bar{f}_a$ in Figure 9.7(a) is projected horizontally onto the curve of the spring characteristic in part (b) of the figure, intersecting it at the operating point, the coordinates of which are $x = \bar{x}, f_K = \bar{f}_K$.

Upon substituting $x = \bar{x} + \hat{x}$ and $f_a(t) = \bar{f}_a + \hat{f}_a(t)$ into (11) and using $\dot{\bar{x}} = \ddot{\bar{x}} = 0$, we obtain

$$M\ddot{\hat{x}} + B\dot{\hat{x}} + f_K(x) = \bar{f}_a + \hat{f}_a(t) \tag{15}$$

which is identical to (12). The first two terms in the Taylor series for the spring force are

$$f_K(\bar{x}) + \left.\frac{df_K}{dx}\right|_{\bar{x}} \hat{x} \tag{16}$$

where $f_K(\bar{x}) = \bar{f}_K$ and where $\bar{x}$ must satisfy (14). Substituting (16) for $f_K(x)$ into (15) and invoking (14) yield the desired linear model:

$$M\ddot{\hat{x}} + B\dot{\hat{x}} + k(\bar{x})\hat{x} = \hat{f}_a(t) \tag{17}$$

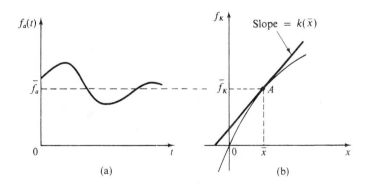

Figure 9.7 (a) Applied force for Example 9.4. (b) Nonlinear spring characteristic with new operating point.

where $k(\bar{x}) = [df_K/dx]|_{\bar{x}}$ and is the slope of the straight line in Figure 9.7(b). Note that the form of the model given by (17) with $\hat{f}_a \neq 0$ is the same as that with $\hat{f}_a = 0$, which is given by (13). The only difference between the two equations is the value of the effective spring constant. The value of $k(\bar{x})$ depends on the value of $\bar{x}$ at which the slope of $f_K(x)$ is measured. Hence the responses of the two linearized models could be rather different, even for the same incremental applied force $\hat{f}_a(t)$.

▶ **EXAMPLE 9.5**

Derive a linear model for the mechanical system and spring characteristic shown in Figure 9.8, where $x = 0$ corresponds to an unstretched spring.

SOLUTION   We obtain the nonlinear model of the system by drawing the free-body diagram shown in Figure 9.9(a) and setting the sum of the vertical forces equal to zero. Because the mass is constrained to move vertically, we must include its weight Mg in the free-body diagram. The resulting nonlinear model is

$$M\ddot{x} + B\dot{x} + f_K(x) = f_a(t) + Mg$$

Note that its form is similar to that given by (11), the nonlinear model for the two preceding examples. By setting $x = \bar{x}$ and $f_a(t) = \bar{f}_a$ and by noting that $\dot{x} = \ddot{x} = 0$, we find the algebraic equation for the operating point to be

$$f_K(\bar{x}) = \bar{f}_a + Mg$$

which is the same as (14) except for inclusion of the force Mg on the right side.

The nonlinear spring characteristic in Figure 9.8(b) is repeated in Figure 9.9(c). Figure 9.9(b) shows $f_a(t)$ and $\bar{f}_a$ as in Example 9.4, but it also shows the total nominal force $\bar{f}_a + Mg$ that must be projected onto the characteristic curve in Figure 9.9(c) to establish the operating point A_1. The point A_2, which is obtained by projecting the force $\bar{f}_a$ from Figure 9.9(b), would be the operating point if the motion of the mass were horizontal, as it is in Example 9.4. For applied forces having the same average values, the two systems will have different linearized spring characteristics if the curve of $f_K(x)$ does not have the same slope at points A_1 and A_2. If the spring were linear, however, the slope of the characteristic in Figure 9.9(c) would be constant and the presence of the weight would have no influence on the effective spring constant, as was observed in Example 2.5. With the provision that

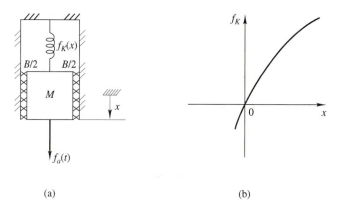

(a)   (b)

Figure 9.8   (a) Mechanical system for Example 9.5. (b) Nonlinear spring characteristic.

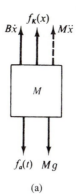

(a)

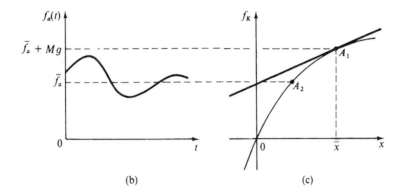

(b)   (c)

Figure 9.9 (a) Free-body diagram for Example 9.5. (b) Input. (c) Nonlinear spring characteristic with two operating points.

$k(\bar{x})$ is the slope of $f_K(x)$ measured at point A_1 rather than at point A_2, the resulting linearized model is again given by (17).

▶ **EXAMPLE 9.6**

A high-speed vehicle of mass M moves along a horizontal track and is subject to a linear retarding force Bv caused by viscous friction associated with the bearings and a nonlinear retarding force $D|v|v$ caused by air drag. Obtain a linear model that is valid when the driving force $f_a(t)$ undergoes variations about a positive nominal value $\bar{f}_a$.

SOLUTION   The nonlinear differential equation governing the vehicle's velocity is

$$M\dot{v} + Bv + D|v|v = f_a(t) \tag{18}$$

Setting $v = \bar{v}$ and $f_a(t) = \bar{f}_a$ and noting that $\dot{\bar{v}} = 0$, we have

$$B\bar{v} + D|\bar{v}|\bar{v} = \bar{f}_a$$

for the operating-point equation. Because $\bar{f}_a$ is positive, we know that $\bar{v}$ is positive and we can replace $D|\bar{v}|\bar{v}$ by $D(\bar{v})^2$. Then

$$D(\bar{v})^2 + B\bar{v} - \bar{f}_a = 0 \tag{19}$$

By inspection we see that (19) will have two real roots, one positive and the other negative. However, we are interested only in the positive root, which is

$$\bar{v} = \frac{-B + \sqrt{B^2 + 4\bar{f}_a D}}{2D} \tag{20}$$

The negative root was introduced when we replaced $|\bar{v}|\bar{v}$ by $(\bar{v})^2$ and is not a root of the original operating-point equation. Provided that v always remains positive, we can replace $D|\bar{v}|\bar{v}$ in (18) by $D v^2$. Then, using $v = \bar{v} + \hat{v}$ and $f_a(t) = \bar{f}_a + \hat{f}_a(t)$ and noting that $\dot{\bar{v}} = 0$, we can rewrite (18) as

$$M\dot{\hat{v}} + B(\bar{v} + \hat{v}) + D v^2 = \bar{f}_a + \hat{f}_a(t) \tag{21}$$

To linearize the term v^2, we replace it by the constant and linear terms in its Taylor series:

$$(\bar{v})^2 + \left.\frac{d}{dv}(v^2)\right|_{\bar{v}} (v - \bar{v}) = (\bar{v})^2 + 2\bar{v}\hat{v} \tag{22}$$

Substituting (22) for v^2 into (21) and regrouping, we have

$$M\dot{\hat{v}} + (B + 2D\bar{v})\hat{v} + B\bar{v} + D(\bar{v})^2 = \bar{f}_a + \hat{f}_a(t)$$

The constant terms cancel because of (19), the operating-point equation. Thus the desired linearized model, which holds for $\bar{f}_a > 0$ and $\bar{v} > 0$, is

$$M\dot{\hat{v}} + b\hat{v} = \hat{f}_a(t)$$

where b denotes the effective damping coefficient

$$b = B + 2D\bar{v}$$

with $\bar{v}$ given by (20) as a function of the average driving force $\bar{f}_a$.

It is worthwhile to observe that in this particular case the Taylor series of the non-linearity has only three terms, and we could have obtained it without differentiation by writing $$v^2 = (\bar{v} + \hat{v})^2 = (\bar{v})^2 + 2\bar{v}\hat{v} + (\hat{v})^2$$
In this example, we can see that $(\hat{v})^2$ is the error introduced by replacing v^2 by $(\bar{v})^2 + 2\bar{v}\hat{v}$. Provided that $\bar{v} \gg |\hat{v}|$, the error will be small compared to the two terms that are retained.

▶ **EXAMPLE 9.7**

A nonlinear system obeys the state-variable equations

$$\dot{x} = y \tag{23a}$$

$$\dot{y} = -|x|x - 2x - 2y^3 - 3 + A\sin t \tag{23b}$$

Find the operating point and develop the linearized model in numerical form.

SOLUTION The operating point, described by $\bar{x}$ and $\bar{y}$, must satisfy the conditions $\dot{\bar{x}} = \dot{\bar{y}} = 0$ with the incremental portion of the input set to zero. Hence the operating-point equations reduce to

$$\bar{y} = 0$$
$$|\bar{x}|\bar{x} + 2\bar{x} + 3 = 0 \tag{24}$$

which have the solution $\bar{x} = -1, \bar{y} = 0$. We replace the two nonlinear elements in (23b) by the first two terms in their respective Taylor-series expansions. For $|x|x$, we write

$$|\bar{x}|\bar{x} + 2|\bar{x}|\hat{x}$$

and we replace y^3 by

$$(\bar{y})^3 + 3(\bar{y})^2 \hat{y}$$

By substituting these approximations into (23) and using $x = \bar{x} + \hat{x}$ and $y = \bar{y} + \hat{y}$, we obtain

$$\dot{\hat{x}} = \bar{y} + \hat{y}$$
$$\dot{\hat{y}} = -(|\bar{x}|\bar{x} + 2|\bar{x}|\hat{x}) - 2(\bar{x} + \hat{x}) - 2[(\bar{y})^3 + 3(\bar{y})^2\hat{y}] - 3 + A\sin t \tag{25}$$

With $\bar{x} = -1$ and $\bar{y} = 0$, (25) reduces to

$$\dot{\hat{x}} = \hat{y}$$
$$\dot{\hat{y}} = -4\hat{x} + A\sin t \tag{26}$$

which is the linearized model in state-variable form. Comparing (26) with (25), we note that all the constant terms have been canceled, the coefficient of $\hat{y}$ in the second equation is zero because $\bar{y} = 0$, and the coefficient of $\hat{x}$ reflects the combined effects of the linear and nonlinear terms.

Circuits with Nonlinear Resistors

We have defined resistors, capacitors, and inductors as elements for which there is an algebraic relationship between the voltage and current, voltage and charge, and current and flux linkage, respectively. If the two variables involved in the algebraic relationship are directly proportional to one another, then the element is linear, as was the case in all the examples in Chapter 6.

We now consider circuits with nonlinear resistors. The general procedure for obtaining a linearized model is the same as that used for the mechanical examples. As in Section 9.1, we express the variables as the sum of a constant portion and a time-varying portion. For example, we write a voltage e_o as

$$e_o = \bar{e}_o + \hat{e}_o$$

where the constant term $\bar{e}_o$ is the nominal value, corresponding to a particular operating point, and where $\hat{e}_o$ is the incremental time-varying component. In the circuit diagrams, we indicate the fact that a resistor is nonlinear by drawing a curved line through its symbol. The procedure is illustrated by two examples of increasing complexity.

▷ **EXAMPLE 9.8**

The circuit shown in Figure 9.10(a) contains a nonlinear resistor that obeys the element law $i_o = 2e_o^3$. Write the differential equation relating e_o and $e_i(t)$. If $e_i(t) = 18 + A\cos\omega t$, find the operating point and derive the linearized input-output differential equation. Also determine the time constant of the linearized model.

SOLUTION The right-hand resistor is nonlinear because the current i_o is not directly proportional to the voltage e_o. Summing the currents leaving the node at the upper right gives

$$\frac{1}{2}\dot{e}_o + [e_o - e_i(t)] + 2e_o^3 = 0$$

Thus the input-output equation is

$$\frac{1}{2}\dot{e}_o + 2e_o^3 + e_o = e_i(t) \tag{27}$$

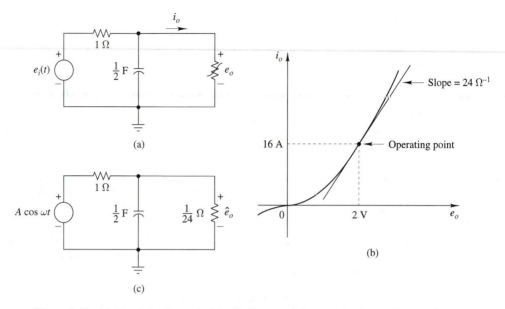

Figure 9.10 (a) Circuit for Example 9.8. (b) Characteristic curve for the nonlinear resistor. (c) Linearized equivalent circuit.

To determine the operating point, we replace $e_i(t)$ by $\bar{e}_i = 18$ V, e_o by $\bar{e}_o$, and $\dot{e}_o$ by zero to obtain

$$2\bar{e}_o^3 + \bar{e}_o = 18$$

The only real value of $\bar{e}_o$ that satisfies this algebraic equation is

$$\bar{e}_o = 2 \text{ V} \tag{28}$$

From the nonlinear element law, we see that $\bar{i}_o = 16$ A, which gives the operating point shown in Figure 9.10(b).

To develop a linearized model, we let

$$e_i(t) = 18 + A\cos\omega t$$

$$e_o = 2 + \hat{e}_o \tag{29}$$

As in (8), we write the first two terms in the Taylor series for the nonlinear term $2e_o^3$, which are

$$\bar{i}_o + \left.\frac{di_o}{de_o}\right|_{\bar{e}_o}(e_o - \bar{e}_o) = \bar{i}_o + (6\bar{e}_o^2)\hat{e}_o$$

$$= 16 + 24\hat{e}_o \tag{30}$$

This approximation describes the tangent to the characteristic curve at the operating point, as shown in Figure 9.10(b). Substituting (29) and (30) into (27) gives

$$\tfrac{1}{2}(\dot{\bar{e}}_o + \dot{\hat{e}}_o) + (16 + 24\hat{e}_o) + (2 + \hat{e}_o) = 18 + A\cos\omega t$$

Because $\dot{\bar{e}}_o = 0$ and because the constant terms cancel (as is always the case), we have for the linearized model

$$\tfrac{1}{2}\dot{\hat{e}}_o + 25\hat{e}_o = A\cos\omega t$$

or

$$\dot{\hat{e}}_o + 50\hat{e}_o = 2A\cos\omega t \tag{31}$$

By inspecting this equation, we see that the time constant of the linearized model is 0.02 s.

In the last example, we can see from (30), as well as from Figure 9.10(b), that $\hat{i}_o = 24\hat{e}_o$. This equation has the form of Ohm's law for a linear resistor: $\hat{i}_o = (1/r)\hat{e}_o$, where $r = 1/24\ \Omega$. Figure 9.10(c) shows the linearized equivalent circuit that relates the time-varying incremental variables. It has the same form as Figure 9.10(a), except that the nonlinear resistor has been replaced by a linearized element. It is straightforward to show that (31) can be obtained directly from part (c) of Figure 9.10. Keep in mind that the value of the linearized element depends on the operating point.

Computer simulations were discussed in Chapter 4. In the next example, we not only develop a linearized model but also use the results of a computer simulation to compare the nonlinear and linearized models. This is usually the only way to determine how large the time-varying component of the input can be and still have the linearized model give good results.

▶ EXAMPLE 9.9

Write the state-variable equations for the circuit shown in Figure 9.11(a), which contains a nonlinear resistor that obeys the element law $i_2 = \frac{1}{8}e_C^3$. Find the operating point when $e_i(t) = 2 + \hat{e}_i(t)$, and derive the linearized state-variable equations in terms of the incremental variables. Plot and compare i_L versus t for the nonlinear and linearized models when $\hat{e}_i(t) = [A\sin t]U(t)$ for $A = 0.1$ V, $A = 1.0$ V, and $A = 10.0$ V.

SOLUTION   We choose e_C and i_L as state variables and note that

$$e_L = e_i(t) - e_C \tag{32}$$

Applying Kirchhoff's current law to the upper right node gives

$$i_L + e_L - \frac{1}{8}e_C^3 - 2\dot{e}_C = 0$$

Inserting (32) into this equation and into the element law $di_L/dt = \frac{1}{3}e_L$ gives

$$\dot{e}_C = \frac{1}{2}\left[e_i(t) + i_L - \frac{1}{8}e_C^3 - e_C\right]$$

$$\frac{di_L}{dt} = \frac{1}{3}[e_i(t) - e_C] \tag{33}$$

which constitute the nonlinear state-variable equations.

At the operating point, the derivatives of the state variables are zero, and (33) reduces to the algebraic equations

$$\bar{e}_i + \bar{i}_L - \frac{1}{8}\bar{e}_C^3 - \bar{e}_C = 0$$

$$\bar{e}_i - \bar{e}_C = 0$$

With $\bar{e}_i = 2$ V, we find that $\bar{e}_C = 2$ V, $\bar{i}_L = 1$ A, and $\bar{i}_2 = 2^3/8 = 1$ A. An alternative way of determining the operating point is to recall from Section 6.5 that when all the voltages and currents are constant, we can replace the capacitors and inductors by open and short circuits, respectively. We do this in Figure 9.11(b), from which we again see that $\bar{e}_C = 2$ V and $\bar{i}_L = \bar{i}_2 = 1$ A.

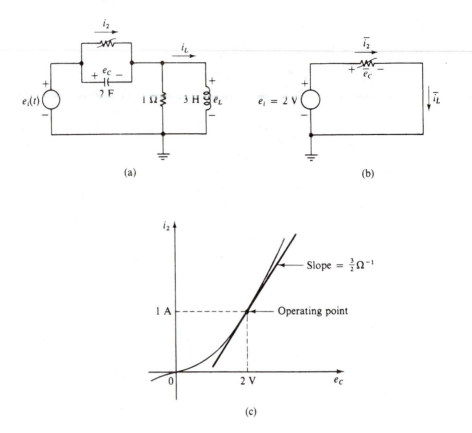

Figure 9.11 (a) Circuit for Example 9.9. (b) Circuit used for determining the operating point. (c) Characteristic curve for the nonlinear resistor.

Next we define the incremental variables $\hat{e}_C, \hat{i}_L$, and $\hat{e}_i(t)$ by the equations

$$e_C = 2 + \hat{e}_C$$
$$i_L = 1 + \hat{i}_L \tag{34}$$
$$e_i(t) = 2 + \hat{e}_i(t)$$

The Taylor-series expansion for the nonlinear resistor is

$$i_2 = \frac{1}{8} e_C^3 = \frac{1}{8} \bar{e}_C^3 + \frac{3}{8} \bar{e}_C^2 (e_C - \bar{e}_C) + \cdots$$

Using the first two terms in the series, with $\bar{e}_C = 2$ V, we have

$$i_2 \simeq 1 + \frac{3}{2} \hat{e}_C \tag{35}$$

which is shown graphically by the straight line in Figure 9.11(c). Substituting (34) and (35) into (33) and canceling the constant terms, we obtain

$$\dot{e}_C = \frac{1}{2} \left[\hat{e}_i(t) + \hat{i}_L - \frac{5}{2} \hat{e}_C \right] \tag{36a}$$

$$\frac{d\hat{i}_L}{dt} = \frac{1}{3} [\hat{e}_i(t) - \hat{e}_C] \tag{36b}$$

as the linearized state-variable equations.

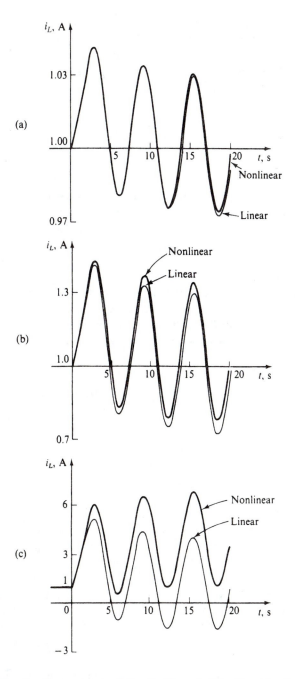

Figure 9.12 Results of computer simulation for Example 9.9 with $e_i(t) = 2 + [A \sin t] U(t)$.
(a) $A = 0.1$ V. (b) $A = 1.0$ V. (c) $A = 10.0$ V.

The results of a computer solution for i_L of the nonlinear model given by (33) with $e_i(t) = 2 + [A\sin t]U(t)$ are shown in Figure 9.12 for three different values of the amplitude A. Plotted on the same axes are curves obtained by calculating $\hat{i}_L$ from the linearized model in (36) with $\hat{e}_i(t) = [A\sin t]U(t)$ and then forming the quantity $\hat{i}_L + \bar{i}_L$. Because $e_i(t) = \bar{e}_i$ for all $t < 0$, we used the initial conditions

$$e_C(0) = \bar{e}_C = 2 \text{ V}, \quad \hat{e}_C(0) = 0$$
$$i_L(0) = \bar{i}_L = 1 \text{ A}, \quad \hat{i}_L(0) = 0$$

Note that the responses of the nonlinear and linearized models are almost identical when $A = 0.1$ V, are in close agreement when $A = 1.0$ V, but differ significantly when $A = 10.0$ V. The steady-state response of the linearized model is always a sinusoidal oscillation about the operating point. For large values of A, however, the steady-state response of the nonlinear model is not symmetrical about the operating point.

We have plotted the three sets of curves with different vertical scales and different origins in order to get a good comparison of the responses of the nonlinear and linearized models for each of the three values of A. If we repeat the example when $\bar{e}_i = 0$ and $\hat{e}_i(t) = [A\sin t]U(t)$, then in the steady state we would expect the response of the nonlinear model to be symmetrical about the operating point. When we carry out a computer run for this case, we obtain not only the expected symmetry but also good agreement of the nonlinear and linearized responses for all three values of A.

▶ 9.3 COMPUTER SIMULATION

We now solve two nonlinear problems using MATLAB and Simulink.

▶ EXAMPLE 9.10

Construct a Simulink diagram to represent (23) and (26) in Example 9.7. Run the simulation for $A = 3$, $x(0) = -1$, and $y(0) = 0$. Plot on the same set of axes x versus t for the nonlinear and linearized models for $0 \le t \le 20$ s.

SOLUTION The top part of the Simulink diagram in Figure 9.13 represents the nonlinear model in (23). Several blocks appear here for the first time. A function block, which appears in the Functions & Tables library, allows the creation of an arbitrary function of the input variable. It is used here to form y^3. Two other blocks are found in the Math library: Abs and Product. Abs forms the absolute value of its input. The Product block has two inputs, and the output is their product. These blocks are used in Figure 9.13 to form the $|x|x$ term in (23). As shown on the diagram, we set the initial condition $x(0)$ equal to -1. The lower part of Figure 9.13 shows the blocks needed to calculate $\hat{x}$ according to (26). The initial conditions on both integrators are set equal to zero. In order to compare the output to that for the nonlinear model, we form $x = \hat{x} + \bar{x}$, where $\bar{x} = -1$. Note that the displacement feedback signal $4\hat{x}$ in the linearized model is the sum of two parts: the $2x$ term in the original nonlinear model plus another $2x$ due to the slope of $|x|x$ at the operating point $\bar{x} = -1$. Although the nonlinear model contains the damping term y^3 fed back to the input summer, the linearized model has no damping. This is because the slope of the characteristic curve for y^3 is zero at the operating point $\bar{y} = 0$.

A graph of the first 20 seconds of the solution is shown in Figure 9.14. Both curves include an oscillation at 1 rad/s caused by the input. However, the linearized model also contains a sinusoidal oscillation at 2 rad/s that persists for all time because of the lack of damping. This combination of two sinusoids at different frequencies in the steady state produces the distinctive curve shown.

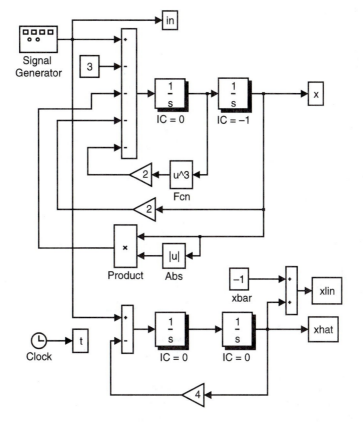

Figure 9.13 Simulink diagram for Example 9.10.

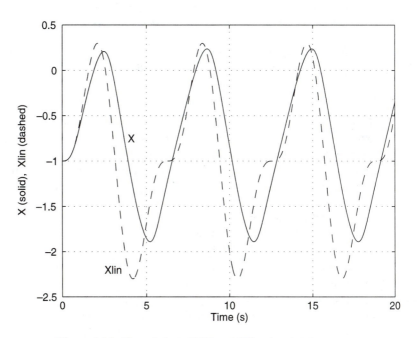

Figure 9.14 The solution of (23) and (26) using $A = 3$ for Example 9.10.

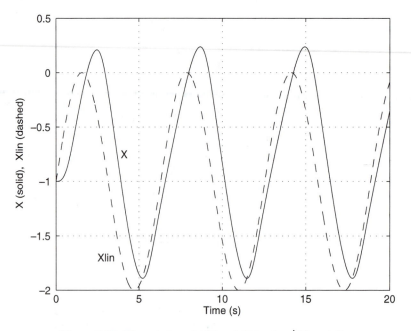

Figure 9.15 The solution of (23) and (26) using $\hat{\dot{x}}(0) = A/3$.

It is possible to choose the initial conditions so that the sinusoidal mode at 2 rad/s in the linearized model is not excited. It turns out that we need $\hat{x}(0) = 0$ and $\hat{\dot{x}}(0) = A/3$ in order for this to happen. When the model in Figure 9.13 is run with this value for $\hat{\dot{x}}(0)$, Figure 9.15 results. This illustrates the effect of failing to excite one of the mode functions of the system, a topic that was discussed in Section 8.2.

We now consider a mechanical system with a nonlinear load. We examine the effect of changing the size of the input, starting from two different operating conditions.

▶ **EXAMPLE 9.11**

The device shown in Figure 9.16 represents a fan driven by a motor. The motor supplies a torque that may be changed instantaneously from one constant value to another. We consider the initial value of that torque to be its operating point, and treat the change as the variable input. The fan blades provide a nonlinear load torque proportional to the cube of the angular velocity. Draw a free-body diagram and write the modeling equation for this system. Construct a Simulink diagram to solve this equation and simulate the response to an applied torque given by $\tau_a(t) = A + BU(t-1)$. Let $A = 0.02$ N·m, $B = 0.05$ N·m, and $C = 10^{-6}$ N·m³. Calculate the steady-state value of ω before the step is applied and use this as the initial condition for ω. Plot the response of ω versus time for $0 < t < 6$ s. Use the step block to produce $\tau_a(t)$, setting the step time to 1 s. Repeat the above when $A = 0.10$ and 0.50 N·m, and plot all the results on the same axes.

Next, linearize the model around the operating point $\bar{\tau}_a = 0.1$ N·m. Make a second plot of the responses of the nonlinear model to steps of $B = 0.05, 0.10$, and 0.20 N·m. Construct a Simulink diagram for the linearized model and plot its response on the same axes as the nonlinear model.

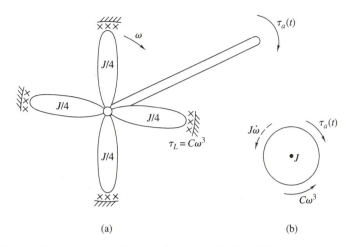

Figure 9.16 (a) A motor-driven fan for Example 9.11. (b) Free-body diagram.

SOLUTION   The applied torque $\tau_a(t)$ is produced by adding a constant term to the output of a step block in the upper left corner of the Simulink diagram in Figure 9.17.

The upper half of the figure is the nonlinear model, and includes a Function block that has been programmed to produce an output that is the cube of the input. The lower half of the figure is the linearized model. The input is only the variable part $\hat{\tau}_a(t)$ of the applied torque. The feedback gain is obtained from the linearized modeling equation $J\dot{\hat{\omega}} + 3(C\bar{\tau}^2)^{(1/3)}\hat{\omega} = \hat{\tau}_a(t)$.

The output of the integrator is $\hat{\omega}$. This is added to $\bar{\omega}$, the operating-point value of ω, to produce the variable called "omegalin," the total value of ω. The following abbreviated M-file establishes the parameter values and runs the simulation to produce the desired responses. Additional commands are needed to plot and label these responses as in Figures 9.18 and 9.19. The reader may access the Web site at www.wiley.com/college/close, where M-files to produce the MATLAB plots in this chapter are available to be downloaded and run.

```
% File is ex9_11_run.m
A = 0.1; B = 0.1; C = 1e-6; J = 0.005;
sim('ex9_11_model')
plot(t,omega);grid
% This produces the middle solid curve in Figure 9.18.
hold
B=0.05;
sim('ex9_11_model'); plot(t,omega, '--')
A=0.5;
sim('ex9_11_model'); plot(t,omega, '--')
B=0.1;
sim('ex9_11_model'); plot(t,omega)
A=0.02; B=0.05;
sim('ex9_11_model'); plot(t,omega, '--')
B=0.1;
sim('ex9_11_model'); plot(t,omega)
hold off
```

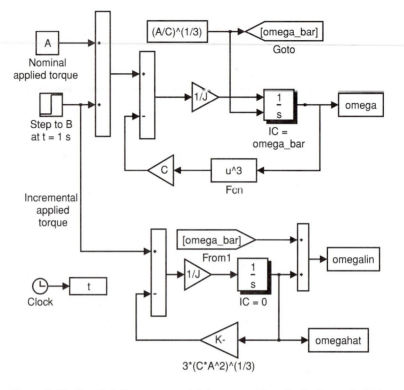

Figure 9.17 Simulink diagram to model the motor-driven fan for Example 9.11.

```
pause
% The next commands produce the responses shown in Figure 9.19.
A=0.1;
sim('ex9_11_model'); plot(t,omega)
hold plot(t,omegalin, '--')
B=0.05;
sim('ex9_11_model'); plot(t,omega)
plot(t,omegalin, '--'); grid
B=0.2;
sim('ex9_11_model'); plot(t,omega)
plot(t,omegalin, '--')
hold off
```

From Figure 9.18 we see that the increase in speed caused by a step of applied torque of 0.05 N·m is much greater if the operating point is at $\bar{\tau}_a = 0.02$ than if the operating point is at $\bar{\tau}_a = 0.50$.

In Figure 9.19 the linear and nonlinear models give similar results when the change from the operating point is small and $B = 0.05$, but the two models give very different results if B increases to 0.2 N·m. Although it is not immediately evident from Figure 9.19, the shape of the response of the nonlinear model is not a simple exponential, as is the case of the dashed lines of the linear model.

Reflecting on the results of this example and the preceding one, we see that a nonlinearity can alter the waveshape and the final value of the output. Its effect may be different

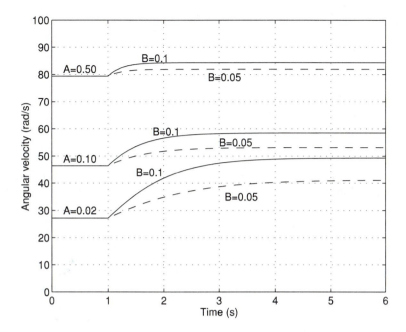

Figure 9.18 Fan speed for two values of B and three levels of starting torque in Example 9.11.

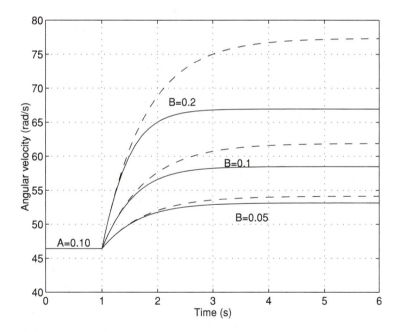

Figure 9.19 Fan speed for the nonlinear model (solid) and the linearized model (dashed) in Example 9.11.

for different operating points, and it is difficult to predict in advance the range of incremental inputs for which a linearized model will give a satisfactory approximation. Because it is difficult to establish general guidelines for approximating nonlinear systems, we must usually rely heavily on computer simulations.

▶ 9.4 PIECEWISE LINEAR SYSTEMS

Many systems are linear over a part of the range of their variables, and then have a different linear model over a different range. When the variables for a particular case stay in one range, the system can be analyzed without regard to the nonlinearity. If the variables cross from one range to another, however, the analytic solution can become difficult, and numerical solutions may be very helpful. We will use MATLAB to illustrate such a numerical solution.

Simulink has several functions that are useful in allowing a simulation to change its modeling equations according to the value of one or more of its variables. The Switch block has inputs u_1 and u_3 at the top and bottom, respectively, plus another input u_2 that determines the state of the switch. The output of the switch is given by

$$y = \begin{cases} u_1 & \text{if } u_2 < 0 \\ u_3 & \text{if } u_2 \geq 0 \end{cases}$$

In the following example, we use this block to build a piecewise linear system in which four of the parameters change their values at different points in the solution.

▶ EXAMPLE 9.12

Consider further the case of the parachute jumper studied in Example 4.7. In that example, we began the simulation with the parachute and jumper in free fall at a known velocity, with the parachute at the end of its risers. At $t = 0$, the parachute opened, and we simulated the displacements and velocities until a constant velocity of fall was achieved. We now extend that study to include the events that occur before the parachute opens, and also after the jumper lands on a large landing pad that has substantial elasticity and viscous damping. The mechanical model of the system, shown in Figure 9.20, is similar to the

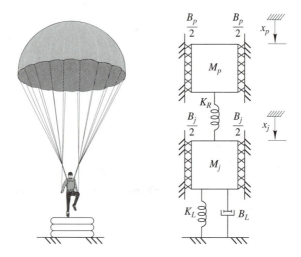

Figure 9.20 A parachute jumper and a mechanical model of the parachute jumper.

one in Figure 4.17, but with the addition of K_L and B_L to represent the properties of the landing pad. Note that the positive direction of the displacements x_j of the jumper and x_p of the parachute are downward.

As in Example 4.7, let $M_p = 10$ kg, $M_j = 60$ kg, and $B_j = 10$ N·s/m. When the parachute is closed, $B_p = 5$ N·s/m and K_R exerts no force on the jumper and parachute. The parachute opens when the length of the risers, which connect the jumper and the parachute, equals or exceeds 20 m, after which $B_p = 300$ N·s/m and $K_R = 400$ N/m.

The jumper and parachute leave the airplane at $t = 0$, with zero initial displacement and zero initial velocity. They both fall freely until the risers extend to their working length of 20 m, at which time the parachute opens. The landing pad is 120 m below the jumper's starting position, and its elastic and damping coefficients are $K_L = 1500$ N/m and $B_L = 300$ N·s/m.

Draw the free-body diagrams and write the modeling equations for the parachute and jumper, including all the elements shown in Figure 9.20. Construct a Simulink diagram that includes the parameter changes that occur at different stages in the jump. Plot the displacements and velocities of the jumper and parachute for $0 \le t \le 15$ s. Also plot the acceleration of the jumper and the length of the risers.

SOLUTION In the Simulink diagram, we need switches to reflect the changes that occur when the parachute opens. We define

$$B_p = \begin{cases} 5 & \text{for } x_R < 20 \\ 100 & \text{for } x_R \ge 20 \end{cases} \qquad \text{where } x_R = x_j - x_p$$

$$F_R = \begin{cases} 0 & \text{for } x_R < 20 \\ 400 x_R & \text{for } x_R \ge 20 \end{cases} \qquad \text{where } x_R = x_j - x_p$$

These functions are implemented by `Switch` and `Switch3`, respectively, which are in the middle left of Figure 9.21. In order to account for the changes that take place after the jumper hits the landing pad, we also define

$$F_{Lj} = \begin{cases} K_L(x_j - 120) + B_L \dot{x}_j & \text{for } x_j \ge 120 \\ 0 & \text{for } x_j < 120 \end{cases}$$

$$F_{Lp} = \begin{cases} K_L(x_p - 120) + B_L \dot{x}_p & \text{for } x_p \ge 120 \\ 0 & \text{for } x_p < 120 \end{cases}$$

which are formed by `Switch2` and `Switch1`, respectively, on the right side of the diagram. The modeling equations, obtained from separate free-body diagrams for the parachute and jumper, are

$$\ddot{x}_p = \frac{1}{M_p} \left[-B_p \dot{x}_p + K_R x_R - F_{Lp} + M_p g \right] \tag{37a}$$

$$\ddot{x}_j = \frac{1}{M_j} \left[M_j g - B_j \dot{x}_j - K_R x_R - F_{Lj} \right] \tag{37b}$$

The complete Simulink diagram is shown in Figure 9.21. It includes near the top of the figure two integrators to relate $\ddot{x}_p$, $\dot{x}_p$, and x_p. A chain of two more integrators near the bottom relates $\ddot{x}_j$, $\dot{x}_j$, and x_j to one another.

With appropriate use of the subplot command we can produce the plots shown in Figure 9.22.

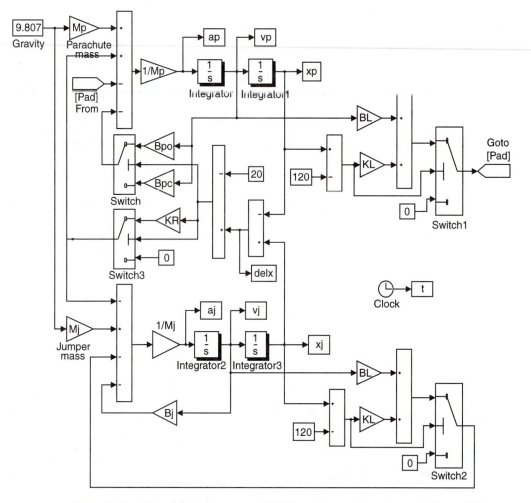

Figure 9.21 A Simulink diagram to model (37) and the parachute jumper in Figure 9.20.

These plots have been labeled with letters A through H to help the reader identify the following stages:

(A) The increase in velocity of the jumper due to gravity before the parachute and risers begin to slow her fall.

(B) The rapid slowing of the parachute as its drag becomes active, but the jumper's weight has not yet extended the risers to pull down hard on the parachute.

(C) A sudden upward acceleration of the jumper as the parachute opens.

(D) The increase in length of the risers as the weight of the jumper is carried up to the parachute.

(E) The increase in parachute velocity as the weight of the jumper becomes applied to it.

(F) The terminal velocity of the jumper and parachute, about 6.24 m/s.

(G) A sudden upward acceleration of the jumper when she hits the landing pad.

(H) The rapid decrease in parachute velocity when it hits the landing pad.

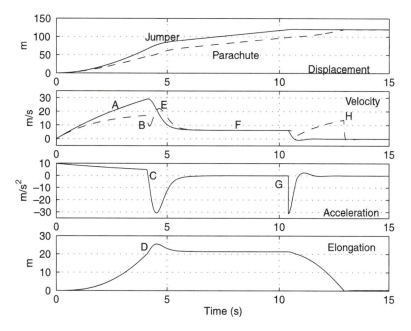

Figure 9.22 Responses for Example 9.12.

At $t = 0$, the jumper and the parachute leave the plane together, with zero velocity and displacement but with an acceleration of g downward. As their velocity increases, the viscous drag increases, lowering their acceleration for about 4 seconds. During this time, the parachute is separating from the heavier jumper, and at about 4 seconds the risers become fully extended, and the parachute opens. This produces a large negative (upward) acceleration on the jumper, lowering her velocity from about 30 m/s to below 8 m/s in about 1.5 seconds. The parachute velocity falls sharply at the instant it opens, but it immediately increases when the weight of the jumper extends the risers and they pull down on the parachute. At 8 to 10 seconds, the parachute and jumper are falling at a terminal velocity of about 6.24 m/s, with a constant riser length of about 21.3 m between them. The acceleration of both is zero. At about 10.4 seconds, the jumper hits the landing pad. This applies an acceleration of around $3g$ to the jumper, and she not only stops but bounces up in the air by a small amount before coming to rest at 12 seconds. When the jumper lands, the parachute velocity slows considerably until the riser length shortens to 20 m. At that moment, the parachute collapses, and with a much lower drag, speeds up until it hits the landing pad, bounces once, and stops. In this respect, the model probably differs from the behavior of a real parachute, which would probably remain open and drift slowly downward under its own weight. Although not visible in these plots, the final displacement of the jumper is 120.39 meters, and that of the parachute is 120.06 meters; each of them sinks into the landing pad a little due to their weight. The difference between these is the final value of the riser length, 0.33 meters.

▶ SUMMARY

To linearize an element law about a particular operating point, we can use the first two terms in its Taylor-series expansion. This is equivalent to approximating its characteristic curve by a straight line tangent to the curve at the operating point.

For a system containing nonlinear elements, we first find the operating point corresponding to a specified constant value of the input. Then, in the nonlinear differential equations, we express the variables as the sum of constant and time-varying components, with nonlinear terms replaced by the first two terms in their Taylor-series expansions. After canceling the constant terms, we obtain linear equations that involve only the time-varying incremental variables. In Section 9.4, we saw how to use MATLAB when some of the parameters change at different points in the solution.

In subsequent chapters, we shall use the same techniques to obtain linearized approximations to some electromechanical, thermal, and hydraulic systems. The basic technique can also be extended to systems whose parameters vary with time and to nonlinearities that depend on two or more independent variables. Such extensions are treated in some of the references in Appendix F.

▶ PROBLEMS

In Problems 9.1 through 9.6, use a Taylor-series expansion to derive the linearized model for the element law and operating point(s) specified. In each case, show the linearized characteristic on a sketch of the nonlinear element law.

9.1. $f(x) = 0.5x^3$ where $\bar{x} = -2, 0$, and 2

*9.2. $f(x) = \begin{cases} -A(1 - \epsilon^x) & \text{for } x < 0 \\ A(1 - \epsilon^{-x}) & \text{for } x \geq 0 \end{cases}$ where $\bar{x} = -1, 0$, and 1

9.3. $f(\theta) = \begin{cases} -\sin^2\theta & \text{for } \theta < 0 \\ \sin^2\theta & \text{for } \theta \geq 0 \end{cases}$ where $\bar{\theta} = 0, \pi/4, \pi/2$, and $3\pi/4$

9.4. $f(\theta) = \begin{cases} -\sin(\theta^2) & \text{for } \theta < 0 \\ \sin(\theta^2) & \text{for } \theta \geq 0 \end{cases}$ where $\bar{\theta} = 0, \sqrt{\pi}/2$, and $\sqrt{\pi/2}$

*9.5. $f(y) = 1/y$ where $y > 0$ and $\bar{y} = 0.5$

9.6. $f(z) = \begin{cases} -\sqrt{|z|} & \text{for } z < 0 \\ \sqrt{z} & \text{for } z \geq 0 \end{cases}$ where $\bar{z} = -2$ and 2

9.7. A nonlinear spring characteristic $f_K(x)$ is shown in Figure P9.7, where x denotes the total length. For each of the operating points specified, determine graphically the force exerted by the spring at the operating point, and evaluate graphically the linearized spring constant.

    a. $\bar{x} = 0.1$ m

    b. $\bar{x} = 0.2$ m

    c. $\bar{x} = 0.3$ m

    d. $\bar{x} = 0.4$ m

*9.8. The nonlinear mechanical system shown in Figure P9.8 has $M = 1.5$ kg, $B = 0.5$ N·s/m, and the spring characteristic $f_K(x)$ plotted in Figure P9.7. The gravitational constant is 9.807 m/s^2. The variable x denotes the total length of the spring.

    a. Verify that the nonlinear input-output differential equation is $1.5\ddot{x} + 0.5\dot{x} + f_K(x) = 14.71$.

    b. Solve for the operating point $\bar{x}$.

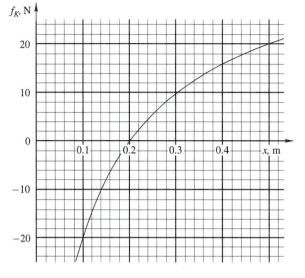

Figure P9.7

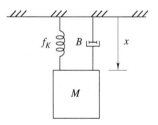

Figure P9.8

c. Derive the linearized differential equation that is valid in the vicinity of the operating point.

d. Give the approximate range of x for which the linearized spring force is within 25% of the nonlinear spring force.

9.9. For the system shown in Figure P9.9, the mass is attached to the wall by a series combination of a linear spring and a nonlinear spring. The parameter values are $M = 4.0\,\text{kg}$, $B = 0.3\,\text{N·s/m}$, and $K = 25\,\text{N/m}$. The applied force is $f_a(t) = 10 + 2\sin 3t$, and the nonlinear spring characteristic $f_K(x)$ is plotted in Figure P9.7. The position $z = 0$ corresponds to $f_a(t) = 0$ with both springs undeflected.

a. Find the nonlinear input-output differential equation relating z to $f_a(t)$.

b. Solve for the operating point $\bar{z}$.

c. Derive the linearized differential equation that is valid in the vicinity of the operating point.

d. Give the approximate range of z for which the linearized spring force is within 25% of the nonlinear spring force.

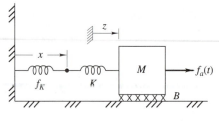

Figure P9.9

*9.10. A mechanical system containing a nonlinear spring obeys the differential equation

$$\ddot{x} + 2\dot{x} + f(x) = A + B \sin 3t$$

where

$$f(x) = \begin{cases} -4\sqrt{|x|} & \text{for } x < 0 \\ 4\sqrt{x} & \text{for } x \geq 0 \end{cases}$$

a. Find the operating point $\bar{x}$ and derive the linearized model in numerical form for $A = 8$.

b. Find the operating point $\bar{x}$ and the linearized spring constant for $A = 4$ and -4.

c. Find $\hat{x}(0)$ and $\dot{\hat{x}}(0)$ when $A = 4$, $x(0) = 1.5$, and $\dot{x}(0) = 0.5$.

9.11. A nonlinear system obeys the equation

$$\ddot{x} + 2\dot{x} + \dot{x}^3 + \frac{4}{x} = A + B \cos t$$

a. Solve for the operating-point conditions on $\bar{x}$ and $\dot{\bar{x}}$. What restriction must be placed on the value of A?

b. Derive the linearized model, evaluating the coefficients in terms of A or numbers.

*9.12. A system is described by the nonlinear equation

$$\ddot{y} + 2(\dot{y} + \dot{y}^3) + 2y + |y|y = A + B \cos t$$

a. For $A = -3$, find the operating point and derive the linearized model, expressing all coefficients in numerical form.

b. Repeat part (a) for $A = 15$.

9.13. The model of a nonlinear system is described by the equation

$$\ddot{x} + \dot{x} + 3x\sqrt{|x|} = A + B \sin \omega t$$

The equilibrium position is known to be $\bar{x} = 4$. Determine the linearized incremental model, and evaluate $\hat{x}(0)$ and $\dot{\hat{x}}(0)$ when $x(0) = 5$ and $\dot{x}(0) = 0$.

9.14. A nonlinear system obeys the equation

$$\dot{x} + 0.5x^2 = 2 + A \sin t$$

a. Sketch the nonlinear term $0.5x^2$ and indicate all possible operating points.

b. For each operating point you found in part (a), derive the linearized model for the system and indicate whether the linear model is stable or unstable.

9.15. A nonlinear system obeys the differential equation
$$\dot{y} + 4\sqrt{y} = 8 + B\cos 2t$$
for $y \geq 0$ and has the initial condition $y(0) = 5$.

a. Derive the linearized incremental model corresponding to the specified input.

b. Sketch the nonlinear element law for $y \geq 0$, and indicate the operating point and the linear approximation.

c. Evaluate the appropriate initial condition for the incremental variable.

9.16. A system obeys the differential equation
$$\ddot{x} + 3|\dot{x}|\dot{x} + 4x^3 = A + B\sin 2t$$
and has the initial conditions $x(0) = 2$ and $\dot{x}(0) = 1$.

a. Find the operating point and derive the linearized model for $A = 4$. Also, find the initial values of $\hat{x}$ and $\dot{\hat{x}}$.

b. Repeat part (a) for $A = 32$.

9.17. In developing the linearized model for a high-speed vehicle in Example 9.6, we assumed that $\bar{f}_a$, the nominal value of the driving force, was positive. Repeat the development for $\bar{f}_a < 0$, and find an expression for b, the effective damping coefficient of the linearized model, that is valid for both cases.

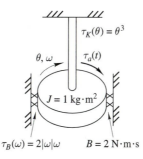

Figure P9.18

***9.18.** The disk shown in Figure P9.18 is supported by a nonlinear torsional spring and is subject to both linear and nonlinear frictional torques. The applied torque is $\tau_a(t) = 8 + \hat{\tau}_a(t)$.

a. Find the operating point $\bar{\theta}$.

b. Derive the linearized input-output equation in terms of $\hat{\theta}(t) = \theta - \bar{\theta}$.

c. Find the initial values of the incremental variables $\hat{\theta}$ and $\dot{\hat{\theta}}$ if $\theta(0) = 0.5$ rad and $\dot{\theta}(0) = -0.5$ rad/s.

9.19. The translational system shown in Figure P9.19 has a linear and a nonlinear spring and is subjected to the applied force $f_u(t)$.

a. Show that the nonlinear model is $2\ddot{x} + 6\dot{x} + 3x + |x|x = f_a(t)$.

b. Solve for the operating point $\bar{x}$ when $\bar{f}_a = 10$ N.

c. Derive the linearized model when $f_a(t) = 10 + \hat{f}_a(t)$.

d. Find the initial values $\hat{x}(0)$ and $\dot{\hat{x}}(0)$ if $x(0) = 3$ m and $\dot{x}(0) = 1$ m/s.

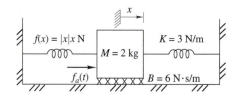

Figure P9.19

9.20. The rotating cylinder shown in Figure P9.20 has damping vanes and a linear frictional torque such that the motion is described by the nonlinear equation

$$\dot{\omega} + 2|\omega|\omega + 2\omega = \tau_a(t)$$

where the applied torque is $\tau_a(t) = 12 + \hat{\tau}_a(t)$. Find the operating point and derive a linearized model in terms of the incremental angular velocity $\hat{\omega}$.

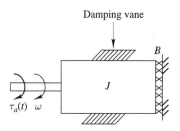

Figure P9.20

*9.21.

a. Verify that the input-output differential equation for the mechanical system modeled in Problem 2.19 is

$$M\ddot{x} + B\dot{x} + x^3 = f_a(t) + Mg$$

b. Find the operating point and derive the linearized input-output equation for the system for the case when $\bar{f}_a = 0$.

c. Repeat part (b) for the case when $\bar{f}_a = Mg$.

9.22. A nonlinear system obeys the state-variable equations

$$\dot{x} = -x + y$$

$$\dot{y} = \frac{1}{y} + 4 + B \sin t$$

Find the operating point and derive the linearized equations in state-variable form. Also, evaluate the initial conditions on the incremental variables when $x(0) = 0$ and $y(0) = -0.5$.

*9.23. A second-order nonlinear system having state variables x and y obeys the equations

$$\dot{x} = -2x + y^3$$

$$\dot{y} = x + 4 + \cos t$$

a. Find the operating-point values $\bar{x}$ and $\bar{y}$.

b. Find the linearized state-variable equations in numerical form.

c. Find the linearized model as an input-output equation relating $\hat{x}$ and its derivatives to the incremental input.

9.24. A nonlinear system with state variables x and y and input $u(t)$ obeys the equations

$$\dot{x} = -2|x|x - y + u(t) - 6$$

$$\dot{y} = x - y - 6$$

a. Verify that when $u(t) = 2 + B \cos 2t$, the operating point is $\bar{x} = 0.7808, \bar{y} = -5.2192$.

b. Evaluate the linearized model about this operating point.

c. Evaluate the initial conditions for the incremental variables when $x(0) = 1$ and $y(0) = -6$.

9.25. The nonlinear resistor in the circuit shown in Figure P9.25 obeys the element law $e_o = 2i^3$.

a. Verify that the nonlinear input-output equation is

$$0.5\frac{di}{dt} + 3i + 2i^3 = 22 + \hat{e}_i(t)$$

b. Find the operating-point values $\bar{i}$ and $\bar{e}_o$.

c. Derive the linearized model that is valid in the vicinity of this operating point.

d. Find the time constant of the linearized model.

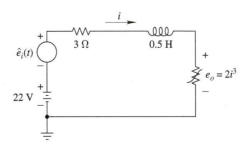

Figure P9.25

9.26. For the circuit shown in Figure 9.10(a), the nonlinear resistor is described by the equation $i_o = e_o/(1 + |e_o|)$.

    a.   Verify that the nonlinear state-variable equation is

$$\dot{e}_o = 2\left[-e_o - \frac{e_o}{1 + |e_o|} + e_i(t) \right]$$

    b.   If the operating point is defined by $\bar{e}_o = 2$ V, what must be $\bar{e}_i$, the nominal value of the input?

    c.   Find the linearized state-variable equation for the operating point defined in part (b).

    d.   Draw the linearized equivalent circuit.

*9.27. The nonlinear resistor in the circuit shown in Figure P9.27 obeys the element law $e_o = 3|i_L|i_L$.

    a.   Verify that the nonlinear state-variable model is

$$\frac{di_L}{dt} = -\frac{10}{7}i_L - \frac{3}{2}|i_L|i_L + \frac{1}{7}e_i(t)$$
$$e_o = 3|i_L|i_L$$

    b.   Find the operating point and derive the linearized model when $e_i(t) = 5 + 0.4\cos t$ for $t > 0$.

    c.   Find the time constant of the linearized model.

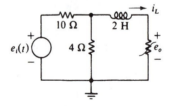

Figure P9.27

*9.28. For the circuit shown in Figure P9.28, $i_o = |e_o|e_o$ and $e_i(t) = 4 + \hat{e}_i(t)$.

    a.   Show that the nonlinear input-output differential equation is

$$2\ddot{e}_o + 6\dot{e}_o + 2e_o + \frac{d}{dt}(|e_o|e_o) = 4\dot{e}_i + 2e_i(t)$$

    b.   Find the operating-point values $\bar{e}_o$ and $\bar{i}_o$.

    c.   Find the linearized input-output equation relating $\hat{e}_o$ and $\hat{e}_i(t)$.

    d.   Draw the linearized equivalent circuit.

9.29. For the circuit shown in Figure P9.29, the element law for the nonlinear resistor is $i_2 = 2|e_c|e_c$.

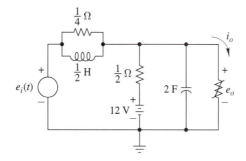

Figure P9.28

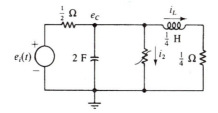

Figure P9.29

a. Verify that the circuit obeys the nonlinear state-variable equations

$$\dot{e}_C = -e_C - |e_C|e_C - \frac{1}{2}i_L + e_i(t)$$

$$\frac{di_L}{dt} = 4e_C - i_L$$

b. Determine the values of $\bar{e}_C$ and $\bar{i}_L$ when $e_i(t) = 10 + \hat{e}_i(t)$.

c. Derive the linearized state-variable equations that are valid in the vicinity of the operating point found in part (b).

d. Find $\hat{e}_C(0)$ and $\hat{i}_L(0)$ when $e_C(0) = 3$ V and $i_L(0) = 1$ A.

9.30. For the circuit shown in Figure P9.30, the element law for the nonlinear resistor is $e_2 = 0.5i_L^3$. The input voltage $e_i(t)$ is 4 V for $t \leq 0$ and is $4 + \hat{e}_i(t)$ for $t > 0$.

a. Verify that the nonlinear state-variable model is

$$\dot{e}_C = -e_C + \frac{1}{2}i_L + \frac{1}{2}e_i(t)$$

$$\frac{di_L}{dt} = -2e_C - i_L^3 + 2e_i(t)$$

$$e_o = -e_C - \frac{1}{2}i_L^3 + e_i(t)$$

b. Find $e_C(0)$ and $i_L(0)$. Assume that steady-state conditions exist at $t = 0-$.

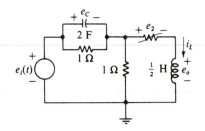

Figure P9.30

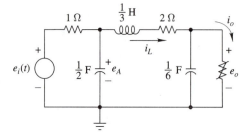

Figure P9.31

   c. Find the values of the state variables at the operating point, and derive a set of linearized state-variable equations.

   d. Find $\hat{e}_C(0)$ and $\hat{i}_L(0)$.

9.31. For the circuit shown in Figure P9.31, $i_o = \frac{1}{4}e_o^3$ and $e_i(t) = 8 + \hat{e}_i(t)$.

   a. Verify that the state-variable equations are

$$\dot{e}_A = 2[-e_A - i_L + e_i(t)]$$

$$\dot{e}_o = 6\left[-\frac{1}{4}e_o^3 + i_L\right]$$

$$\frac{di_L}{dt} = 3[e_A - e_o - 2i_L]$$

   b. Verify that the operating point is defined by $\bar{e}_o = 2$ V.

   c. Find the linearized state-variable equations for the operating point in part (b) and draw the linearized equivalent circuit.

In Problems 9.32 through 9.34, draw a block diagram for the nonlinear system that was modeled in the example indicated, using the equation cited. Also draw a block diagram for the linearized model.

9.32. Equations (23) in Example 9.7.

9.33. Equation (27) in Example 9.8.

9.34. Equations (33) in Example 9.9.

9.35. Draw a Simulink diagram for Problem 9.25. Solve the full nonlinear model and the linearized model with incremental variables. Let $\hat{e}_i(t)$ be a train of square waves with peak amplitude 1 V and with a frequency of 2 Hz. Plot the following three curves: $e_i(t), i$ for the nonlinear model, and $\hat{i} + \bar{i}$ for the linearized model. Repeat with a peak amplitude of 10 V for $\hat{e}_i(t)$.

9.36. Draw a Simulink diagram for Problem 9.18. Solve the full nonlinear model and the linearized model with incremental variables. Let $\hat{\tau}_a(t)$ be a sinusoid with amplitude 10 N·m and with a frequency of 1 rad/s. Plot θ for the nonlinear model and also $\hat{\theta} + \bar{\theta}$ on the same set of axes.

9.37. Draw Simulink diagrams for each of the following nonlinear models.

    a. $\dot{v}_1 = -3v_1 + 2|v_2|v_2 + 3\cos 2t$
         $\dot{v}_2 = v_1 - v_2$

    b. $\ddot{y} + 2|\dot{y}|\dot{y} + y^3 = 5u(t)$

10

ELECTROMECHANICAL SYSTEMS

We can construct a wide variety of very useful devices by combining electrical and mechanical elements. Among the electromechanical devices that we shall consider are potentiometers, the galvanometer, the microphone, and motors and generators.

We first discuss coupling the electrical and mechanical parts of the system by mechanically varying a resistance. In the following section we consider the coupling associated with the movement of a current-carrying conductor through a magnetic field. Two new laws are needed to describe the additional forces and voltages caused by this action.

A number of magnetically coupled devices are considered in detail. The examples incorporate many of the analytical tools developed in previous chapters, such as damping ratio, steady-state response, transfer functions, and linearization. A comprehensive example points out some of the important features that must be considered in the design of an electromechanical measuring device. The final section illustrates the use of MATLAB and Simulink.

▶ 10.1 RESISTIVE COUPLING

We can control a variable resistance by mechanical motion either continuously by moving an electrical contact or discretely by opening and closing a switch. Because resistors cannot store energy, this method of coupling electrical and mechanical parts of a system, in contrast to coupling by magnetic and electrical fields, does not involve mechanical forces that depend on electrical variables.

Figure 10.1(a) shows a strip of conducting material having resistivity ρ and cross-sectional area A. One terminal is fixed to the left end of the conductor. The other terminal, known as the **wiper**, is free to slide along the bar while maintaining a good electrical connection at all times—that is, zero resistance at the contact point. The resistance per unit length of the bar is ρ/A, so the resistance between the terminals is

$$R = \left[\frac{\rho}{A}\right] x(t) \tag{1}$$

where $x(t)$, the displacement of the wiper, can vary with time. The rotational equivalent of this device is shown in Figure 10.1(b). Here the resistance between the two terminals is a function of the angle $\theta(t)$, which defines the angular orientation of the wiper with respect to the fixed terminal.

A very useful device known as a **potentiometer** is obtained by adding a third terminal to the right ends of the variable resistors shown in Figure 10.1. For the translational po-

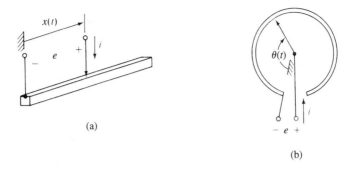

Figure 10.1 Variable resistors. (a) Translational. (b) Rotational.

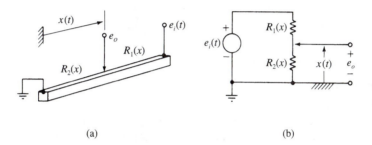

Figure 10.2 (a) Translational potentiometer. (b) Equivalent circuit.

tentiometer shown in Figure 10.2(a), the two end terminals are normally connected across a voltage source, and the voltage at the wiper is considered the output. The resistances R_1 and R_2 depend on the wiper position and are therefore labeled $R_1(x)$ and $R_2(x)$ in the figure. In our mathematical development, however, we shall omit the x in parentheses. We can regard either the position $x(t)$ or the voltage source $e_i(t)$, or both, as inputs.

The circuit diagram for the potentiometer is shown in Figure 10.2(b). Provided that no current flows through the wiper, we can apply the voltage-divider rule of Equations (6.31) to obtain

$$e_o = \left[\frac{R_2}{R_1 + R_2}\right] e_i(t) \tag{2}$$

If the distance and the resistance between the fixed terminals are denoted by $x_{\max}$ and R_T, respectively, then $R_2 = [R_T / x_{\max}] x(t)$ and $R_1 + R_2 = R_T$. Substituting these two expressions into (2), we can write the output voltage as

$$e_o = \left[\frac{1}{x_{\max}}\right] x(t) e_i(t) \tag{3}$$

where the ratio $x(t)/x_{\max}$ must be between 0 and 1, inclusive.

We can interpret (3) as saying that the output voltage is proportional to the product of the input voltage $e_i(t)$ and the mechanical variable $x(t)$. For a constant input voltage E, the output voltage is proportional only to the mechanical displacement $x(t)$:

$$e_o = \left[\frac{E}{x_{\max}}\right] x(t) \tag{4}$$

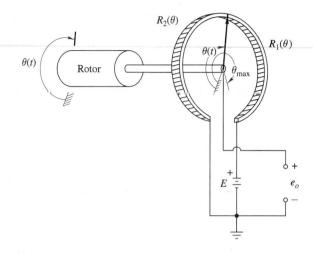

Figure 10.3 Potentiometer used to measure the angular orientation of a rotor.

The rotary potentiometer shown in Figure 10.3 has a constant voltage E across its fixed terminals. The wiper, which divides the total potentiometer resistance R_T into R_1 and R_2, is attached to a rotor that has an angular displacement $\theta(t)$. If the resistance per unit length of the potentiometer is constant, $R_2 = [R_T/\theta_{max}]\theta(t)$ and $R_1 + R_2 = R_T$.

Equation (2) remains valid when these expressions are used for R_2 and R_T and when $e_i(t)$ is replaced by E. Thus

$$e_o = \left[\frac{E}{\theta_{max}}\right]\theta(t) \tag{5}$$

Loading Effect

The output terminals of a potentiometer are frequently connected to some electrical component, such as an amplifier, voltmeter, or recording device. The **loading effect** of such a component can often be represented by an external resistor R_o connected across the output of the potentiometer. This is shown in Figure 10.4 for the case of a rotary potentiometer, although the discussion applies equally well to a translational one. Equation (2) is no longer valid, because R_1 and R_2 are no longer in series. Hence, we cannot use (4) and (5), and the simple relationship between e_o and the mechanical displacement has been destroyed. In such a case, we say that R_o *loads* the potentiometer.

In order to avoid the loading effect and in order to use the simple potentiometer characterizations in (4) and (5), we can try to make R_o very large compared to all possible values

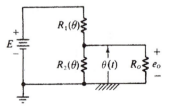

Figure 10.4 Equivalent circuit for a loaded potentiometer.

of R_2. Since $R_2 \leq R_T$, we want

$$R_o \gg R_T \tag{6}$$

When (6) applies, the current through R_o will be negligible compared to that through R_2. Then the current through R_2 will be essentially the same as the current in R_1, and R_1 and R_2 can still be regarded as a series combination. In that case, we can neglect the loading effect and still use (4) and (5).

▶ 10.2 COUPLING BY A MAGNETIC FIELD

A great variety of electromechanical devices contain current-carrying wires that can move within a magnetic field. The physical laws governing this type of electromechanical coupling are given in introductory physics textbooks such as Halliday, Resnick, and Walker (see Appendix F). These laws state that (1) a wire in a magnetic field that carries a current will have a force exerted on it, and (2) a voltage will be induced in a wire that moves relative to the magnetic field. The variables we need to model such devices are

> f_e, the force on the conductor in newtons (N)
>
> v, the velocity of the conductor with respect to the magnetic field in meters per second (m/s)
>
> ℓ, the length of the conductor in the magnetic field in meters (m)
>
> ϕ, the flux in webers (Wb)
>
> $\mathcal{B}$, the flux density of the magnetic field in webers per square meter (Wb/m^2)
>
> i, the current in the conductor in amperes (A)
>
> e_m, the voltage induced in the conductor in volts (V)

In all of our examples the variables will be scalar quantities. However, we shall first introduce the basic laws in a very general way, where the force, velocity, length, and flux density can have any spatial orientation. For the general case we will represent these four quantities as vectors and use the boldface symbols $\mathbf{f_e}$, $\mathbf{v}$, $\boldsymbol{\ell}$, and $\boldsymbol{\mathcal{B}}$.

The force on a conductor of differential length $d\boldsymbol{\ell}$ carrying a current i in a magnetic field of flux density $\boldsymbol{\mathcal{B}}$ is

$$d\mathbf{f_e} = i(d\boldsymbol{\ell} \times \boldsymbol{\mathcal{B}}) \tag{7}$$

The cross in (7) represents the vector cross product. To obtain the total electrically induced force $\mathbf{f_e}$, we must integrate (7) along the length of the conductor.

In our applications, the wires will be either straight conductors that are perpendicular to a unidirectional magnetic field or circular conductors in a radial magnetic field. In either case, the differential length $d\boldsymbol{\ell}$ will be perpendicular to a uniform flux density $\boldsymbol{\mathcal{B}}$. Then (7) simplifies to the scalar relationship

$$f_e = \mathcal{B}\ell i \tag{8}$$

where the direction of the force is perpendicular to both the wire and the magnetic field and can be found by the following rule: If the bent fingers of the right hand are pointed from the positive direction of the current toward the positive direction of the magnetic field (through the right angle), the thumb will point in the positive direction of the force. Figure 10.5(a) shows that for a positive current into the page and a positive flux to the left, the force on a straight conductor is upward. The position of the right hand corresponding to this situation is shown in Figure 10.5(b).

The voltage induced in a conductor of differential length $d\boldsymbol{\ell}$ moving with velocity $\mathbf{v}$ in a field of flux density $\boldsymbol{\mathcal{B}}$ is

$$de_m = (\mathbf{v} \times \boldsymbol{\mathcal{B}}) \cdot d\boldsymbol{\ell} \tag{9}$$

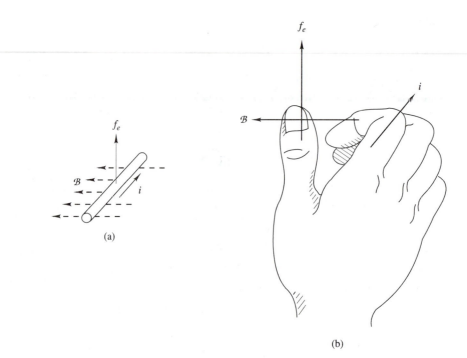

Figure 10.5 Right-hand rule for the force on a conductor.

where the dot denotes the scalar product, or dot product, used with vector notation. We obtain the total induced voltage for a conductor by integrating (9) between the ends of the conductor.

In practice, the three vectors in (9) will be mutually perpendicular, so integrating (9) yields the scalar relationship

$$e_m = \mathcal{B}\ell v \tag{10}$$

If the bent fingers of the right hand are pointed from the positive direction of the velocity toward the positive direction of the magnetic field (through the right angle), the thumb will point in the direction in which the current caused by the induced voltage tends to flow. In Figure 10.6(a), a straight conductor is shown moving downward in a magnetic field directed to the left. As indicated by the polarity signs and by the sketch shown in Figure 10.6(b), the positive sense of the induced voltage is into the page. If the conductor were part of a complete circuit with no external sources, the current in the conductor would be into the page. According to Figure 10.5, this current would cause there to be exerted on the conductor an upward force that would oppose the downward motion.

Equations (8) and (10), which describe the force and induced voltage associated with a wire moving perpendicularly to a magnetic field, can be incorporated in a schematic representation of a translational electromechanical system as shown in Figure 10.7. The induced voltage is represented by a source in the electrical circuit, whereas the magnetically induced force is shown acting on the mass M to which the conductor is attached. We can determine the proper polarity marks for e_m and the reference direction for f_e by examining the specific device under consideration, but the figure indicates two combinations of reference directions that are consistent with the conservation of energy.

In Figure 10.7(a), the polarity of the electrical source is such that power is absorbed from the remainder of the circuit when both e_m and i are positive. Likewise, when f_e and v

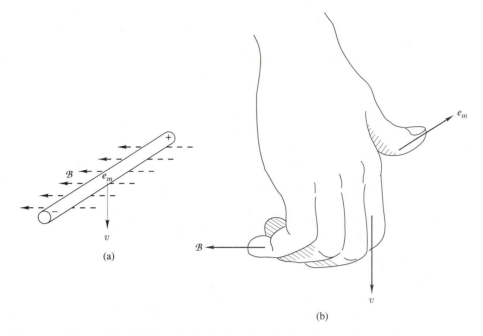

Figure 10.6 Right-hand rule for the voltage induced in a moving conductor.

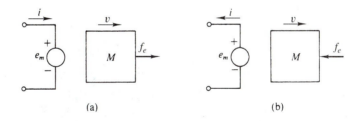

Figure 10.7 Representations of a translational electromechanical system.
(a) Electrical-to-mechanical power flow. (b) Mechanical-to-electrical power flow.

are positive, there is a transfer of power into the mechanical part of the system. Conversely, for Figure 10.7(b), power flows from the mechanical side to the electrical side when all four variables are positive.

It is instructive to evaluate the power involved in the coupling mechanism. The external power delivered to the electrical part of Figure 10.7(a) is

$$p_e = e_m i = (\mathcal{B}\ell v)i$$

whereas the power available to whatever mechanical elements are attached to the coil is

$$p_m = f_e v = (\mathcal{B}\ell i)v$$

Hence,

$$p_m = p_e$$

which says that any power delivered to the coupling mechanism in electrical form will be passed on undiminished to the mechanical portion. Of course, any practical system has losses resulting from the resistance of the conductor and the friction between the moving mechanical elements. However, any such dissipative elements can be modeled separately by a resistor in the electrical circuit or a viscous-friction element acting on the mass. In a similar way, you can demonstrate that for the coupling shown in Figure 10.7(b), the mechanical power supplied by the force f_e is transmitted to whatever electrical elements are connected across e_m.

▶ 10.3 DEVICES COUPLED BY MAGNETIC FIELDS

Having introduced the basic laws that govern the behavior of a single conductor in a magnetic field, we now describe several of the most common types of electromechanical systems and derive their mathematical models. In each case, we consider idealized versions of the device that omit certain aspects that may be important from a design standpoint but are not essential to understanding its operation as part of an overall system. We shall examine in turn the galvanometer, the microphone, and a motor.

The Galvanometer

The galvanometer is a device that produces an angular deflection dependent on the current passing through a coil attached to a pointer. It is widely used in electrical measurement devices. As shown in Figure 10.8(a), a permanent magnet supplies a radial magnetic field, and the flux passes through a stationary iron cylinder between the poles of the magnet. A coil of wire whose terminals can be connected to an external circuit is suspended by bearings so that it can rotate about a horizontal axis passing through the center of the cylinder. A torsional spring mounted on its axis restrains the coil.

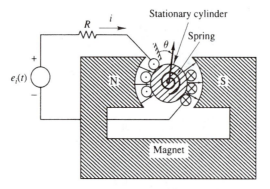

(a)

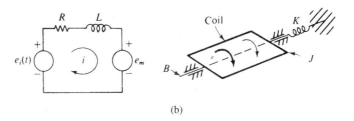

(b)

Figure 10.8 Galvanometer. (a) Physical device. (b) Diagrams used for analysis.

The magnet provides a uniform flux density $\mathcal{B}$ within the air gaps between it and the iron cylinder directed from the north to the south pole. The moment of inertia of the coil is J, and the combination of bearing friction and damping due to the air is represented by the viscous-damping coefficient B. The torsional spring has the rotational spring constant K. It is assumed that the electrical connection between the external circuit and the movable coils is made in such a way that the connection exerts no torque on the coil.

The coil consists of N rectangular turns, each of which has a radius of a and a length of ℓ along the direction of the axis of rotation, and it has a total inductance L. The dots in the wires on the left side of the coil and the crosses on the right side indicate that when i is positive, the current flows out of the page in the left conductors and into the page in the right conductors. The remainder of the circuit consists of a voltage source $e_i(t)$ and a resistance R that accounts for any resistance external to the galvanometer as well as for the resistance of the coils.

For purposes of analysis, we represent the idealized system by the circuit and mechanical diagrams shown in Figure 10.8(b). We have used the rotational equivalent of Figure 10.7(a) rather than Figure 10.7(b) to represent the electromechanical coupling, because the purpose of the device is to convert an electrical variable (current) into a mechanical variable (angular displacement). It remains to use (8) and (10), with the appropriate right-hand rules, to determine the expressions for e_m and τ_e, the electrically induced torque.

Assume that the rotation of the coil is sufficiently small that all the conductors remain in the region of constant flux density. Then we can obtain the torque τ_e acting about the axis of the coil by summing the torques on the $2N$ individual conductors of length ℓ. Because the magnetic field is confined to the iron cylinder as it passes between the air gaps, there is no contribution to the torque from the ends of the coil outside the gaps. If the current is positive, each conductor on the left side of the coil has a force of $f_e = \mathcal{B}\ell i$ acting upward, whereas the conductors on the right side have forces of the same magnitude acting downward. Because the arrow denoting the electrically induced torque τ_e was taken as clockwise in Figure 10.8(b), the expression for the torque is

$$\tau_e = (2N\mathcal{B}\ell a)i \tag{11}$$

Next we express the voltage induced in the coil in terms of the angular velocity $\dot{\theta}$. Because there are $2N$ conductors in series, which move with the velocity $a\dot{\theta}$ with respect to the magnetic field, a voltage of $2N\mathcal{B}\ell a\dot{\theta}$ is induced in the coil. We find the sign of the mechanically induced voltage e_m by applying the right-hand rule illustrated in Figure 10.6(b). The conductors on the right side of the coil in Figure 10.8(a) move downward when $\dot{\theta} > 0$ and the flux-density vector $\boldsymbol{\mathcal{B}}$ points to the right, toward the south pole. Thus the positive sense of e_m is directed toward the viewer, and the corresponding current direction is out of the paper, opposite to the reference arrow for i in the figure. The circuit shown in Figure 10.8(b) was drawn with this assumed polarity for e_m, so we have

$$e_m = (2N\mathcal{B}\ell a)\dot{\theta} \tag{12}$$

Summing torques on the coil and using (11) give

$$J\ddot{\theta} + B\dot{\theta} + K\theta = (2N\mathcal{B}\ell a)i \tag{13}$$

Using (12) in a voltage-law equation for the single loop that makes up the electrical part of the system gives

$$L\frac{di}{dt} + Ri + (2N\mathcal{B}\ell a)\dot{\theta} = e_i(t) \tag{14}$$

Equations (13) and (14) constitute the complete model of the galvanometer, which is a third-order model. From these two equations we can determine the state-variable equations, the transfer function, and the input-output differential equation. In practical situations, the inductance of the coil is often sufficiently small to allow us to neglect the term $L\,di/dt$ in (14). This simplification is particularly helpful in solving for the response to an input voltage, because the model becomes second-order. When this term is dropped, we can solve (14) for the current, obtaining

$$i = \frac{1}{R}\left[e_i(t) - (2N\mathcal{B}\ell a)\dot{\theta}\right]$$

By substituting this expression into (13), we have, for the input-output equation,

$$\ddot{\theta} + \left(\frac{B}{J} + \frac{\alpha^2}{JR}\right)\dot{\theta} + \frac{K}{J}\theta = \frac{\alpha}{JR}e_i(t) \tag{15}$$

where $\alpha = 2N\mathcal{B}\ell a$ is the electromechanical coupling coefficient. Comparing the left side of (15) with the left side of Equation (8.31), we find that the undamped natural frequency ω_n and the damping ratio ζ of the galvanometer are

$$\omega_n = \sqrt{\frac{K}{J}}$$

$$\zeta = \frac{1}{2\sqrt{KJ}}\left(B + \frac{\alpha^2}{R}\right)$$

Hence the undamped natural frequency depends on the mechanical parameters K and J, and the damping ratio depends on both the mechanical and electrical parameters and the electromechanical coupling coefficient α.

We can find the sensitivity of the galvanometer in radians per volt by solving for the steady-state response to the constant excitation $e_i(t) = A$. The steady-state value of θ can be found by setting the derivatives in (15) equal to zero, so

$$\theta_{ss} = \left(\frac{\alpha}{KR}\right)A \tag{16}$$

The galvanometer sensitivity is $\theta_{ss}/A = \alpha/KR$. Thus a higher flux density will make the device more sensitive, and increasing either the spring stiffness or the electrical resistance will reduce its sensitivity.

We can arrive at the same result for the sensitivity by the following argument: Because in the steady state the coils will be stationary, no voltage will be induced in the coils. Thus the steady-state current will be $i_{ss} = A/R$. The steady-state torque exerted on the coils through the action of the magnetic field will be $(\tau_e)_{ss} = \alpha A/R$, which must be balanced entirely by the torsional spring that exerts the steady-state torque $K\theta_{ss}$. Equating these two torques gives (16).

If we wish to characterize the system by a set of state-variable equations, we can choose θ and ω as the state variables and use (15) to write

$$\dot{\theta} = \omega$$

$$\dot{\omega} = -\frac{K}{J}\theta - \left(\frac{B}{J} + \frac{\alpha^2}{JR}\right)\omega + \frac{\alpha}{JR}e_i(t)$$

Had we retained the inductance of the coil, the current i would have become the third state variable.

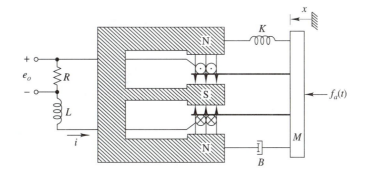

Figure 10.9 Representation of a microphone.

The Microphone

The microphone shown in cross section in Figure 10.9 consists of a diaphragm attached to a circular coil of wire that moves back and forth through a magnetic field when sound waves impinge on the diaphragm. The magnetic field is supplied by a cylindrical permanent magnet having concentric north and south poles, which result in radial lines of flux directed inward toward the axis of the magnet. The coil has N turns with a radius of a and is connected in series with an external resistor R, across which the output voltage is measured. The positive direction for the current i is assumed to be counterclockwise from the perspective of viewing the magnet from the diaphragm. In the figure, the resistance of the coil has been neglected, but the inductance of the coil is represented by the inductor L located externally to the coil. If the coil resistance were not negligible, another resistor could be added in series with the external resistor in order to account for it.

A single stiffness element K has been used to represent the stiffness of the entire diaphragm, and the viscous-friction element B has been used to account for the energy dissipation due to air resistance. The net force of the impinging sound waves is represented by $f_a(t)$, which is the input to the system. Although the forces acting on the diaphragm are certainly distributed in nature, it is a justifiable simplification to consider them as point forces associated with the lumped elements K, B, and M.

The first step in modeling the microphone is to construct idealized diagrams of the electrical and mechanical portions. In developing the equations for the galvanometer, we assumed at the outset positive senses for τ_e and e_m. Then we wrote expressions for these quantities, determining their signs by using the right-hand rules. This time, we shall first determine the positive directions of f_e, the electrically induced force on the diaphragm, and the mechanically induced voltage e_m when the other variables are positive. After that, we shall label the diagram with senses that agree with the positive directions of f_e and e_m. When we use this approach, we know in advance that the expressions for f_e and e_m will have plus signs.

Figure 10.10(a) shows the upper portion of a single turn of the coil as viewed from the diaphragm looking toward the magnet. Because the flux arrow points downward from the north to the south pole and the current arrow points to the left, the right-hand rule shown in Figure 10.5(b) indicates that the positive sense of f_e is toward the diaphragm. Because the velocity arrow points away from the diaphragm, the right-hand rule shown in Figure 10.6(b) indicates that the positive sense of the induced voltage is the same as that of the current. We can reach similar conclusions by examining any other part of the coil. Thus we can draw the complete diagram of the system, as shown in Figure 10.10(b).

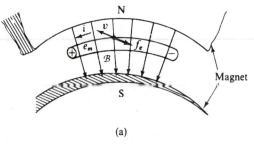

(a)

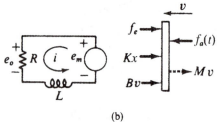

(b)

Figure 10.10 Microphone. (a) Relationships for a portion of a single coil. (b) Diagrams used for analysis.

Because of the radial symmetry of the coil and the flux lines in the air gap, the entire coil of length $2\pi aN$ is perpendicular to the flux. Thus

$$f_e = \alpha i \tag{17a}$$

$$e_m = \alpha v \tag{17b}$$

where $\alpha = 2\pi aN\mathcal{B}$ is the electromechanical coupling coefficient for the system.

Summing forces on the free-body diagram for the diaphragm and using (17a), we obtain

$$M\dot{v} + Bv + Kx = -\alpha i + f_a(t) \tag{18}$$

We find the circuit equation by applying Kirchhoff's voltage law and using (17b), which yields

$$L\frac{di}{dt} + Ri = \alpha v \tag{19}$$

The system has three independent energy-storing elements (L, K, and M), so an appropriate set of state variables is i, x, and v. By rewriting (18) and (19), we obtain the state-variable equations

$$\frac{di}{dt} = \frac{1}{L}(-Ri + \alpha v)$$

$$\dot{x} = v$$

$$\dot{v} = \frac{1}{M}[-\alpha i - Kx - Bv + f_a(t)]$$

The output e_o is not a state variable, but we can find it from

$$e_o = Ri \tag{20}$$

In one of the end-of-chapter problems, you are asked to verify that the corresponding transfer function is

$$H(s) = \frac{E_o(s)}{F_a(s)} = \frac{\alpha R s}{MLs^3 + (MR + BL)s^2 + (KL + BR + \alpha^2)s + KR}$$ (21)

and that the input-output equation is

$$ML\ddot{e}_o + (MR + BL)\ddot{e}_o + (KL + BR + \alpha^2)\dot{e}_o + KRe_o = \alpha R \dot{f}_a$$ (22)

It is interesting to note that because the right side of (22) is proportional to $\dot{f}_a$ rather than to $f_a(t)$, the forced response for a constant-force input is zero. Thus a constant force applied to the diaphragm yields no output voltage in the steady state. This same conclusion can be reached by noting from (21) that $H(0) = 0$. In Figure 10.9, the constant applied force will be balanced by the steady-state spring force Kx_{ss}, the diaphragm will become stationary, and no voltage will be induced in the coil.

A Direct-Current Motor

A direct-current (dc) motor is somewhat similar to the galvanometer but differs from it in several significant respects. In all but the smallest motors, the magnetic field is established not by a permanent magnet but by a current in a separate field winding on the iron core that constitutes the stationary part of the motor, the **stator**. Figure 10.11 indicates the manner in which the field winding establishes the flux when the field current i_F flows. Because of the saturation effects of magnetic fields in iron, the flux ϕ is not necessarily proportional to i_F at high currents.

In a dc motor, the iron cylinder between the poles of the magnet is free to rotate and is called the **rotor**. The rotating coils are embedded in the surface of the rotor and are known as the **armature winding**. There is no restraining torsional spring, and the rotor is free to rotate through an indefinite number of revolutions. However, this fact requires a significant deviation from the galvanometer in the construction of the rotor and the armature winding. First, if the rotor shown in Figure 10.11 rotates through 180° without a change in the direction of the currents in the individual armature conductors, then the torque exerted on it through the magnetic field will undergo a change in direction. Second, if there were a direct connection of the external armature circuit to the rotating armature, the wires would soon become tangled and halt the machine.

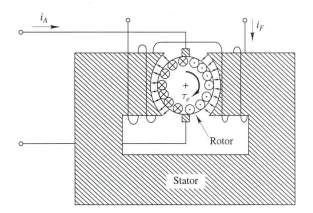

Figure 10.11 DC motor showing field and armature windings.

To solve both of these problems, we use a **commutator**, which consists of a pair of low-resistance carbon brushes that are fixed with respect to the stator and make contact with the ends of the armature windings on the rotor (see the references in Appendix F for details). As indicated in Figure 10.11, when a conductor is located to the right of the commutator brushes, a positive value of i_A implies that the individual conductor current will be directed toward the reader. When the conductor is located to the left of the brushes, its current flows away from the reader when $i_A > 0$. As the ends of a particular conductor pass under the brushes, the direction of the current in that conductor changes sign. Under the arrangement just described, each conductor exerts a unidirectional torque on the rotor as it passes through a complete revolution. Hence the sliding contact at the brushes solves the mechanical problem of connecting the stationary and rotating parts of the armature circuit.

For modeling purposes, it is convenient to represent the important characteristics of the motor as shown in Figure 10.12. Representing the armature by a stationary circuit having resistance R_A, inductance L_A, and induced-voltage source e_m is justified by the presence of the commutator, which makes the armature behave as if it were stationary even though the individual conductors are indeed rotating. Likewise, the field circuit has resistance R_F and inductance L_F but no induced voltage. The rotor has moment of inertia J, rotational viscous-damping coefficient B, a driving torque τ_e caused by the forces acting on the individual conductors, and a load torque τ_L. The electrical inputs to the motor may be considered to be the currents i_A and i_F or the applied voltages e_A and e_F. The output is ω, the angular velocity of the rotor.

We begin the modeling process by expressing the voltage e_m and the torque τ_e in terms of the other system variables. We then write voltage-law equations for both the armature and field circuits, unless i_A or i_F is an input, and apply D'Alembert's law to the rotor.

When we are modeling dc motors and generators, it is convenient to express the flux density $\mathcal{B}$ as

$$\mathcal{B} = \frac{1}{A}\phi(i_F) \tag{23}$$

where $\phi(i_F)$ is the total flux established by the field current and A is the effective cross-sectional area of the flux path in the air gap between the rotor and stator. If ℓ denotes the total length of the armature conductors within the magnetic field and a denotes the radius of the armature, the electromechanical torque exerted on the rotor is

$$\tau_e = \left(\frac{\phi}{A}\right)\ell a i_A \tag{24}$$

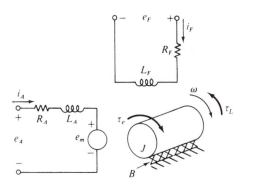

Figure 10.12 Diagram of the dc motor used for analysis.

Because the parameters ℓ, a, and A depend only on the geometry of the motor, we can define the motor parameter

$$\gamma = \frac{\ell a}{A} \tag{25}$$

and rewrite (24) in terms of the flux and armature current as

$$\tau_e = [\gamma \, \phi(i_F)] i_A \tag{26}$$

Similarly, the voltage induced in the armature is

$$e_m = [\gamma \, \phi(i_F)] \omega \tag{27}$$

Keep in mind that the flux $\phi(i_F)$ in (26) and (27) is a function of the field current i_F and, for that matter, is generally a nonlinear function.

Having established the model for the internal behavior of a dc motor represented by Figure 10.12 and (25), (26), and (27), we are prepared to develop models of complete electromechanical systems involving dc motors. We shall consider two such systems in the following examples.

▶ **EXAMPLE 10.1**

Derive the state-variable equations for a dc motor that has a constant field voltage E_F, an applied armature voltage $e_i(t)$, and a load torque $\tau_L(t)$. Also obtain the input-output equation with ω as the output, and determine the steady-state angular velocities corresponding to the following sets of inputs: $e_i(t) = E$, $\tau_L(t) = 0$ and $e_i(t) = 0$, $\tau_L(t) = T$.

SOLUTION   The basic motor diagram in Figure 10.12 is repeated in Figure 10.13, with the specified field and armature input voltages added. The field voltage E_F is constant, so the field current will be the constant $i_F = E_F/R_F$. We can write the electromechanical driving torque τ_e and the induced voltage e_m as

$$\begin{aligned}
\tau_e &= \alpha i_A \\
e_m &= \alpha \omega
\end{aligned} \tag{28}$$

where α is a constant defined by

$$\alpha = \gamma \, \phi(i_F) \tag{29}$$

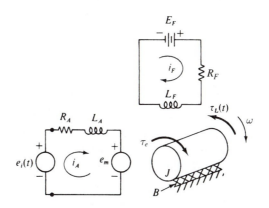

Figure 10.13 DC motor with a constant field current.

and γ is given by (25). We select i_A and ω as the state variables and write a voltage equation for the armature circuit and a torque equation for the rotor. Then, using (28) and solving for the derivatives of the state variables, we find the state-variable equations to be

$$\frac{di_A}{dt} = \frac{1}{L_A}[-R_A i_A - \alpha\omega + e_i(t)]$$

$$\dot{\omega} = \frac{1}{J}[\alpha i_A - B\omega - \tau_L(t)] \tag{30}$$

In order to find the system's transfer functions, we apply the Laplace transform to (30) and assume that there is no initial stored energy. With $i_A(0) = 0$ and $\omega(0) = 0$, we have

$$L_A s I_A(s) = -R_A I_A(s) - \alpha\Omega(s) + E_i(s)$$

$$Js\Omega(s) = \alpha I_A(s) - B\Omega(s) - \tau_L(s)$$

Eliminating $I_A(s)$ from this pair of algebraic equations and solving for the transformed output $\Omega(s)$, we find that

$$\Omega(s) = H_1(s)E_i(s) + H_2(s)\tau_L(s)$$

where

$$H_1(s) = \frac{\alpha/JL_A}{P(s)} \tag{31a}$$

$$H_2(s) = \frac{-(1/J)s - (R_A/JL_A)}{P(s)} \tag{31b}$$

$$P(s) = s^2 + \left(\frac{R_A}{L_A} + \frac{B}{J}\right)s + \left(\frac{R_A B + \alpha^2}{JL_A}\right) \tag{31c}$$

The quantity $H_1(s)$ is the transfer function relating the output velocity and input voltage when $\tau_L(t) = 0$. $H_2(s)$ relates $\Omega(s)$ and $\tau_L(s)$ when $e_i(t) = 0$. The general input-output differential equation is

$$\ddot{\omega} + \left(\frac{R_A}{L_A} + \frac{B}{J}\right)\dot{\omega} + \left(\frac{R_A B + \alpha^2}{JL_A}\right)\omega = \frac{\alpha}{JL_A}e_i(t) - \frac{1}{J}\dot{\tau}_L - \frac{R_A}{JL_A}\tau_L(t) \tag{32}$$

As expected, both the electrical and the mechanical parameters contribute to the system's undamped natural frequency ω_n and to the damping ratio ζ.

To solve for the steady-state motor speed when the voltage source has the constant value $e_i(t) = E$ and when $\tau_L(t) = 0$, we omit all derivative terms in (32) and substitute these values for $e_i(t)$ and $\tau_L(t)$, obtaining

$$\omega_{ss} = \frac{\alpha E}{R_A B + \alpha^2} \tag{33}$$

In physical terms, the motor will run at a constant speed such that the driving torque $\tau_e = \alpha i_A$ exactly balances the viscous-frictional torque $B\omega_{ss}$. However, the steady-state armature current is $i_A = (E - e_m)/R_A$, where $e_m = \alpha\omega_{ss}$. Making the appropriate substitutions, we again obtain (33). We get the same expression by examining $H_1(0)$.

When $e_i(t) = 0$ and $\tau_L(t)$ has the constant value T, the steady-state solution to (32) is

$$\omega_{ss} = -\frac{R_A T}{R_A B + \alpha^2} \tag{34}$$

which indicates that the motor will be driven backward at a constant angular velocity. We can also obtain (34) by noting that $\omega_{ss} = TH_2(0)$. To understand the behavior of the motor

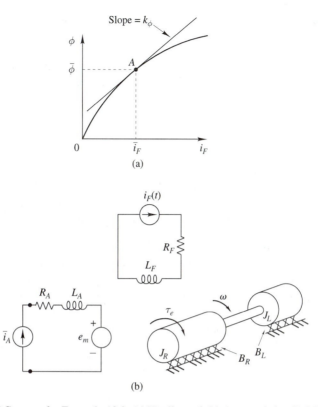

Figure 10.14 DC motor for Example 10.2. (a) Nonlinear field characteristic. (b) Diagram used for analysis.

under this condition, we observe that the electromechanical torque, $\tau_e = \alpha i_A$, must balance the sum of the load torque T and the viscous-frictional torque $B\omega_{ss}$. Thus an armature current must flow that will make $\alpha i_A = T + B\omega_{ss}$. Furthermore, because the applied armature voltage is zero and $e_m = \alpha\omega_{ss}$, it follows that $i_A R_A = -\alpha\omega_{ss}$. As anticipated, solving these two equations for ω_{ss} results in (34).

In this situation, the motor is acting as a **generator** connected to a load of zero resistance. Part of the mechanical power supplied by the load torque is being converted to electrical form and dissipated in the armature resistance R_A. Basically, we may think of a generator as a motor that is being driven mechanically and that delivers a portion of the power to an electrical load connected across the armature terminals.

You can verify that when the constant applied voltage is $e_i(t) = E$ and the constant load torque is $\tau_L(t) = \alpha E / R_A$, the steady-state motor speed is $\omega_{ss} = 0$. In this condition, the electromechanical driving torque τ_e exactly matches the load torque and the motor is stalled.

▶ **EXAMPLE 10.2**

A dc motor has a constant armature current $\bar{i}_A$ but a variable current source $i_F(t)$ supplying the field winding. The relationship between the flux ϕ and the field current $i_F(t)$ is nonlinear and is shown in Figure 10.14(a). A load having moment of inertia J_L and viscous-damping coefficient B_L is connected to the rotor by a rigid shaft. Find a linear model suitable for analyzing small perturbations about the operating point A indicated on the flux-versus-current curve.

SOLUTION In Figure 10.14(b), we have repeated the basic motor diagram in Figure 10.12 and have also added the specified current sources and the mechanical load. From (26), the electromechanical torque is

$$\tau_e = \gamma \, \phi \bar{i}_A \tag{35}$$

where, in this case, γ and $\bar{i}_A$ are constants and $\phi = \phi(i_F)$. Because the armature current is constant and the field current is specified, there is no need to write an equation for either of the electrical circuits. Rather, we obtain the system model by summing torques on the rotor and load and using (35). This yields

$$(J_R + J_L)\dot{\omega} + (B_R + B_L)\omega = \gamma \, \bar{i}_A \phi(i_F) \tag{36}$$

To obtain the linearized model, we write the field current and angular velocity as $i_F(t) = \bar{i}_F + \hat{i}_F(t)$ and $\omega = \bar{\omega} + \hat{\omega}$, respectively. The operating-point values $\bar{i}_F$ and $\bar{\omega}$ must satisfy (36), which reduces to

$$\bar{\omega} = \frac{\gamma \bar{i}_A \phi(\bar{i}_F)}{B_R + B_L} \tag{37}$$

The flux is approximated by the first two terms in the Taylor-series expansion for $\phi(i_F)$ about the operating point, namely

$$\phi(\bar{i}_F) + k_\phi \hat{i}_F \tag{38}$$

where

$$k_\phi = \left. \frac{d\phi}{di_F} \right|_{\bar{i}_F} \tag{39}$$

Substituting (38) into (36), writing $\omega = \bar{\omega} + \hat{\omega}$, and using (37) to cancel the constant terms, we find the linearized model to be

$$(J_R + J_L)\dot{\hat{\omega}} + (B_R + B_L)\hat{\omega} = \gamma \bar{i}_A k_\phi \hat{i}_F(t) \tag{40}$$

where k_ϕ is the slope of the curve of ϕ versus i_F evaluated at the operating point.

In the last example note that (40) is a first-order equation, which implies that the system is only first order. Although there are two inductors that can store energy, their currents are those of the current sources $\bar{i}_A$ and $i_F(t)$ and thus cannot be state variables.

Furthermore, it is apparent that the moments of inertia J_R and J_L and the friction coefficients B_R and B_L are merely added to obtain equivalent parameters for the system as a whole. If the motor had been connected to its load through a gear train (as would usually be the case), the equivalent parameters, when reflected to the motor, would be

$$J_{\text{eq}} = J_R + \frac{1}{N^2} J_L$$

$$B_{\text{eq}} = B_R + \frac{1}{N^2} B_L$$

The symbol N denotes the motor-to-load gear ratio; that is, the motor rotates at N times the angular velocity of the load.

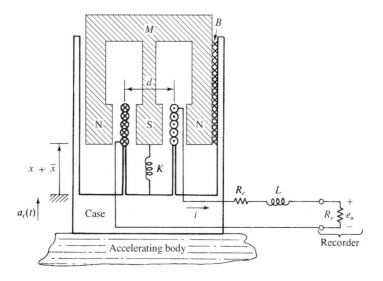

Figure 10.15 Acceleration-measuring device.

▶ 10.4 A DEVICE FOR MEASURING ACCELERATION

In many areas of technology, it is important to be able to measure and record the acceleration of a moving body as a function of time. For example, deceleration measurements are required in the testing of automobiles for crash resistance. Accelerometers are important components in ship, aircraft, and rocket navigation and guidance systems.

Figure 10.15 shows an electromechanical device whose response depends on the acceleration of its case relative to an inertially fixed reference frame. The device is not intended to be representative of commonly used accelerometers, but modeling and analyzing its response will provide practice in using many of the techniques that we have discussed in this chapter and in Chapter 8.

System Description

Basically, the device shown in Figure 10.15 consists of a case attached to the body or material whose motion is to be measured, a circular coil fixed to the case, and a permanent magnet supported on the case by a spring, with viscous damping between the magnet and the case. As the case moves vertically as a result of motion of the body supporting it, a voltage is induced in the coil because of the relative motion of the coil and the magnetic field. A recorder is attached to the terminals of the coil and draws a graph of the coil voltage e_o as a function of time.

The magnet has mass M and provides a flux density $\mathcal{B}$ in the annular space between its north and south poles. Although several springs would be used to support the mass, the single spring shown with spring constant K may be considered equivalent to whatever springs are present. The viscous-damping coefficient B accounts for all viscous effects between the magnet and the case.

The coil has N turns of diameter d, and the parameter $\ell = \pi d N$ denotes the length of the coil. The total coil resistance and inductance are modeled by lumped elements having values R_c and L, respectively. The dots and crosses associated with the coil in Figure 10.15 indicate that the assumed positive direction of current is clockwise as viewed from above.

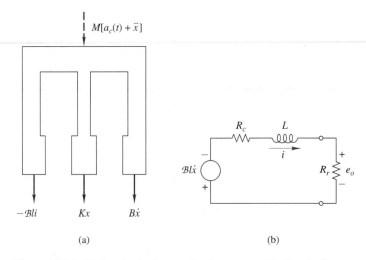

Figure 10.16 (a) Free-body diagram for the magnet. (b) Circuit diagram.

The recorder is attached directly to the terminals of the coil and is assumed to provide a resistance R_r in the coil circuit.

The acceleration of the case relative to a fixed inertial reference frame is denoted by $a_c(t)$ and is the input to the system. The vertical distance between the magnet and the case is $x + \bar{x}$, so the variable x denotes the incremental displacement of the magnet relative to the case, with the value $x = 0$ corresponding to the equilibrium condition. Likewise, the relative velocity of the magnet with respect to the case is $\dot{x}$.

System Model

We derive the differential equations describing the behavior of the system by drawing a free-body diagram for the magnet, drawing a circuit diagram for the coil and recorder, and expressing the electromechanically induced force and voltage in terms of the appropriate system variables. Three aspects of this task deserve specific mention.

First, the inertial force shown on the free-body diagram must use the acceleration of the mass relative to the inertial reference frame. Hence this force on the diagram shown in Figure 10.16(a) is $M[a_c(t) + \ddot{x}]$ in the downward direction. Second, recall that the electrically induced force on the coil is given by $f_e = \mathcal{B}\ell i$. In this instance, however, we must show on the free-body diagram the force on the magnet, which, by the law of reaction forces, is $-\mathcal{B}\ell i$ in the downward direction. Finally, in order to determine the proper sign for the voltage induced in the coil, we note that when $\dot{x} > 0$, the coil is moving downward relative to the magnetic field. This is because the magnet is moving upward relative to the case, and the coil is attached to the case.

Taking these points into consideration, we can draw the free-body and circuit diagrams shown in Figure 10.16. Summing forces on the magnet and writing a voltage equation for the circuit, we obtain

$$L\frac{di}{dt} + (R_c + R_r)i = -\mathcal{B}\ell\dot{x} \tag{41a}$$

$$M\ddot{x} + B\dot{x} + Kx = \mathcal{B}\ell i - Ma_c(t) \tag{41b}$$

$$e_o = R_r i \tag{41c}$$

Transfer Function

To determine a single overall transfer function relating the input transform $A_c(s)$ to the output transform $E_o(s)$, we shall transform (41) with zero initial conditions. Doing this, we obtain the following set of three algebraic transform equations:

$$(Ls + R_c + R_r)I(s) = -\mathcal{B}\ell sX(s) \tag{42a}$$

$$(Ms^2 + Bs + K)X(s) = \mathcal{B}\ell I(s) - MA_c(s) \tag{42b}$$

$$E_o(s) = R_r I(s) \tag{42c}$$

We can combine these equations to eliminate the variables $X(s)$ and $I(s)$, obtaining a single equation relating the input transform $A_c(s)$ and the output transform $E_o(s)$.

First, if we combine (42a) and (42b) to eliminate $X(s)$, the result is

$$(Ms^2 + Bs + K)\left(\frac{Ls + R}{-\mathcal{B}\ell s}\right)I(s) - \mathcal{B}\ell I(s) = -MA_c(s)$$

where $R = R_c + R_r$. Combining the two terms involving $I(s)$ in this equation and then using (42c) to express $I(s)$ in terms of $E_o(s)$, we obtain the single transformed equation

$$\frac{1}{R_r}[(Ms^2 + Bs + K)(Ls + R) + \mathcal{B}^2\ell^2 s]E_o(s) = \mathcal{B}\ell MsA_c(s)$$

Solving for the ratio $E_o(s)/A_c(s)$ yields the overall transfer function

$$\frac{E_o(s)}{A_c(s)} = \frac{R_r \mathcal{B}\ell Ms}{P(s)} \tag{43}$$

where $P(s)$ is the characteristic polynomial of the device and is

$$P(s) = MLs^3 + (BL + MR)s^2 + (BR + KL + \mathcal{B}^2\ell^2)s + KR \tag{44}$$

Because the acceleration, velocity, and displacement of the case are related by $a_c(t) = \dot{v}_c = \ddot{x}_c$, and because for zero initial conditions $A_c(s) = sV_c(s) = s^2X_c(s)$, we can also write the transfer functions corresponding to a velocity or displacement input:

$$\frac{E_o(s)}{V_c(s)} = \frac{R_r \mathcal{B}\ell Ms^2}{P(s)}$$

$$\frac{E_o(s)}{X_c(s)} = \frac{R_r \mathcal{B}\ell Ms^3}{P(s)}$$

The characteristic polynomial $P(s)$ is a cubic, so it is difficult to be very specific about the behavior of the system as a measuring device without substituting numerical values for the parameters and calculating the frequency response or simulating the response to specific inputs, such as the impulse and step function.

Rather than doing this, we shall make the approximation that the coil inductance can be neglected. In terms of the frequency response, this approximation should be well justified for low frequencies but not for high frequencies. Because our interest is principally in the response at low frequencies, we are justified in setting L equal to zero in (44). With this change $P(s)$ becomes quadratic. In essence, we have eliminated the current as a state variable, and (43) reduces to

$$\frac{E_o(s)}{A_c(s)} = \frac{R_r \mathcal{B}\ell Ms}{MRs^2 + (BR + \mathcal{B}^2\ell^2)s + KR}$$

$$= \frac{(\mathcal{B}\ell R_r/R)s}{s^2 + \left(\dfrac{B}{M} + \dfrac{\mathcal{B}^2\ell^2}{MR}\right)s + \dfrac{K}{M}} \tag{45}$$

Inspection of (45) indicates that the transfer function $E_o(s)/A_c(s)$ has a zero at $s = 0$ and a pair of poles that may be real or complex. To determine the undamped natural frequency ω_n and the damping ratio ζ associated with these poles, we compare (45) to

$$H(s) = \frac{Cs}{s^2 + 2\zeta\omega_n s + \omega_n^2} \tag{46}$$

which can be viewed as the standard form for a transfer function having a zero at $s = 0$ and two poles. Comparing the coefficients in (45) and (46), we obtain

$$C = \frac{\mathcal{B}\ell R_r}{R}$$

$$\omega_n = \sqrt{\frac{K}{M}} \tag{47}$$

$$\zeta = \frac{1}{2}\sqrt{\frac{M}{K}}\left(\frac{B}{M} + \frac{\mathcal{B}^2\ell^2}{MR}\right)$$

If it were known what values of ζ and ω_n would result in a device that would perform well as a measuring instrument, a designer could attempt to select the physical parameters in order to achieve these values of ζ and ω_n. Frequency-response plots can be used to determine suitable values for ζ and ω_n.

Frequency Response

The steady-state response to a sinusoidal input is formed by examining $H(j\omega)$, as discussed in Section 8.4. A frequency-response analysis of (45) would be cumbersome unless specific numerical values were given for all but one or two of the physical parameters, so we shall work with (46). When s is replaced by $j\omega$, (46) becomes

$$\begin{aligned} H(j\omega) &= \frac{jC\omega}{\omega_n^2 - \omega^2 + j2\zeta\omega_n\omega} \\ &= \frac{j(C/\omega_n)(\omega/\omega_n)}{1 - (\omega/\omega_n)^2 + j2\zeta(\omega/\omega_n)} \end{aligned} \tag{48}$$

where the second form of the equation results from dividing both the numerator and the denominator by ω_n^2. The quantity ω/ω_n can be thought of as a normalized frequency. Because the factor C/ω_n in the numerator of (48) is a multiplying constant that does not affect the variation of $H(j\omega)$ with ω, we shall also normalize the magnitude of the transfer function by defining

$$\begin{aligned} H_N(j\omega) &= \frac{\omega_n}{C}H(j\omega) \\ &= \frac{j(\omega/\omega_n)}{1 - (\omega/\omega_n)^2 + j2\zeta(\omega/\omega_n)} \end{aligned} \tag{49}$$

We obtain the magnitude of $H_N(j\omega)$ by dividing the magnitude of its numerator by that of its denominator. This results in

$$|H_N(j\omega)| = \frac{\omega/\omega_n}{\sqrt{(\omega/\omega_n)^4 + (4\zeta^2 - 2)(\omega/\omega_n)^2 + 1}} \tag{50}$$

Comparing $|H_N(j\omega)|$ for different values of ζ indicates the relative shapes of the frequency-response magnitudes corresponding to different values of the damping ratio. If the device

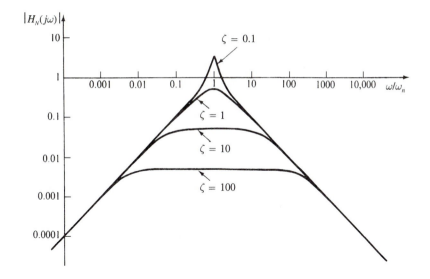

Figure 10.17 Frequency response of the acceleration-measuring device for several damping ratios.

is to be of value in measuring the acceleration of the case, there should be a range of frequencies for which $|H_N(j\omega)|$ is fairly flat—that is, independent of frequency.

From (50), we see that $|H_N(j\omega)| \simeq \omega/\omega_n$ for small values of ω/ω_n and that it approaches $1/(\omega/\omega_n)$ for large values of ω/ω_n. For $\omega/\omega_n = 1$, $|H_N(j\omega)| = 1/2\zeta$. Using this information and calculating a few additional points, we can draw the plots shown in Figure 10.17. Logarithmic scales are commonly used for frequency-response plots, and they enable us to include a wide range of values of $|H_N(j\omega)|$ and ω/ω_n.

It is apparent that the device must be heavily damped ($\zeta \gg 1$) if a range of frequencies is to be achieved for which $|H_N(j\omega)|$ is essentially constant. If $\zeta = 100$, $|H_N(j\omega)|$ will remain between 0.0045 and 0.0050 for $0.01 < \omega/\omega_n < 100$, which is a four-decade range of frequencies. In contrast, if $\zeta \leq 1$, there is no range of frequencies over which $|H_N(j\omega)|$ is essentially constant.

Because $|H_N(j\omega)| \simeq \omega/\omega_n$, as ω approaches zero, a constant acceleration will result in a zero steady-state output. We can also see this from (46) by noting that the steady-state response to a unit step-function input is $H(0) = 0$. When the acceleration $a_c(t)$ is constant, the velocity of the magnet will become equal to the velocity of the coil. There will then be no voltage induced in the coil, and the output of the recorder will be zero.

In order to read the relative displacement of the magnet with respect to the case, we could attach a pointer to the mass and put scale markings on the case. There would then be a steady-state output reading even for a constant acceleration of the case, and the device could sense acceleration of arbitrarily low frequencies. To obtain an electrical output for recording purposes, we can replace the pointer by the wiper of a potentiometer and use the voltage at the wiper arm as the output.

▶ 10.5 COMPUTER SIMULATION

In this section, we simulate two problems and examine the responses to particular inputs.

▶ EXAMPLE 10.3

Construct a Simulink diagram to solve Equations (30) from Example 10.1. Assume the load torque $\tau_L(t) = K|\omega|\omega$. The applied armature voltage $e_i(t)$ is a square wave having

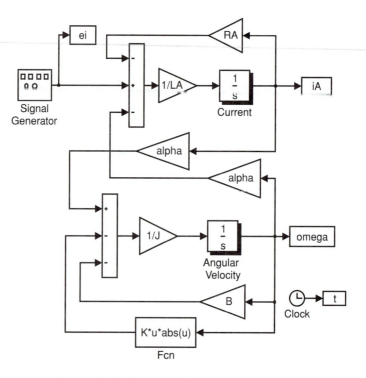

Figure 10.18 Simulink diagram for Example 10.3.

zero mean and amplitude 5 V, at a frequency of 0.05 Hz. The parameter values are: $R_A = 1\ \Omega$, $L_A = 0.5$ H, $J = 50$ kg·m^2, $B = 25$ N·m·s, and $K = 50$ N·m. Make two plots of ω on the same axes, one with $\alpha = 2$ N·m/A and the other with $\alpha = 5$ N·m/A.

SOLUTION The Simulink diagram in Figure 10.18 represents the second-order nonlinear system described by (30). The `Fcn` block comes from the `Functions & Tables` Library, and has been set up to implement the nonlinear term representing the load torque. The signal generator is set to produce the specified square waves. When the correct parameter values are entered into MATLAB, preferably using an M-file, the plot shown as Figure 10.19 results. Since the angular velocity never exceeds 0.4 rad/s, and the input voltage is 5 V, we have divided the voltage by 10, in order to make all the curves easily visible on the same axes.

One effect of increasing α is to increase the steady-state motor speed. Another effect is to cause speed changes to occur more rapidly when the input voltage changes. This is because the α^2 term in the expression for $P(s)$ in (31) is not negligible compared to the $R_A B$ term.

We now consider a system in linear motion.

▶ EXAMPLE 10.4

A loudspeaker produces sound waves by moving a diaphragm in response to an electrical input. In the cross-sectional view shown in Figure 10.20, $e_i(t)$ is the input voltage and the output is the displacement x. A coil of wire with N turns and radius a is attached to the diaphragm. Let $\alpha = 2\pi a N \mathcal{B}$, where $\mathcal{B}$ is the flux density in the air gap of the permanent

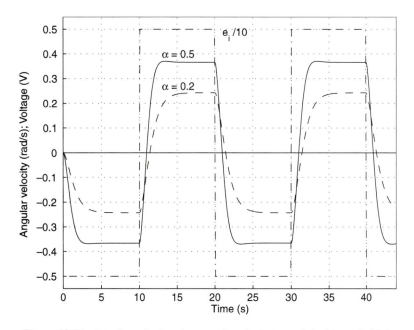

Figure 10.19 Angular velocity when $\alpha = 2$ and when $\alpha = 5$ for Example 10.3.

magnet. It can be shown that the equations

$$\dot{x} = v$$
$$\dot{v} = \frac{1}{M}(-Kx - Bv + \alpha i)$$
$$\frac{di}{dt} = \frac{1}{L}[-\alpha v - Ri + e_i(t)]$$

(51)

constitute a valid state-variable model. Construct a Simulink diagram to represent these equations. Let $e_i(t)$ be a sinusoidally varying voltage of amplitude 1 V with frequencies of 10, 100, 1000, and 5000 Hz. For each frequency, find the steady-state amplitude of the displacement x. The values of the system parameters are: $L = 0.02$ mH, $K = 1.25 \times 10^5$ N/m, $R = 2\ \Omega$, $B = 30$ N·s/m, $M = 2$ g $= 2 \times 10^{-3}$ kg, and $\alpha = 2.5$ V·s/m.

SOLUTION   Figure 10.21 shows a Simulink diagram representing the modeling equations. The top row of blocks models the first-order subsystem comprising the electrical

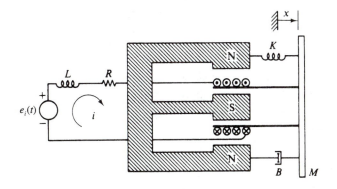

Figure 10.20 Cross-section of a loudspeaker.

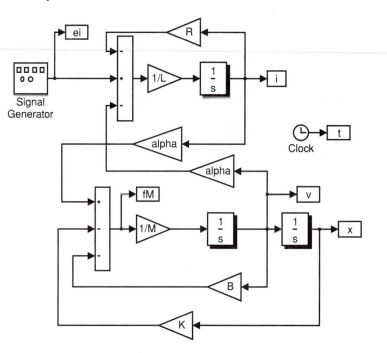

Figure 10.21 Simulink diagram to solve Example 10.4.

components of the speaker included in the third modeling equation. The lower row containing two integrators is a second-order model of the mechanical properties of the system given in the first two modeling equations. These two subsystems are interconnected by two blocks with gains of α, representing the electromechanical coupling between the electrical and mechanical elements of the speaker. We have included a block to send to the MATLAB workspace a variable called fM, the net force that accelerates the mass M. This force does not appear explicitly in the modeling equations, but it is available for plotting or other analysis. If the frequency of the sinusoidal signals produced by the Signal Generator in the upper left corner of the diagram is specified as a variable f, the model can be run with $f = 10$, and then rerun with new values of f without altering the model.

After the given parameter values, including the desired frequency, have been entered into MATLAB, the simulation is run for 0.05 s. During this time, any transient has time to decay, and the output contains only the steady-state response. Observation of the magnitudes of the peak outputs in the steady state yields the results shown in Table 10.1.

At low frequencies, when the input voltage varies slowly, the displacement is able to follow the input. As the input frequency increases, the speaker diaphragm is not able to move with sufficient speed, and the amplitude of its oscillation becomes smaller. Above about 800 Hz, the amplitude begins to drop off rapidly with frequency. This speaker might be satisfactory for bass sounds, but would not work well for sounds at the high end of a normal person's hearing range, which is about 15,000 Hz.

▷ SUMMARY

In this chapter, we discussed two mechanisms for coupling the electrical and mechanical parts of a system. The potentiometer, which is an example of resistive coupling, can

TABLE 10.1 **Steady-State Amplitude versus Frequency for Example 10.4**

Frequency Hz	Peak amplitude mm
10	0.0500
100	0.0496
1000	0.0292
5000	0.0028

produce an output voltage that is proportional to a mechanical displacement, provided that the loading effect is avoided.

The coupling for most of the electromechanical devices we considered was through a magnetic field. The applications used a current-carrying wire that moved perpendicularly to the current and to the magnetic field. When summing forces, we included the force on the wire, $f_e = \mathcal{B}\ell i$. When summing voltages, we included the induced voltage, $e_m = \mathcal{B}\ell v$. The characteristics of the overall systems depended on the electrical parameters, the mechanical parameters, and the strength of the magnetic coupling.

We presented several examples to illustrate the basic steps in modeling electromechanical devices. The references in Appendix F contain further information about nonlinear effects and construction details, including the commutator needed for a direct-current motor. They also discuss other possible types of electromechanical coupling, such as mechanically varying the characteristics of the flux path.

▶ PROBLEMS

*10.1. The potentiometer model shown in Figure 10.2(b) includes the resistances R_1 and R_2, both of which depend on $x(t)$. However, many potentiometers contain some inductance because they are constructed by winding many turns of wire about a core. The circuit shown in Figure P10.1 represents this inductance by the lumped elements whose values also depend on $x(t)$.

    a. Find the differential equation relating e_o to $x(t)$ and $e_i(t)$.

    b. Assume that R_2 and L_2 are proportional to the displacement $x(t)$. Let $L_2 = L_T[x(t)/x_{\max}]$ and $R_2 = R_T[x(t)/x_{\max}]$, where $L_1 + L_2 = L_T$ and $R_1 + R_2 = R_T$. Find the differential equation relating e_o to $x(t)$ and $e_i(t)$. Compare your answer to (3).

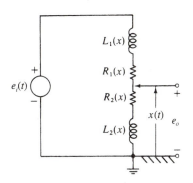

Figure P10.1

10.2.

    a. Find the state-variable equations, the transfer function, and the input-output differential equation for the galvanometer shown in Figure 10.8 when the inductance L of the coil is not neglected.

    b. Find the steady-state response to a constant input by examining $H(0)$, and compare the result to (16).

10.3. Figure P10.3 shows a galvanometer whose flux is obtained from the same current i that passes through the movable coil, rather than from a permanent magnet as shown in Figure 10.8. The flux density in the air gaps is $\mathcal{B} = k_{\mathcal{B}}i$. The coil has moment of inertia J and viscous-damping coefficient B, and it is restrained by a torsional spring with spring constant K. The total length of the coil in the magnetic field is d, and its radius is a. Let $\alpha = adk_{\mathcal{B}}$.

    a. Verify that a valid state-variable model is

$$\dot{\theta} = \omega$$

$$\dot{\omega} = \frac{1}{J}(-K\theta - B\omega + \alpha i^2)$$

$$\frac{di}{dt} = \frac{1}{L}[-\alpha i\omega - Ri + e_i(t)]$$

    b. Find $\bar{\theta}$ corresponding to the operating point $\bar{e}_i$, where $e_i(t) = \bar{e}_i + \hat{e}_i(t)$.

    c. Find a set of linearized state-variable equations valid in the vicinity of the operating point you found in part (b). To do this, first let $\theta = \bar{\theta} + \hat{\theta}$, $\omega = \bar{\omega} + \hat{\omega}$, and $i = \bar{i} + \hat{i}$. Then assume that the incremental variables are small, so that $(\hat{i})^2$ and the product $\hat{i}\hat{\omega}$ can be neglected.

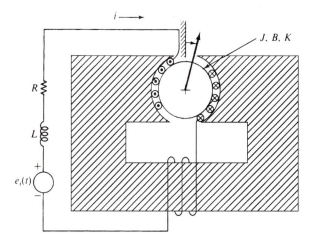

Figure P10.3

10.4.

    a. Verify (21) and (22) for the microphone shown in Figure 10.9 by transforming (18), (19), and (20) with zero initial conditions.

b. Write the input-output equation for the case $L = 0$. Find expressions for the damping ratio ζ and the undamped natural frequency ω_n in terms of the physical parameters M, K, B, R, and α.

10.5.

A transducer to measure translational motion is shown in Figure P10.5. The permanent magnet produces a uniform magnetic field in the air gap with flux density $\mathcal{B}$ and can move with a displacement x and velocity v. A wire that is fixed in space has a length d within the magnetic field. The inductance and resistance of the wire are included in the lumped elements L and R, and the output of the system is the voltage e_o across the resistor R_o.

a. Verify that the transfer function is

$$H(s) = \frac{E_o(s)}{X(s)} = \frac{d\mathcal{B}R_o s}{sL + (R + R_o)}$$

b. If $i(0) = 0$, find the response for all $t > 0$ when $x(t) = U(t)$. What is the steady-state response? What is the time constant?

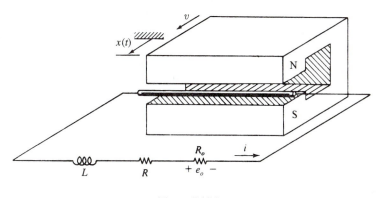

Figure P10.5

*10.6. The magnet shown in Figure P10.5 and considered in Problem 10.5 has mass M and is separated from a fixed horizontal surface by an oil film with viscous friction coefficient B. Instead of a displacement input, a force $f_a(t)$ is applied to the magnet in the positive x direction. The wire of length d remains fixed in space. Find the transfer function $H(s) = E_o(s)/F_a(s)$.

10.7. A wire of length ℓ, rigidly attached to a mass M, is in a magnetic field with flux density $\mathcal{B}$. In the cross-sectional view shown in Figure P10.7, the wire is perpendicular to the page, with point a connected to the front of the wire and point b to the back. The input to the system is a 12-V battery. With no energy initially stored within the system, the switch closes at $t = 0$.

a. Which way will the wire—and hence the mass—move when the switch is closed?

b. What will be the steady-state displacement of the mass from its original position?

c. Define a set of state variables and write the state-variable equations describing the behavior of the system for $t > 0$.

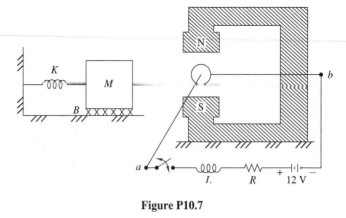

Figure P10.7

*10.8. The conductor of mass M shown in Figure P10.8 can move vertically through a uniform magnetic field of flux density $\mathcal{B}$ whose positive sense is into the page. There is no friction, the effective length of the conductor in the field is d, and $e_i(t)$ is a voltage source. The lumped elements $e_i(t), L$, and R are outside the magnetic field.

   a. Write a set of state-variable equations.

   b. For what constant voltage $\bar{e}_i$ will the conductor remain stationary?

   c. What will be the steady-state velocity of the conductor if $e_i(t)$ is always zero?

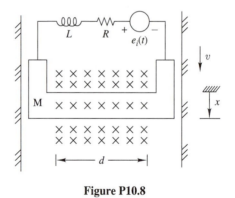

Figure P10.8

10.9. A plunger is made to move horizontally through the center of a fixed cylindrical permanent magnet by the application of a voltage source $e_i(t)$. Attached to the plunger is a coil having N turns and radius a. Figure P10.9 shows the system, including a cross-sectional view of the magnet and plunger, and indicates typical paths for the magnetic flux by dashed lines. The magnetic field between the plunger and the south pole is assumed to have a constant flux density $\mathcal{B}$. The resistance and inductance of the coil are represented by the lumped elements R and L, respectively, and the plunger has mass M.

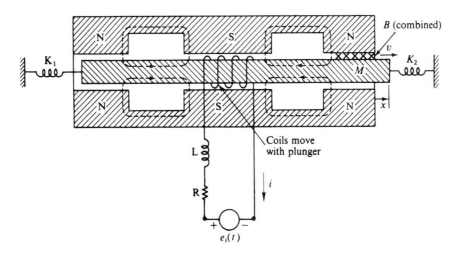

Figure P10.9

a. Verify that the following is a valid set of state-variable equations:

$$\dot{x} = v$$

$$\dot{v} = \frac{1}{M}[-(K_1 + K_2)x - Bv + \alpha i]$$

$$\frac{di}{dt} = \frac{1}{L}[-\alpha v - Ri + e_i(t)]$$

where $\alpha = 2\pi a N \mathcal{B}$.

b. Determine the direction in which the plunger will move if there is no initial stored energy and if $e_i(t) = U(t)$.

c. Find the steady-state displacement of the plunger for the input in part (b).

10.10. For the electric motor shown in Figure 10.14(b), let $i_F(t)$ have the constant value $\bar{i}_F$.

a. Write the differential equation describing the system and identify the time constant when $i_A(t)$ is the input and ω the output.

b. If $i_A(t) = i_{A_1}$ for all $t < 0$ and $i_A(t) = i_{A_2}$ for all $t > 0$, sketch ω versus t. Assume that steady-state conditions exist at $t = 0-$.

c. Repeat part (b) when i_{A_2} is replaced by $-i_{A_1}$. Find the value of t for which $\omega = 0$.

*10.11.

a. Write the differential equation describing the motor shown in Figure 10.14(b) when $i_A(t)$ and $i_F(t)$ are separate inputs and when $\phi = k_\phi i_F(t)$.

b. Find an expression for $\bar{\omega}$ at the operating point corresponding to $\bar{i}_A$ and $\bar{i}_F$.

c. If $i_A(t) = \bar{i}_A$ and $i_F(t) = \bar{i}_F + \hat{i}_F(t)$, find a linearized model that is valid in the vicinity of the operating point you found in part (b).

10.12. Let the shaft connecting J_R and J_L in Figure 10.14(b) have a stiffness constant K rather than being rigid. Also replace the current source $\bar{i}_A$ by a voltage source $e_i(t)$, with its positive sense upward. Assume that $\phi = k_\phi i_F$, and denote the angular displacements of the rotor and the load by θ_R and θ_L, respectively. Choose as state variables i_A, ω_R, ω_L, and $\theta = \theta_R - \theta_L$. Write the state-variable equations.

10.13. Assume that $L_A = 0$ for the motor shown in Figure 10.14(b). Replace the current source $\bar{i}_A$ by a voltage source that has a constant value of E_A volts.

a. Write the differential equation describing the system if the input is $i_F(t)$ and if $\phi = k_\phi i_F(t)$.

b. Find the operating point corresponding to $\bar{i}_F$.

c. Find a linearized model that is valid about the operating point you found in part (b). Identify its time constant. Let $\omega = \bar{\omega} + \hat{\omega}$ and $i_F = \bar{i}_F + \hat{i}_F$, and assume that terms involving $(\hat{i}_F)^2$ and the product $\hat{i}_F \hat{\omega}$ can be neglected.

*10.14. For the motor depicted in Figure 10.13, replace $e_i(t)$ by a constant voltage source E_A, and replace the source in the field winding by a time-varying voltage $e_F(t)$. Assume that $\phi = k_\phi i_F$.

a. Using i_A, i_F, and ω as state variables, write the state-variable equations.

b. Find $\bar{\omega}$ for the operating point corresponding to $e_F(t) = \bar{e}_F$ and $\tau_L(t) = 0$.

c. Derive a linearized model valid about the operating point you found in part (b). Let $\omega = \bar{\omega} + \hat{\omega}$, $i_A = \bar{i}_A + \hat{i}_A$, and $i_F = \bar{i}_F + \hat{i}_F$. Assume that terms involving the products $\hat{i}_F \hat{\omega}$ and $\hat{i}_A \hat{\omega}$ can be neglected.

10.15. The field and armature windings of an electric motor are connected in parallel directly across a voltage source $e_i(t)$, as shown in Figure P10.15. The resistances of the field and armature windings are R_F and R_A, respectively, and the inductances of both windings are negligible.

a. Verify that the differential equation relating ω to $e_i(t)$ and $\tau_L(t)$ is

$$J\dot{\omega} + B\omega = \frac{\gamma k_\phi}{R_F} \left(\frac{R_F - \gamma k_\phi \omega}{R_A R_F} \right) e_i^2(t) - \tau_L(t)$$

b. Find an expression for $\bar{\omega}$ at the operating point corresponding to $e_i(t) = \bar{e}_i$ and $\tau_L(t) = 0$.

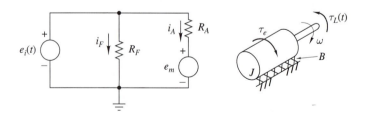

Figure P10.15

10.16. For the motor described in Problem 10.15, add an armature inductance L_A and a field inductance L_F in series with R_A and R_F, respectively.

  a. Verify that

$$\frac{di_A}{dt} = \frac{1}{L_A}[-R_A i_A - \gamma k_\phi i_F \omega + e_i(t)]$$

$$\frac{di_F}{dt} = \frac{1}{L_F}[-R_F i_F + e_i(t)]$$

$$\dot{\omega} = \frac{1}{J}[\gamma k_\phi i_F i_A - B\omega - \tau_L(t)]$$

  constitute a suitable set of state-variable equations.

  b. Find an expression for $\bar{\omega}$ at the operating point corrresponding to $e_i(t) = \bar{e}_i$ and $\tau_L(t) = 0$.

10.17. The rotor shown in Figure P10.17 is driven by a torque $\tau_a(t)$. The rotor has a moment of inertia J, but the friction is negligible. The field winding is excited by a constant voltage source, resulting in a constant magnetic flux. The armature resistance is denoted by R_A, and the armature inductance is negligible. Use (28) to obtain expressions for e_m and τ_e.

  a. With the rotor initially at rest and with the switch in the left-hand position connecting the armature to a short circuit, the applied torque is $\tau_a(t) = U(t)$. Find and sketch ω as a function of time.

  b. After the rotor has reached a steady-state speed under the conditions of part (a), the switch is thrown to the right, thereby connecting the armature to the battery. Find and sketch ω versus t if a constant unit torque continues to be applied to the rotor.

  c. After a new steady-state speed has been reached under the conditions of part (b), the applied torque is removed, with the switch still in the right-hand position. Again find and sketch ω versus t.

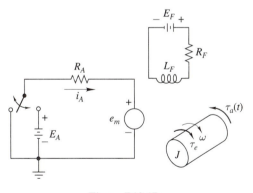

Figure P10.17

*10.18. The magnetic field of a small motor is supplied by a permanent magnet. The following two sets of measurements are made in the steady state with a 6-V battery connected across the armature. When there is no load and negligible friction, the motor rotates at 100 rad/s. When the motor is mechanically blocked to prevent its movement, the armature current is 2 A. The rotor is then attached to a propeller, which creates a viscous-friction load of 2.0×10^{-3} N·m·s, and the 6-V battery is again connected across the armature. Find the steady-state speed of the propeller.

10.19. Consider the electric motor shown in Figure 10.13 and discussed in Example 10.1. Taking i_A and ω as the state variables, write the state-variable equations in matrix form. Also write a matrix output equation for the output vector $\mathbf{y} = [\tau_e \ e_m]^T$. Identify the matrices $\mathbf{A}$, $\mathbf{B}$, $\mathbf{C}$, and $\mathbf{D}$.

10.20. Derive the transfer function $H(s) = X(s)/E_i(s)$ for the loudspeaker described in Example 10.4 and modeled by Equations (51). Then simplify this third-order system by letting $L = 0$. For the resulting second-order transfer function, evaluate the characteristic equation with the given parameter values, and find the damping ratio ζ and the undamped natural frequency ω_n.

11

THERMAL SYSTEMS

Thermal systems are systems in which the storage and flow of heat are involved. Their mathematical models are based on the fundamental laws of thermodynamics. Examples of thermal systems include a thermometer, an automobile engine's cooling system, an oven, and a refrigerator. Generally thermal systems are distributed, and thus they obey partial rather than ordinary differential equations. We shall restrict our attention to lumped mathematical models by making approximations where necessary. Our purpose is to obtain linear ordinary differential equations that are capable of describing the dynamic response to a good approximation. We shall not examine the steady-state analysis of thermodynamic cycles that might be required in the design of a chemical process. Although systems that involve changes of phase (such as boiling or condensation) can be modeled, such treatment is beyond the scope of this book.

As in the previous chapters on modeling, we first introduce the variables and the element laws used to describe the dynamic behavior of thermal systems. Then we present a number of examples illustrating their application. A comprehensive example of the analysis of a thermal system appears in Section 11.4.

▶ 11.1 VARIABLES

The variables used to describe the behavior of a thermal system are

θ, temperature in kelvins (K)[1]

q, heat flow rate in joules per second (J/s) or in watts (W) where 1 watt = 1 joule per second.

The temperatures at various points in a distributed body usually differ from one another. For modeling and analysis, however, it is desirable to assume that all points in the body have the same temperature, presumably the average temperature of the body. If the temperature deviations from the average at various points do affect the validity of the single-temperature model, then the body may be partitioned into segments. Each of the segments can have a different average temperature associated with it, as illustrated in a later example. Unless otherwise noted, we shall use average temperatures for individual bodies. Furthermore, because the temperature is a measure of the energy stored in a body (provided there are no changes of phase), we normally select the temperatures as the state variables of a thermal system.

---

[1]Although the kelvin is the SI temperature unit, degrees Celsius (°C) may be more familiar. A temperature expressed in kelvins can be converted to degrees Celsius by subtracting 273.15 from its value.

For most thermal systems, an equilibrium condition exists that defines the nominal operation. Generally, only deviations of the variables from their nominal values are of interest from a dynamic point of view. In these cases, **incremental temperatures** and **incremental heat flow rates** are defined by relationships of the form

$$\hat{\theta}(t) = \theta(t) \quad \bar{\theta}$$
$$\hat{q}(t) = q(t) - \bar{q}$$

where $\bar{\theta}$ and $\bar{q}$ are the nominal values. The ambient temperature of the environment surrounding the system is considered constant and is denoted by θ_a. In some problems, the nominal values of the temperature variables may be equal to θ_a, in which case we may refer to the incremental temperatures as **relative temperatures**.

▶ 11.2 ELEMENT LAWS

A consequence of the laws of thermodynamics is that there are only two types of passive thermal elements: thermal capacitance and thermal resistance. Strictly speaking, thermal capacitance and thermal resistance are characteristics associated with bodies that are distributed in space and are not lumped elements. However, because we seek to describe the dynamic behavior of thermal systems by lumped models, we shall refer to them as elements. These elements are described next, and a brief discussion of thermal sources follows.

Thermal Capacitance

An algebraic relationship exists between the temperature of a physical body and the heat stored within it. Provided that there is no change of phase and that the range of temperatures is not excessive, this relationship can be considered linear.

If $q_{in}(t) - q_{out}(t)$ denotes the net heat flow rate into the body as a function of time, then the net heat supplied between time t_0 and time t is

$$\int_{t_0}^{t} [q_{in}(\lambda) - q_{out}(\lambda)] \, d\lambda$$

We assume that the heat supplied during this time interval equals a constant C times the change in temperature. If the temperature of the body at the reference time t_0 is denoted by $\theta(t_0)$, then

$$\theta(t) = \theta(t_0) + \frac{1}{C} \int_{t_0}^{t} [q_{in}(\lambda) - q_{out}(\lambda)] \, d\lambda \tag{1}$$

The constant C is known as the **thermal capacitance** and has units of joules per kelvin (J/K). For a body having a mass M and specific heat σ, with units of joules per kilogram-kelvin, the thermal capacitance is $C = M\sigma$.

Differentiating (1), we have

$$\dot{\theta} = \frac{1}{C}[q_{in}(t) - q_{out}(t)] \tag{2}$$

which relates the rate of temperature change to the instantaneous net heat flow rate into the body. Because we generally select the temperatures of the bodies that constitute a thermal system as the state variables, we shall use (2) extensively to write the state-variable equations. As indicated earlier, we can use (1) and (2) only when the temperature of the body is assumed to be uniform. If the thermal gradients within the body are so great that we cannot make this assumption, then the body should be divided into two or more parts with separate thermal capacitances.

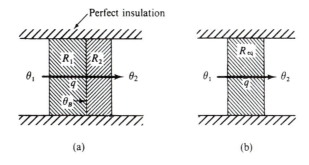

Figure 11.1 (a) Two thermal resistances in series. (b) Equivalent resistance.

Thermal Resistance

Heat can flow between points by three different mechanisms: conduction, convection, and radiation. We shall consider only conduction, whereby heat flows from one body to another through the medium connecting them at a rate proportional to the temperature difference between the points. Specifically, the flow of heat by conduction from a body at temperature θ_1 to a body at temperature θ_2 obeys the relationship

$$q(t) = \frac{1}{R}[\theta_1(t) - \theta_2(t)] \tag{3}$$

where R is the **thermal resistance** of the path between the bodies, with units of kelvin-seconds per joule (K·s/J) or kelvins per watt (K/W). For a path of cross-sectional area A and length d composed of material having a thermal conductivity α (with units of watts per meter-kelvin), the thermal resistance is

$$R = \frac{d}{A\alpha} \tag{4}$$

We can use (3) only when the material or body being treated as a thermal resistance does not store any heat. Should it become important to account for the heat stored in the resistance, then we must also include a thermal capacitance in the model.

In developing lumped models of thermal systems, we often find it convenient to combine two or more thermal resistances into a single equivalent resistance. The following two examples illustrate the techniques for doing this for combinations of two resistances.

▶ **EXAMPLE 11.1**

Figure 11.1(a) shows two bodies at temperatures θ_1 and θ_2 separated by two resistances R_1 and R_2. Heat flows through each of the resistances at the rate q but cannot flow through the perfect insulating material above and below the resistances. Find the value of the equivalent thermal resistance R_{eq} in Figure 11.1(b) and solve for the interface temperature θ_B.

SOLUTION   We can use (3) for each of the thermal resistances to express the heat flow rate q in terms of the resistance and the temperature difference. Specifically,

$$q = \frac{1}{R_1}(\theta_1 - \theta_B) \tag{5a}$$

$$q = \frac{1}{R_2}(\theta_B - \theta_2) \tag{5b}$$

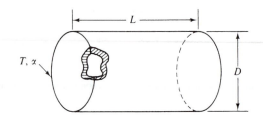

Figure 11.2 Cylindrical vessel for Example 11.2.

Combining the equations to eliminate θ_B and arranging the result in the form of (3), we find that

$$q = \frac{1}{R_1 + R_2}(\theta_1 - \theta_2)$$

Hence the equivalent thermal resistance is

$$R_{eq} = R_1 + R_2 \tag{6}$$

where R_1 and R_2 are said to be in **series** because the heat flow rate is the same through each.

To calculate the interface temperature θ_B, we combine (5a) and (5b) to eliminate q, getting

$$\theta_B = \frac{R_2\theta_1 + R_1\theta_2}{R_1 + R_2} \tag{7}$$

You should verify that equivalent forms of (7) are

$$\theta_B = \theta_1 - \frac{R_1}{R_{eq}}(\theta_1 - \theta_2)$$

$$\theta_B = \theta_2 + \frac{R_2}{R_{eq}}(\theta_1 - \theta_2)$$

▶ **EXAMPLE 11.2**

Figure 11.2 shows a hollow cylindrical vessel whose walls have thickness T and a material whose thermal conductivity is α. Calculate the thermal resistances of the side of the cylinder (R_c) and of each end (R_e). Then find the equivalent resistance of the entire vessel in terms of R_c and R_e.

SOLUTION   From (4), with d replaced by T and A replaced by $\pi D^2/4$, the resistance of each end of the vessel is

$$R_e = \frac{4T}{\pi D^2 \alpha} \tag{8}$$

Similarly, the resistance of the cylindrical portion is

$$R_c = \frac{T}{\pi D L \alpha} \tag{9}$$

The rate at which heat flows through each end is

$$q_e = \frac{\Delta\theta}{R_e}$$

where $\Delta\theta$ denotes the interior temperature minus the exterior temperature. Likewise, the heat flow rate through the cylindrical wall is

$$q_c = \frac{\Delta\theta}{R_c}$$

The total heat flow rate is

$$q_T = 2q_e + q_c$$
$$= \left(\frac{2}{R_e} + \frac{1}{R_c}\right)\Delta\theta \tag{10}$$

Because the equivalent thermal resistance R_{eq} must satisfy the relationship

$$q_T = \frac{\Delta\theta}{R_{eq}}$$

it follows from (10) that

$$\frac{1}{R_{eq}} = \frac{2}{R_e} + \frac{1}{R_c}$$

and

$$R_{eq} = \frac{R_c R_e}{2R_c + R_e} \tag{11}$$

Because the two ends and the cylindrical wall present independent paths for the heat to flow between the interior and exterior of the vessel, with the same temperature difference existing across each path, the three thermal resistances are said to be in **parallel**.

Thermal Sources

There are two types of ideal thermal sources. Figure 11.3 represents a source that adds or removes heat at a specified rate. Heat is added to the system when $q_i(t)$ is positive and removed when $q_i(t)$ is negative. On occasion, we shall consider the temperature of a body to be an input, in which case the temperature is a known function of time regardless of the rate at which heat flows between that body and the rest of the system.

▶ 11.3 DYNAMIC MODELS OF THERMAL SYSTEMS

We shall demonstrate how to construct and analyze dynamic models of thermal systems by considering several examples. The general technique is to select the temperature of each thermal capacitance as a state variable and use (2) to obtain the corresponding state-variable equation. The net heat flow rate into a thermal capacitance depends on heat sources and

Figure 11.3 Representation of an ideal thermal source.

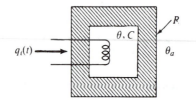

Figure 11.4 Thermal system with one capacitance.

heat flow rates through thermal resistances. By using (3), we can express the heat flow rates through the resistances in terms of the system's state variables, the temperatures of the thermal capacitances.

We first consider systems that have thermal capacitances from which heat can escape to the environment through thermal resistances. We then illustrate approximating a distributed system by a lumped model by analyzing two possible models for heating a bar. In the final example, a thermal capacitance is heated by both a heater and an incoming liquid stream.

▷ **EXAMPLE 11.3**

Figure 11.4 shows a thermal capacitance C enclosed by insulation that has an equivalent thermal resistance R. The temperature within the capacitance is θ and is assumed to be uniform. Heat is added to the interior of the system at the rate $q_i(t)$. The nominal values of $q_i(t)$ and θ are denoted by $\bar{q}_i$ and $\bar{\theta}$, respectively. The ambient temperature surrounding the exterior of the insulation is θ_a, a constant. Find the system model in terms of θ, $q_i(t)$, and θ_a and also in terms of incremental variables. Solve for the transfer function and the unit step response.

SOLUTION The appropriate state variable is θ. We obtain an expression for its derivative by using (2) with

$$q_{in}(t) = q_i(t)$$

and, from (3),

$$q_{out}(t) = \frac{1}{R}(\theta - \theta_a)$$

Thus the state-variable model is

$$\dot{\theta} = \frac{1}{C}\left[q_i(t) - \frac{1}{R}(\theta - \theta_a)\right]$$

where we consider the ambient temperature θ_a an input to the system, along with $q_i(t)$. Rewriting the model, we have

$$\dot{\theta} + \frac{1}{RC}\theta = \frac{1}{C}q_i(t) + \frac{1}{RC}\theta_a \tag{12}$$

which is readily recognized as the differential equation of a linear first-order system with the time constant $\tau = RC$ and the inputs $q_i(t)$ and θ_a. At the operating point, (12) reduces to

$$\frac{1}{RC}\bar{\theta} = \frac{1}{C}\bar{q}_i + \frac{1}{RC}\theta_a \tag{13}$$

so

$$\bar{\theta} = \theta_a + R\bar{q}_i \tag{14}$$

When the system is in equilibrium, the temperature of the thermal capacitance is constant, and the heat flow rate $\bar{q}_i$ supplied by the heater must equal the rate of heat flow through the thermal resistance. Then the temperature difference across the resistance is $R\bar{q}_i$, which agrees with (14).

To obtain a model in terms of incremental variables, we define

$$\hat{\theta}(t) = \theta(t) - \bar{\theta}$$
$$\hat{q}_i(t) = q_i(t) - \bar{q}_i$$

Substituting these expressions into (12) gives

$$\dot{\hat{\theta}} + \frac{1}{RC}(\hat{\theta} + \bar{\theta}) = \frac{1}{C}[\hat{q}_i(t) + \bar{q}_i] + \frac{1}{RC}\theta_a$$

By using (13), we can cancel the constant terms in the last equation, giving

$$\dot{\hat{\theta}} + \frac{1}{RC}\hat{\theta} = \frac{1}{C}\hat{q}_i(t) \tag{15}$$

Examining (14) shows that if $\bar{q}_i > 0$, then $\bar{\theta} > \theta_a$ and the capacitance is being heated. If $\bar{q}_i < 0$, then $\bar{\theta} < \theta_a$ and the capacitance is being cooled. If $\bar{q}_i = 0$, then the nominal value of the temperature is $\bar{\theta} = \theta_a$ and $\hat{q}_i(t) = q_i(t)$. Because the system is linear, the incremental model given by (15) has the same coefficients regardless of the operating point.

Recall from Chapter 8 that the transfer function is $H(s) = Y(s)/U(s)$, where $U(s)$ is the transformed input and $Y(s)$ is the transform of the zero-state response. Thus we transform (15) with $\hat{\theta}(0) = 0$ to obtain

$$s\hat{\Theta}(s) + \frac{1}{RC}\hat{\Theta}(s) = \frac{1}{C}\hat{Q}_i(s)$$

We can rearrange this transformed equation to give the transfer function $H(s) = \hat{\Theta}(s)/\hat{Q}_i(s)$ as

$$H(s) = \frac{\dfrac{1}{C}}{s + \dfrac{1}{RC}}$$

which has a single pole at $s = -1/RC$. The response of $\hat{\theta}$ to a unit step function for $\hat{q}_i(t)$ will be

$$\hat{\theta} = R(1 - \epsilon^{-t/RC}) \quad \text{for } t > 0$$

which approaches a steady-state value of R with the time constant RC. Note that we can also find the steady-state value of $\hat{\theta}$ by evaluating $H(s)$ at $s = 0$. This is an application of the result given in Equation (82) in Chapter 8, where the steady-state response to a constant input was shown to be the input times $H(0)$. To find the response in terms of the actual temperature θ, we merely add $\bar{\theta}$ to $\hat{\theta}$, getting

$$\theta = \bar{\theta} + R(1 - \epsilon^{-t/RC}) \quad \text{for } t > 0$$

which is shown in Figure 11.5, along with the input $q_i(t)$.

▶ **EXAMPLE 11.4**

A system with two thermal capacitances is shown in Figure 11.6. Heat is supplied to the left capacitance at the rate $q_i(t)$ by a heater, and heat is lost at the right end to the environment,

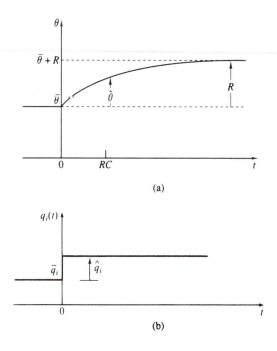

(a)

(b)

Figure 11.5 (a) Temperature response for Example 11.3. (b) Heat input rate.

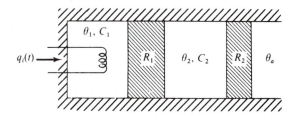

Figure 11.6 Thermal system with two capacitances.

which has the constant ambient temperature θ_a. Except for the thermal resistances R_1 and R_2, the enclosure is assumed to be perfectly insulated. Find the transfer function relating the transforms of the incremental variables $\hat{q}_i(t)$ and $\hat{\theta}_2$.

SOLUTION   Taking θ_1 and θ_2 as the two state variables and using (2) for each thermal capacitance, we can write the differential equations

$$
\begin{aligned}
\dot{\theta}_1 &= \frac{1}{C_1}\left[q_i(t) - \frac{1}{R_1}(\theta_1 - \theta_2)\right] \\
\dot{\theta}_2 &= \frac{1}{C_2}\left[\frac{1}{R_1}(\theta_1 - \theta_2) - \frac{1}{R_2}(\theta_2 - \theta_a)\right]
\end{aligned}
\tag{16}
$$

At the operating point corresponding to the nominal input $\bar{q}_i$, (16) reduces to

$$\bar{q}_i - \frac{1}{R_1}(\bar{\theta}_1 - \bar{\theta}_2) = 0$$

$$\frac{1}{R_1}(\bar{\theta}_1 - \bar{\theta}_2) - \frac{1}{R_2}(\bar{\theta}_2 - \theta_a) = 0$$

(17)

from which

$$\bar{\theta}_2 = \theta_a + R_2\bar{q}_i \tag{18a}$$

$$\bar{\theta}_1 = \theta_a + (R_1 + R_2)\bar{q}_i \tag{18b}$$

At the operating point, where equilibrium conditions exist and the temperatures are constant, the heat flow rates through R_1 and R_2 must both equal $\bar{q}_i$. Because the rates of heat flow are the same, we can regard the two resistances as being in series. The equivalent resistance is $R_{eq} = R_1 + R_2$, as in (6), so $\bar{\theta}_1$ must be the temperature difference $(R_1 + R_2)\bar{q}_i$ plus the ambient temperature θ_a, which agrees with (18b).

We define the incremental temperatures $\hat{\theta}_1 = \theta_1 - \bar{\theta}_1$ and $\hat{\theta}_2 = \theta_2 - \bar{\theta}_2$ and substitute these expressions into (16). After canceling the constant terms by using (17), we obtain

$$\dot{\hat{\theta}}_1 + \frac{1}{R_1 C_1}\hat{\theta}_1 = \frac{1}{R_1 C_1}\hat{\theta}_2 + \frac{1}{C_1}\hat{q}_i(t)$$

$$\dot{\hat{\theta}}_2 + \left(\frac{1}{R_1 C_2} + \frac{1}{R_2 C_2}\right)\hat{\theta}_2 = \frac{1}{R_1 C_2}\hat{\theta}_1$$

(19)

where the ambient temperature θ_a no longer appears. To find the transfer function $H(s) = \hat{\Theta}_2(s)/\hat{Q}_i(s)$, we transform (19) with $\hat{\theta}_1(0) = \hat{\theta}_2(0) = 0$, getting

$$\left(s + \frac{1}{R_1 C_1}\right)\hat{\Theta}_1(s) = \frac{1}{R_1 C_1}\hat{\Theta}_2(s) + \frac{1}{C_1}\hat{Q}_i(s)$$

$$\left[s + \left(\frac{1}{R_1 C_2} + \frac{1}{R_2 C_2}\right)\right]\hat{\Theta}_2(s) = \frac{1}{R_1 C_2}\hat{\Theta}_1(s)$$

Combining these equations to eliminate $\hat{\Theta}_1(s)$ and rearranging to form the ratio $\hat{\Theta}_2(s)/\hat{Q}_i(s)$, we get

$$H(s) = \frac{\dfrac{1}{R_1 C_1 C_2}}{s^2 + \left(\dfrac{1}{R_1 C_1} + \dfrac{1}{R_1 C_2} + \dfrac{1}{R_2 C_2}\right)s + \dfrac{1}{R_1 C_1 R_2 C_2}} \tag{20}$$

Observe that $H(0) = R_2$. This reflects the fact that in the steady-state, all the heat supplied must leave the region at temperature θ_2 by passing through the barrier with resistance R_2. Note that the denominator of $H(s)$ is a quadratic function of s, which implies that it will have two poles. In fact, for any combination of numerical values for R_1, R_2, C_1, and C_2, these poles will be real, negative, and distinct. As a consequence, the transient response of the system consists of two decaying exponential functions.

▶ **EXAMPLE 11.5**

Consider a bar of length L and cross-sectional area A that is perfectly insulated on all its boundaries except at the left end, as shown in Figure 11.7. The temperature at the left end

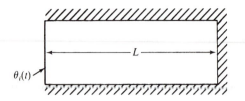

Figure 11.7 Insulated bar considered in Example 11.5.

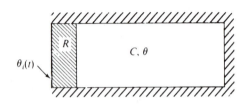

Figure 11.8 Single-capacitance approximation to the insulated bar shown in Figure 11.7.

of the bar is $\theta_i(t)$, a known function of time that is the system's input. The interior of the bar is initially at the ambient temperature θ_a. The specific heat of the material is σ with units of joules per kilogram-kelvin, its density is ρ expressed in kilograms per cubic meter, and its thermal conductivity is α expressed in watts per meter-kelvin. Although the system is distributed and can be modeled exactly only by a partial differential equation, develop a lumped model consisting of a single thermal capacitance and resistance and then find the step response.

SOLUTION Assume that all points within the bar have the same temperature θ except the left end, which has the prescribed temperature $\theta_i(t)$. Then the thermal capacitance is

$$C = \sigma \rho A L \tag{21}$$

with units of joules per kelvin. To complete the single-capacitance approximation, we assume that the thermal resistance of the entire bar, which from (4) is

$$R = \frac{L}{A\alpha} \tag{22}$$

with units of kelvins per watt, is all lumped together at the left end of the bar. The lumped approximation is shown in Figure 11.8, with C and R given by (21) and (22), respectively. From (2), with $q_{\text{out}} = 0$ because of the perfect insulation, and with

$$q_{\text{in}} = \frac{1}{R}[\theta_i(t) - \theta]$$

it follows that the single-capacitance model is

$$\dot{\theta} = \frac{1}{RC}[\theta_i(t) - \theta]$$

or

$$\dot{\theta} + \frac{1}{RC}\theta = \frac{1}{RC}\theta_i(t) \tag{23}$$

where $\theta(0) = \theta_a$.

If the input $\theta_i(t)$ is the sum of the constant ambient temperature θ_a and a step function of height B, we can write

$$\theta_i(t) = \theta_a + BU(t)$$

In this example, it is convenient to define the nominal values of the temperatures to be the ambient temperature θ_a and to let the incremental variables be the temperatures relative to θ_a. The relative input temperature is

$$\hat{\theta}_i(t) = \theta_i(t) - \theta_a$$
$$= BU(t)$$

and the relative bar temperature is

$$\hat{\theta} = \theta - \theta_a$$

With these definitions, (23) becomes

$$\dot{\hat{\theta}} + \frac{1}{RC}(\hat{\theta} + \theta_a) = \frac{1}{RC}[\theta_a + BU(t)]$$

which reduces to

$$\dot{\hat{\theta}} + \frac{1}{RC}\hat{\theta} = \frac{B}{RC}U(t) \tag{24}$$

with the initial condition $\hat{\theta}(0) = 0$. The response of the relative temperature is

$$\hat{\theta} = B(1 - \epsilon^{-t/RC}) \qquad \text{for } t > 0$$

which indicates that the temperature of the bar will rise from the ambient temperature to that of its left end with the time constant

$$RC = \frac{\sigma \rho L^2}{\alpha}$$

► **EXAMPLE 11.6**

Analyze the step response of the insulated bar shown in Figure 11.7, using the two-capacitance approximation shown in Figure 11.9.

SOLUTION   As indicated in Figure 11.9, we take the value of each thermal capacitance as $0.5C$, where C is given by (21). Likewise, we take the value of each thermal resistance

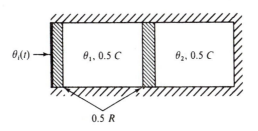

Figure 11.9   A two-capacitance approximation to the insulated bar shown in Figure 11.7.

as $0.5R$, where R is given by (22). Applying (2) to the left capacitance with

$$q_{in} = \frac{1}{0.5R}[\theta_i(t) - \theta_1]$$

and

$$q_{out} = \frac{1}{0.5R}(\theta_1 - \theta_2)$$

gives the state-variable equation

$$\dot{\theta}_1 = \frac{1}{0.25RC}[\theta_i(t) - 2\theta_1 + \theta_2] \qquad (25)$$

Applying (2) to the right capacitance with

$$q_{in} = \frac{1}{0.5R}(\theta_1 - \theta_2)$$

and $q_{out} = 0$ gives the second state-variable equation

$$\dot{\theta}_2 = \frac{1}{0.25RC}(\theta_1 - \theta_2) \qquad (26)$$

To evaluate the responses of θ_1 and θ_2 to the input $\theta_i(t) = \theta_a + BU(t)$, we can define the relative temperatures $\hat{\theta}_1 = \theta_1 - \theta_a$, $\hat{\theta}_2 = \theta_2 - \theta_a$, and $\hat{\theta}_i(t) = \theta_i(t) - \theta_a$ and then derive the transfer functions $H_1(s) = \hat{\Theta}_1(s)/\hat{\Theta}_i(s)$ and $H_2(s) = \hat{\Theta}_2(s)/\hat{\Theta}_i(s)$. Knowing $H_1(s)$ and $H_2(s)$, we can write, for the step responses,

$$\hat{\theta}_1(t) = \mathcal{L}^{-1}\left[H_1(s)\frac{B}{s}\right]$$

$$\hat{\theta}_2(t) = \mathcal{L}^{-1}\left[H_2(s)\frac{B}{s}\right]$$

Rewriting (25) and (26) in terms of the relative temperatures, we have

$$\dot{\hat{\theta}}_1 + \frac{8}{RC}\hat{\theta}_1 = \frac{4}{RC}\hat{\theta}_2 + \frac{4}{RC}\hat{\theta}_i(t)$$

$$\dot{\hat{\theta}}_2 + \frac{4}{RC}\hat{\theta}_2 = \frac{4}{RC}\hat{\theta}_1 \qquad (27)$$

Transforming (27) with $\hat{\theta}_1(0) = \hat{\theta}_2(0) = 0$ gives the pair of algebraic equations

$$\left(s + \frac{8}{RC}\right)\hat{\Theta}_1(s) = \frac{4}{RC}[\hat{\Theta}_2(s) + \hat{\Theta}_i(s)] \qquad (28a)$$

$$\left(s + \frac{4}{RC}\right)\hat{\Theta}_2(s) = \frac{4}{RC}\hat{\Theta}_1(s) \qquad (28b)$$

Solving (28b) for $\hat{\Theta}_2(s)$, substituting the result into (28a), and rearranging terms, we find that

$$\left[\left(s + \frac{8}{RC}\right)\left(s + \frac{4}{RC}\right) - \left(\frac{4}{RC}\right)^2\right]\hat{\Theta}_1(s) = \left[\frac{4}{RC}s + \left(\frac{4}{RC}\right)^2\right]\hat{\Theta}_i(s)$$

Simplifying the quadratic factor on the left side and solving for $H_1(s) = \hat{\Theta}_1(s)/\hat{\Theta}_i(s)$, we obtain

$$H_1(s) = \frac{4}{RC} \left[\frac{s + \dfrac{4}{RC}}{s^2 + \dfrac{12}{RC}s + \dfrac{16}{(RC)^2}} \right]$$

$$= \frac{4}{RC} \left[\frac{s + \dfrac{4}{RC}}{\left(s + \dfrac{1.528}{RC}\right)\left(s + \dfrac{10.472}{RC}\right)} \right] \tag{29}$$

The transfer function has poles at $s_1 = -1.528/RC$ and $s_2 = -10.472/RC$, and the free response has two terms that decay exponentially with the time constants

$$\tau_1 = \frac{RC}{1.528} = 0.6545RC$$

$$\tau_2 = \frac{RC}{10.472} = 0.0955RC$$

Because $H_1(0) = 1$ from (29), the steady-state value of $\hat{\theta}_1$ equals B when $\hat{\theta}_i(t) = BU(t)$. For this input,

$$\hat{\Theta}_1(s) = \frac{4}{RC} \left[\frac{s + \dfrac{4}{RC}}{\left(s + \dfrac{1.528}{RC}\right)\left(s + \dfrac{10.472}{RC}\right)} \right] \cdot \frac{B}{s}$$

$$= \frac{A_1}{s} + \frac{A_2}{s + \dfrac{1.528}{RC}} + \frac{A_3}{s + \dfrac{10.472}{RC}} \tag{30}$$

You can verify that the numerical values of the coefficients in the partial-fraction expansion are

$$A_1 = B$$
$$A_2 = -0.7235B$$
$$A_3 = -0.2764B$$

Substituting these values into (30) and taking the inverse transform, we find that for $t > 0$,

$$\hat{\theta}_1(t) = B\left(1 - 0.7235\epsilon^{-1.528t/RC} - 0.2764\epsilon^{-10.472t/RC}\right)$$

From (28b),

$$\hat{\Theta}_2(s) = \frac{4}{RCs + 4}\hat{\Theta}_1(s)$$

$$= \frac{16}{(RC)^2} \left[\frac{B}{s\left(s + \dfrac{1.528}{RC}\right)\left(s + \dfrac{10.472}{RC}\right)} \right]$$

whose inverse transform can be shown to be

$$\hat{\theta}_2(t) = B\left(1 - 1.1708\epsilon^{-1.528t/RC} + 0.1708\epsilon^{-10.472t/RC}\right)$$

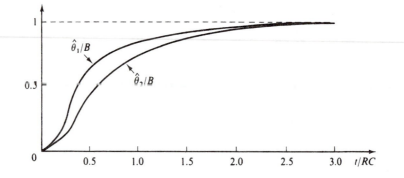

Figure 11.10 Response to a step function in $\hat{\theta}_i(t)$ for the two-capacitance approximation to the insulated bar shown in Figure 11.9.

The ratios $\hat{\theta}_1(t)/B$ and $\hat{\theta}_2(t)/B$ are shown in Figure 11.10 versus the normalized time variable t/RC. Other lumped-element approximations that lead to more accurate results for the insulated bar are investigated in some of the problems at the end of the chapter.

▶ **EXAMPLE 11.7**

The insulated vessel shown in Figure 11.11 is filled with liquid at a temperature θ, which is kept uniform throughout the vessel by perfect mixing. Liquid enters at a constant volumetric flow rate of $\bar{w}$, expressed in units of cubic meters per second, and at a temperature $\theta_i(t)$. It leaves at the same rate and at the temperature θ_o. Because of the perfect mixing, the exit temperature θ_o is the same as the liquid temperature θ. The thermal resistance of the vessel and its insulation is R, and the ambient temperature is θ_a, a constant.

Heat is added to the liquid in the vessel by a heater at a rate $q_h(t)$. The volume of the vessel is V, and the liquid has a density of ρ (with units of kilograms per cubic meter) and a specific heat of σ (with units of joules per kilogram-kelvin). Derive the system model and find the appropriate transfer functions.

SOLUTION   The thermal capacitance of the liquid is the product of the liquid's volume, density, and specific heat. The heat entering the vessel is the sum of that due to the heater and that contained in the incoming stream. The heat leaving the vessel is the sum of that taken out by the outgoing stream and that lost to the ambient through the vessel walls and insulation.

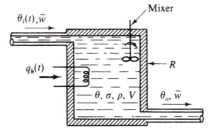

Figure 11.11 Insulated vessel with liquid flowing through it.

From (2),

$$\dot{\theta} = \frac{1}{C}(q_{\text{in}} - q_{\text{out}}) \tag{31}$$

where

$$q_{\text{in}} = q_h(t) + \bar{w}\rho\sigma\,\theta_i(t)$$

$$q_{\text{out}} = \bar{w}\rho\sigma\,\theta + \frac{1}{R}(\theta - \theta_a)$$

$$C = \rho\sigma V$$

Substituting these expressions into (31) and rearranging, we obtain the first-order differential equation

$$\dot{\theta} + \left(\frac{\bar{w}}{V} + \frac{1}{RC}\right)\theta = \frac{\bar{w}}{V}\theta_i(t) + \frac{1}{C}q_h(t) + \frac{1}{RC}\theta_a \tag{32}$$

The time constant of the system is

$$\tau = \frac{1}{\dfrac{\bar{w}}{V} + \dfrac{1}{RC}} \tag{33}$$

which approaches RC as $\bar{w}$ approaches zero (no liquid flow). As R approaches infinity (perfect insulation), the time constant becomes $V/\bar{w}$, which is the time required to replace the total tank volume at the volumetric flow rate $\bar{w}$.

Because the initial values of the state variables must be zero when we calculate the transfer functions, we first rewrite the model in terms of incremental variables defined with respect to the operating point. Setting $\dot{\theta}$ equal to zero in (32) yields the relationship between the nominal values $\bar{q}_h$, $\bar{\theta}_i$, $\bar{\theta}$, and the ambient temperature θ_a. It is

$$\bar{q}_h + \frac{C\bar{w}}{V}(\bar{\theta}_i - \bar{\theta}) = \frac{1}{R}(\bar{\theta} - \theta_a) \tag{34}$$

Rewriting (32) in terms of the incremental variables $\hat{\theta} = \theta - \bar{\theta}$, $\hat{\theta}_i(t) = \theta_i(t) - \bar{\theta}_i$, and $\hat{q}_h(t) = q_h(t) - \bar{q}_h$ and using (34), we obtain for the incremental model

$$\dot{\hat{\theta}} + \frac{1}{\tau}\hat{\theta} = \frac{\bar{w}}{V}\hat{\theta}_i(t) + \frac{1}{C}\hat{q}_h(t) \tag{35}$$

The equilibrium condition, or operating point, corresponds to the initial condition $\hat{\theta}(0) = 0$ and to incremental inputs $\hat{q}_h(t) = \hat{\theta}_i(t) = 0$. To find the transfer function $H_1(s) = \hat{\Theta}(s)/\hat{Q}_h(s)$, we transform (35) with $\hat{\theta}(0) = 0$ and $\hat{\theta}_i(t) = 0$, obtaining

$$\left(s + \frac{1}{\tau}\right)\hat{\Theta}(s) = \frac{1}{C}\hat{Q}_h(s)$$

Solving for the ratio $\hat{\Theta}(s)/\hat{Q}_h(s)$ yields

$$H_1(s) = \frac{\dfrac{1}{C}}{s + \dfrac{1}{\tau}}$$

which has a single pole at $s = -1/\tau$. The steady-state value of $\hat{\theta}$ in response to a unit step change in the heater input is

$$H_1(0) = \frac{\tau}{C} = \frac{1}{\bar{w}\rho\sigma + \dfrac{1}{R}}$$

Hence, either a high flow rate $\bar{w}$ or a low thermal resistance R tends to reduce the steady-state effect of a change in the heater input.

In similar fashion, you can verify that the transfer function $H_2(s) = \hat{\Theta}(s)/\hat{\Theta}_i(s)$ is

$$H_2(s) = \frac{\dfrac{\bar{w}}{V}}{s + \dfrac{1}{\tau}}$$

The steady-state response of $\hat{\theta}$ to a unit step change in the inlet temperature is

$$H_2(0) = \frac{\bar{w}\tau}{V} = \frac{\bar{w}\rho\sigma}{\bar{w}\rho\sigma + \dfrac{1}{R}}$$

$$= \frac{1}{1 + \dfrac{1}{\bar{w}\rho\sigma R}}$$

Thus a step change in the inlet temperature with constant heater input affects the steady-state value of the liquid temperature by only some fraction of the change. Either a high flow rate or a high thermal resistance will tend to make that fraction approach unity.

In an actual process for which θ must be maintained constant in spite of variations in $\theta_i(t)$, a feedback control system may be used to sense changes in θ and make corresponding adjustments in q_h. The type of mathematical modeling and transfer-function analysis that we have demonstrated here plays an important part in the design of such control systems.

▶ 11.4 A THERMAL SYSTEM

Producing chemicals almost always requires control of the temperature of liquids contained in vessels. In a continuous process, a vessel within which a reaction is taking place typically has liquid flowing into and out of it continuously, and a control system is needed to maintain the liquid at a constant temperature. Such a system was modeled in Example 11.7. In a batch process, a vessel would typically be filled with liquid, sealed, and then heated to a prescribed temperature. In the design and operation of batch processes, it is important to be able to calculate in advance the time required for the liquid to reach the desired temperature. Such a batch system is modeled and analyzed in the following case study.

System Description

Figure 11.12 shows a closed, insulated vessel, filled with liquid, that contains an electrical heater immersed in the liquid. The heating element is contained within a metal jacket that has a thermal resistance of R_{HL}. The thermal resistance of the vessel and its insulation is R_{La}. The heater has a thermal capacitance of C_H, and the liquid has a thermal capacitance of C_L. The heater temperature is θ_H and the temperature of the liquid is θ_L, which is assumed to be uniform because of the mixer in the vessel. The rate at which energy is supplied to the heating element is $q_i(t)$.

The heater and the liquid are initially at the ambient temperature θ_a, with the heater turned off. At time $t = 0$, the heater is connected to an electrical source that supplies energy at a constant rate. We wish to determine the response of the liquid temperature θ_L

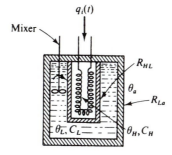

Figure 11.12 Vessel with heater.

and to calculate the time required for the liquid to reach a desired temperature, denoted by θ_d. The numerical values of the system parameters are

Heater capacitance: $C_H = 20.0 \times 10^3$ J/K

Liquid capacitance: $C_L = 1.0 \times 10^6$ J/K

Heater-liquid resistance: $R_{HL} = 1.0 \times 10^{-3}$ s·K/J

Liquid-ambient resistance: $R_{La} = 5.0 \times 10^{-3}$ s·K/J

Ambient temperature: $\theta_a = 300$ K

Desired temperature: $\theta_d = 365$ K

We will derive the system model for an arbitrary input $q_i(t)$ and for an arbitrary ambient temperature θ_a, using θ_L and θ_H as the state variables. Then we will define a set of variables relative to the ambient conditions and find the transfer function relating the transforms of the relative liquid temperature and the input. Finally, we will calculate the time required for the liquid to reach the desired temperature.

System Model

Because θ_H and θ_L represent the energy stored in the system, we can write the state-variable model as

$$\dot{\theta}_H = \frac{1}{C_H}[q_i(t) - q_{HL}]$$

$$\dot{\theta}_L = \frac{1}{C_L}[q_{HL} - q_{La}]$$

(36)

where $q_{HL} = (\theta_H - \theta_L)/R_{HL}$ and $q_{La} = (\theta_L - \theta_a)/R_{La}$. Substituting these expressions for q_{HL} and q_{La} and the appropriate numerical parameter values into (36) leads to the pair of state-variable equations

$$\dot{\theta}_H = -0.050\theta_H + 0.050\theta_L + [0.50 \times 10^{-4}]q_i(t)$$

$$\dot{\theta}_L = -[1.20 \times 10^{-3}]\theta_L + 10^{-3}\theta_H + [0.20 \times 10^{-3}]\theta_a$$

(37)

where the initial conditions are $\theta_H(0) = \theta_L(0) = \theta_a$.

At this point, we could transform (37) for a specific ambient temperature and a specific input $q_i(t)$. Then we could solve for $\Theta_L(s)$ and take its inverse transform to find $\theta_L(t)$.

Instead, we define the relative variables

$$\hat{\theta}_H = \theta_H - \theta_a \tag{38a}$$

$$\hat{\theta}_L = \theta_L - \theta_a \tag{38b}$$

$$\hat{q}_i(t) = q_i(t) \tag{38c}$$

where the last equation implies that $\bar{q}_i = 0$. Using (38) to rewrite (37), we find that the system model in terms of the relative variables is

$$\dot{\hat{\theta}}_H = -0.050\hat{\theta}_H + 0.050\hat{\theta}_L + [0.50 \times 10^{-4}]\hat{q}_i(t)$$
$$\dot{\hat{\theta}}_L = -[1.20 \times 10^{-3}]\hat{\theta}_L + 10^{-3}\hat{\theta}_H \tag{39}$$

where the initial conditions are $\hat{\theta}_H(0) = \hat{\theta}_L(0) = 0$.

When transformed and rearranged, (39) becomes the pair of algebraic equations

$$(s + 0.050)\hat{\Theta}_H(s) = 0.050\hat{\Theta}_L(s) + (0.50 \times 10^{-4})\hat{Q}_i(s)$$
$$(s + 1.20 \times 10^{-3})\hat{\Theta}_L(s) = 10^{-3}\hat{\Theta}_H(s) \tag{40}$$

Combining the two equations in (40) to eliminate $\hat{\Theta}_H(s)$, we find that the transfer function $H(s) = \hat{\Theta}_L(s)/\hat{Q}_i(s)$ is

$$H(s) = \frac{0.50 \times 10^{-4}}{1000s^2 + 51.20s + 0.010}$$

$$= \frac{0.50 \times 10^{-7}}{s^2 + 0.05120s + 10^{-5}}$$

$$= \frac{0.50 \times 10^{-7}}{(s + 0.0510)(s + 0.000196)} \tag{41}$$

System Response

Having found the transfer function $\hat{\Theta}_L(s)/\hat{Q}_i(s)$ given by (41), we can now solve for the response to various inputs. Specifically, we shall find $\theta_L(t)$ for the ambient temperature $\theta_a = 300$ K (approximately 27 °C or 80 °F) and for a step-function input $\hat{q}_i(t) = 1.50 \times 10^4$ W for $t > 0$.

To find $\hat{\Theta}_L(s)$, we multiply $\hat{Q}_i(s) = [1.50 \times 10^4](1/s)$ by $H(s)$ as given by (41), getting

$$\hat{\Theta}_L(s) = \frac{0.750 \times 10^{-3}}{s(s + 0.0510)(s + 0.000196)} \tag{42}$$

Next we expand $\hat{\Theta}_L(s)$ in partial fractions to obtain

$$\hat{\Theta}_L(s) = \frac{75.0}{s} + \frac{0.289}{s + 0.0510} - \frac{75.29}{s + 0.000196}$$

Thus for $t > 0$, the relative liquid temperature is

$$\hat{\theta}_L = 75.0 + 0.289\epsilon^{-0.0510t} - 75.29\epsilon^{-0.000196t} \tag{43}$$

We obtain the actual liquid temperature by adding the ambient temperature of 300 K to $\hat{\theta}_L$, getting

$$\theta_L = 375.0 + 0.289\epsilon^{-0.0510t} - 75.29\epsilon^{-0.000196t} \tag{44}$$

which has a steady-state value of 375 K and is shown in Figure 11.13.

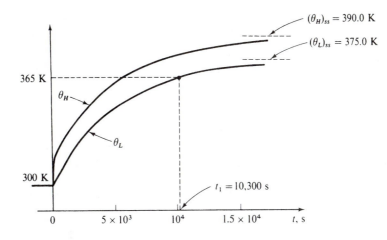

Figure 11.13 Responses of liquid and heater temperatures.

From inspection of either (43) or (44), we see that the transient response of the system has two exponentially decaying modes with time constants of

$$\tau_1 = \frac{1}{0.0510} = 19.61 \text{ s}$$

$$\tau_2 = \frac{1}{0.000196} = 5102 \text{ s}$$

Because τ_2 exceeds τ_1 by several orders of magnitude, the transient having the longer time constant dominates the response. The principal effect of the shorter time constant is to give θ_L a zero slope at $t = 0+$, which it would not have if the transient response were a single decaying exponential.

Finally, we must calculate the value of t_1, the time required for the liquid to reach the desired temperature $\theta_d = 365$ K. Because two exponential terms are present, an explicit solution of (44) is not possible. However by writing (44) in terms of the time constants τ_1 and τ_2 as

$$\theta_L = 375.0 + 0.289\epsilon^{-t/\tau_1} - 75.29\epsilon^{-t/\tau_2}$$

and noting that the solution for t_1 must be much greater than τ_1, we can see that the term $0.289\epsilon^{-t/\tau_1}$ is negligible when $t = t_1$. Hence we can use the simpler approximate expression

$$\theta_L = 375.0 - 75.29\epsilon^{-t/5102} \tag{45}$$

where the value of 5102 s has been substituted for τ_2. Because (45) contains only one exponential function, we can solve it explicitly for t_1. Replacing the left side of (45) by the value 365.0 and replacing t by t_1 on the right side, we have

$$365.0 = 375.0 - 75.29\epsilon^{-t_1/5102}$$

which leads to

$$t_1 = 5102 \ln \left(\frac{75.29}{375.0 - 365.0} \right)$$

$$= 10{,}300 \text{ s}$$

which is approximately 2.86 hours.

Though we have completed our original task, we can use the preceding modeling and analysis results to carry out a variety of additional tasks. For instance, we can find the response of the heater temperature θ_H for the specified $q_i(t)$ by eliminating $\Theta_L(s)$ from (40). The heater temperature θ_H is shown in Figure 11.13, and you are encouraged to evaluate the analytical expression for it. In practice, it might be important to evaluate the time t_1 for constant values of $\hat{q}_i(t)$ other than the value 1.50×10^4 W that we used. For example, if $\hat{q}_i(t)$ is doubled to 3.0×10^4 W, the desired liquid temperature $\theta_d = 365$ K will be reached in only 2916 s (48.6 minutes). On the other hand, if the constant energy-input rate is less than 1.30×10^4 W, the liquid will never reach 365 K. If we want to investigate the effects of replacing the constant power source by a variable source, we can multiply the transfer function $H(s)$ by $\hat{Q}_i(s)$ to give the transform of the relative temperature. After we find the inverse transform, we can add it to the ambient temperature to get θ_L.

▶ **EXAMPLE 11.8**

Consider the situation that might arise if this heated chamber needed to be maintained at a temperature near 365 K, but the only power source was a heater that was twice as large as was needed. In order to keep the desired temperature while using this heater, it is decided to switch it off half of the time. The question we want to address is: How fast does the liquid temperature approach 365 K, and how much does it change around its final average value if a heater delivers 26 kW for 25 minutes, followed by 25 minutes off, and this on-off cycle continues indefinitely?

SOLUTION A Simulink diagram to simulate the system in Figure 11.12 and solve (36) is shown in Figure 11.14. It uses the same parameter values given above. The heater power $q_i(t)$ is zero for 25 minutes, then increases to 26 kW for 25 minutes, and then returns to

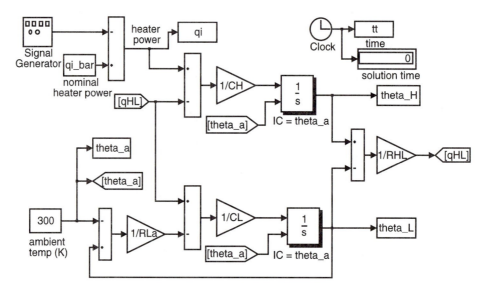

Figure 11.14 A Simulink diagram to solve (36).

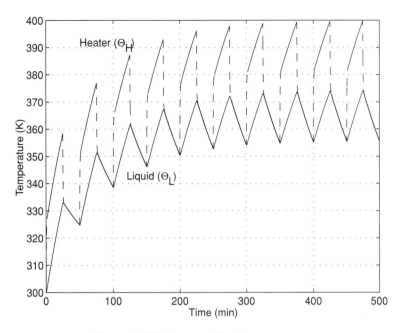

Figure 11.15 Heater and liquid temperatures.

zero and repeats this cycle. This is achieved by adding a steady level of $\bar{q}_i$ to the output of the `Signal Generator`, which varies from $-\bar{q}_i$ to $+\bar{q}_i$. The result is a signal that varies between zero and $2\bar{q}_i$.

This diagram uses several new Simulink features: `Goto`, `From`, and `Display` blocks. The `Goto` and `From` blocks are used together to avoid the need to draw long lines across Simulink diagrams. A signal which is introduced to any `Goto` block with a particular name appears automatically at all `From` blocks that have that same name. This feature is used in Figure 11.14 for the variables θ_a and qHL. Because of the lengthy solution time required by the 30,000 s stopping time (500 minutes), we include a `Display` block whose input is the time variable `tt`. This allows the user to monitor the problem time as the simulation progresses, without having to be concerned if something has gone wrong. `Display` blocks could also be connected to the outputs of the two integrators to allow monitoring of the heater and liquid temperatures. This diagram also uses the integrator block which shows the initial condition, θ_a, explicitly.

Figure 11.15 contains plots of the temperature of the heater, θ_H, and that of the liquid, θ_L. Both start at the ambient temperature of 300 K, and increase while the heater is on. It is during the fourth "on" cycle of the heater that the liquid temperature first reaches 365 K. The average liquid temperature reaches a steady state only after about 400 minutes. Thereafter, the liquid temperature varies between about 373 K and 357 K. If this range were unacceptable, it would be relatively easy to change the frequency of the signal generator to switch more rapidly, which would reduce the size of the temperature fluctuations, but at the cost of placing more demands on the heater controller.

▶ SUMMARY

We first introduced the basic variables—temperature and heat flow rate—and then presented the element laws. In contrast to the three types of passive elements in mechanical and electrical systems, there are only two types of passive elements in thermal systems: thermal capacitance and thermal resistance. Also, a greater degree of approximation is often necessary in order to represent a thermal system by a lumped-element model.

The temperature of each body that can store heat is usually taken to be a state variable. Because there is only one type of energy-storing passive element, the poles of the transfer functions are real numbers, and the mode functions that characterize the free response are exponential rather than sinusoidal terms. Of course, adding mechanical or other elements to the thermal part of a system can change the nature of this response.

We investigated many examples, which culminated in the case study presented in Section 11.4. The procedures for finding a state-variable model, an input-output differential equation, or a transfer function are similar to those used for mechanical and electrical examples. In fact, the general methods of modeling and analysis introduced in earlier chapters are applicable to a wide variety of systems. In the next chapter, we shall apply them to hydraulic systems.

▶ PROBLEMS

11.1. Find the equivalent thermal resistance for the three resistances shown in Figure P11.1. Also express the temperatures θ_A and θ_B at the interfaces in terms of θ_1 and θ_2.

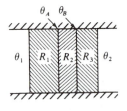

Figure P11.1

*11.2. The hollow enclosure shown in Figure P11.2 has four sides, each with resistance R_a, and two ends, each with resistance R_b. Find the equivalent thermal resistance between the interior and the exterior of the enclosure. Also calculate the total heat flow rate from the interior to the exterior when the interior of the enclosure is at a constant temperature θ_1 and the ambient temperature is θ_a.

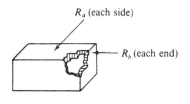

Figure P11.2

11.3. A perfectly insulated enclosure containing a heater is filled with a liquid having thermal capacitance C. The temperature of the liquid is assumed to be uniform and is denoted by θ. The heat supplied by the heater is $q_i(t)$.

   a. Write the differential equation obeyed by the liquid temperature θ.

   b. Solve the equation you found in part (a) for θ in terms of the arbitrary initial temperature $\theta(0)$.

   c. Sketch θ versus t when $q_i(t) = 1$ for $10 < t \leq 20$ and zero otherwise.

11.4. Figure P11.4 shows a volume for which the temperature is θ_1 and the thermal capacitance is C. The volume is perfectly insulated from the environment except for the thermal resistances R_1 and R_2. Heat is supplied at the rate $q_i(t)$. The ambient temperature is θ_a.

   a. Write the system model.

   b. Find and sketch θ_1 versus t when $q_i(t) = AU(t)$ and $\theta_1(0) = \theta_a$. (*Hint:* First solve for the relative temperature $\hat{\theta}_1 = \theta_1 - \theta_a$.)

   c. Evaluate the transfer function $\hat{\Theta}_1(s)/Q_i(s)$ and sketch the magnitude of the frequency response versus ω.

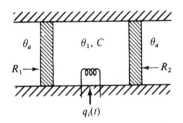

Figure P11.4

*11.5.

   a. Repeat part (a) of Problem 11.4 when the temperature to the right of R_2 is $\theta_2(t)$ rather than the constant ambient temperature θ_a.

   b. Using the relative temperatures $\hat{\theta}_1 = \theta_1 - \theta_a$ and $\hat{\theta}_2(t) = \theta_2(t) - \theta_a$, rewrite the model and evaluate the transfer functions $H_1(s) = \hat{\Theta}_1(s)/\hat{\Theta}_2(s)$ and $H_2(s) = \hat{\Theta}_1(s)/Q_i(s)$.

11.6. The system shown in Figure P11.6 is composed of two thermal capacitances, three thermal resistances, and two heat sources.

   a. Verify that the model can be written in state-variable form as

$$\dot{\theta}_1 = \frac{1}{C_1}\left[-\left(\frac{1}{R_1} + \frac{1}{R_2} \right)\theta_1 + \frac{1}{R_2}\theta_2 + \frac{1}{R_1}\theta_a + q_1(t) \right]$$

$$\dot{\theta}_2 = \frac{1}{C_2}\left[\frac{1}{R_2}\theta_1 - \left(\frac{1}{R_2} + \frac{1}{R_3} \right)\theta_2 + \frac{1}{R_3}\theta_a + q_2(t) \right]$$

   b. Derive the input–output differential equation relating θ_1, the inputs $q_1(t)$ and $q_2(t)$, and the ambient temperature θ_a.

   c. Defining the relative temperature $\hat{\theta}_1$ as $\hat{\theta}_1 = \theta_1 - \theta_a$, write the transfer function $\hat{\Theta}_1(s)/\hat{Q}_1(s)$, and evaluate the steady-state response to a unit step-function input.

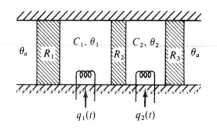

Figure P11.6

Identify the three other transfer functions associated with the system, but do not solve for them.

*11.7.

   a. Find the state-variable model for Figure P11.6 when the temperature to the right of R_3 is $\theta_3(t)$ rather than the constant ambient temperature θ_a.

   b. Rewrite the model in terms of the relative temperatures $\hat{\theta}_1 = \theta_1 - \theta_a$, $\hat{\theta}_2 = \theta_2 - \theta_a$, and $\hat{\theta}_3(t) = \theta_3(t) - \theta_a$, and derive the transfer functions $H_1(s) = \hat{\Theta}_1(s)/\hat{\Theta}_3(s)$ and $H_2(s) = \hat{\Theta}_2(s)/\hat{\Theta}_3(s)$.

11.8. Figure P11.8 shows an electronic amplifier and a fan that can be turned on to cool the amplifier. The electronic equipment has a thermal capacitance C and generates heat at the constant rate $\bar{q}$ when it is operating. The amplifier's thermal conductivity to the ambient temperature due to convection is K with the fan off and $2K$ with it on. The ambient temperature is θ_a, and the temperature of the amplifier is θ_1.

   a. Verify that, with the fan off, the differential equation obeyed by θ_1 is

$$\dot{\theta}_1 + \frac{K}{C}\theta_1 = \frac{1}{C}(K\theta_a + \bar{q})$$

and that when the fan is on, the equation is

$$\dot{\theta}_1 + \frac{2K}{C}\theta_1 = \frac{1}{C}(2K\theta_a + \bar{q})$$

   b. Solve for and sketch θ_1 when the system undergoes the following sequence of operations. In each case, indicate the steady-state temperature and the time constant.

     (i) With $\theta_1(t_1) = \theta_a$ and the fan off, the amplifier is turned on at $t = t_1$.

     (ii) After the system reaches steady-state conditions, the fan is turned on at $t = t_2$.

     (iii) After the system reaches steady-state conditions, the amplifier is turned off at $t = t_3$ with the fan running.

   c. Show how the response for $t > t_3$ in part (b) is affected when the fan also is turned off at $t = t_3$.

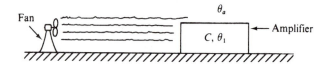

Figure P11.8

*11.9. Figure P11.9 shows a piece of hot metal that has been immersed in a water bath to cool it. We can develop a simplified model by assuming that the temperatures of the metal and water, denoted by θ_m and θ_w, respectively, are uniform and that the rate of heat loss from the metal is proportional to the temperature difference $\theta_m - \theta_w$. The thermal capacitances are C_m and C_w, and the thermal resistance is R. We can neglect initially any heat lost to the environment at the surface.

    a. Verify that the mathematical model of the system subject to these assumptions can be written as

$$\dot{\theta}_m = \frac{1}{RC_m}(-\theta_m + \theta_w)$$

$$\dot{\theta}_w = \frac{1}{RC_w}(\theta_m - \theta_w)$$

    b. Taking the initial metal temperature and water temperature as $\theta_m(0)$ and $\theta_w(0)$, respectively, solve for and sketch θ_m versus t and θ_w versus t.

    c. Modify your answer to part (a) to allow for heat flow from the water to the environment. Denote the ambient temperature by θ_a and the thermal resistance between the water and the environment by R_a.

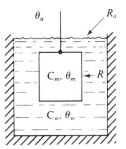

Figure P11.9

11.10. A heat exchanger in a chemical process uses steam to heat a liquid flowing in a pipe. We can apply an approximate linear model to relate changes in the temperature of the liquid leaving the heat exchanger to changes in the rate of steam flow. This model is given by the transfer function

$$\frac{\hat{\Theta}(s)}{\hat{W}(s)} = \frac{Ae^{-sT_d}}{(\tau_1 s + 1)(\tau_2 s + 1)}$$

where $\hat{\Theta}(s)$ and $\hat{W}(s)$ are the Laplace transforms of the incremental outlet temperature and the incremental steam flow rate, respectively. We can determine the coefficient A, the time delay T_d, and the time constants τ_1 and τ_2 experimentally by recording the response to a step change in the steam flow rate with all other process conditions held constant. For parameter values $A = 0.5$ K·s/kg, $T_d = 20$ s, $\tau_1 = 15$ s, and $\tau_2 = 150$ s, evaluate and sketch the following:

    a. The response to a step in $\hat{w}(t)$ of 20 kg/s

    b. The unit impulse response $h(t)$

    c. The magnitude of the frequency response $H(j\omega)$ versus ω

*11.11. Write the state-variable equations, using relative temperatures with respect to the ambient temperature, for a three-capacitance model of the insulated bar considered in Examples 11.5 and 11.6. Use three equal capacitances of value $C/3$ and three equal resistances of value $R/3$. Find the characteristic polynomial in terms of the parameter RC.

11.12. Repeat Problem 11.11, using nonuniform lengths for the three lumped segments of the bar. Specifically, take segments of length $0.2L$, $0.3L$, and $0.5L$, from left to right, where L is the length of the bar. Explain why such an approximation should yield greater accuracy than the equal-segment approximation of Problem 11.11.

11.13.

a. Solve part (a) and part (b) of Problem 11.9 in numerical form for the following parameter values and initial conditions. The parameter values are

> Metal mass $= 50$ kg
> Metal specific heat $= 460$ J/(kg·K)
> Liquid volume $= 0.15$ m³
> Liquid density $= 1000$ kg/m³
> Liquid specific heat $= 4186$ J/(kg·K)
> Thermal resistance between the metal and the liquid $= 10^{-3}$ s·K/J
> Thermal resistance between the liquid and the environment $= 10^{-2}$ s·K/J

The initial conditions are

> Initial metal temperature $= 750$ K
> Initial liquid temperature $= 320$ K
> Ambient temperature $= 300$ K

b. Do part (c) of Problem 11.9 and solve for θ_m. Sketch θ_m versus t and show qualitatively the shape of the curve for θ_w.

11.14. Using a digital computer, simulate the response of the insulated bar considered in Examples 11.5 and 11.6 by using N elements of equal length, as shown in Figure P11.14. Normalize the model by taking $RC = 1$ and finding the response to $\hat{\theta}_i(t) = U(t)$, where the initial relative temperature of each element is taken as zero.

a. Use $N = 3$ and plot θ_a, θ_b, and θ_c versus time.

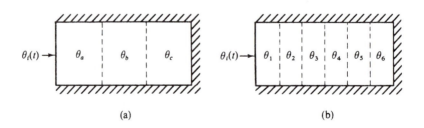

(a)   (b)

Figure P11.14

b. Use $N = 6$ and plot the temperatures.

$$\theta_a^* = 0.5(\theta_1 + \theta_2)$$
$$\theta_b^* = 0.5(\theta_3 + \theta_4)$$
$$\theta_c^* = 0.5(\theta_5 + \theta_6)$$

Compare the results for the two cases and comment on the difference.

11.15. The temperature of a uniform bar is analyzed in Example 11.5 and Example 11.6, and extensions of the analysis are proposed in Problems 11.11 and 11.12. For each of the following lumped models, using relative temperatures with respect to the ambient temperature, write the state-variable equations in matrix form and write an output equation for the average temperature of the bar:

   a. The two-capacitance model with uniform segment lengths analyzed in Example 11.6

   b. The three-capacitance model with uniform segment lengths described in Problem 11.11

   c. The three-capacitance model with nonuniform segment lengths described in Problem 11.12

11.16. Use the Simulink diagram in Figure 11.14 to examine the effect of increasing the frequency with which the heater is turned on and off. Change the frequency of the square-wave heater input to 0.0012 Hz, so that the period is 13.9 minutes. What is the peak-to-peak change in bath temperature in the steady-state?

11.17. Consider the heated chamber of Example 11.8. In the example, the heater was on for 25 minutes, then off for 25 minutes. This would be described as a "duty cycle" of 50 percent. Replace the `Signal Generator` and the `Constant` which produce the heater power signal used in Figure 11.14 with a `Pulse Generator` from the `Sources` library. Plot the heater and liquid temperatures if the heater size is increased to 39 kW, and the duty cycle is reduced to 33.33 percent.

12

FLUID SYSTEMS

A hydraulic system is one in which liquids, which are generally considered incompressible, flow. Hydraulic systems commonly appear in chemical processes, automatic control systems, and actuators and drive motors for manufacturing equipment. Such systems are usually interconnected to mechanical systems through pumps, valves, and movable pistons. A turbine driven by water and used for driving an electric generator is an example of a system with interacting hydraulic, mechanical, and electrical elements. We will not discuss here systems that include compressible fluids such as air and other gases.

An exact analysis of hydraulic systems is usually not feasible because of their distributed nature and the nonlinear character of the resistance to flow. For our dynamic analysis, however, we can obtain satisfactory results by using lumped elements and linearizing the resulting nonlinear mathematical models. On the other hand, the design of chemical processes requires a more exact analysis wherein static, rather than dynamic, models are used.

In most cases, hydraulic systems operate with the variables remaining close to a specific operating point. Thus we are generally interested in models involving incremental variables. This fact is particularly helpful because such models are usually linear, although the model in terms of the total variables may be quite nonlinear.

In the next two sections, we shall define the variables to be used and introduce and illustrate the element laws. Then we will present a variety of examples to demonstrate the modeling process and the application of the analytical techniques discussed in previous chapters, including Laplace transforms and transfer functions.

▶ 12.1 VARIABLES

Because hydraulic systems involve the flow and accumulation of liquid, the variables used to describe their dynamic behavior are

> w, flow rate in cubic meters per second (m^3/s)
>
> v, volume in cubic meters (m^3)
>
> h, liquid height in meters (m)
>
> p, pressure in newtons per square meter (N/m^2)

Unless otherwise noted, a pressure will be the **absolute pressure**. In addition, we shall sometimes find it convenient to express pressures in terms of gauge pressures. A **gauge pressure**, denoted by p^*, is defined to be the difference between the absolute pressure and the atmospheric pressure p_a:

$$p^*(t) = p(t) - p_a \tag{1}$$

A pressure difference, denoted by Δp, is the difference between the pressures at two points.

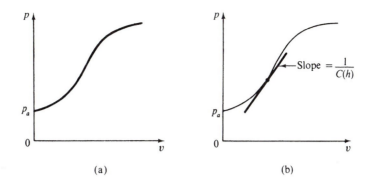

Figure 12.1 Pressure versus liquid volume for a vessel with variable cross-sectional area $A(h)$.

▶ ## 12.2 ELEMENT LAWS

Hydraulic systems exhibit three types of characteristics that can be approximated by lumped elements: capacity, resistance to flow, and inertance. In this section we shall discuss the first two. The inertance, which accounts for the kinetic energy of a moving fluid stream, is usually negligible, and we will not consider it. A brief discussion of centrifugal pumps that act as hydraulic sources appears at the end of this section.

Capacitance

When liquid is stored in an open vessel, there is an algebraic relationship between the volume of the liquid and the pressure at the base of the vessel. If the cross-sectional area of the vessel is given by the function $A(h)$, where h is the height of the liquid level above the bottom of the vessel, then the liquid volume v is the integral of the area from the base of the vessel to the top of the liquid. Hence,

$$v = \int_0^h A(\lambda)\, d\lambda \tag{2}$$

where λ is a dummy variable of integration. For a liquid of density ρ expressed in kilograms per cubic meter, the absolute pressure p and the liquid height h are related by

$$p = \rho g h + p_a \tag{3}$$

where g is the gravitational constant (9.807 m/s^2) and where p_a is the atmospheric pressure, which is taken as 1.013×10^5 N/m^2.

Equations (2) and (3) imply that for any vessel geometry, liquid density, and atmospheric pressure, there is a unique algebraic relationship between the pressure p and the liquid volume v. A typical characteristic curve describing this relationship is shown in Figure 12.1(a).

If the tangent to the pressure-versus-volume curve is drawn at some point, as shown in Figure 12.1(b), then the reciprocal of the slope is defined to be the **hydraulic capacitance**, denoted by $C(h)$. As indicated by the h in parentheses, the capacitance depends on the point on the curve being considered and hence on the liquid height h. Now

$$C(h) = \frac{1}{dp/dv} = \frac{dv}{dp}$$

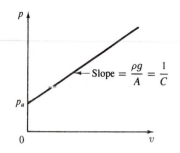

Figure 12.2 Pressure versus liquid volume for a vessel with constant A.

and, from the chain rule of differentiation,

$$C(h) = \frac{dv}{dh}\frac{dh}{dp}$$

We see that $dv/dh = A(h)$ from (2) and that $dh/dp = 1/\rho g$ from (3). Thus for a vessel of arbitrary shape,

$$C(h) = \frac{A(h)}{\rho g} \tag{4}$$

which has units of $m^4 \cdot s^2/kg$ or, equivalently, m^5/N.

For a vessel with constant cross-sectional area A, (2) reduces to $v = Ah$. We can substitute the height $h = v/A$ into (3) to obtain the pressure in terms of the volume:

$$p = \frac{\rho g}{A}v + p_a \tag{5}$$

Equation (5) yields a linear plot of pressure versus volume, as shown in Figure 12.2. The slope of the line is the reciprocal of the capacitance C, where

$$C = \frac{A}{\rho g} \tag{6}$$

The volume of liquid in a vessel at any instant is the integral of the net flow rate into the vessel plus the initial volume. Hence we can write

$$v(t) = v(0) + \int_0^t [w_{in}(\lambda) - w_{out}(\lambda)]\, d\lambda$$

which can be differentiated to give the alternative form

$$\dot{v} = w_{in}(t) - w_{out}(t) \tag{7}$$

To obtain expressions for the time derivatives of the pressure p and the liquid height h that are valid for vessels with variable cross-sectional areas, we use the chain rule of differentiation to write

$$\frac{dv}{dt} = \frac{dv}{dh}\frac{dh}{dt}$$

where dv/dt is given by (7) and where $dv/dh = A(h)$. Thus the rate of change of the liquid height depends on the net flow rate according to

$$\dot{h} = \frac{1}{A(h)}[w_{in}(t) - w_{out}(t)] \tag{8}$$

Alternatively, we can write dv/dt as

$$\frac{dv}{dt} = \frac{dv}{dp}\frac{dp}{dt}$$

where $dv/dp = C(h)$. Hence the rate of change of the pressure at the base of the vessel is

$$\dot{p} = \frac{1}{C(h)}[w_{in}(t) - w_{out}(t)] \tag{9}$$

where $C(h)$ is given by (4).

Because any of the variables v, h, and p can be used as a measure of the amount of liquid in a vessel, we generally select one of them as a state variable. Then (7), (8), or (9) will yield the corresponding state-variable equation when w_{in} and w_{out} are expressed in terms of the state variables and inputs.

If the cross-sectional area of the vessel is variable, then the coefficient $A(h)$ in (8) will be a function of h, and the system model will be nonlinear. To develop a linearized model, we must find the operating point, define the incremental variables, and retain the first two terms in the Taylor-series expansion. Likewise, the term $C(h)$ in (9) will cause the differential equation to be nonlinear because the capacitance varies with h, which in turn is a function of the pressure.

▶ **EXAMPLE 12.1**

Consider a vessel formed by a circular cylinder of radius R and length L that contains a liquid of density ρ in units of kilograms per cubic meter. Find the hydraulic capacitance of the vessel when the cylinder is vertical, as shown in Figure 12.3(a). Then evaluate the capacitance when the cylinder is on its side, as shown in Figure 12.3(b).

SOLUTION For the configuration shown in Figure 12.3(a), the cross-sectional area is πR^2 and so is independent of the liquid height. Thus we can use (6), and the vessel's hydraulic capacitance is $C_a = \pi R^2 / \rho g$.

When the vessel is on its side, as shown in Figure 12.3(b), the cross-sectional area is a function of the liquid height h. You can verify that the width of the liquid surface is $2\sqrt{R^2 - (R-h)^2}$, which is zero when $h = 0$ and $h = 2R$ and which has a maximum value of $2R$ when $h = R$. Using (4), we find that the capacitance is

$$C_b = \frac{2L}{\rho g}\sqrt{R^2 - (R-h)^2}$$

which is shown in Figure 12.4.

Resistance

As liquid flows through a pipe, there is a drop in the pressure of the liquid over the length of pipe. There is likewise a pressure drop if the liquid flows through a valve or through an orifice. The change in pressure associated with a flowing liquid results from the dissipation of energy and usually obeys a nonlinear algebraic relationship between the flow rate w and the pressure difference Δp. The symbol for a valve is shown in Figure 12.5; it can also be used for other energy-dissipating elements. A positive value of w indicates that liquid is flowing in the direction of the arrow; a positive value of Δp indicates that the pressure at the end marked $+$ is higher than the pressure at the other end. The expression

$$w = k\sqrt{\Delta p} \tag{10}$$

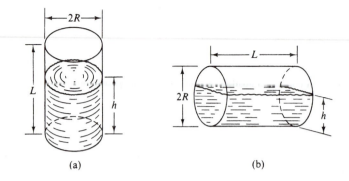

Figure 12.3 Cylindrical vessel for Example 12.1. (a) Cylinder vertical. (b) Cylinder horizontal.

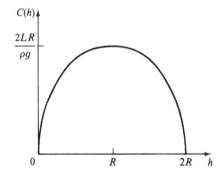

Figure 12.4 Capacitance of the vessel shown in Figure 12.3(b).

describes an orifice and a valve and is a good approximation for turbulent flow through pipes. We can treat all situations of interest to us by using a nonlinear element law of the form of (10). In this equation, k is a constant that depends on the characteristics of the pipe, valve, or orifice. A typical curve of flow rate versus pressure difference is shown in Figure 12.6(a).

Because (10) is a nonlinear relationship, we must linearize it about an operating point in order to develop a linear model of a hydraulic system. If we draw the tangent to the curve of w versus Δp at the operating point, the reciprocal of its slope is defined to be the **hydraulic resistance** R. Figure 12.6(b) illustrates the geometric interpretation of the resistance, which has units of newton-seconds per meter[5].

Expanding (10) in a Taylor series about the operating point gives

$$w = \overline{w} + \left.\frac{dw}{d\,\Delta p}\right|_{\overline{\Delta p}} (\Delta p - \overline{\Delta p}) + \cdots$$

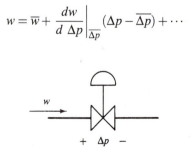

Figure 12.5 Symbol for a hydraulic valve.

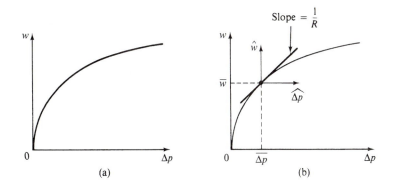

Figure 12.6 (a) Flow rate versus pressure difference given by (10). (b) Geometric interpretation of hydraulic resistance.

The incremental variables $\hat{w}$ and $\widehat{\Delta p}$ are defined by

$$\hat{w} = w - \overline{w} \tag{11a}$$

$$\widehat{\Delta p} = \Delta p - \overline{\Delta p} \tag{11b}$$

and the second- and higher-order terms in the expansion are dropped. Thus the incremental model becomes

$$\hat{w} = \frac{1}{R}\widehat{\Delta p} \tag{12}$$

where

$$\frac{1}{R} = \frac{dw}{d\,\Delta p}\bigg|_{\overline{\Delta p}}$$

We can express the resistance R in terms of either $\overline{\Delta p}$ or $\overline{w}$ by carrying out the required differentiation using (10). Specifically,

$$\frac{1}{R} = \frac{d}{d\,\Delta p}(k\,\Delta p^{1/2})\bigg|_{\overline{\Delta p}}$$

$$= \frac{k}{2\sqrt{\overline{\Delta p}}}$$

so

$$R = \frac{2\sqrt{\overline{\Delta p}}}{k} \tag{13}$$

To express the resistance in terms of $\overline{w}$, we note from (10) that

$$\overline{w} = k\sqrt{\overline{\Delta p}} \tag{14}$$

Substituting (14) into (13) gives the alternative equation for the hydraulic resistance as

$$R = \frac{2\overline{w}}{k^2} \tag{15}$$

Because liquids typically flow through networks composed of pipes, valves, and orifices, we must often combine several relationships of the form of (10) into a single

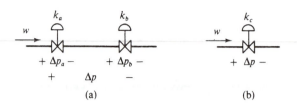

Figure 12.7 (a) Two valves in series. (b) Equivalent valve.

equivalent expression. We use linearized models in much of our analysis of hydraulic systems, so it is important to develop rules for combining the resistances of linearized elements that occur in series and parallel combinations. In the following example, we consider the relationship of flow versus pressure difference and the equivalent resistance for two valves in series. The parallel-flow situation is treated in one of the end-of-chapter problems.

▶ **EXAMPLE 12.2**

Figure 12.7(a) shows a series combination of two valves through which liquid flows at a rate of w and across which the pressure difference is Δp. An equivalent valve is shown in Figure 12.7(b). The two valves obey the relationships $w = k_a\sqrt{\Delta p_a}$ and $w = k_b\sqrt{\Delta p_b}$, respectively. Find the coefficient k_c of an equivalent valve that obeys the relationship $w = k_c\sqrt{\Delta p}$. Also evaluate the resistance R_c of the series combination in terms of the individual resistances R_a and R_b.

SOLUTION   Because the two valves are connected in series, they have the same flow rate w, and the total pressure difference is $\Delta p = \Delta p_a + \Delta p_b$. To determine k_c, we write Δp in terms of w as

$$\Delta p = \Delta p_a + \Delta p_b = \left(\frac{1}{k_a^2} + \frac{1}{k_b^2}\right)w^2$$

and then solve for w in terms of Δp. After some manipulations, we find that

$$w = \left(\frac{k_a k_b}{\sqrt{k_a^2 + k_b^2}}\right)\sqrt{\Delta p} \tag{16}$$

Comparing (16) with (10) reveals that the equivalent valve constant is

$$k_c = \frac{k_a k_b}{\sqrt{k_a^2 + k_b^2}} \tag{17}$$

Using (15) for the resistance of the linearized model of the equivalent valve, we can write

$$R_c = \frac{2\overline{w}}{k_c^2} = 2\overline{w}\left(\frac{1}{k_a^2} + \frac{1}{k_b^2}\right) \tag{18}$$

However, by applying (15) to the individual valves, we see that their resistances are $R_a = 2\overline{w}/k_a^2$ and $R_b = 2\overline{w}/k_b^2$, respectively. Using these expressions for R_a and R_b, we can rewrite (18) as

$$R_c = R_a + R_b \tag{19}$$

which is identical to the result for a linear electrical circuit.

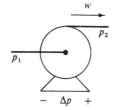

Figure 12.8 Symbolic representation of a pump.

Sources

In most hydraulic systems, the source of energy is a pump that derives its power from an electric motor. Here we shall consider the centrifugal pump driven at a constant speed, which is widely used in chemical processes. The symbolic representation of a pump is shown in Figure 12.8. Typical input-output relationships for a centrifugal pump being driven at three different constant speeds are shown in Figure 12.9(a). Pump curves of Δp versus w are determined experimentally under steady-state conditions and are quite nonlinear. To include a pump being driven at constant speed in a linear dynamic model, we first determine the operating point for the particular pump speed by calculating the values of $\overline{\Delta p}$ and $\overline{w}$. Then we find the slope of the tangent to the pump curve at the operating point and define it to be $-K$, which has units of newton-seconds per meter[5]. Having done this, we can express the incremental pressure difference $\widehat{\Delta p}$ in terms of the incremental flow rate $\hat{w}$ as

$$\widehat{\Delta p} = -K\hat{w} \tag{20}$$

where the constant K is always positive. Solving (20) for $\hat{w}$ yields

$$\hat{w} = -\frac{1}{K}\widehat{\Delta p} \tag{21}$$

Figure 12.9(b) illustrates the relationship of the linearized approximation to the nonlinear pump curve.

We can write the Taylor-series expansion for a pump driven at a constant speed as

$$w = \overline{w} + \frac{dw}{d\,\Delta p}\bigg|_{\overline{\Delta p}} (\Delta p - \overline{\Delta p}) + \cdots$$

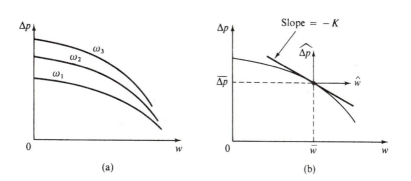

Figure 12.9 Typical centrifugal pump curves where $\Delta p = p_2 - p_1$. (a) For three different pump speeds ($\omega_1 < \omega_2 < \omega_3$). (b) Showing the linear approximation.

where the coefficient $(dw/d\,\Delta p)|_{\overline{\Delta p}}$ is the slope of the tangent to the curve of w versus Δp, measured at the operating point, and has the value $-1/K$. By dropping the second- and higher-order terms in the expansion and using the incremental variables $\hat{w}$ versus $\widehat{\Delta p}$, we obtain the linear relationship (21).

The manner in which a constant-speed pump can be incorporated into the dynamic model of a hydraulic system is illustrated in Example 12.4 in the following section. References in Appendix F contain more comprehensive discussions of pumps and their models, including ones where variations in the pump speed are significant.

12.3 DYNAMIC MODELS OF HYDRAULIC SYSTEMS

In this section, we apply the element laws presented in Section 12.2 and use many of the analytical techniques from previous chapters. We will develop and analyze dynamic models for a single vessel with a valve; a combination of a pump, vessel, and valve; and finally two vessels with two valves and a pump. In each case, we will derive the nonlinear model and then develop and analyze a linearized model.

▶ **EXAMPLE 12.3**

Figure 12.10 shows a vessel that receives liquid at a flow rate $w_i(t)$ and loses liquid through a valve that obeys the nonlinear flow-pressure relationship $w_o = k\sqrt{p_1 - p_a}$. The cross-sectional area of the vessel is A and the liquid density is ρ. Derive the nonlinear model obeyed by the absolute pressure p_1 at the bottom of the vessel. Then develop the linearized version that is valid in the vicinity of the operating point, and find the transfer function relating the transforms of the incremental input $\hat{w}_i(t)$ and the incremental pressure $\hat{p}_1$.

Having developed the system models in literal form, determine in numerical form the operating point, the transfer function, and the response to a 10 percent step-function increase in the input flow rate for the following parameter values:

$$A = 2\,\text{m}^2$$
$$\rho = 1000\,\text{kg/m}^3$$
$$k = 5.0 \times 10^{-5}\,\text{m}^4/\text{s} \cdot \text{N}^{1/2}$$
$$\overline{w}_i = 6.0 \times 10^{-3}\,\text{m}^3/\text{s}$$
$$p_a = 1.013 \times 10^5\,\text{N/m}^2$$
$$g = 9.807\,\text{m/s}^2$$

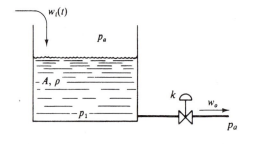

Figure 12.10 Hydraulic system for Example 12.3.

SOLUTION   Taking the pressure p_1 as the single state variable, we use (9), with $C(h)$ replaced by the constant $C = A/\rho g$, to write

$$\dot{p}_1 = \frac{1}{C}[w_{in}(t) - w_{out}(t)] \tag{22}$$

where

$$w_{in}(t) = w_i(t) \tag{23a}$$

$$w_{out}(t) = k\sqrt{p_1 - p_a} \tag{23b}$$

Substituting (23) into (22) gives the nonlinear system model as

$$\dot{p}_1 = \frac{1}{C}[-k\sqrt{p_1 - p_a} + w_i(t)] \tag{24}$$

To develop a linearized model, we rewrite (24) in terms of the incremental variables $\hat{p}_1 = p_1 - \overline{p}_1$ and $\hat{w}_i(t) = w_i(t) - \overline{w}_i$. The nominal values $\overline{p}_1$ and $\overline{w}_i$ must satisfy the algebraic equation

$$k\sqrt{\overline{p}_1 - p_a} = \overline{w}_i$$

or

$$\overline{p}_1 = p_a + \frac{1}{k^2}\overline{w}_i^2 \tag{25}$$

which corresponds to an outflow rate equal to the inflow rate, resulting in a constant liquid level and pressure. The nominal height of the liquid is

$$\overline{h} = \frac{(\overline{p}_1 - p_a)}{\rho g} \tag{26}$$

Making the appropriate substitutions into (24) and using (12) with (15) for the linearized valve equation, we obtain

$$\dot{\hat{p}}_1 + \frac{1}{RC}\hat{p}_1 = \frac{1}{C}\hat{w}_i(t) \tag{27}$$

where $R = 2\overline{w}_i/k^2$. Transforming (27) with $\hat{p}_1(0) = 0$, which corresponds to $p_1(0) = \overline{p}_1$, we find that the system's transfer function $H(s) = \hat{P}_1(s)/\hat{W}_i(s)$ is

$$H(s) = \frac{\dfrac{1}{C}}{s + \dfrac{1}{RC}} \tag{28}$$

which has a single pole at $s = -1/RC$.

For the parameter values specified, the operating point given by (25) reduces to

$$\overline{p}_1 = 1.013 \times 10^5 + \left(\frac{6.0 \times 10^{-3}}{5.0 \times 10^{-5}}\right)^2 = 1.157 \times 10^5 \ \text{N/m}^2$$

and from (26), the nominal liquid height is

$$\overline{h} = \frac{1.440 \times 10^4}{1000 \times 9.807} = 1.468 \ \text{m}$$

The numerical values of the hydraulic resistance and capacitance are, respectively,

$$R = \frac{2 \times 6.0 \times 10^{-3}}{(5.0 \times 10^{-5})^2} = 4.80 \times 10^6 \text{ N} \cdot \text{s/m}^5$$

$$C = \frac{2.0}{1000 \times 9.807} = 2.039 \times 10^{-4} \text{ m}^5/\text{N}$$

Substituting these values of R and C into (28), we obtain the numerical form of the transfer function as

$$H(s) = \frac{4904.}{s + 1.0216 \times 10^{-3}} \tag{29}$$

If $w_i(t)$ is originally equal to its nominal value of $\overline{w}_i = 6.0 \times 10^{-3}$ m^3/s and undergoes a 10 percent step-function increase, then $\hat{w}_i(t) = [0.60 \times 10^{-3}]U(t)$ m^3/s and $\hat{W}_i(s) = 0.60 \times 10^{-3}$ (1/s). Thus

$$\hat{P}_1(s) = \frac{4904. \times 0.60 \times 10^{-3}}{s(s + 1.0216 \times 10^{-3})}$$

$$= \frac{2.942}{s(s + 1.0216 \times 10^{-3})}$$

From the final-value theorem, the steady-state value of $\hat{p}_1$ is $s\hat{P}_1(s)$ evaluated at $s = 0$, namely

$$\lim_{t \to \infty} \hat{p}_1(t) = \frac{2.942}{1.0216 \times 10^{-3}} = 2880. \text{ N/m}^2$$

The time constant of the linearized model is $\tau = RC$, which becomes

$$\tau = (4.80 \times 10^6)(2.039 \times 10^{-4})$$

$$= 978.7 \text{ s}$$

which is slightly over 16 minutes. Thus the response of the incremental pressure is

$$\hat{p}_1 = 2880.(1 - \epsilon^{-t/978.7})$$

The change in the incremental level is $\hat{p}_1/\rho g$, which becomes

$$\hat{h} = 0.2937(1 - \epsilon^{-t/978.7})$$

To obtain the responses of the actual pressure and liquid level, we merely add the nominal values $\overline{p}_1 = 1.157 \times 10^5$ N/m^2 and $\overline{h} = 1.468$ m to these incremental variables. In the steady state, $p_1 = 1.186 \times 10^5$ N/m^2 and $h = 1.76$ m. It is interesting to note that because of the nonlinear valve, the 10 percent increase in the flow rate results in a 20 percent increase in both the gauge pressure $p_1 - p_a$ and the height h.

▶ **EXAMPLE 12.4**

Find the linearized model of the hydraulic system shown in Figure 12.11(a), which consists of a constant-speed centrifugal pump feeding a vessel from which liquid flows through a pipe and valve obeying the relationship $w_o = k\sqrt{p_1 - p_a}$. The pump characteristic for the specified pump speed $\overline{\omega}$ is shown in Figure 12.11(b).

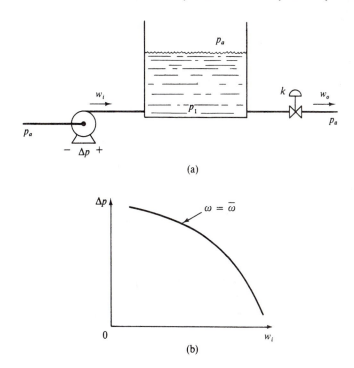

(a)

(b)

Figure 12.11 (a) System for Example 12.4. (b) Pump curve.

SOLUTION   The equilibrium condition for the system corresponds to

$$\overline{w}_i = \overline{w}_o \tag{30}$$

where $\overline{w}_i$ and $\overline{\Delta p} = \overline{p}_1 - p_a$ must be one of the points on the pump curve in Figure 12.11(b), and where $\overline{w}_o$ obeys the nonlinear flow relationship

$$\overline{w}_o = k\sqrt{\overline{\Delta p}} \tag{31}$$

To determine the operating point, we find the solution to (30) graphically by plotting the valve characteristic (31) on the pump curve. Doing this gives Figure 12.12(a), where the operating point is the intersection of the valve curve and the pump curve, designated as point A in the figure. Once we have located the operating point, we can draw the tangent to the pump curve as shown in Figure 12.12(b) and determine its slope $-K$ graphically.

Following this preliminary step, we can use (9) to write the model of the system as

$$\dot{p}_1 = \frac{1}{C}(w_i - w_o) \tag{32}$$

where, from (12), the approximate flow rate through the valve is

$$w_o = \overline{w}_o + \frac{1}{R}\widehat{\Delta p} \tag{33}$$

and where, from (21), the approximate flow rate through the pump is

$$w_i = \overline{w}_i - \frac{1}{K}\widehat{\Delta p} \tag{34}$$

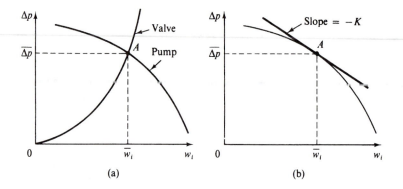

Figure 12.12 (a) Combined pump and valve curves for Example 12.4. (b) Pump curve with linear approximation.

Substituting (33) and (34) into (32), using $\dot{p}_1 = \hat{\dot{p}}_1$ and (30), and noting that $\widehat{\Delta p} = \hat{p}_1$ because p_a is constant, we find the incremental model to be

$$\hat{\dot{p}}_1 = \frac{1}{C}\left(-\frac{1}{K}-\frac{1}{R}\right)\hat{p}_1$$

which we can write as the homogeneous first-order differential equation

$$\hat{\dot{p}}_1 + \frac{1}{C}\left(\frac{1}{K}+\frac{1}{R}\right)\hat{p}_1 = 0 \tag{35}$$

Inspection of (35) indicates that the magnitude of the slope of the pump curve at the operating point enters the equation in exactly the same manner as the resistance associated with the valve. Hence, if we evaluate the equivalent resistance R_{eq} according to

$$R_{eq} = \frac{RK}{R+K}$$

(35) is the same as (27), which was derived for a vessel and a single valve, except for the absence of an input flow rate.

▶ **EXAMPLE 12.5**

The valves in the hydraulic system shown in Figure 12.13 obey the flow-pressure relationships $w_1 = k_1\sqrt{p_1 - p_2}$ and $w_2 = k_2\sqrt{p_2 - p_a}$. The atmospheric pressure is p_a, and the capacitances of the vessels are C_1 and C_2. Find the equations that determine the operating point, and show how the pump curve is used to solve them. Derive a linearized model that is valid about the operating point.

SOLUTION Because the pump and the two vessels are in series at equilibrium conditions, we define the operating point by equating the three flow rates $\overline{w}_p$, $\overline{w}_1$, and $\overline{w}_2$. The flow rates through the two valves are given by

$$\overline{w}_1 = k_1\sqrt{\overline{p}_1 - \overline{p}_2} \tag{36a}$$

$$\overline{w}_2 = k_2\sqrt{\overline{p}_2 - p_a} \tag{36b}$$

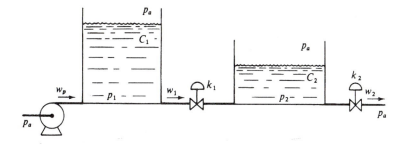

Figure 12.13 Hydraulic system with two vessels considered in Example 12.5.

The flow rate $\overline{w}_p$ through the pump and the pressure difference $\overline{\Delta p}_1 = \overline{p}_1 - p_a$ must correspond to a point on the pump curve. Equations (36a) and (36b) are those of two valves in series, and, as shown in Example 12.2, we can replace such a combination of valves by an equivalent valve specified by (16). Hence

$$\overline{w}_p = k_{eq}\sqrt{\overline{\Delta p}_1} \tag{37}$$

where, from (17),

$$k_{eq} = \frac{k_1 k_2}{\sqrt{k_1^2 + k_2^2}}$$

Plotting (37) on the pump curve as shown in Figure 12.12(a) yields the values of $\overline{\Delta p}_1$ and $\overline{w}_p$, from which we can find the other nominal values.

With this information, we can develop the incremental model. Using (9), (12), and (21), we can write the pair of linear differential equations

$$\dot{\hat{p}}_1 = \frac{1}{C_1}\left[-\frac{1}{K}\hat{p}_1 - \frac{1}{R_1}(\hat{p}_1 - \hat{p}_2)\right]$$
$$\dot{\hat{p}}_2 = \frac{1}{C_2}\left[\frac{1}{R_1}(\hat{p}_1 - \hat{p}_2) - \frac{1}{R_2}\hat{p}_2\right] \tag{38}$$

where the valve resistances are given by $R_1 = 2\overline{w}_1/k_1^2$ and $R_2 = 2\overline{w}_2/k_2^2$, and where $-K$ is the slope of the pump curve at the operating point. As indicated by (38), the incremental model has no inputs and hence can respond only to nonzero initial conditions—that is, $\hat{p}_1(0) \neq 0$ and/or $\hat{p}_2(0) \neq 0$.

In practice, there might be additional liquid streams entering either vessel, or the pump speed might be changed. It is also possible for either of the two valves to be opened or closed slightly. Such a change would modify the respective hydraulic resistance.

In the absence of an input, we can transform (38) to find the Laplace transform of the zero-input response. Doing this, we find that after rearranging,

$$\left[C_1 s + \left(\frac{1}{K} + \frac{1}{R_1}\right)\right]\hat{P}_1(s) = \frac{1}{R_1}\hat{P}_2(s) + C_1\hat{p}_1(0) \tag{39a}$$

$$\left[C_2 s + \left(\frac{1}{R_1} + \frac{1}{R_2}\right)\right]\hat{P}_2(s) = \frac{1}{R_1}\hat{P}_1(s) + C_2\hat{p}_2(0) \tag{39b}$$

We can find either $\hat{P}_1(s)$ or $\hat{P}_2(s)$ by combining these two equations into a single transform equation. The corresponding inverse transform will yield the zero-input response in terms of $\hat{p}_1(0)$ and $\hat{p}_2(0)$. The denominator of either $\hat{P}_1(s)$ or $\hat{P}_2(s)$ will be the characteristic

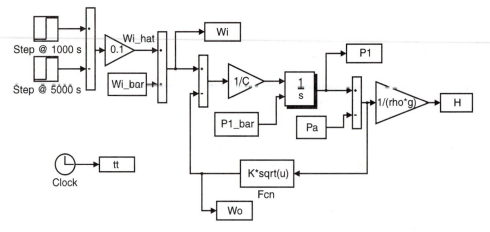

Figure 12.14 A Simulink diagram to solve (24).

polynomial of the system, which, as you may verify, is

$$s^2 + \left[\frac{1}{C_1} \left(\frac{1}{K} + \frac{1}{R_1} \right) + \frac{1}{C_2} \left(\frac{1}{R_1} + \frac{1}{R_2} \right) \right] s + \frac{1}{C_1 C_2} \left(\frac{1}{KR_1} + \frac{1}{KR_2} + \frac{1}{R_1 R_2} \right)$$

▶ **EXAMPLE 12.6**

Draw a Simulink diagram to simulate the system in Example 12.3 and solve Equation (24). Use the parameter values in Example 12.3, and allow $w_i(t)$ to increase step-wise by 10 percent at $t = 1000$ s and decrease back to its initial value at $t = 5000$ s. Simulate 10,000 seconds of the response. Use MATLAB to make three plots, one above the other, with the flows $w_i(t)$ and w_o on the upper axes, the pressure p_1 on the middle axes, and the liquid height h on the lower axes. Show the time variable in minutes.

SOLUTION The Simulink diagram is shown in Figure 12.14. This diagram uses the integrator block which shows the initial condition, $\bar{p}_1$, explicitly. Two step function blocks are used to generate the changes in inflow rate, $\hat{w}_i$, and these changes are added to the steady-state value, $\bar{w}_i$, to form w_i.

Figure 12.15 contains plots of the inflow and outflow, the pressure at the bottom of the tank, p_1, and the height of liquid in the tank, h. The steady-state values of p_1 and h achieved just before the reduction in $w_i(t)$ agree with the values calculated in Example 12.3.

▶ SUMMARY

The basic variables in a hydraulic system are flow rate and pressure. Other variables that are equivalent to the pressure at the bottom of a container are the volume of the liquid and the liquid height.

Because hydraulic systems are generally nonlinear, especially in the resistance to fluid flow, we developed linearized models valid in the vicinity of an operating point. We

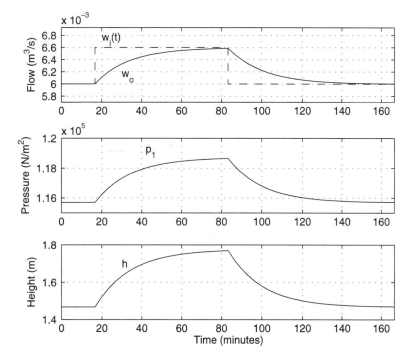

Figure 12.15 Inflow (dashed) and responses (solid) for Example 12.6.

introduced the passive elements of hydraulic capacitance and hydraulic resistance in constructing such models. The former is associated with the potential energy of a fluid in a
vessel, the latter with the energy dissipated when fluid flows through valves, orifices, and
pipes. Another possible passive element is inertance, which is associated with the kinetic
energy of fluids in motion. Because the inertance is normally negligible, we did not include
it in our models.

A source may consist of a specified flow rate into a vessel. However, most practical hydraulic energy sources are mechanically driven pumps. We generally describe such pumps
by a nonlinear relationship between pressure and flow rate, rather than specifying one of
these two variables independently. This contrasts with the ideal force, velocity, voltage,
and current sources used in earlier chapters. Because of the algebraic relationship between
pressure and flow rate, a pump enters into the linearized system equations in a way somewhat similar to hydraulic resistance.

The pressure in each vessel (or else the volume or the liquid height) is normally chosen
to be a state variable. The procedures for finding and solving hydraulic models, including
state-variable equations and transfer functions, are generally the same as those used in the
earlier chapters.

▶ PROBLEMS

*12.1. Figure P12.1 shows a conical vessel that has a circular cross section and contains
a liquid. Evaluate and sketch the hydraulic capacitance as a function of the liquid height
h. Also evaluate and sketch the gauge pressure $p^* = p - p_a$ at the base of the vessel as a
function of the liquid volume v.

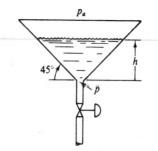

Figure P12.1

12.2. Find the equivalent hydraulic resistances for the linear models of hydraulic networks shown in Figure P12.2. Express your answers as single fractions.

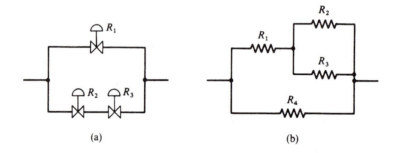

(a)   (b)

Figure P12.2

12.3. Two valves that obey the relationships $w_a = k_a\sqrt{\Delta p}$ and $w_b = k_b\sqrt{\Delta p}$ are connected in parallel, as indicated in Figure P12.3.

    a.   Determine the equivalent valve coefficient k_c in terms of k_a and k_b such that the total flow rate is given by $w = k_c\sqrt{\Delta p}$.

    b.   Show that the hydraulic resistance of the equivalent linearized model is $R_c = 2\sqrt{\overline{\Delta p}}/k_c$, where $\overline{\Delta p}$ denotes the nominal pressure drop across the valves.

    c.   Show that $R_c = R_a R_b/(R_a + R_b)$, where R_a and R_b are the hydraulic resistances of the individual valves evaluated at the nominal pressure difference $\overline{\Delta p}$.

    d.   Assume curves for w_a versus Δp and w_b versus Δp, and sketch the corresponding curve for w_c. Indicate the linearized approximations at a typical value of $\overline{\Delta p}$.

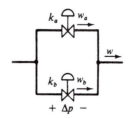

Figure P12.3

12.4. Consider the hydraulic system that was modeled in Example 12.3, consisting of a single vessel and a valve.

    a.   Obtain the linearized model in terms of the incremental pressure that is valid for the nominal flow rate $\overline{w}_i = 3.0 \times 10^{-3}$ m³/s, and evaluate the transfer function $\hat{P}_1(s)/\hat{W}_i(s)$. Also find $\overline{p}_1$.

    b.   Rewrite the model and transfer function you found in part (a) in terms of the incremental volume $\hat{v}$. Also find $\overline{v}$.

    c.   Solve for and sketch the incremental volume versus time when $\hat{w}_i(t) = 0.5 \times 10^{-3}U(t)$ m³/s and when the system starts at the nominal conditions found in part (a).

*12.5. Write the state-variable equations for the system shown in Figure P12.5, using the incremental pressures $\hat{p}_1$ and $\hat{p}_2$ as state variables, where the pump obeys the relationship

$$w_p = \overline{w}_p - \frac{1}{K}(\hat{p}_2 - \hat{p}_1)$$

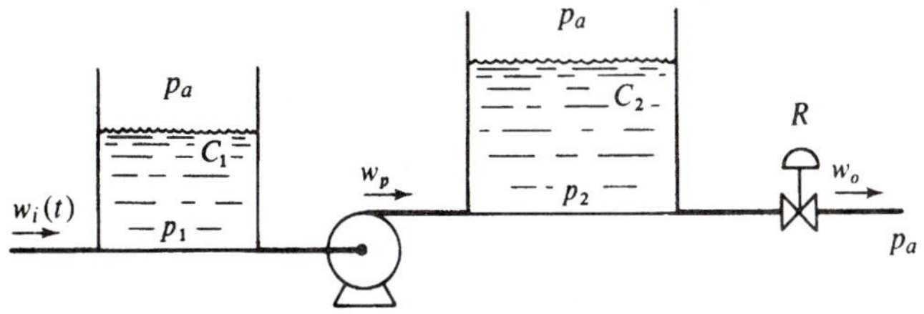

Figure P12.5

12.6. For the hydraulic system shown in Figure P12.6, the incremental input is $\hat{w}_i(t)$ and the incremental output is $\hat{w}_o$.

    a.   Verify that the following equations represent an appropriate state-variable model in terms of the incremental state variables $\hat{p}_1$ and $\hat{p}_2$.

$$\dot{\hat{p}}_1 = \frac{1}{C_1}\left[-\frac{1}{R_1}\hat{p}_1 + \hat{w}_i(t)\right]$$

$$\dot{\hat{p}}_2 = \frac{1}{C_2}\left[\frac{1}{R_1}\hat{p}_1 - \frac{1}{R_2}\hat{p}_2\right]$$

$$\hat{w}_o = \frac{1}{R_2}\hat{p}_2$$

    b.   Derive the system transfer function $\hat{W}_o(s)/\hat{W}_i(s)$.

    c.   Define the time constants $\tau_1 = R_1C_1$ and $\tau_2 = R_2C_2$, and evaluate the unit step response for the case $\tau_1 \neq \tau_2$. Sketch the response for the case where $\tau_1 = 2\tau_2$.

    d.   Evaluate the transfer function $\hat{H}_2(s)/\hat{W}_i(s)$ that relates the transform of $\hat{w}_i(t)$ to the transform of the incremental liquid height $\hat{h}_2$.

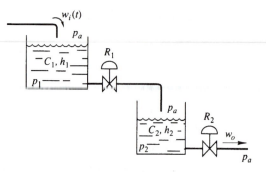

Figure P12.6

*12.7. The hydraulic system shown in Figure P12.7 has the incremental pressure $\hat{p}_i(t)$ as its input and the incremental flow rate $\hat{w}_o$ as its output.

    a.   Verify that the following equations represent an appropriate state-variable model in terms of the incremental state variables $\hat{p}_1$ and $\hat{p}_2$.

$$\dot{\hat{p}}_1 = -\frac{1}{R_3 C_1}\hat{p}_1 + \frac{1}{R_1 C_1}\hat{p}_i(t)$$

$$\dot{\hat{p}}_2 = \frac{1}{R_3 C_2}\hat{p}_1 - \frac{1}{R_4 C_2}\hat{p}_2 + \frac{1}{R_2 C_2}\hat{p}_i(t)$$

$$\hat{w}_o = \frac{1}{R_4}\hat{p}_2$$

    b.   Find the transfer function $\hat{W}_o(s)/\hat{P}_i(s)$.

    c.   Solve for the steady-state value of the unit step response.

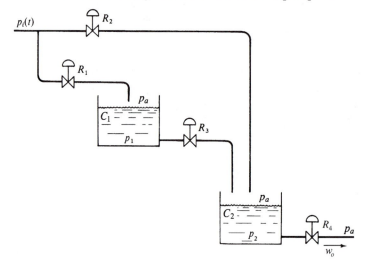

Figure P12.7

12.8. Figure P12.8 shows three identical tanks having capacitance C connected by identical lines having resistance R. An input stream flows into tank 1 at the incremental flow rate $\hat{w}_i(t)$, and an output stream flows from tank 3 at the incremental flow rate $\hat{w}_o(t)$.

a. Verify that the following equations represent an appropriate state-variable model in terms of the incremental state variables $\hat{v}_1$, $\hat{v}_2$, and $\hat{v}_3$.

$$\dot{\hat{v}}_1 = -\frac{2}{RC}\hat{v}_1 + \frac{1}{RC}\hat{v}_2 + \frac{1}{RC}\hat{v}_3 + \hat{w}_i(t)$$

$$\dot{\hat{v}}_2 = \frac{1}{RC}\hat{v}_1 - \frac{2}{RC}\hat{v}_2 + \frac{1}{RC}\hat{v}_3$$

$$\dot{\hat{v}}_3 = \frac{1}{RC}\hat{v}_1 + \frac{1}{RC}\hat{v}_2 - \frac{2}{RC}\hat{v}_3 - \hat{w}_o(t)$$

b. Let $RC = 1$ and find $\hat{V}_2(s)$ in terms of $\hat{W}_1(s)$ and $\hat{W}_o(s)$ by transforming the three state-variable equations and eliminating $\hat{V}_1(s)$ and $\hat{V}_3(s)$.

c. Solve for $\hat{v}_2$ as a function of time when $\hat{w}_i(t) = U(t)$ and $\hat{w}_o(t) = 0$. Repeat the solution when $\hat{w}_i(t) = 0$ and $\hat{w}_o(t) = U(t)$. Sketch the responses.

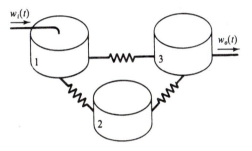

Figure P12.8

12.9. The valve and pump characteristics for the hydraulic system shown in Figure P12.9(a) are plotted in Figure P12.9(b). Curves are given for the two pump speeds 100 rad/s and 150 rad/s. The cross-sectional area of the vessel is 2.0 m^2, and the liquid density is 1000 kg/m^3.

a. Determine the steady-state flow rate, gauge pressure, and liquid height for each of the pump speeds for which curves are shown.

b. Derive the linearized models for each of the pump speeds in numerical form.

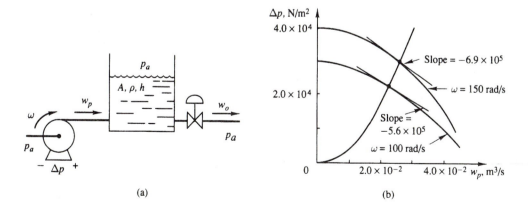

(a)   (b)

Figure P12.9

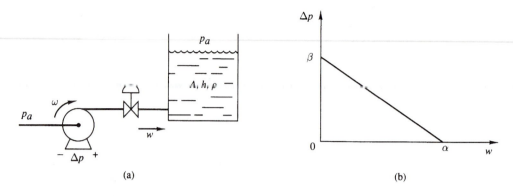

Figure P12.10

*12.10. As shown in Figure P12.10(a), liquid can flow into a vessel through a pump and a valve. The assumed pump characteristic is shown in Figure P12.10(b), where α and β are the maximum flow rate and the maximum pressure difference, respectively. The valve is shut for all $t < 0$, is opened at $t = 0$, and presents no resistance to the flow of the liquid for $t > 0$.

    a. Verify that the differential equation obeyed by the liquid height h after the valve is opened is

$$\dot{h} + \left(\frac{\alpha\rho g}{\beta A}\right) h = \frac{\alpha}{A}$$

    b. Write expressions for the time constant and the steady-state height.

    c. Solve for $h(t)$ and sketch it versus time.

12.11. Consider the hydraulic system described in Problem 12.10 with a pump having the nonlinear pressure-flow relationship shown in Figure P12.11.

    a. Verify that the differential equation obeyed by the liquid height h after the valve is opened is

$$\dot{h} + \frac{\alpha}{A}\left(\frac{\rho g}{\beta}\right)^2 h^2 = \frac{\alpha}{A}$$

    b. Use Simulink and MATLAB to simulate this system, and plot $h(t)$ and $w(t)$ versus time. Use the parameter values: $A = 1.50 \text{ m}^2$, $\rho = 1000 \text{ kg/m}^3$, $\alpha = 0.120 \text{ m}^3/\text{s}$, and $\beta = 3.0 \times 10^4 \text{ N/m}^2$.

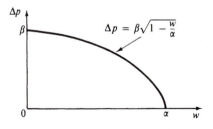

Figure P12.11

13

BLOCK DIAGRAMS FOR DYNAMIC SYSTEMS

In this chapter, we build upon the block diagrams that were introduced in Chapter 4 for using SIMULINK to simulate system responses, and upon the transfer functions from Chapter 8 that allow us to represent linear dynamic systems as ratios of polynomials. By combining these two important concepts, we will be in a position to analyze and design linear dynamic systems that can be used for control and signal processing.

We begin by developing several techniques for simplifying block diagrams, usually to obtain the transfer function(s) of a linear system as a ratio of polynomials in the Laplace transform variable s. In order to be able to do this for systems involving feedback, as will often be the case, we develop some specific techniques. We also need methods for dealing with models in input-output form when the differential equation involves derivatives of the input.

In order to introduce the reader to some of the typical concerns and to provide background for the material to be covered in Chapters 14 and 15, we conclude the chapter by examining in detail a particular electromechanical system. We construct a block-diagram model, find the transfer function, and then improve the performance by adjusting those parameters that are under the designer's control.

▶ 13.1 RULES FOR ALTERING DIAGRAM STRUCTURE

We begin this chapter by considering several ways of modifying the structure of an existing block diagram that will simplify finding the overall transfer function. In addition to the summers, integrators, and gain blocks that were used in Chapter 4, we include blocks that are characterized by their transfer functions. Such blocks might represent subsystems whose inner workings are not of concern. The individual transfer functions, which are generally the ratio of two polynomials, are often denoted by $F(s)$, $G(s)$, and other symbols in addition to $H(s)$, which was used in Chapter 8. When the transfer function is a constant, then that block reduces to a gain block.

Recall that transfer functions are formed by assuming that there is no initial stored energy, and they therefore give only the zero-state response. This restriction is not a severe one for stable systems, because any initial energy does not affect the steady-state response at all and does not affect the form of the transient response. If the size of the transient response is important, we can show additional inputs in our diagram to represent the initial energy.

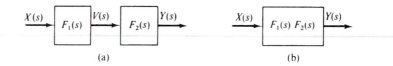

Figure 13.1 (a) Two blocks in series. (b) Equivalent diagram.

Series Combination

Two blocks are said to be in **series** when the output of one goes *only* to the input of the other, as shown in Figure 13.1(a). The transfer functions of the individual blocks in the figure are $F_1(s) = V(s)/X(s)$ and $F_2(s) = Y(s)/V(s)$.

When we evaluate the individual transfer functions, it is essential that we take any **loading effects** into account. This means that $F_1(s)$ is the ratio $V(s)/X(s)$ when the two subsystems are connected, so any effect the second subsystem has on the first is accounted for in the mathematical model. The same statement holds for calculating $F_2(s)$. As shown in Section 10.1, for example, the input-output relationship for a potentiometer would be changed by an external resistor connected from its wiper to the ground node.

In Figure 13.1(a), $Y(s) = F_2(s)V(s)$ and $V(s) = F_1(s)X(s)$. It follows that

$$Y(s) = F_2(s)[F_1(s)X(s)]$$
$$= [F_1(s)F_2(s)]X(s)$$

Thus the transfer function relating the input transform $X(s)$ to the output transform $Y(s)$ is $F_1(s)F_2(s)$, the product of the individual transfer functions. The equivalent block diagram is shown in Figure 13.1(b).

Parallel Combination

Two systems are said to be in **parallel** when they have a common input and their outputs are combined by a summing junction. If, as indicated in Figure 13.2(a), the individual blocks have the transfer functions $F_1(s)$ and $F_2(s)$ and the signs at the summing junction are both positive, the overall transfer function $Y(s)/X(s)$ will be the sum $F_1(s) + F_2(s)$, as shown in Figure 13.2(b). To prove this statement, we note that

$$Y(s) = V_1(s) + V_2(s)$$

where $V_1(s) = F_1(s)X(s)$ and $V_2(s) = F_2(s)X(s)$. Substituting for $V_1(s)$ and $V_2(s)$, we have

$$Y(s) = [F_1(s) + F_2(s)]X(s)$$

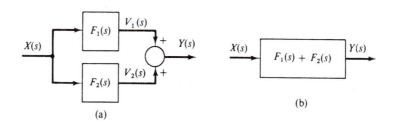

Figure 13.2 (a) Two blocks in parallel. (b) Equivalent diagram.

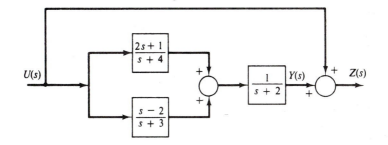

Figure 13.3 Block diagram for Example 13.1.

If either of the summing-junction signs associated with $V_1(s)$ or $V_2(s)$ is negative, we must change the sign of the corresponding transfer function in forming the overall transfer function. The following example illustrates the rules for combining blocks that are in parallel or in series.

▷ **EXAMPLE 13.1**

Evaluate the transfer functions $Y(s)/U(s)$ and $Z(s)/U(s)$ for the block diagram shown in Figure 13.3, giving the results as rational functions of s.

SOLUTION   Because $Z(s)$ can be viewed as the sum of the outputs of two parallel blocks, one of which has $Y(s)$ as its output, we first evaluate the transfer function $Y(s)/U(s)$. To do this, we observe that $Y(s)$ can be considered the output of a series combination of two parts, one of which is a parallel combination of two blocks. Starting with this parallel combination, we write

$$\frac{2s+1}{s+4} + \frac{s-2}{s+3} = \frac{3s^2+9s-5}{s^2+7s+12}$$

and redraw the block diagram as shown in Figure 13.4(a). The series combination in this version has the transfer function

$$\frac{Y(s)}{U(s)} = \frac{3s^2+9s-5}{s^2+7s+12} \cdot \frac{1}{s+2}$$

$$= \frac{3s^2+9s-5}{s^3+9s^2+26s+24}$$

which leads to the diagram shown in Figure 13.4(b). We can reduce the final parallel combination to the single block shown in Figure 13.4(c) by writing

$$\frac{Z(s)}{U(s)} = 1 + \frac{Y(s)}{U(s)}$$

$$= 1 + \frac{3s^2+9s-5}{s^3+9s^2+26s+24}$$

$$= \frac{s^3+12s^2+35s+19}{s^3+9s^2+26s+24}$$

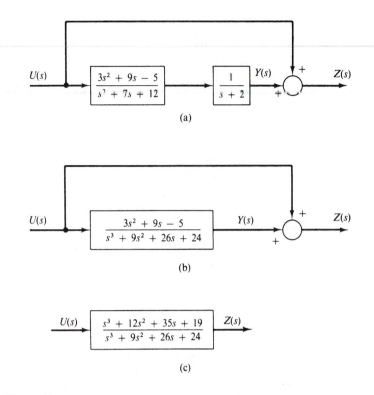

(a)

(b)

(c)

Figure 13.4 Equivalent block diagrams for the diagram shown in Figure 13.3.

Moving a Pick-Off Point

A **pick-off point** is a point where an incoming variable in the diagram is directed into more than one block. In the partial diagram of Figure 13.5(a), the incoming signal $X(s)$ is used not only to provide the output $Q(s)$ but also to form the signal $W(s)$, which in practice might be fed back to a summer that appears elsewhere in the complete diagram. The pick-off point can be moved to the right of $F_1(s)$ if the transfer function of the block leading to $W(s)$ is modified as shown in Figure 13.5(b). Both parts of the figure give the same equations:

$$Q(s) = F_1(s)X(s)$$
$$W(s) = F_2(s)X(s)$$

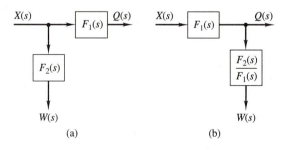

Figure 13.5 Moving a pick-off point.

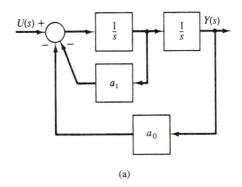

(a)

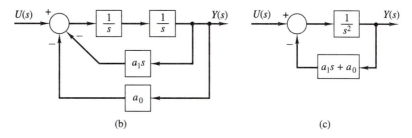

(b)   (c)

Figure 13.6 (a) Block diagram for Example 13.2. (b), (c) Equivalent block diagrams.

▶ **EXAMPLE 13.2**

Figure 13.6(a) shows a system having two feedback loops. Use Figure 13.5 and the rules for combining blocks in series and parallel to redraw the block diagram so that it has a single feedback loop.

SOLUTION   The pick-off point leading to the gain block a_1 can be moved to the output $Y(s)$ by replacing a_1 by a_1s, as shown in Figure 13.6(b). Then the two integrator blocks, which are now in series, can be combined to give the transfer function $1/s^2$. The two feedback blocks are now in parallel and can be combined into the single transfer function $a_1s + a_0$. With these changes, the revised diagram has a single feedback loop, as shown in Figure 13.6(c).

Moving a Summing Junction

Suppose that, in the partial diagram of Figure 13.7(a), we wish to move the summing junction to the left of (before) the block that has the transfer function $F_2(s)$. We can do this by modifying the transfer function of the block whose input is $X_1(s)$, as shown in part (b) of Figure 13.7. For each part of this figure,

$$Q(s) = F_1(s)X_1(s) + F_2(s)X_2(s) \tag{1}$$

We could also move the summing junction to the left of $F_1(s)$ by using the diagram shown in part (c) of the figure. As before, $Q(s)$ obeys the expression in (1).

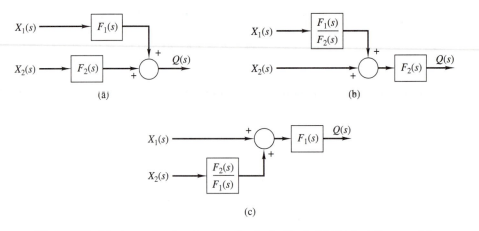

Figure 13.7 Moving a summing junction ahead of a block. (a) Original diagram. (b), (c) Equivalent diagrams.

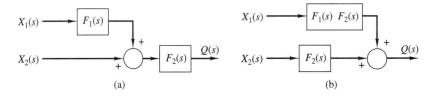

Figure 13.8 Moving a summing junction after a block. (a) Original diagram. (b) Equivalent diagram.

Alternatively, suppose that in the partial diagram of Figure 13.8(a), we wish to move the summing junction to the right of (after) the block that has the transfer function $F_2(s)$. We can do this by modifying the transfer function of the block whose input is $X_1(s)$, as shown in part (b) of Figure 13.8. For each part of this figure,

$$Q(s) = F_1(s)F_2(s)X_1(s) + F_2(s)X_2(s) \qquad (2)$$

▶ **EXAMPLE 13.3**

Use Figure 13.7 and the rule for a parallel combination to modify the block diagram in Figure 13.9(a) to remove the right summing junction, leaving only the left summing junction.

SOLUTION   We use Figure 13.7 to move the right summer from *after* the integrator block to *before* it, resulting in Figure 13.9(b). Since the two summing junctions now have no blocks between them, they can be combined as a single summer, as shown in part (c) of Figure 13.9. The final step is to combine the two parallel paths from $U(s)$ to the summing junction, resulting in part (d), where the transfer function $bs + 1$ appears explicitly.

The diagram in part (d) of Figure 13.9 may not appear to be much simpler than the original diagram. However, it has the desirable property that the summing junction that entered the feedback loop after the integrator has been removed from the loop. As will be

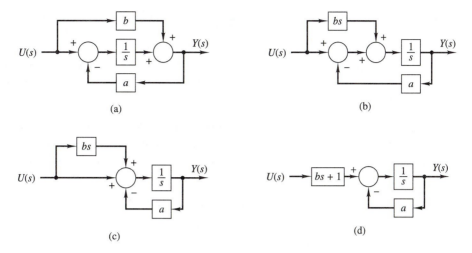

Figure 13.9 (a) Block diagram for Example 13.3. (b), (c), (d) Equivalent diagrams.

discussed in the following section, it is a simple task to collapse the remaining feedback loop into a single block whose transfer function can be written as a ratio of polynomials. Once that has been done, the rule for combining blocks in series can be used to obtain a single overall transfer function from the input $U(s)$ to the output $Y(s)$.

▶ 13.2 REDUCING DIAGRAMS FOR FEEDBACK SYSTEMS

The block diagrams developed in Section 4.2 contained only integrators, gain blocks, and summers. Feedback paths were formed around the integrators by multiplying the transformed state variables by constants and then feeding these signals back to an input summer. In this section, we shall consider the more general situation where the individual blocks can have arbitrary transfer functions.

Figure 13.10(a) shows the block diagram of a general feedback system that has a forward path from the summing junction to the output and a feedback path from the output back to the summing junction. The transforms of the system's input and output are $U(s)$ and $Y(s)$, respectively. The transfer function $G(s) = Y(s)/V(s)$ is known as the **forward transfer function**, and $H(s) = Z(s)/Y(s)$ is called the **feedback transfer function**. We

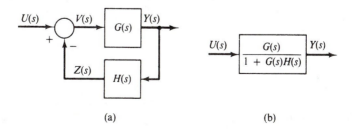

Figure 13.10 (a) Block diagram of a feedback system. (b) Equivalent diagram.

must evaluate both of these transfer functions with the system elements connected in order to account properly for any possible loading effects of the interconnections. The product $G(s)H(s)$ is referred to as the **open-loop transfer function**. The minus sign is shown to be associated with the feedback signal from the block $H(s)$ at the summing junction because a minus sign naturally occurs in the majority of feedback systems, particularly in control systems.

Given the model of a feedback system in terms of its forward and feedback transfer functions $G(s)$ and $H(s)$, it is often necessary to determine the **closed-loop transfer function** $T(s) = Y(s)/U(s)$. We do this by writing the algebraic transform equations corresponding to the block diagram shown in Figure 13.10(a) and solving them for the ratio $Y(s)/U(s)$. We can write the following transform equations directly from the block diagram.

$$V(s) = U(s) - Z(s)$$
$$Y(s) = G(s)V(s)$$
$$Z(s) = H(s)Y(s)$$

If we combine these equations in such a way as to eliminate $V(s)$ and $Z(s)$, we find that
$$Y(s) = G(s)[U(s) - H(s)Y(s)]$$

which can be rearranged to give
$$[1 + G(s)H(s)]Y(s) = G(s)U(s)$$

Hence the closed-loop transfer function $T(s) = Y(s)/U(s)$ is

$$T(s) = \frac{G(s)}{1 + G(s)H(s)} \tag{3}$$

where it is implicit that the sign of the feedback signal at the summing junction is negative. It is readily shown that when a plus sign is used at the summing junction for the feedback signal, the closed-loop transfer function becomes

$$T(s) = \frac{G(s)}{1 - G(s)H(s)} \tag{4}$$

A commonly used simplification occurs when the feedback transfer function is unity— when $H(s) = 1$. Such a system is referred to as a **unity-feedback system**, and (3) reduces to

$$T(s) = \frac{G(s)}{1 + G(s)} \tag{5}$$

We now consider three examples that make use of (3) and (4). The first two illustrate determining the closed-loop transfer function by reducing the block diagram. They also show the effects of feedback gains on the closed-loop poles, time constant, damping ratio, and undamped natural frequency. In the third example, a block diagram is drawn directly from the system's state-variable equations and then reduced to give the system's transfer functions.

▷ **EXAMPLE 13.4**

Find the closed-loop transfer function for the feedback system shown in Figure 13.11(a), and compare the locations of the poles of the open-loop and closed-loop transfer functions in the s-plane.

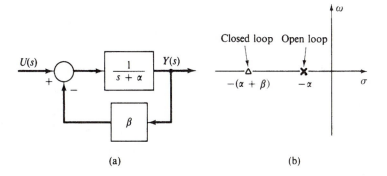

Figure 13.11 Single-loop feedback system for Example 13.4. (a) Block diagram. (b) Pole locations in the s-plane.

SOLUTION   By comparing the block diagram shown in Figure 13.11(a) with that shown in Figure 13.10(a), we see that $G(s) = 1/(s+\alpha)$ and $H(s) = \beta$. Substituting these expressions into (3) gives

$$T(s) = \frac{\dfrac{1}{s+\alpha}}{1 + \left(\dfrac{1}{s+\alpha}\right)\beta}$$

which we can write as a rational function of s by multiplying the numerator and denominator by $s + \alpha$. Doing this, we obtain the closed-loop transfer function

$$T(s) = \frac{1}{s+\alpha+\beta}$$

This result illustrates an interesting and useful property of feedback systems: the fact that the poles of the closed-loop transfer function differ from the poles of the open-loop transfer function $G(s)H(s)$. In this case, the single open-loop pole is at $s = -\alpha$, whereas the single closed-loop pole is at $s = -(\alpha + \beta)$. These pole locations are indicated in Figure 13.11(b) for positive α and β. Hence, in the absence of feedback, the pole of the transfer function $Y(s)/U(s)$ is at $s = -\alpha$, and the free response will be of the form $\epsilon^{-\alpha t}$. With feedback, however, the free response will be $\epsilon^{-(\alpha+\beta)t}$. Thus the time constant of the open-loop system is $1/\alpha$, whereas that of the closed-loop system is $1/(\alpha + \beta)$.

▷ **EXAMPLE 13.5**

Find the closed-loop transfer function of the two-loop feedback system shown in Figure 13.12. Also express the damping ratio and the undamped natural frequency of the closed-loop system in terms of the gains a_0 and a_1.

SOLUTION   Because the system's block diagram contains one feedback path inside another, we cannot use (3) directly to evaluate $Y(s)/U(s)$. However, we can redraw the block diagram such that the summing junction is split into two summing junctions, as shown in Figure 13.13(a). Then it is possible to use (3) to eliminate the inner loop by calculating the transfer function $W(s)/V(s)$. Taking $G(s) = 1/s$ and $H(s) = a_1$ in (3), we

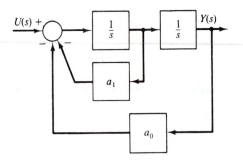

Figure 13.12 System with two feedback loops for Example 13.5.

obtain

$$\frac{W(s)}{V(s)} = \frac{\dfrac{1}{s}}{1 + \dfrac{a_1}{s}} = \frac{1}{s + a_1}$$

Redrawing Figure 13.13(a) with the inner loop replaced by a block having $1/(s+a_1)$ as its transfer function gives Figure 13.13(b). The two blocks in the forward path of this version are in series and can be combined by multiplying their transfer functions, as in Figure 13.1, which gives the block diagram shown in Figure 13.13(c). Then we can apply (3) again to find the overall closed-loop transfer function $T(s) = Y(s)/U(s)$ as

$$T(s) = \frac{\dfrac{1}{s(s+a_1)}}{1 + \dfrac{1}{s(s+a_1)} \cdot a_0} = \frac{1}{s^2 + a_1 s + a_0} \tag{6}$$

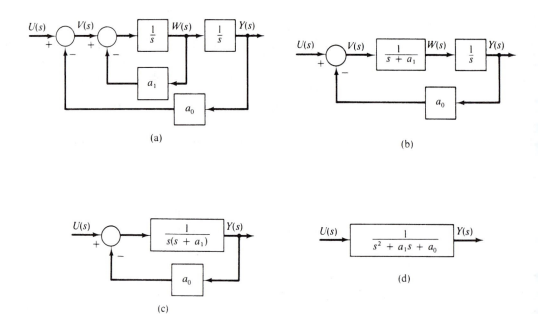

Figure 13.13 Equivalent block diagrams for the system shown in Figure 13.12.

The block-diagram representation of the feedback system corresponding to (6) is shown in Figure 13.13(d). Note that the same answer can be obtained by using the results of Example 13.2 as shown in Figure 13.6(c).

The poles of the closed-loop transfer function are the roots of the equation

$$s^2 + a_1 s + a_0 = 0 \tag{7}$$

which we obtain by setting the denominator of $T(s)$ equal to zero and which is the characteristic equation of the closed-loop system. Equation (7) has two roots, which may be real or complex, depending on the sign of the quantity $a_1^2 - 4a_0$. However, the roots of (7) will have negative real parts and the closed-loop system will be stable provided that a_0 and a_1 are both positive.

If the poles are complex, it is convenient to rewrite the denominator of $T(s)$ in terms of the damping ratio ζ and the undamped natural frequency ω_n, which were introduced in Section 8.2. Comparing the characteristic equation in Equation (8.28) to (7), we see that

$$a_0 = \omega_n^2 \tag{8a}$$

$$a_1 = 2\zeta\omega_n \tag{8b}$$

Solving (8a) for ω_n and substituting it into (8b) give the damping ratio and the undamped natural frequency of the closed-loop system as

$$\zeta = \frac{a_1}{2\sqrt{a_0}}$$

$$\omega_n = \sqrt{a_0}$$

We see from these expressions that a_0, the gain of the outer feedback path in Figure 13.12, determines the undamped natural frequency ω_n and that a_1, the gain of the inner feedback path, affects only the damping ratio. If we can specify both a_0 and a_1 at will, then we can attain any desired values of ζ and ω_n for the closed-loop transfer function.

▶ **EXAMPLE 13.6**

Draw a block diagram for the translational mechanical system studied in Example 3.7, whose state-variable equations are given by Equations (3.14). Reduce the block diagram to determine the transfer functions $X_1(s)/F_a(s)$ and $X_2(s)/F_a(s)$ as rational functions of s.

SOLUTION   Transforming (3.14) with zero initial conditions, we have

$$sX_1(s) = V_1(s) \tag{9a}$$

$$MsV_1(s) = -K_1X_1(s) - B_1V_1(s) + K_1X_2(s) + F_a(s) \tag{9b}$$

$$B_2sX_2(s) = K_1X_1(s) - (K_1 + K_2)X_2(s) \tag{9c}$$

We use (9b) to draw a summing junction that has $MsV_1(s)$ as its output. After the summing junction, we insert the transfer function $1/Ms$ to get $V_1(s)$, which, from (9a), equals $sX_1(s)$. Thus an integrator whose input is $V_1(s)$ has $X_1(s)$ as its output. Using (9c), we form a second summing junction that has $B_2sX_2(s)$ as its output. Following this summing junction by the transfer function $1/B_2s$, we get $X_2(s)$ and can complete the four feedback paths required by the summing junctions. The result of these steps is the block diagram shown in Figure 13.14(a).

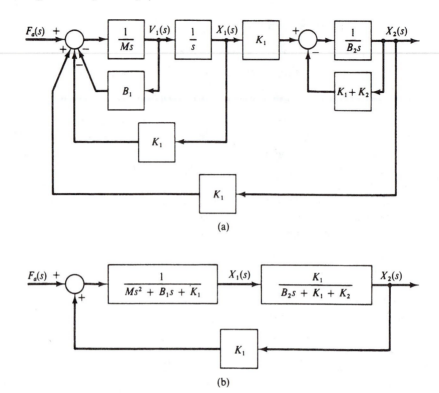

(a)

(b)

Figure 13.14 Block diagrams for the system in Example 13.6. (a) As drawn from (9). (b) With the three inner feedback loops eliminated.

To simplify the block diagram, we use (3) to reduce each of the three inner feedback loops, obtaining the version shown in Figure 13.14(b). To evaluate the transfer function $X_1(s)/F_a(s)$, we can apply (4) to this single-loop diagram because the sign associated with the feedback signal at the summing junction is positive rather than negative. Doing this with

$$G(s) = \frac{1}{Ms^2 + B_1 s + K_1}$$

and

$$H(s) = \frac{K_1^2}{B_2 s + K_1 + K_2}$$

we find

$$
\begin{aligned}
\frac{X_1(s)}{F_a(s)} &= \frac{\dfrac{1}{Ms^2 + B_1 s + K_1}}{1 - \dfrac{1}{Ms^2 + B_1 s + K_1} \cdot \dfrac{K_1^2}{B_2 s + K_1 + K_2}} \\[2mm]
&= \frac{B_2 s + K_1 + K_2}{(Ms^2 + B_1 s + K_1)(B_2 s + K_1 + K_2) - K_1^2} \\[2mm]
&= \frac{B_2 s + K_1 + K_2}{P(s)}
\end{aligned}
\tag{10}
$$

where

$$P(s) = MB_2s^3 + [(K_1 + K_2)M + B_1B_2]s^2$$
$$+ [B_1(K_1 + K_2) + B_2K_1]s + K_1K_2$$

To obtain $X_2(s)/F_a(s)$, we can write

$$\frac{X_2(s)}{F_a(s)} = \frac{X_1(s)}{F_a(s)} \cdot \frac{X_2(s)}{X_1(s)}$$

where $X_1(s)/F_a(s)$ is given by (10) and, from Figure 13.14(b),

$$\frac{X_2(s)}{X_1(s)} = \frac{K_1}{B_2s + K_1 + K_2} \tag{11}$$

The result of multiplying (10) and (11) is a transfer function with the same denominator as (10) but with a numerator of K_1. Note that the two transfer functions are consistent with the corresponding input-output differential equations found in Equations (3.27) and (3.28) for this system.

In the previous examples, we used the rules for combining blocks that are in series or in parallel, as shown in Figures 13.1 and 13.2. We also repeatedly used the rule for simplifying the basic feedback configuration given in Figure 13.10(a). To conclude this chapter, we present an example in which the feedback loops have an additional summing junction that must be moved outside the loop before (3) can be used to eliminate the loop.

▶ **EXAMPLE 13.7**

Find the closed-loop transfer function $T(s) = Y(s)/U(s)$ for the feedback system shown in Figure 13.15(a), which has two feedback loops, each of which has multiple summing junctions.

SOLUTION   We cannot immediately apply (3) to the inner feedback loop consisting of the first integrator and the gain block a_1, because the output of gain block b_1 enters a summer within that loop. We therefore use Figure 13.7 to move this summer to the left of the first integrator block, where it can be combined with the first summer. The resulting diagram is given in Figure 13.15(b).

Now (3) can be applied to the inner feedback loop to give the transfer function

$$G_1(s) = \frac{1/s}{1 + a_1/s} = \frac{1}{s + a_1}$$

The equivalent block with the transfer function $G_1(s)$ is then in series with the remaining integrator, which results in a combined transfer function of $1/[s(s + a_1)]$. Also, the two blocks with transfer functions of sb_1 and 1 are in parallel and can be combined into a single block. These simplifications are shown in Figure 13.15(c).

We can now repeat the procedure and move the right summer to the left of the block labeled $1/[s(s + a_1)]$, where it can again be combined with the first summer. This is done in part (d) of Figure 13.15. The two blocks in parallel at the left can now be combined by adding their transfer functions, and (3) can be applied to the right part of the diagram to

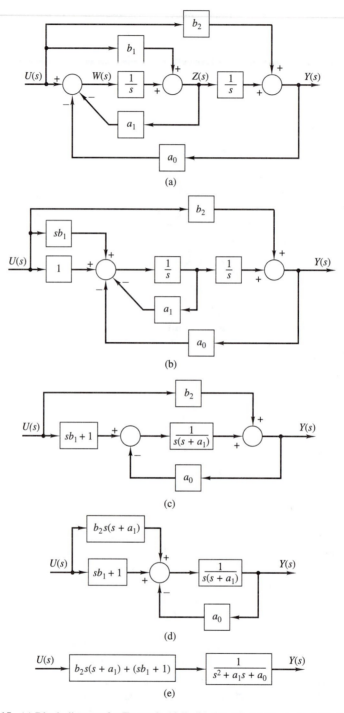

Figure 13.15 (a) Block diagram for Example 13.7. (b), (c), (d), (e) Equivalent block diagrams.

give

$$\frac{\dfrac{1}{s(s+a_1)}}{1+\dfrac{a_0}{s(s+a_1)}} = \frac{1}{s^2+a_1s+a_0}$$

These steps yield Figure 13.15(e), from which we see that

$$T(s) = \frac{b_2s^2 + (a_1b_2+b_1)s + 1}{s^2+a_1s+a_0} \tag{12}$$

Because performing operations on a given block diagram is equivalent to manipulating the algebraic equations that describe the system, it may sometimes be easier to work with the equations themselves. As an alternative solution to the last example, suppose that we start by writing the equations for each of the three summers in Figure 13.15(a):

$$W(s) = U(s) - a_0Y(s) - a_1Z(s) \tag{13a}$$

$$Z(s) = \frac{1}{s}W(s) + b_1U(s) \tag{13b}$$

$$Y(s) = \frac{1}{s}Z(s) + b_2U(s) \tag{13c}$$

Substituting (13a) into (13b) to eliminate $W(s)$, we see that

$$Z(s) = \frac{1}{s}[U(s) - a_0Y(s) - a_1Z(s)] + b_1U(s)$$

from which

$$Z(s) = \frac{1}{s+a_1}[-a_0Y(s) + (b_1s+1)U(s)] \tag{14}$$

Using (14) to eliminate $Z(s)$ in (13c), we get

$$Y(s) = \frac{1}{s(s+a_1)}[-a_0Y(s) + (b_1s+1)U(s)] + b_2U(s)$$

Rearranging this equation, we obtain the transfer function

$$\frac{Y(s)}{U(s)} = \frac{b_2s^2 + (a_1b_2+b_1)s + 1}{s^2+a_1s+a_0}$$

which agrees with (12), as found in Example 13.7.

▶ 13.3 DIAGRAMS FOR INPUT-OUTPUT MODELS

The input-output form of the model of an nth-order fixed linear system with input $u(t)$ and output y is the single differential equation

$$a_ny^{(n)} + a_{n-1}y^{n-1} + \cdots + a_0y = b_mu^{(m)} + \cdots + b_0u(t) \tag{15}$$

where $y^{(k)}$ denotes d^ky/dt^k, and where in practice $m \le n$. Transforming this equation with the initial-condition terms set equal to zero gives

$$(a_ns^n + a_{n-1}s^{n-1} + \cdots + a_0)Y(s) = (b_ms^m + \cdots + b_0)U(s)$$

corresponding to the transfer function

$$H(s) = \frac{Y(s)}{U(s)} = \frac{b_m s^m + \cdots + b_0}{a_n s^n + a_{n-1} s^{n-1} + \cdots + a_0} \tag{16}$$

In this section we shall present a general method of representing these equations by a block diagram consisting only of summers, gains, and integrators. We continue our practice of avoiding the use of differentiator blocks, because of the inaccuracies that can result from using them in numerical solutions. We also want to be able to write a state-variable model corresponding to a given input-output equation. Because the choice of state variables is not unique, our method will result in only one of several possible state-variable models.

We shall develop the method in two stages. First we consider systems for which (15) contains no derivatives of the input, and then we extend the method to include input derivatives. For a first-order system, the state-variable and input-output models involve a single first-order differential equation and are essentially identical. Thus we start with second-order systems and generalize the results to apply to higher-order systems.

Second-Order Systems

We shall consider the input-output differential equation

$$a_2 \ddot{y} + a_1 \dot{y} + a_0 y = f(t)$$

for three choices of the input function:

1. $f(t) = u(t)$
2. $f(t) = b_1 \dot{u} + b_0 u(t)$
3. $f(t) = b_2 \ddot{u} + b_1 \dot{u} + b_0 u(t)$

We shall construct the corresponding block diagrams in the following three examples, the last one of which will constitute the most general case of a second-order fixed linear system.

▶ **EXAMPLE 13.8**

Construct the block diagram for the system described by the differential equation

$$a_2 \ddot{y} + a_1 \dot{y} + a_0 y = u(t) \tag{17}$$

which involves no input derivatives in its input function. Then use the block diagram to find a state-variable model for the system.

SOLUTION With $y(0) = \dot{y}(0) = 0$, the transformed equation is

$$a_2 s^2 Y(s) + a_1 s Y(s) + a_0 Y(s) = U(s)$$

so

$$s^2 Y(s) = \frac{1}{a_2}[-a_1 s Y(s) - a_0 Y(s) + U(s)] \tag{18}$$

We begin by drawing a series combination of two integrators in Figure 13.16(a), which shows the transform variables $Y(s)$, $sY(s)$, and $s^2 Y(s)$. Using gain blocks and a summer, we then form $s^2 Y(s)$ according to (18). The complete diagram is shown in Figure 13.16(b).

Although the block diagram is labeled in terms of the Laplace-transformed variables $U(s), Y(s), sY(s)$, and $s^2 Y(s)$, the reader can easily translate these transformed variables

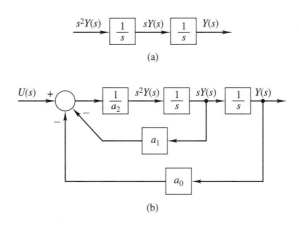

Figure 13.16 Diagrams for Example 13.8. (a) Partial diagram. (b) Complete block diagram.

into their time-domain counterparts, namely $u(t), y(t), \dot{y}(t)$, and $\ddot{y}(t)$. Likewise, the two $1/s$ blocks represent integrators, as first discussed in Chapter 4.

In order to write the state-variable equations, we take the output of each integrator as a state variable, so the integrator input must be the derivative of the corresponding state variable. Referring to Figure 13.16(b), we assign the output of the right integrator (labeled $Y(s)$ in the figure) to be the state variable q_1, which is also the output y. We assign the output of the left integrator (labeled $sY(s)$ in the figure) to be the state variable q_2. With these definitions, we can write the state-variable model from Figure 13.16(b) as

$$\dot{q}_1 = q_2 \tag{19a}$$

$$\dot{q}_2 = \frac{1}{a_2}[-a_0 q_1 - a_1 q_2 + u(t)] \tag{19b}$$

$$y = q_1 \tag{19c}$$

In the last example, note that (19a) and (19b) do have the correct form for the state-variable equations describing a second-order fixed linear system having $u(t)$ as its input. It is easy to check (19) by showing that these equations satisfy the input-output equation in (17). To do this, we use (19a) to replace q_2 with $\dot{q}_1$ in (19b) and then use (19c) to replace q_1 with y. Transforming the resulting equations, with zero initial conditions, and solving for the ratio $Y(s)/U(s)$, we obtain

$$H(s) = \frac{Y(s)}{U(s)} = \frac{1}{a_2 s^2 + a_1 s + a_0} \tag{20}$$

which is the transfer function of the system given by (17).

We started the solution to Example 13.8 by solving the transformed equation for the output term containing the highest power of s. The reader may wish to compare this procedure to Example 4.1, where we solved the time-domain equation for the highest derivative of the output. In fact, the diagram in Figure 4.6(b) is the same as the one in Figure 13.16(b), with $a_2 = M, a_1 = B, a_0 = K$, and $u(t) = f_a(t)$. However, we next wish to extend the approach used in Example 13.8 to the case for which the input function involves both $u(t)$ and its first derivative.

▶ **EXAMPLE 13.9**

Draw the block diagram and write the state-variable model when

$$a_2\ddot{y} + a_1\dot{y} + a_0 y = b_1\dot{u} + b_0 u(t) \tag{21}$$

SOLUTION   With zero initial conditions, the transformed output can be written as

$$Y(s) = \frac{b_1 s + b_0}{a_2 s^2 + a_1 s + a_0} U(s) \tag{22}$$

If we define $Q(s)$ to be

$$Q(s) = \left[\frac{1}{a_2 s^2 + a_1 s + a_0} \right] U(s) \tag{23}$$

we can rewrite (22) as

$$Y(s) = (b_1 s + b_0) Q(s)$$
$$= b_1 s Q(s) + b_0 Q(s) \tag{24}$$

The quantity inside the brackets in (23) is the transfer function corresponding to (17) in Example 13.8. Thus (23) is described by the block diagram in Figure 13.16(b) if we replace the symbol $Y(s)$ by $Q(s)$. This is done in Figure 13.17(a). Once $Q(s)$ and $sQ(s)$ are available in the block diagram, we use two additional gain blocks and a summer to satisfy (24), which results in the complete diagram in Figure 13.17(b).

To write state-variable equations, we introduce the transformed variables $Q_1(s) = Q(s)$ and $Q_2(s) = sQ(s)$. We can then write the transformed state-variable model directly from

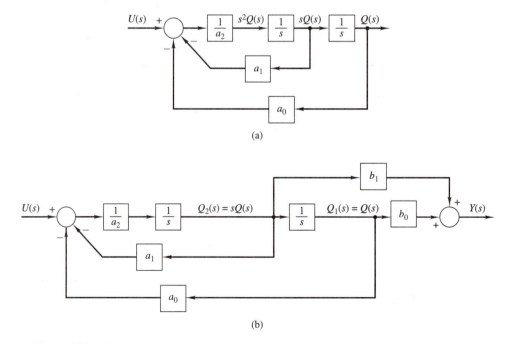

(a)

(b)

Figure 13.17   Diagram for Example 13.9. (a) Diagram for $Q(s)$. (b) Complete block diagram.

Figure 13.17(b) as

$$sQ_1(s) = Q_2(s)$$

$$sQ_2(s) = \frac{1}{a_2}[-a_0Q_1(s) - a_1Q_2(s) + U(s)] \tag{25}$$

$$Y(s) = b_0Q_1(s) + b_1Q_2(s)$$

Recall that when constructing block diagrams, we always assume the initial conditions to be zero. Because $\mathcal{L}[\dot{q}_1] = sQ_1(s)$ and $\mathcal{L}[\dot{q}_2] = sQ_2(s)$ for zero initial conditions, (25) leads directly to the time-domain state-variable model, namely,

$$\dot{q}_1 = q_2$$

$$\dot{q}_2 = \frac{1}{a_2}[-a_0q_1 - a_1q_2 + u(t)] \tag{26}$$

$$y = b_0q_1 + b_1q_2$$

With a little practice, it is possible to write the state-variable model directly from the block diagram, and we shall do this in subsequent examples. Finally, comparing these equations with the results of Example 13.8, we see that the only effect of the more complicated input function has been to change the algebraic output equation to include both state variables, rather than just q_1.

The third of these second-order examples involves an input function consisting of the input and both its first and second derivatives.

▶ **EXAMPLE 13.10**

Repeat the previous example when

$$a_2\ddot{y} + a_1\dot{y} + a_0y = b_2\ddot{u} + b_1\dot{u} + b_0u(t) \tag{27}$$

SOLUTION   We can write the zero-state transformed output as

$$Y(s) = \left[\frac{b_2s^2 + b_1s + b_0}{a_2s^2 + a_1s + a_0}\right]U(s) \tag{28}$$

or, with $Q(s)$ defined by (23),

$$Y(s) = (b_2s^2 + b_1s + b_0)Q(s)$$
$$= b_2s^2Q(s) + b_1sQ(s) + b_0Q(s) \tag{29}$$

The definition of $Q(s)$ is the same as for Example 13.9, so the relationship between $Q(s)$ and $U(s)$ is still described by Figure 13.17(a). To satisfy (29), we can use three gain blocks and a summer to obtain the complete block diagram in Figure 13.18(a). In order to facilitate writing the state-variable equations, we have replaced $Q(s)$ and $sQ(s)$, respectively, by $Q_1(s)$ and $Q_2(s)$ when labeling the outputs of the integrators. From this diagram we see that

$$\dot{q}_1 = q_2 \tag{30a}$$

$$\dot{q}_2 = \frac{1}{a_2}[-a_0q_1 - a_1q_2 + u(t)] \tag{30b}$$

$$y = b_0q_1 + b_2\dot{q}_2 + b_1q_2 \tag{30c}$$

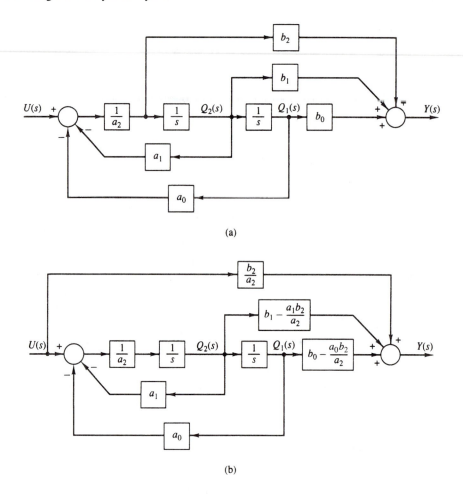

(a)

(b)

Figure 13.18 Diagrams for Example 13.10. (a) Block diagram corresponding to (29). (b) Block diagram corresponding to (31).

Note, however, that (30c) does not have the standard form of the output equation in a state-variable model, because y is given as a function of the *derivative* of one of the state variables, in addition to the state variables themselves. In spite of this feature, (30c) is a useful form of the output equation for many purposes, particularly because each of the six gains appearing in the block diagram is a coefficient of the input-output differential equation. Thus we can draw the diagram without performing any calculations to evaluate the gains.

The basic reason for the form of (30) is that one of the signals entering the output summer in Figure 13.18(a) does not come from the output of an integrator. This in turn is caused by the fact that the numerator of the transfer function in (28) is not of lower order than the denominator. When this situation occurs, we can carry out a preliminary step of long division, as was done in Section 7.6 (Transform Inversion) whenever a function of s was not a strictly proper rational function. Then (28) becomes

$$Y(s) = \left[\frac{b_2}{a_2} + \frac{(b_1 - a_1 b_2/a_2)s + (b_0 - a_0 b_2/a_2)}{a_2 s^2 + a_1 s + a_0} \right] U(s)$$

$$= \frac{b_2}{a_2} U(s) + \left[\left(b_1 - \frac{a_1 b_2}{a_2} \right) s + \left(b_0 - \frac{a_0 b_2}{a_2} \right) \right] Q(s) \qquad (31)$$

The quantity $Q(s)$ is still given by (23), and the relationship between $Q(s)$ and $U(s)$ is still shown in Figure 13.17(a). Starting with that figure, adding the extra blocks needed to implement (31), and replacing $Q(s)$ by $Q_1(s)$ and $sQ(s)$ by $Q_2(s)$, we obtain the diagram in Figure 13.18(b). The state-variable equations corresponding to the diagram are again given by (30a) and (30b), but the output equation becomes

$$y = \left(b_0 - \frac{a_0 b_2}{a_2} \right) q_1 + \left(b_1 - \frac{a_1 b_2}{a_2} \right) q_2 + \frac{b_2}{a_2} u(t) \tag{32}$$

which includes the input $u(t)$ in addition to the state variables q_1 and q_2.

Higher-Order Systems

Extending the three previous examples to systems of order higher than two is straightforward. The general input-output differential equation and the corresponding transfer function are given by (15) and (16), respectively. We first consider the related differential equation

$$a_n q^{(n)} + a_{n-1} q^{(n-1)} + \cdots + a_0 q = u(t) \tag{33}$$

where the input function is assumed to be $u(t)$. Transforming (33) with zero initial conditions, we have

$$a_n s^n Q(s) + a_{n-1} s^{n-1} Q(s) + \cdots + a_0 Q(s) = U(s) \tag{34}$$

and

$$Q(s) = \left[\frac{1}{a_n s^n + a_{n-1} s^{n-1} + \cdots + a_0} \right] U(s) \tag{35}$$

We rearrange (34) as

$$s^n Q(s) = \frac{1}{a_n} \left[-a_{n-1} s^{n-1} Q(s) - \cdots - a_0 Q(s) + U(s) \right] \tag{36}$$

which is described by the block diagram shown in Figure 13.19.

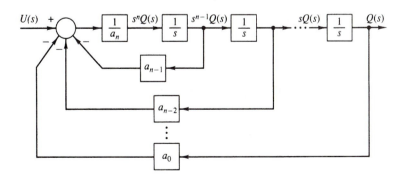

Figure 13.19 Block diagram for (33) through (36).

Returning to the more general model given by (15), we rewrite (16) as

$$Y(s) = \left[\frac{b_m s^m + \cdots + b_0}{a_n s^n + a_{n-1} s^{n-1} + \cdots + a_0} \right] U(s)$$

$$= (b_m s^m + \cdots + b_0) Q(s)$$

$$= b_m s^m Q(s) + \cdots + b_0 Q(s) \tag{37}$$

where $Q(s)$ is defined by (35) and where the relationship between $Q(s)$ and $U(s)$ is represented by the diagram in Figure 13.19. To form the block diagram corresponding to (15) and (16), we merely add to Figure 13.19 additional gain and summer blocks according to (37).

As long as $m < n$, this procedure will give an entirely satisfactory diagram. For the case of $m = n$, we may choose to carry out a preliminary step of long division on the transfer function $H(s)$ in (16). The following two examples illustrate the method for third-order systems.

▶ **EXAMPLE 13.11**

Draw the block diagram for the system described by

$$\dddot{y} + 5\ddot{y} + 2\dot{y} + y = 3\ddot{u} + 4u(t)$$

Also find a state-variable model.

SOLUTION   The block diagram corresponding to

$$\dddot{q} + 5\ddot{q} + 2\dot{q} + q = u(t)$$

is shown in Figure 13.20(a). Noting that

$$Y(s) = (3s^2 + 4)Q(s) = 3s^2 Q(s) + 4Q(s)$$

we add to that figure two gain blocks and a summer in order to obtain the block diagram in Figure 13.20(b).

The outputs of the integrators are labeled $Q_1(s)$, $Q_2(s)$, and $Q_3(s)$ from right to left. Then we can write the state-variable model as

$$\begin{aligned}
\dot{q}_1 &= q_2 \\
\dot{q}_2 &= q_3 \\
\dot{q}_3 &= -q_1 - 2q_2 - 5q_3 + u(t) \\
y &= 4q_1 + 3q_3
\end{aligned} \tag{38}$$

▶ **EXAMPLE 13.12**

Draw the block diagram for the system described by

$$0.5\dddot{y} + 2\ddot{y} + \dot{y} + y = 2\ddot{u} - 3\dot{u} + u(t)$$

Also find a state-variable model.

SOLUTION   We can write

$$Y(s) = \left[\frac{2s^3 - 3s + 1}{0.5s^3 + 2s^2 + s + 1} \right] U(s) \tag{39}$$

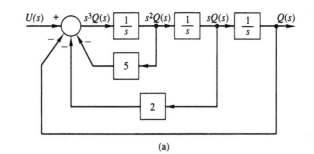

(a)

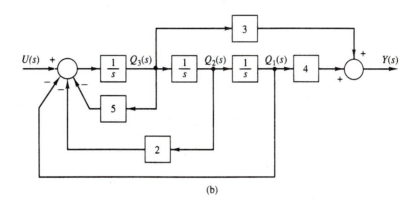

(b)

Figure 13.20 Diagrams for Example 13.11. (a) Partial block diagram. (b) Complete block diagram.

where the quantity inside the bracket is the system's transfer function $H(s)$. Then

$$Y(s) = (2s^3 - 3s + 1)Q(s)$$
$$= 2s^3 Q(s) - 3sQ(s) + Q(s) \tag{40}$$

where

$$Q(s) = \left[\frac{1}{0.5s^3 + 2s^2 + s + 1} \right] U(s) \tag{41}$$

The block diagram for (41) is shown in Figure 13.21(a). Adding the gain and summer blocks called for by (40) gives the complete diagram in Figure 13.21(b). Then we have

$$\dot{q}_1 = q_2 \tag{42a}$$
$$\dot{q}_2 = q_3 \tag{42b}$$
$$\dot{q}_3 = 2[-q_1 - q_2 - 2q_3 + u(t)] \tag{42c}$$
$$y = q_1 - 3q_2 + 2\dot{q}_3 \tag{42d}$$

Because of the presence of the term $2\dot{q}_3$, the last equation does not have the standard form for the output equation in a state-variable model. For an alternative block diagram, we perform a step of long division on the transfer function within the brackets in (39). Then

$$Y(s) = \left[4 + \frac{-8s^2 - 7s - 3}{0.5s^3 + 2s^2 + s + 1} \right] U(s)$$
$$= 4U(s) - 8s^2 Q(s) - 7sQ(s) - 3Q(s)$$

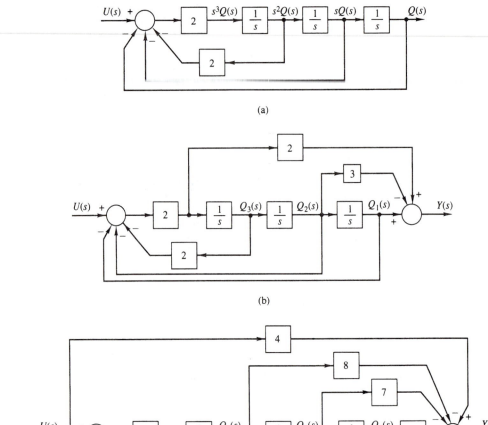

(a)

(b)

(c)

Figure 13.21 Diagrams for Example 13.12. (a) Partial block diagram. (b) Complete block diagram. (c) Alternative block diagram.

Again starting with Figure 13.21(a), we add four gain blocks and a summer to obtain the diagram in part (c) of the figure. The corresponding state-variable model consists of (42a), (42b), and (42c), together with the new output equation

$$y = -3q_1 - 7q_2 - 8q_3 + 4u(t) \tag{43}$$

which does not contain any derivatives. This output equation can also be obtained by substituting (42c) into (42d) to eliminate the derivative on the right side of (42d).

13.4 APPLICATION TO A CONTROL SYSTEM

Most control systems use feedback to force the output variable to follow a reference input while remaining relatively insensitive to the effects of one or more disturbance inputs. A common type of control system is the **servomechanism**. Here a mechanical input variable, such as the position or angular rotation of an element, is required to follow a reference input, such as the orientation of a knob or dial that is varied by a human operator. For example, a mechanical manipulator used for working with radioactive materials would require several servomechanisms to translate the operator's hand motion into equivalent physical motion at a remote location behind the shielding material.

In this section, we model and analyze a simple servomechanism whose purpose is to make the angular orientation of an output shaft follow that of a manually adjusted dial. First, we will develop a mathematical model of the feedback system by writing the algebraic and differential equations that describe the individual elements, transforming these equations and drawing a block diagram, and then reducing the block diagram to give a single transfer function. Finally, we shall analyze the system's performance and consider means of improving it.

System Description

The positional servomechanism we will consider is shown in Figure 13.4.1. The manually set input potentiometer at the left has its two ends connected to constant voltage sources, and its wiper voltage e_1 obeys the algebraic relationship

$$e_1 = K_\theta \theta_i(t) \tag{44}$$

where K_θ is a constant, provided that the potentiometer is linear and that no current is drawn by the amplifier—that is, there is no loading. The output potentiometer at the right of the figure is identical to the input potentiometer, except that its wiper is mechanically connected to the output shaft. Hence the voltage of the wiper of the output potentiometer obeys the equation

$$e_2 = K_\theta \theta_o \tag{45}$$

If both potentiometers are constrained to one full revolution and if the constant voltage sources are $\pm A$ volts, then $K_\theta = A/\pi$ volts per radian.

The amplifier's output voltage is

$$e_a = K_A(e_1 - e_2) \tag{46}$$

where K_A is the amplifier gain in volts per volt. It is assumed that the amplifier is powerful enough to ensure that (46) holds regardless of the current i flowing in the armature circuit of the motor and that the amplifier draws no current from the wipers of the input and output potentiometers. Note that the amplifier is described by an algebraic rather than a differential equation, which implies that the amplifier responds rapidly enough to ensure that (46) holds, regardless of how rapidly e_1 or e_2 may vary.

The motor is assumed to have a constant field current and negligible inductance in the armature winding. The electromechanical driving torque τ_e and the induced voltage e_m are given, as in Equation (10.28), by

$$\tau_e = \alpha i$$

$$e_m = \alpha \dot{\phi}$$

where the coupling coefficient α has units of volt-seconds or, equivalently, newton-meters per ampere and is dependent on the field current $\bar{i}_f$. The symbol ϕ denotes the angular displacement of the motor shaft. If the armature resistance is denoted by R, the viscous-friction coefficient by B, and the moment of inertia by J, the motor can be modeled by the

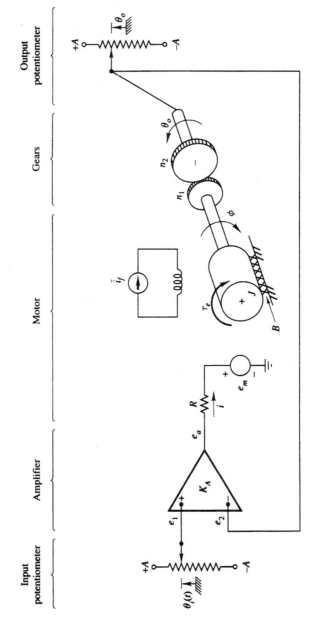

Figure 13.22 Servomechanism components

pair of equations

$$i = \frac{1}{R}(e_a - \alpha\dot{\phi}) \tag{47a}$$

$$J\ddot{\phi} + B\dot{\phi} = \alpha i \tag{47b}$$

The motor shaft is connected to the output shaft through a pair of gears having the gear ratio $N = n_2/n_1$. Hence the motor angle ϕ and the output wiper angle θ_o are related by

$$\theta_o = \frac{1}{N}\phi \tag{48}$$

Although Figure 13.4.1 does not indicate a moment of inertia attached to the right gear or moments of inertia for the gears themselves, such moments of inertia could be referred to the motor shaft and incorporated in the value of J if they were not negligible. Likewise, any viscous friction associated with the output shaft could be referred to the motor and combined with B.

System Model

In order to obtain a block diagram for the system, we transform (44) through (48) with zero initial conditions. For the combination of the input and output potentiometers and the amplifier, we substitute (44) and (45) into (46) and take the Laplace transform to obtain

$$E_a(s) = K_A K_\theta [\Theta_i(s) - \Theta_o(s)] \tag{49}$$

which is represented by the two gains of K_θ, the summing junction, and the gain of K_A within the dashed rectangle shown in Figure 13.23.

When (47) is transformed and $I(s)$ is eliminated, we obtain the single transformed equation

$$\left[Js^2 + \left(B + \frac{\alpha^2}{R} \right) s \right] \Phi(s) = \frac{\alpha}{R} E_a(s)$$

for the motor, which yields the transfer function

$$\frac{\Phi(s)}{E_a(s)} = \frac{\dfrac{\alpha}{RJ}}{s \left[s + \left(\dfrac{B}{J} + \dfrac{\alpha^2}{JR} \right) \right]}$$

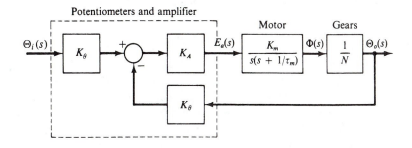

Figure 13.23 Servomechanism block diagram.

If we define the parameters

$$K_m = \frac{\alpha}{RJ}$$

$$\tau_m = \frac{1}{\dfrac{B}{J} + \dfrac{\alpha^2}{JR}}$$

the transfer function of the motor becomes

$$\frac{\Phi(s)}{E_a(s)} = \frac{K_m}{s\left(s + \dfrac{1}{\tau_m}\right)} \tag{50}$$

which results in a single block located in the forward path of the diagram shown in Figure 13.23. Finally, we describe the gears by the gain $1/N$ according to

$$\Theta_o(s) = \frac{1}{N}\Phi(s) \tag{51}$$

and draw the feedback path from the output to the output-potentiometer block.

Closed-Loop Transfer Function

To calculate $T(s) = \Theta_o(s)/\Theta_i(s)$, the transfer function of the closed-loop system, we note that the first block with gain K_θ is in series with the feedback-loop portion of the system. We can obtain the transfer function of the feedback loop by applying (3) with

$$G(s) = \frac{K_A K_m/N}{s(s + 1/\tau_m)}$$

$$H(s) = K_\theta$$

Then

$$T(s) = K_\theta \left[\frac{G(s)}{1 + G(s)H(s)}\right] = \frac{K_A K_m K_\theta/N}{s^2 + (1/\tau_m)s + K_A K_m K_\theta/N} \tag{52}$$

By inspecting (52), we can observe several important aspects of the behavior of the closed-loop system. First, the system model is second order, and its closed-loop transfer function $T(s)$ has two poles and no zeros in the finite s-plane. Assuming that the values of all the parameters appearing in $T(s)$ are positive, the poles of $T(s)$ will be in the left half of the complex plane, and the closed-loop system will be stable regardless of the specific numerical values of the parameters.

We obtain the steady-state value of the unit step response of the closed-loop system by setting s equal to zero in (52), in accordance with Equation (8.82).

$$T(0) = \frac{K_A K_m K_\theta/N}{K_A K_m K_\theta/N} = 1$$

Hence a step input for $\theta_i(t)$ will result in an output angle θ_o that is identical to the input angle in the steady state. This steady-state condition of zero error will occur regardless of the specific numerical values for the system parameters. This property is a result of the feedback structure of the servomechanism, whereby a signal that is proportional to the error signal $\theta_i(t) - \theta_o$ is used to drive the motor. Because the motor acts like an integrator (its transfer function $\Phi(s)/E_a(s)$, given by (50), has a pole at $s = 0$), the armature voltage e_a must become zero if the system is to reach steady state with a constant input; otherwise, $\dot\phi$ would not be zero. Because of (46), the difference of the wiper voltages e_1 and e_2 must be zero in the steady state, and this condition requires that $(\theta_o)_{ss} = \theta_i$.

TABLE 13.1 **Numerical Values of Servomechanism Parameters**

Parameter	Value
A (magnitude of potentiometer voltages)	15 V
K_m (for the motor)	150 V $\cdot$ rad/s^2
τ_m (for the motor)	0.40 s
N (gear ratio)	10

Design for Specified Damping Ratio

In practice, all the parameter values except one are often fixed and we must select the remaining one to yield some specific response characteristic such as the damping ratio, undamped natural frequency, or steady-state response. Suppose that all the parameters except the amplifier gain K_A are fixed at the values listed in Table 13.1 and that we must select K_A to yield a value of 0.50 for the damping ratio.

Because the potentiometer gains are $K_\theta = A/\pi$, it follows that $K_\theta = 15.0/\pi = 4.775$ V/rad. Substituting the known parameter values into (52) results in the closed-loop transfer function

$$T(s) = \frac{(150 \times 4.775/10)K_A}{s^2 + (1/0.40)s + (150 \times 4.775/10)K_A}$$

$$= \frac{71.62K_A}{s^2 + 2.50s + 71.62K_A} \tag{53}$$

Comparing the denominator of (53) to the polynomial $s^2 + 2\zeta\omega_n s + \omega_n^2$, which is used to define the damping ratio ζ and the undamped natural frequency ω_n, we see that

$$2\zeta\omega_n = 2.50 \tag{54a}$$

$$\omega_n^2 = 71.62K_A \tag{54b}$$

Setting ζ equal to its specified value of 0.50 in (54a), we get $\omega_n = 2.50$ rad/s. Substituting for ω_n^2 in (54b) gives the required amplifier gain as

$$K_A = 0.08727 \text{ V/V}$$

Substituting this value of K_A into (53) yields the closed-loop transfer function in numerical form as

$$T(s) = \frac{6.25}{s^2 + 2.50s + 6.25} \tag{55}$$

You can verify that the denominator of $T(s)$ results in a pair of complex poles having $\zeta = 0.50$ as specified.

Proportional-Plus-Derivative Feedback

The system designed in the foregoing paragraphs is constrained in several ways that turn out to be undesirable in practice. For example, although we obtained the specified damping ratio of 0.50, the value of ω_n was dictated by the requirement on ζ and could not have been specified independently. In practice, one might want to speed up the response by increasing ω_n while retaining the damping ratio at $\zeta = 0.5$. This would require a higher K_A, which would decrease ζ unless an additional term can be introduced to increase the coefficient of s in the denominator of (52). We shall demonstrate that this can be done by providing a signal proportional to the angular velocity of the motor shaft (or output shaft) that is fed back and added to the amplifier input.

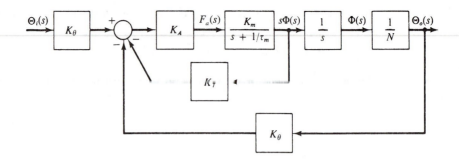

Figure 13.24 Block diagram of servomechanism with tachometer feedback added.

A tachometer is a device with a shaft that rotates and produces a voltage that is proportional to the angular velocity. It is therefore just a type of electric generator, as studied in Chapter 10. Assume that a tachometer is attached directly to the motor and a potentiometer is placed across the output terminals of the tachometer, such that the voltage on the potentiometer wiper is

$$e_3 = K_T \dot{\phi}$$

where K_T has units of volts/rad/second = volt-seconds, and may have any positive value. The signal e_3 is added to e_2, such that the armature voltage is now

$$\begin{aligned} e_a &= K_A(e_1 - e_2 - e_3) \\ &= K_A[K_\theta \theta_i(t) - K_\theta \theta_o - K_T \dot{\phi}] \end{aligned}$$

Then the transform of the armature voltage becomes

$$E_a(s) = K_A[K_\theta \Theta_i(s) - K_\theta \Theta_o(s) - K_T s \Phi(s)] \tag{56}$$

The modified system can be represented by the block diagram shown in Figure 13.24, which we obtain by separating the motor transfer function $K_m/[s(s + 1/\tau_m)]$ into the pair of blocks in series shown in the figure and then adding the inner feedback path corresponding to the term $K_T s \Phi(s)$. Because we assume zero initial conditions when evaluating transfer functions, the input to the gain block K_T is $s\Phi(s) = \mathcal{L}[\dot{\phi}(t)]$. Thus a signal proportional to the angular velocity of the motor shaft is being fed through the inner feedback path to the summing junction.

To determine the effect of the tachometer feedback on the closed-loop transfer function, we can reduce the inner loop shown in Figure 13.24 by using (3) with $G(s) = K_A K_m/(s + 1/\tau_m)$ and $H(s) = K_T$ to give

$$T'(s) = \frac{K_A K_m}{s + (1/\tau_m) + K_A K_m K_T}$$

Applying (3) again with $G(s) = T'(s)/sN$ and $H(s) = K_\theta$ and then multiplying the result by K_θ, we obtain the overall transfer function

$$T(s) = \frac{K_A K_m K_\theta/N}{s^2 + [(1/\tau_m) + K_A K_m K_T]s + K_A K_m K_\theta/N} \tag{57}$$

Comparing (57) to (52) indicates that the only effect of the tachometer feedback is to modify the coefficient of s in the denominator of the transfer function. Hence the damping ratio and undamped natural frequency now satisfy the relationships

$$\begin{aligned} 2\zeta\omega_n &= (1/\tau_m) + K_A K_m K_T \\ \omega_n^2 &= K_A K_m K_\theta/N \end{aligned} \tag{58}$$

If the values of both K_A and K_T can be selected with τ_m, K_m, K_θ, and N fixed as before, we can specify values for both ζ and ω_n. For example, using the parameter values given in

Table 13.1 with $\zeta = 0.50$, we find that (55) reduces to

$$\omega_n = 2.50 + 150 K_A K_T \tag{59a}$$

$$\omega_n^2 = 71.62 K_A \tag{59b}$$

If, for example, we want ω_n to be 5.0 rad/s, instead of the 2.5 rad/s it was previously, it follows that

$$K_A = 0.3491 \text{ V/V}$$

$$K_T = 0.04775 \text{ V·s/rad}$$

Thus, K_A has been increased without changing ζ by adding the K_T term in (57). When we substitute these parameter values and those listed in Table 13.1 into (57), the numerical form of the closed-loop transfer function with tachometer feedback is

$$T(s) = \frac{25.0}{s^2 + 5.0s + 25.0}$$

It is easy to verify that the two poles of $T(s)$ are complex and have $\zeta = 0.50$ and $\omega_n = 5.0$ rad/s, as required.

In concluding this example, we note that the control law given by (56) uses a combination of the output-shaft angle θ_o and the motor angular velocity $\dot{\phi}$ as the feedback signals. It is also possible to think in terms of having the output-shaft angular velocity $\dot{\theta}_o$ fed back by noting that the two angular velocities in question are proportional to one another: $\dot{\theta}_o = \dot{\phi}/N$. Hence we can rewrite (56) in the time domain and in terms of the output-shaft angle as

$$e_a = K_A [K_\theta \theta_i(t) - K_\theta \theta_o - K_{\dot{\theta}} \dot{\theta}_o] \tag{60}$$

where $K_{\dot{\theta}} = N K_T$, which is the gain of the angular-velocity term. The block diagram corresponding to this version of the control law can take either of the equivalent forms shown in Figure 13.25. If we combine the two parallel feedback paths shown in Figure 13.25(b)

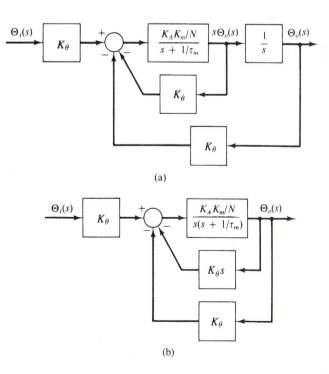

(a)

(b)

Figure 13.25 Equivalent block diagrams for servomechanism with tachometer feedback.

to yield the single feedback transfer function $K_\theta + K_{\dot\theta}s$, using (3) gives the same expression for $T(s) = \Theta_o(s)/\Theta_i(s)$ as found in (57).

▶ SUMMARY

In this chapter we have developed a number of tools that will be useful for modifying and simplifying block diagrams that represent linear systems, in particular those involving feedback. We have shown how to combine two blocks in series or parallel into a single block with its transfer function expressed as a rational function of the Laplace transform variable s. We also showed how to move a pick-off point and a summing junction, both of which are essential steps in reducing the diagrams of feedback systems.

The simplification of block diagrams for linear systems involving feedback was considered in detail, followed by a discussion of methods for dealing with input derivatives in the modeling equations. Finally, we used a number of these methods to develop a block diagram for an electromechanical servomechanism, determine its closed-loop transfer function as a rational function, and determine the adjustable gains with both position and velocity feedback. In the remaining two chapters, we will expand on these ideas by introducing a number of concepts and methods that are useful for the analysis and design of feedback control systems.

▶ PROBLEMS

13.1. Find the transfer functions of the blocks shown in parts (b) and (c) of Figure P13.1 such that the transfer functions $Y(s)/X_1(s)$ and $Y(s)/X_2(s)$ are identical to those for part (a) of the figure.

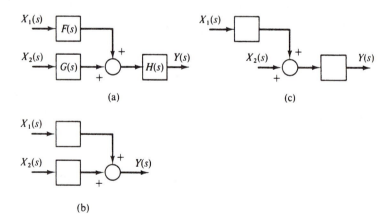

Figure P13.1

13.2. Find the transfer functions of the blocks shown in parts (b) and (c) of Figure P13.2 such that the transfer functions $Y_1(s)/X(s)$ and $Y_2(s)/X(s)$ are identical to those for part (a) of the figure.

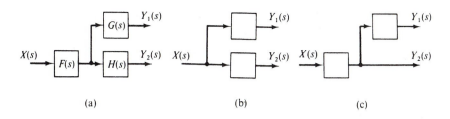

Figure P13.2

*13.3. Evaluate the transfer functions $T_1(s) = Y(s)/U(s)$ and $T_2(s) = Z(s)/U(s)$ as rational functions for the block diagram shown in Figure P13.3.

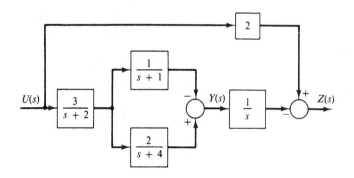

Figure P13.3

13.4. Draw block diagrams for each of the following input-output models.

    a. $\dot{y} + 3y = 2u(t)$

    b. $\dot{y} + 3y = 2\dot{u} + u(t)$

    c. $\ddot{z} + 4\dot{z} + 2z = u(t)$

    d. $\ddot{z} + 4\dot{z} + 2z = 3\dot{u} + 2u(t)$

*13.5. Draw block diagrams for each of the following input-output models. Also write a set of state-variable equations for each of the models.

    a. $\ddot{y} + 5\dot{y} + 6y = 12u(t)$

    b. $\ddot{y} + 5\dot{y} + 6y = 2\ddot{u} + \dot{u} + 2u(t)$

    c. $\dddot{y} + 2\ddot{y} + 3\dot{y} + 4y = 2\ddot{u} - \dot{u} + 4u(t)$

13.6. Repeat Problem 13.5 for each of the following models.

    a. $2\ddot{y} + 3\dot{y} + 6y = 4u(t)$

    b. $2\ddot{y} + 3\dot{y} + 6y = -2\dot{u} + 3u(t)$

    c. $\ddot{z} + 4\dot{z} + 3z = \ddot{u} + 2u(t)$

In Problems 13.7 and 13.8, draw block diagrams for the systems that were modeled in the examples indicated, using the equations cited.

13.7. Equation (5.29) in Example 5.2

13.8. Equation (6.21) in Example 6.3

In Problems 13.9 and 13.10, determine the closed-loop transfer function $Y(s)/U(s)$ as a rational function of s for the block diagram shown in the figure cited.

*13.9. Figure P13.9

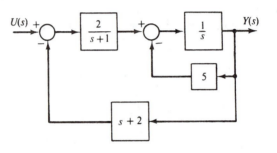

Figure P13.9

13.10. Figure P13.10

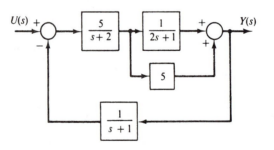

Figure P13.10

13.11.

a. Draw a block diagram for the rotational mechanical system modeled in Example 5.5 by transforming Equations (5.42) and drawing the corresponding diagram.

b. Set $\tau_L(s)$ equal to zero and reduce the block diagram to find the transfer function $T_1(s) = \Omega_2(s)/\tau_a(s)$.

c. Set $\tau_a(s)$ equal to zero and reduce the block diagram you found in part (a) to obtain the transfer function $T_2(s) = \Omega_2(s)/\tau_L(s)$.

d. Transform Equation (5.43) and use the result to verify your answers to parts (b) and (c).

13.12.

a. Draw a block diagram for the electromechanical system modeled in Example 10.1 by Equations (10.30).

b. Obtain the transfer functions $T_1(s) = \Omega(s)/E_i(s)$ and $T_2(s) = \Omega(s)/\tau_L(s)$ by reducing the block diagram, first with $\tau_L(s) = 0$ and then with $E_i(s) = 0$. Compare your answers with the transfer functions $H_1(s)$ and $H_2(s)$ derived in Example 10.1.

13.13.

a. Draw a block diagram for the thermal system modeled in Example 11.4 by Equation (11.19).

b. Obtain the transfer function $T(s) = \hat{Q}_2(s)/\hat{Q}_i(s)$ by reducing the block diagram. Compare your answer with Equation (11.20).

*13.14.

a. For the block diagram shown in Figure P13.14, find the closed-loop transfer function $T(s) = Y(s)/U(s)$ as a ratio of polynomials.

b. Determine the steady-state response to a unit step-function input in terms of K.

c. Write the damping ratio ζ and the undamped natural frequency ω_n in terms of K. Solve for the value of K for which $\zeta = 1/\sqrt{2}$, and find the corresponding value of ω_n.

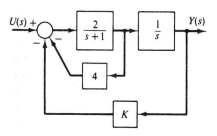

Figure P13.14

13.15.

a. Find the closed-loop transfer function $T(s) = Y(s)/U(s)$ as a ratio of polynomials for the block diagram shown in Figure P13.15.

b. Determine the undamped natural frequency ω_n and the damping ratio ζ of the closed-loop system. Solve for the value of K for which $\omega_n = 5$ rad/s, and find the corresponding value of ζ.

c. Find the steady-state value of the response when the input is a step function of height 5 and when K has the value determined in part (b).

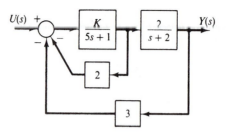

Figure P13.15

*13.16. Consider the feedback system shown in Figure P13.16, which has inputs $U(s)$ and $V(s)$ and output $Y(s)$.

a. Find the transfer function $T_1(s) = Y(s)/U(s)$ as a ratio of polynomials. *Note:* $V(s)$ should be set equal to zero for this calculation.

b. Find the transfer function $T_2(s) = Y(s)/V(s)$ as a ratio of polynomials, taking $U(s) = 0$. *Hint:* Define the output of the right summing junction as $Z(s)$, and write the algebraic transform equations relating $V(s)$, $Z(s)$, and $Y(s)$.

c. Determine the damping ratio ζ and the undamped natural frequency ω_n of the closed-loop system. Determine the value of K for which $\zeta = 1$.

d. Find the steady-state response when $u(t)$ is the unit step function and $v(t) = 0$. Repeat the solution when $u(t) = 0$ and $v(t)$ is the unit step function, and comment on the difference in the results.

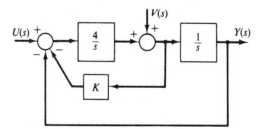

Figure P13.16

13.17.

a. Find the closed-loop transfer function $T(s) = Y(s)/U(s)$ as a ratio of polynomials for the block diagram shown in Figure P13.17.

b. Express the damping ratio ζ and the undamped natural frequency ω_n in terms of K. Show that both poles of $T(s)$ are on the negative real axis for $0 \leq K \leq 1/4$.

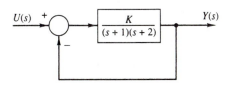

Figure P13.17

*13.18.

a. Find the closed-loop transfer function $T(s) = Y(s)/U(s)$ as a ratio of polynomials for the block diagram shown in Figure P13.18.

b. Express the damping ratio ζ and the undamped natural frequency ω_n in terms of the gains K_2 and K_3. Assume that K_3 is positive but that K_2 can be either positive or negative. Explain why ζ and ω_n are not affected by K_1.

c. For what values of K_2 is the system stable?

d. For what values of K_2 will the zero-input response contain decaying oscillations?

e. Determine the values of K_2 and K_3 such that $\omega_n = 2$ rad/s and $\zeta = 1/2$.

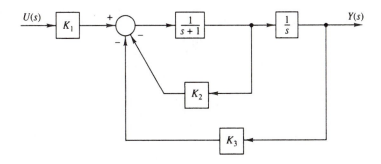

Figure P13.18

13.19.

a. Write a set of state-variable equations for the system represented by the block diagram in Figure P13.19.

b. Determine the closed-loop transfer function $T(s) = \Theta(s)/F_a(s)$ as a ratio of polynomials.

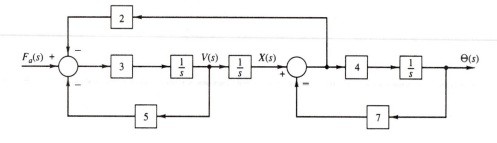

Figure P13.19

*13.20. Find the closed-loop transfer functions $T_1(s) = Y(s)/U(s)$ and $T_2(s) = Z(s)/U(s)$ in terms of the individual transfer functions $A(s), \ldots, E(s)$ for the block diagram shown in Figure P13.20. Give your answer as a ratio of terms that involve only sums, differences, and products of the individual transfer functions.

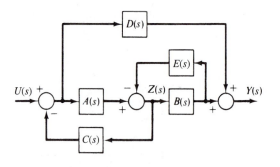

Figure P13.20

13.21. Repeat Problem 13.20 for the block diagram shown in Figure P13.21.

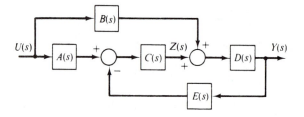

Figure P13.21

14

MODELING, ANALYSIS, AND DESIGN TOOLS

In Chapter 4, we showed how Simulink and MATLAB could be used to simulate dynamic systems. In Chapters 7 and 8, we used the Laplace transform and transfer functions to analyze linear models. For nonlinear systems, we showed in Chapter 9 how to develop a linear approximation that is valid when the variables remain within some region.

For most linear models of third and higher order, and for nearly all nonlinear models, computers are used to obtain solutions for the time responses. They can also carry out other types of analyses for linear models. In this chapter we show how linear models of dynamic systems can be constructed and analyzed using MATLAB and its Control System Toolbox. These tools will be used in Chapter 15 to design feedback controllers for linear systems.

▶ 14.1 BUILDING LINEAR MODELS

With MATLAB and the Control System Toolbox, we can easily construct three different types of models for linear, continuous-time systems. These are the transfer-function, zero-pole-gain, and state-space models. Although the capability exists for building and analyzing these same types of discrete-time linear models, we restrict our attention to continuous-time systems.

For each of the three types we illustrate how the model can be created and converted from one type to another. We also show how to extract important properties of the model, such as getting the transfer function from a state-space model, without having to convert the model from state-space to transfer-function form.

Then we show how to combine the models for subsystems in series or parallel, and will consider the feedback configuration in Section 14.4. Methods for analyzing these models, such as computing and plotting step and impulse responses, are considered in Section 14.2.

Most of the commands that we will use are contained in the Control System Toolbox, which is used in conjunction with MATLAB. This toolbox consists of a set of M-files, all having extension .m, that contain MATLAB commands for implementing the model-building and analytical operations that one commonly encounters when working with control systems. Other toolboxes have been developed for such areas as signal processing and system identification. The interested reader is referred to the Web site of The MathWorks at http://www.mathworks.com for additional information on MATLAB, the Control System Toolbox, and Simulink. MATLAB and its toolboxes have both online and Web-based help facilities that can be used to obtain detailed information about commands while the program is running.

We introduce some of the notation and capabilities of MATLAB, but do not attempt to be comprehensive. All our examples are for continuous-time single-input, single-output (SISO) systems. However, the procedures can readily be extended to include discrete-time systems, as well as systems with multiple inputs and/or multiple outputs.

Transfer-Function Form

The usual form for the transfer function of a single-input, single-output linear model is a ratio of polynomials—that is, a rational function of the variable s. To create the model in MATLAB, we represent the numerator and denominator polynomials by row vectors whose elements are the coefficients of the corresponding powers of s in the polynomials. For example, if the numerator of $G_1(s)$ is the polynomial $N(s) = 3s^2 + 4s + 5$, it can be expressed by entering the command numG1 = [3 4 5], where the symbol numG1 is the name we have assigned to the numerator polynomial.

If the denominator of $G_1(s)$ is the polynomial $D(s) = s^3 + 2s^2 + 4s + 8$, we can give it the name denG1 and define it by entering denG1 = [1 2 4 8]. If we enter these two commands exactly as shown here, and follow them with the command G1 = tf(numG1, denG1), we will have created a linear system with the transfer function

$$G_1(s) = \frac{3s^2 + 4s + 5}{s^3 + 2s^2 + 4s + 8}$$

without any further keystrokes. The computer will automatically respond with the display

```
Transfer function:
  3 s^2 + 4 s + 5
---------------------
s^3 + 2 s^2 + 4 s + 8
```

This model will exist in the MATLAB workspace as a linear time-invariant (LTI) object in transfer-function (TF) form. To obtain a plot of its step response, we need enter only the command step(G1).

Almost any other names can be used for the numerator and denominator vectors, in place of numG1 and denG1. For example, nn and dd will work, as well as numerator_-of_G1 and denominator_of_G1. One should keep in mind that MATLAB is case-sensitive, so numG and numg are two distinct entities. MATLAB will normally display the names and values of the arguments on the screen, which facilitates verifying that they have been entered correctly. This echoing by MATLAB can, however, be suppressed by appending a semicolon to the line. Suppressing the echo is helpful if the response is extensive, for example when there are elements of a large array. Systems can also be created as LTI objects in zero-pole-gain (ZPK) form and in state-space (SS) form, depending on what information is known about the system. Also, LTI systems can be converted from one form to another, and sometimes this conversion is done automatically when two or more systems are being interconnected.

Zero-Pole-Gain Form

We know from Chapter 8 that a transfer function $G(s) = N(s)/D(s)$ can be expressed in terms of a gain constant, its zeros [the solutions of $N(s) = 0$], and its poles [the roots of $D(s) = 0$]. To create a linear time-invariant model for a single-input, single-output linear system in this form, we enter the gain as a scalar and the zeros and poles as column vectors, and then use the zpk command.

For example, to build a MATLAB model for a SISO system whose transfer function has a gain of 5.5, zeros at $s = -3$ and -5, and poles at $s = -1$, $-2 + j3$, and $-2 - j3$, we could enter the lines

```
k = 5.5
zro = [-3; -5]
pol = [-1; -2+3*i; -2-3*i]
G2 = zpk(zro,pol,k)
```

To inform the user that the model had been built in ZPK form, MATLAB will respond with

```
Zero/pole/gain:
    5.5 (s+3) (s+5)
  ---------------------
  (s+1) (s^2 + 4s + 13)
```

Again, different names could have been selected for the input arguments, such as `kk`, `zz`, and `pp` or something more expressive. The symbols `i` and `j` are predefined by MAT-LAB to represent $\sqrt{-1}$, which we have denoted by j throughout this book. The semicolons within the brackets defining `zro` and `pol` indicate to MATLAB that the element that follows is on a new row. Thus the quantities `zro` and `pol` are column vectors. An alternative method of specifying them is to omit the semicolons and transpose the resulting row vector by appending the symbol `'`, as in

```
zro = [-3   -5]'
pol = [-1   -2+3*i   -2-3*i]'
```

In either case, the transfer function defined by the foregoing commands is

$$G_2(s) = \frac{5.5(s+3)(s+5)}{(s+1)(s+2-j3)(s+2+j3)}$$

Note that the numerical values for the zeros and poles can be entered in any order without affecting the result.

State-Space Form

For a fixed linear system, the matrix form of the state-variable model was given in Equations (3.34) as

$$\begin{aligned} \dot{\mathbf{q}} &= \mathbf{Aq} + \mathbf{Bu} \\ \mathbf{y} &= \mathbf{Cq} + \mathbf{Du} \end{aligned} \tag{1}$$

For example, the four matrices corresponding to the third-order system described by

$$\begin{aligned} \dot{q}_1 &= q_2 \\ \dot{q}_2 &= q_3 \\ \dot{q}_3 &= -q_1 - 4q_2 - 2q_3 + 2u \\ y &= q_1 - 3q_2 + 2q_3 + 0.5u \end{aligned}$$

are

$$\mathbf{A} = \begin{bmatrix} 0 & 1 & 0 \\ 0 & 0 & 1 \\ -1 & -4 & -2 \end{bmatrix} \qquad \mathbf{B} = \begin{bmatrix} 0 \\ 0 \\ 2 \end{bmatrix} \qquad \mathbf{C} = \begin{bmatrix} 1 & -3 & 2 \end{bmatrix} \qquad \mathbf{D} = 0.5$$

$$\tag{2}$$

We define this model in MATLAB as the LTI object G3 by entering the lines

```
A = [0 1 0; 0 0 1; -1 -4 -2]
B = [0; 0; 2]
C = [1 -3 2]
D = 0.5
G3 = ss(A,B,C,D)
```

MATLAB will respond with

a =

	x1	x2	x3
x1	0	1	0
x2	0	0	1
x3	-1	-4	-2

b =

	u1
x1	0
x2	0
x3	2

c =

	x1	x2	x3
y1	1	-3	2

d =

	u1
y1	0.5

```
Continuous-time model.
```

Note that the generic lowercase letters a, b, c, and d are used in the display, rather than the names assigned by the user, namely, A, B, C, and D. Also, we see that the rows and columns of the four matrices have been labeled with the corresponding state, input, or output variable. MATLAB uses the symbol x rather than q to represent a state variable.

To analyze a model that exists in SS form within MATLAB, we need only specify the name that has been assigned to the LTI object as the argument of the appropriate command.

Extracting Data and Changing Forms

Once a system model has been created as an LTI object in any one of the three forms (TF, ZPK, or SS), the user can extract useful properties and change from one form to another. For example, if the model G1 exists in TF form, its zeros, poles, and gain can be found by using the command [zz,pp,kk] = zpkdata(G1,'v'). We include the second argument, namely, the character 'v', so that the zeros and poles will be displayed as column vectors, rather than as cell arrays.[1] This feature will work only for SISO systems.

---

[1]Cell arrays are a special class of MATLAB arrays whose elements may themselves contain MATLAB arrays; they allow for the storage of dissimilar objects, such as numbers and strings, in the same array.

Note that the zeros, poles, and gain of $G_1(s)$ can be determined without changing its LTI object from TF to ZPK form.

Likewise, state-space matrices can be found, but an additional step is necessary. To find the state-space matrices for $G_1(s)$, enter the command [a,b,c,d] = ssdata(G1, 'v'). These matrices will be returned as cell arrays, as indicated by the display

```
a =   [3x3 double]
b =   [3x1 double]
c =   [1x3 double]
d = 0
```

where the notation a = [3x3 double] indicates that the matrix a is a 3×3 double precision matrix. To see it in numerical form, namely, as a MATLAB matrix, we can enter a{:,:}, where the use of the braces is required because a is a cell array. The colons tell MATLAB to display all the rows and columns.

Once a model has been defined as an LTI object, it is a simple matter to transform it to either of the other two forms. For example, entering the command G1zpk = zpk(G1) will result in the display

```
Zero/pole/gain:
 3 (s^2 + 1.333s + 1.667)
-----------------------
   (s+2) (s^2 + 4)
```

where the numerator and denominator contain quadratic polynomials in order to avoid having to display the complex zeros and poles as complex numbers. Likewise, the transformation to state-space form can be done via the command G1ss = ss(G1), and the numerical values for the four state-space matrices will be displayed.

The required syntax for the creation of models as LTI objects, the extraction of their key properties, and the conversion from one form to another are shown in Table 14.1. In each case, the names of the quantities to be determined are entered in square brackets on the left side of the equals sign, and the names of the known quantities appear as arguments of the command.

We now illustrate the use of these commands in the following example by entering a system model in zero-pole-gain form, transforming it to the two other forms, and finally returning it to the original form.

TABLE 14.1 Commands for Creating LTI Objects, Extracting Data, and Changing Forms

——— creating LTI objects ———
G = tf(num, den)	Transfer-function form
G = zpk(zeros, poles, gain)	Zero-pole-gain form
G = ss(a, b, c, d)	State-space form

——— extracting data ———
[num,den] = tfdata(G, 'v')	Numerator and denominator
[z, p, k] = zpkdata(G, 'v')	Zeros, poles, and gain
[a,b,c,d] = ssdata(G, 'v')	State-space matrices

— conversion from one form to another —
H = tf(G)	From SS or ZPK form to TF form
H = zpk(G)	From SS or TF form to ZPK form
H = ss(G)	From TF or ZPK form to SS form

▶ **EXAMPLE 14.1**

For a linear system whose transfer function $H(s)$ is

$$H(s) = \frac{4(s+1)(s+4)(s+12)}{s(s+5)(s+3-j6)(s+3+j6)}$$

create an LTI object in ZPK form. Then extract its numerator and denominator polynomials without changing its form. Finally, convert the model to state-space form, extract the zeros, poles, and gain from the SS form, and compare them with those used to create the original ZPK form.

SOLUTION   We begin by entering the parameter values, with the zeros and poles entered as column vectors, via the following commands:

```
z = [-1; -4; -12]
p = [0; -5; -3+6*i; -3-6*i]
k = 4
```

Because there are no semicolons at the ends of the lines, MATLAB echoes the values as

```
z =   -1
      -4
     -12


p =   0
     -5.0000
     -3.0000 + 6.0000i
     -3.0000 - 6.0000i


k =   4
```

To create the LTI object, we enter H = zpk(z,p,k), and MATLAB responds with $H(s)$ as a built-up fraction, with a quadratic polynomial for the two complex poles. The result is

```
Zero/pole/gain:
 4 (s+1) (s+4) (s+12)
-----------------------
s (s+5) (s^2 + 6s + 45)
```

To obtain the coefficients of the numerator and denominator polynomials of $H(s)$, we enter the command

```
[num,den] = tfdata(H, 'v')
```

which results in the output

```
num =     0     4    68   256   192
den =     1    11    75   225     0
```

From these two row vectors of polynomial coefficients, we can write the transfer function as

$$H(s) = \frac{4s^3 + 68s^2 + 256s + 192}{s^4 + 11s^3 + 75s^2 + 225s}$$

To obtain the state-space form of the model as the LTI object HH, as distinct from just extracting its state-space matrices, we enter the line HH = ss(H), which yields the following four matrices:

```
A =            -6    -5.625    0.46211     1.3863
                8        0          0          0
                0        0         -5          0
                0        0          4          0

B =             0
                0
           6.9282
                0

C =       -1.2494    -6.4031    0.57735     1.7321

D =        0
```

We see that A has four rows and four columns, corresponding to four state variables. This also implies that B will have four rows and that C will have four columns. Because there is one input, B and D have one column, and because there is one output, C and D have one row.

As the final step in this example, we extract the zeros, poles, and gain of the state-space model. To distinguish their values from the original set, we refer to them as zz, pp, and kk. To do this, we enter the command

```
[zz,pp,kk] = zpkdata(HH,'v')
```

and the following results are obtained:

```
zz =   -12.0000
        -4.0000
        -1.0000

pp =    -3.0000 + 6.0000i
        -3.0000 - 6.0000i
              0
        -5.0000

kk =     4.0000
```

These values agree with the original values of z, p, and k, although the ordering of the zeros and poles differs. The fact that there are only three zeros in the list indicates that the order of the numerator polynomial is one less than that of the denominator. Hence $H(s)$ approaches zero as s approaches infinity.

Series Connection

Figure 14.1(a) shows two single-input, single-output subsystems that have been defined in transfer-function form and that are to be connected in series, resulting in the equivalent system shown in part (b) of the figure. We know from Chapter 8 that the transfer function

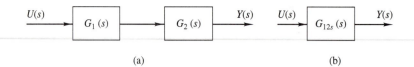

Figure 14.1 (a) Series connection of two subsystems. (b) Equivalent system.

of the resulting system can be written as the product of the individual transfer functions. For the general case we write the individual transfer functions in reverse order,[2] namely

$$G_{12s}(s) = G_2(s)G_1(s)$$

where we use the extra subscript s to denote the series combination. Provided that $G_1(s)$ and $G_2(s)$ exist in the MATLAB workspace as the LTI objects G1 and G2, their series combination G12s can be created as an LTI object with the command G12s = G2*G1. The symbol * is an **overloaded operator** that produces the series combination, because it is being used with the two LTI objects G1 and G2, rather than with two MATLAB variables that represent scalars, vectors, or matrices. We assume that the outputs of the first subsystem are the same as the inputs of the second.

▶ **EXAMPLE 14.2**

Using MATLAB, create as LTI objects the subsystems that have the transfer functions

$$G_1(s) = \frac{2(s+1)}{s(s+4)}$$

and

$$G_2(s) = \frac{3}{s^2 + 2s + 5}$$

and connect them in series to form a new system that has the transfer function

$$G_{12s}(s) = G_2(s)G_1(s)$$

Then obtain the zeros, poles, gain, and also the state-space matrices of $G_{12s}(s)$.

SOLUTION   Because we know the zeros, poles, and gain of $G_1(s)$, we enter them with the commands kG1 = 2, zerG1 = -1, and polG1 = [0; -4] and create an LTI object for $G_1(s)$ in ZPK form with the command

G1 = zpk(zerG1,polG1,kG1)

Because $G_2(s)$ is given in polynomial form, we enter the numerator and denominator polynomials as

numG2 = 3
denG2 = [1 2 5]

We now create an LTI object for $G_2(s)$ with the command

G2 = tf(numG2, denG2)

---

[2]If we restrict our attention to single-input, single-output (SISO) systems, the transfer function can also be written as $G_{12s}(s) = G_1(s)G_2(s)$, but this expression is not valid otherwise.

As an aside, if we wanted to know the poles of $G_2(s)$, we could extract the zeros, poles, and gain of G2 by entering

```
[zerG2,polG2,kG2] = zpkdata(G2, 'v')
```

Doing so yields

```
    zerG2 =   Empty matrix: 0-by-1

    polG2 =   -1.0000 + 2.0000i
              -1.0000 - 2.0000i

     kG2 =    3
```

We see that $G_2(s)$ has a pair of complex poles at $s = -1 + j2$ and $s = -1 - j2$ and a gain of 3. Because $G_2(s)$ has no zeros, the display shows zerG2 = Empty matrix: 0-by-1. This means that the symbol zerG2 has been defined but at present contains no values. Because the numerator of $G_2(s)$ is just the constant 3, the transfer function has no zeros in the finite s-plane, i.e., the set of finite zeros of $G_2(s)$ is empty.

Now that both subsystems have been defined as LTI objects, we make the series connection by giving the command G12s = G2*G1, which results in

```
Zero/pole/gain:
       6 (s+1)
---------------------
s (s+4) (s^2 + 2s + 5)
```

This indicates that G12s has been built as an LTI object in ZPK form, although G1 exists in ZPK form, and G2 exists in TF form. There is an order of precedence for determining the form of an LTI object that is the combination of two or more objects of different forms. This rule is based on always ending up with the form that has the best numerical properties of the constituent forms. For reasons that are beyond the scope of this book, the state-space form has preference over the other two, and the zero-pole-gain form has precedence over the transfer-function form. In this example, the fact that G1 is in ZPK form causes the series combination G12s to also be in ZPK form. The user can verify that this is the case by entering G12s as a command and observing that the transfer function is displayed in ZPK form, as shown above.

This is MATLAB's representation of the rational function

$$G_{12s}(s) = \frac{6s + 6}{s^4 + 6s^3 + 13s^2 + 20s}$$

To verify that the poles and zeros of the series combination are the union of those of the two subsystems, we enter the command

```
[zer_G12s,pol_G12s,k_G12s] = zpkdata(G12s, 'v')
```

and obtain the result

```
zer_G12s =   -1

pol_G12s =    0
             -4.0000
             -1.0000 + 2.0000i
             -1.0000 - 2.0000i

k_G12s =      6
```

which agrees with the form

$$G_{12s}(s) = \frac{6(s+1)}{s(s+4)(s^2+2s+5)}$$

where the complex poles correspond to the quadratic term in the denominator.

To obtain the state-space matrices for the series connection from G12s, we enter

```
[A_G12s,B_G12s,C_G12s,D_G12s] = ssdata(G12s,'v')
```

and obtain a display that says that A_G12s, B_G12s, and C_G12s are cell arrays, and D_G12s is the scalar 0. Recall that to determine the numerical values of the three cell arrays, we can enter A_G12s{:,:}, B_G12s{:,:}, and C_G12s{:,:}, getting

```
A_G12s = -2.0000   -2.5000          0    1.0000
          2.0000         0          0         0
               0         0    -4.0000         0
               0         0     2.0000         0

B_G12s =        0
                0
           2.4495
                0

C_G12s =   1.2247    0.6124          0         0

D_G12s =        0
```

From the dimensions of these matrices, we can see that the resulting system has four state variables, one input, and one output.

Parallel Connection

Figure 14.2(a) shows two single-input, single-output subsystems that have been defined in transfer-function form and that are to be connected in parallel, resulting in the equivalent system shown in part (b) of the figure. We know from Chapter 8 that the transfer function of the resulting system can be written as the sum of the individual transfer functions, namely

$$G_{12p}(s) = G_2(s) + G_1(s)$$

Provided that $G_1(s)$ and $G_2(s)$ exist in the MATLAB workspace as the LTI objects G1 and G2, their parallel combination G12p can be created as an LTI object with the command

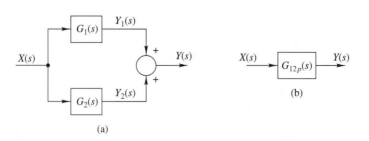

(a)

Figure 14.2 (a) Parallel connection of two subsystems. (b) Equivalent system.

G12p = G2+G1 or G12p = G1+G2, where the symbol + is an **overloaded operator** that produces the parallel combination, because it is being used with the two LTI objects G1 and G2, rather than with two MATLAB variables that represent scalars, vectors, or matrices. As indicated in part (a) of Figure 14.2, each of the subsystems has the same input $U(s)$, and their respective outputs $Y_1(s)$ and $Y_2(s)$ are added together to form the single output $Y(s)$ of the combined system. Should we require that the output $Y_1(s)$ be summed with a negative sign, we could use G12p = G2 - G1.

▷ **EXAMPLE 14.3**

If the same two subsystems $G_1(s)$ and $G_2(s)$ that are defined in Example 14.2 are connected in parallel, use MATLAB to form a new system that has the transfer function

$$G_{12p}(s) = G_2(s) + G_1(s)$$

Then obtain the zeros, poles, gain, and also the state-space matrices of $G_{12p}(s)$.

SOLUTION   We use the same commands as in Example 14.2 to construct the two subsystems as LTI objects in the MATLAB workspace. Having done this, we make the parallel connection by giving the command G12p = G2+G1, which results in

```
Zero/pole/gain:
2 (s+0.4469) (s^2 + 4.053s + 11.19)
-----------------------------------
      s (s+4) (s^2 + 2s + 5)
```

which indicates that G12p has been built as an LTI object in ZPK form. As in the previous example, the fact that G1 is in ZPK form causes the parallel combination G12p to also be in ZPK form. The user can verify that this is the case by entering G12p as a command and observing that the transfer function is displayed in ZPK form, as shown above.

This is MATLAB's representation of the rational function

$$G_{12p}(s) = \frac{2(s+0.4469)(s^2+4.053s+11.19)}{s^4+6s^3+13s^2+20s}$$

To verify that the poles of the parallel combination are the union of those of the two subsystems, but that its zeros are different from those of either subsystem, we enter the command

```
[zer_G12p,pol_G12p,k_G12p] = zpkdata(G12p, 'v')
```

and obtain the result

```
zer_G12p =   -2.0266 + 2.6612i
             -2.0266 - 2.6612i
             -0.4469

pol_G12p =        0
             -4.0000
             -1.0000 + 2.0000i
             -1.0000 - 2.0000i

k_G12p =     2.0000
```

which agrees with the form given above, where the complex zeros and poles correspond to the quadratic terms in the numerator and denominator, respectively.

To obtain the state-space matrices for the parallel connection from G12p, we enter

```
[A_G12p,B_G12p,C_G12p,D_G12p] = ssdata(G12p,'v')
```

and obtain a display that says that A_G12p, B_G12p, and C_G12p are cell arrays, and D_G12p is the scalar 0. Recall that to determine the numerical values of the three cell arrays, we can enter A_G12p{:,:}, B_G12p{:,:}, and C_G12p{:,:}, getting

```
A_G12p =   -2.0000    -2.5000     1.2439     0.5559
            2.0000          0          0          0
                 0          0    -4.0000          0
                 0          0     1.0000          0


B_G12p =          0
                  0
             2.8284
                  0


C_G12p =     1.1672     1.7591     0.7071     0.3160

D_G12p =          0
```

From the dimensions of these matrices, we can see that the resulting system has four state variables, one input, and one output.

▷ 14.2 TIME-DOMAIN ANALYSIS

Having illustrated how the model of a fixed linear system can be expressed in a variety of forms, we now wish to use MATLAB to analyze such models. Complete coverage of MATLAB's varied capabilities is well beyond our scope here, but a number of the most basic and most commonly used commands will be presented. We show how to compute and plot the responses to step functions and impulses, and then the responses to arbitrary inputs.

The commands we discuss can be used either with or without arguments on the left side of an equals sign. When one or more arguments are present, as in the command [y,x,t] = step(G), MATLAB supplies the numerical values of the variables listed inside the square brackets to the left of the equals sign. If a plot is to be drawn, such as y versus t, the user must enter the appropriate plotting command(s). Alternatively, when a command is given without the equals sign and any arguments to the left of it, as in step(G), MATLAB generates the plot automatically but does not supply any numerical values.

Each of the commands we illustrate can be done with the model expressed as an LTI object in any of the three forms: transfer-function (TF), zero-pole-gain (ZPK), or state-space (SS). For the most part, we will use the form in which the model is given. For systems of relatively low order, say 10 or less, the choice is not critical. However, for systems of higher order, the state-space form is preferred, because it is less susceptible to numerical errors than the transfer-function and zero-pole-gain forms.

Step and Impulse Responses

There are specific commands for computing the step and impulse responses, and they will yield the values of both the outputs and the state variables, if desired. The user can have the time range and number of points be chosen automatically or can enter a vector of uniformly spaced time points as an input argument.

A more general command called lsim exists for obtaining the responses to rather arbitrary inputs; it will be demonstrated shortly. In the following example, we illustrate the use of the step and impulse commands.

▶ **EXAMPLE 14.4**

Compute and plot the unit step and impulse responses for the fourth-order system whose transfer function is

$$G(s) = \frac{2(s^3 + 3s^2 + 7s + 5)}{s^4 + 6s^3 + 13s^2 + 26s + 6} \tag{3}$$

Let MATLAB and the Control System Toolbox determine an appropriate time interval and step size for the plots.

SOLUTION   Because we are going to let the time variable for the plot be determined automatically, we need only build the system as an LTI object and issue the step and impulse commands, as follows:

```
num = 2*[1   3   7   5]      % numerator of G(s)
den = [1   6   13   26   6]  % denominator of G(s)
G = tf(num,den)              % build G(s) as LTI in transfer-
                             % function form
step(G)                      % compute and plot step response
impulse(G)                   % compute and plot impulse response
```

The resulting plots appear in Figure 14.3. The titles and the labels for the axes are produced automatically by MATLAB, although it is possible for the user to change them.

Responses to Arbitrary Inputs and Initial Conditions

In addition to finding the step and impulse responses of a linear model, MATLAB can also compute the responses to arbitrary inputs, both with and without initial conditions on the state variables. The command that accomplishes this is called lsim, which stands for *linear simulation*.

The model can be an LTI object in any of the usual three forms, but the user must define both the time vector and the input. For single-input models, the input variable must be a column vector that has one element per time point. For multi-input models, it must be a matrix that has one column for each of the model's inputs. The time vector is defined as for the step and impulse responses—that is, as a row or column vector of uniformly spaced time points. If the initial state is not zero, the state-space form of the model must be used.

▶ **EXAMPLE 14.5**

Obtain the zero-state response of the fourth-order system used in Example 14.4 over the interval $0 \le t \le 50$ seconds. The input is the signal shown in Figure 14.4, where $A = 1, t_1 = 1$ s, $t_2 = 16$ s, and $t_3 = 31$ s.

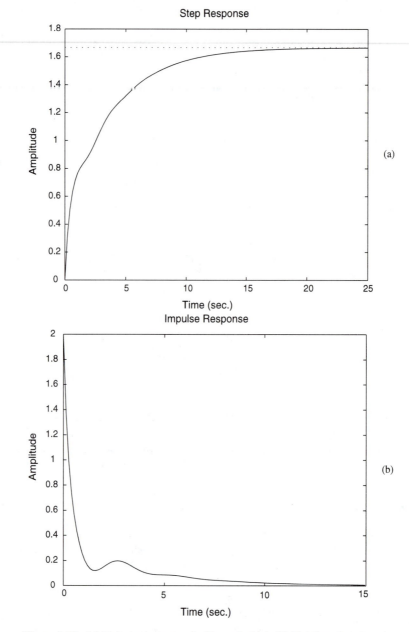

Figure 14.3 (a) Unit step response for Example 14.4. (b) Unit impulse response.

SOLUTION The system G is built as an LTI object in TF form, exactly as in the previous example. To create a column vector for the time variable, with the points separated by 0.1 s, we enter

```
time = [0:0.1:50]';
```

which will give us 501 time points. The apostrophe after the closing bracket denotes the transpose of the matrix.

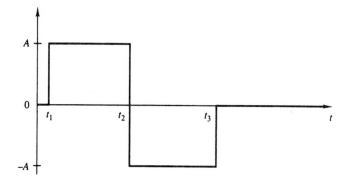

Figure 14.4 Input signal for Example 14.5.

The length of the input vector must always be the same as the length of the time vector. We shall first initialize the input vector in to be a column of 501 zeros. Then we shall change the input values to 1 for $1 < t \leq 16$ and to -1 for $16 < t \leq 31$, as required by Figure 14.4.

```
in = 0*time;
for j=11:160, in(j) = 1.0;end;
for j=161:310, in(j) = -1.0;end;
```

Because the initial states are zero, we are ready to compute the response of the output y and plot it, along with the input, via the commands

```
y = lsim(G,in,time);
plot(time,y,time,in,'-.');grid
```

The resulting plot is shown in Figure 14.5. The labels for the input and output curves have been added by hand using the `gtext` command.

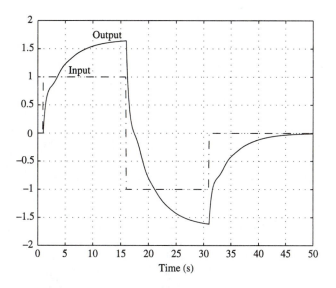

Figure 14.5 Response to pulse input for Example 14.5.

As a check on the nature of the output plot, we can compute several important properties of the system directly from the LTI object. We can use the command damp(G) to determine that the pair of complex poles has a damping ratio of $\zeta = 0.310$ and an undamped natural frequency of $\omega_n = 2.31$ rad/s. The steady-state value of the unit step response is $G(0)$, which can be computed via the command dcgain(G). The result is $G(0) = 1.6667$, which is consistent with the response shown in Figure 14.5.

▶ **EXAMPLE 14.6**

Use the lsim command to compute and plot the zero-input response for the third-order system whose state-space matrices are

$$\mathbf{A} = \begin{bmatrix} 0 & 1 & 0 \\ 0 & 0 & 1 \\ -1 & -4 & -2 \end{bmatrix} \qquad \mathbf{B} = \begin{bmatrix} 0 \\ 0 \\ 2 \end{bmatrix} \qquad \mathbf{C} = \begin{bmatrix} 1 & -3 & 2 \end{bmatrix} \qquad \mathbf{D} = 0.5$$

Take the initial state vector as

$$\mathbf{x}(0) = \begin{bmatrix} 1.0 \\ 2.0 \\ 3.0 \end{bmatrix}$$

SOLUTION　We define this model in MATLAB as the LTI object G by entering the lines

```
A = [0 1 0; 0 0 1; -1 -4 -2]
B = [0; 0; 2]
C = [1 -3 2]
D = 0.5
G = ss(A,B,C,D)
```

The time, input, and initial state vectors can be specified by entering the following commands:

```
time = [0:0.1:15]';
input = 0*time;
x0 = [1; 2; 3]
```

where the input has been defined to be a vector of zeros, having the same dimensions as the time vector. We can compute the zero-input response y by issuing the lsim command with four arguments, namely, y = lsim(G,input,time,x0).

The plot that results is shown in Figure 14.6, which shows transients that decay to zero, as one would expect for a stable system. If we wish to obtain the complete response of a linear system for which the input and the initial conditions are both nonzero, the lsim command should be used with the model in state-space form with the input and initial state vectors defined appropriately.

▶ **14.3 FREQUENCY-DOMAIN ANALYSIS**

A designer often characterizes parts of a feedback control system by their sinusoidal-steady-state responses. The basic ideas were presented in Section 8.4 and can be summarized by Equations (8.64), (8.69), and (8.71). We shall start this section with these same

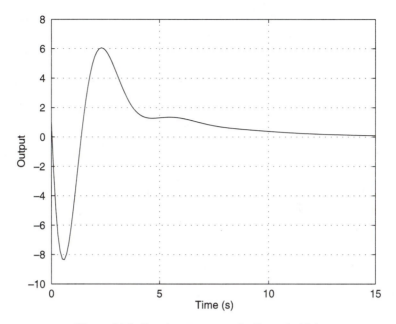

Figure 14.6 Zero-input response for Example 14.6.

equations. As in Chapter 13, however, we shall use $T(s)$ rather than $H(s)$ for a general transfer function in order to avoid confusion with the use of $H(s)$ for the feedback transfer function in feedback systems. To obtain an expression for the frequency-response function $T(j\omega)$, we replace s by $j\omega$ in $T(s)$ and then write the resulting complex quantity in polar form as

$$T(j\omega) = M(\omega)\epsilon^{j\theta(\omega)} \tag{4}$$

where $M(\omega)$ is the magnitude of $T(j\omega)$ and $\theta(\omega)$ is its angle. If the input of a stable system is the sinusoidal function

$$u(t) = \sin\omega t \tag{5}$$

then the steady-state response is

$$y_{ss}(t) = M\sin(\omega t + \theta) \tag{6}$$

More generally, the steady-state response to the sinusoidal input $u(t) = B\sin(\omega t + \phi_1)$ is

$$y_{ss}(t) = BM\sin(\omega t + \phi_1 + \theta)$$

If we draw curves of $M(\omega)$ and $\theta(\omega)$ versus ω, we can see how the magnitude and angle of the steady-state response change as the frequency of the input is changed, as was illustrated in Section 8.4. Later in the present section, we show how to create these curves using MATLAB. In many cases, however, it is more useful to express the magnitude of $T(j\omega)$ in decibels, to use semilog paper, and to present the two curves in the form of a Bode diagram.

Bode diagrams have several advantages. A wide range of values of $M(\omega)$ and of ω can be included (although the point corresponding to $\omega = 0$ can never be shown because of the logarithmic frequency scale). A number of rules can be developed to enable a designer quickly to sketch reasonable approximations to the Bode diagrams, although we shall emphasize computer-generated plots. Conversely, the transfer function $T(s)$ can be approximated from Bode diagrams that have been constructed from experimental measurements. For subsystems in series, the corresponding Bode diagrams can be added to obtain the diagram for the combination. For feedback systems, the diagrams for the open-loop transfer functions provide important information about the stability of the overall systems.

To express a real, positive, dimensionless quantity in decibels (usually abbreviated dB), we take its logarithm (to the base 10) and then multiply by 20. Thus

$$M(\omega)|_{dB} = 20 \log_{10} M(\omega) \tag{7}$$

Note that the decibel gain $M(\omega)|_{dB}$ is positive when $M(\omega) > 1$ and negative when $M(\omega) < 1$. When the magnitude of the transfer function is 1, 10, 100, and 1000, respectively, the corresponding decibel gain is 0, 20, 40, and 60.

We begin the detailed discussion by examining two simple diagrams produced by MATLAB. The commands used to obtain such diagrams are explained later in this section.

▶ EXAMPLE 14.7

Figure 14.7 shows the Bode diagram for $T(s) = 1/(1 + \tau s)$, which has the frequency-response function

$$T(j\omega) = \frac{1}{1 + j\omega\tau} \tag{8}$$

The dashed lines have been added by hand in order to emphasize the high-frequency asymptotes. Explain the low- and high-frequency asymptotes and calculate for both curves the values at $\omega = 1/\tau$.

SOLUTION　For very small values of ω, $T(j\omega)$ approaches 1, so the magnitude curve approaches a constant value of 0 dB, and the angle curve approaches 0°. At very high

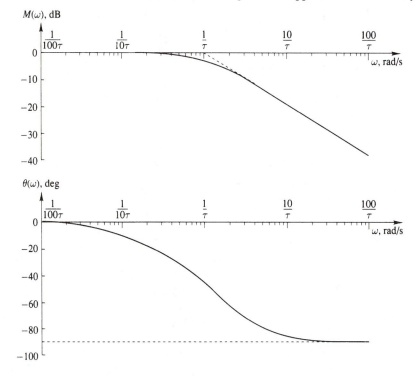

Figure 14.7　Bode diagram for $T(s) = 1/(1 + \tau s)$.

frequencies, where $\omega\tau \gg 1$, $T(j\omega)$ approaches $1/(j\omega\tau)$. Thus the curve of $\theta(\omega)$ must approach $-90°$ for large values of ω.

For the high-frequency asymptote to the magnitude curve, where $T(j\omega) = 1/(j\omega\tau)$, we write $M(\omega)|_{dB} = 20\log_{10}(1/\omega\tau) = -20\log_{10}(\omega\tau) = -20\log_{10}\tau - 20\log_{10}\omega$, which describes a straight line when plotted versus $\log_{10}\omega$. The slope of the line is usually expressed in units of decibels per decade or decibels per octave. A decade corresponds to increasing the frequency by a multiplying factor of 10, an octave to doubling the frequency. For this asymptote, when the frequency is increased from ω_a to $10\omega_a$ the resulting magnitude change is

$$20\log_{10}(1/10\omega_a\tau) - 20\log_{10}(1/\omega_a\tau) = -20\log_{10}10 = -20 \text{ dB}$$

If we double the frequency, the magnitude change is

$$20\log_{10}(1/2\omega_a\tau) - 20\log_{10}(1/\omega_a\tau) = -20\log_{10}2 \simeq -6 \text{ dB}$$

Thus the slope of the line in Figure 14.7 is -20 dB per decade and -6 dB per octave.

We see from (8) that $T(j\omega)$ becomes $1/(1 + j1)$ when $\omega = 1/\tau$. At this frequency, $M(\omega)|_{dB}$ becomes $20\log_{10}(1/\sqrt{2}) \simeq -3$ dB and $\theta = -45°$.

Notice that when the low- and high-frequency asymptotes to the magnitude curve are extended, they meet at $\omega = 1/\tau$, which is sometimes called the **corner frequency** or **break frequency**. Together, these asymptotes might serve as a rough approximation to the exact curve, although there would be a sizeable error near the break frequency. The references in Appendix F contain guidelines for drawing approximate curves by hand, although we shall rely on computer-generated plots.

The results of the last example can be generalized to help us draw low- and high-frequency asymptotes for more complicated functions. For very small or very large values of ω, frequency-response functions usually reduce to one of three cases: $T(j\omega)$ approaches K, $K/(j\omega)^n$, or $K(j\omega)^n$, where n is a positive integer.

When $T(j\omega)$ is a positive constant, the magnitude curve becomes a horizontal line, and $\theta(\omega)$ is zero. When $T(j\omega) = K/(j\omega)$, as for the high-frequency asymptote in the last example, $M(\omega)|_{dB}$ becomes a straight line with a slope of -20 dB per decade. Using arguments similar to those in the last example, we can show that for $T(j\omega) = j\omega K$, the magnitude curve has a slope of $+20$ dB per decade. More generally, the slope of the magnitude curve is $-20n$ dB per decade for $T(j\omega) = K/(j\omega)^n$ and is $+20n$ dB per decade for $T(j\omega) = (j\omega)^n K$.

In addition to knowing the slope, we need to locate one point on the magnitude asymptote when $T(j\omega)$ is $K/(j\omega)^n$ or $K(j\omega)^n$. One way to do so is to note that when $\omega = 1$, $M(\omega)|_{dB} = 20\log_{10}K$. Another easy way is to note that the line (extended if necessary) crosses the zero-dB axis when $|T(j\omega)| = 1$. For $T(j\omega) = K/(j\omega)^n$, this occurs when $\omega = \sqrt[n]{K}$; for $T(j\omega) = (j\omega)^n K$, it occurs when $\omega = \sqrt[n]{1/K}$.

▶ **EXAMPLE 14.8**

Figure 14.8 shows the Bode diagram for $T(s) = \omega_n^2/(s^2 + 2\zeta\omega_n s + \omega_n^2)$, for which

$$T(j\omega) = \frac{\omega_n^2}{-\omega^2 + j(2\zeta\omega_n)\omega + \omega_n^2} = \frac{1}{1 + j2\zeta(\omega/\omega_n) - (\omega/\omega_n)^2} \qquad (9)$$

The magnitude and angle curves are plotted when the damping ratio $\zeta = 1$, 0.5, and 0.1. Note that in order to make the curves more generally applicable, we have made the abscissa the normalized frequency ω/ω_n. Explain the low- and high-frequency asymptotes and calculate for both sets of curves the values when $\omega/\omega_n = 1$.

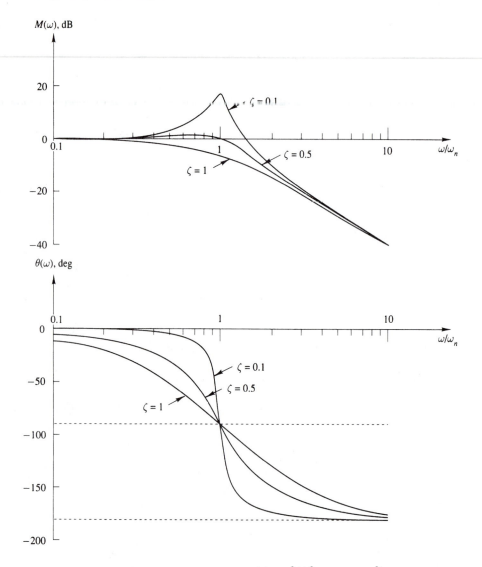

Figure 14.8 Bode diagram for $T(s) = \omega_n^2/(s^2 + 2\zeta\omega_n s + \omega_n^2)$.

SOLUTION From (9), we see that for very small values of ω, $T(j\omega)$ approaches 1. Thus the low-frequency asymptotes are 0 dB and 0°. For large values of ω, $T(j\omega)$ approaches $-(\omega_n/\omega)^2$. Thus the high-frequency magnitude asymptote must have a slope of -40 dB per decade and must cross the zero-dB axis at $\omega = \omega_n$. The high-frequency asymptote for the angle is $-180°$.

We could try to approximate the magnitude curve by using only the low- and high-frequency asymptotes. If we do this, however, very large errors could occur near the break frequency, especially if ζ is very small. Replacing ω by ω_n in (9), we see that $T(j\omega_n) = 1/(j2\zeta)$. Thus $\theta(\omega_n) = -90°$ for all values of ζ, while

$$M(\omega_n)|_{dB} = 20\log_{10}\frac{1}{2\zeta}$$

which gives 13.98 dB when $\zeta = 0.1$, 0 dB when $\zeta = 0.5$, and -6.021 dB when $\zeta = 1$.

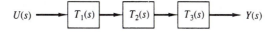

Figure 14.9 Blocks in series corresponding to (10).

Frequently it is desirable to express a transfer function as the product of several factors. For example, let

$$T(s) = T_1(s)T_2(s)T_3(s) \tag{10}$$

in which case

$$T(j\omega) = T_1(j\omega)T_2(j\omega)T_3(j\omega) \tag{11}$$

Using the appropriate subscripts, we let $M(\omega)$ and $\theta(\omega)$ denote the magnitude and angle of the individual frequency-response functions in the usual way. By the rules for complex numbers, we can write, for the overall function $T(j\omega)$,

$$M(\omega) = M_1(\omega)M_2(\omega)M_3(\omega) \tag{12a}$$

$$\theta(\omega) = \theta_1(\omega) + \theta_2(\omega) + \theta_3(\omega) \tag{12b}$$

Furthermore, $M(\omega)|_{dB} = 20\log_{10}[M_1(\omega)M_2(\omega)M_3(\omega)]$, so

$$M(\omega)|_{dB} = M_1(\omega)|_{dB} + M_2(\omega)|_{dB} + M_3(\omega)|_{dB} \tag{13}$$

Thus the individual magnitude curves can be added, as can the individual angle curves.

One application of (12b) and (13) occurs when a block diagram contains several blocks in series. The overall transfer function for the partial diagram in Figure 14.9 is given by (10). To construct the Bode diagram for the overall frequency-response function, we merely add the magnitude and angle curves for the individual blocks.

As will be discussed in Chapter 15, it is sometimes necessary to insert additional components in series with the original open-loop system in order to meet the design specifications. Using (12b) and (13) enables us to determine, for these additional components, the magnitude and angle characteristics needed in order to allow the complete Bode diagram to be appropriately modified. As a simple example, adding a gain block will not affect the angle curve, but will raise the $M(\omega)|_{dB}$ curve by a constant amount (if the gain is greater than one) or lower it (if the gain is less than one).

Using MATLAB to Construct Bode Diagrams

MATLAB's Control System Toolbox has a `bode` command that can produce magnitude and phase angle curves for models in either state-space or transfer-function form. The user has the option of either specifying the frequency values for which the calculations are to be done or letting MATLAB make the selection on the basis of the poles and zeros of the system's transfer function.

If the frequency values are to be specified, the user must first create a row vector of frequency values, expressed in radians per unit of time (typically, seconds). Because we use the logarithm of the frequency as the independent variable, it is customary to have the elements of the frequency list spaced logarithmically so that adjacent values in the list have the same ratio rather than the same difference. This is easily done in MATLAB by means of the `logspace` command, for which the first two arguments are the beginning and ending frequencies, expressed as powers of 10. The optional third argument is the number of points in the list, 50 being the default value.

For each of the frequencies in the list, the `bode` command calculates the magnitude and phase angle of the frequency response, expressed as a magnitude ratio and in degrees, respectively. To obtain a plot with the magnitude expressed in decibels, we must convert the magnitude ratios to decibels by taking $20 \log_{10}$ of each element. This is done by operating on the entire vector of magnitude ratios, as illustrated in Example 14.9. For plotting with a logarithmic horizontal scale and a linear vertical scale, MATLAB has the command `semilogx`, which is used in just the same way as the `plot` command.

▶ **EXAMPLE 14.9**

Draw plots of the frequency response magnitude (in decibels) and phase angle (in degrees) when $T(s)$ is the transfer function given in (3) for the LTI system used in Examples 14.4 and 14.5.

SOLUTION   The steps to be taken are: (1) build $T(s)$ in TF form, (2) create the list of logarithmically-spaced frequencies, (3) compute the magnitude ratio and the phase angle at each frequency value as three-dimensional arrays, (4) convert the magnitude ratios to a two-dimensional array of decibel values, (5) convert the array of phase-angle values to a two-dimensional array, and (6) construct the magnitude and phase plots versus frequency. The commands that will accomplish these steps, along with some comments (on lines beginning with percent signs), follow.

```
% (1)   ---- build T in TF form -------------
num = 2*[1   3   7   5]
den = [1   6   13   26   6]
T = tf(num,den)
% (2) ---- 100 log-spaced freq values from 0.01 to 100 rad/s
freq = logspace(-2,2,100);
nn = length(freq);
% (3) ---- compute magnitude ratio and phase as 3-dimensional
% arrays ---
[mag_ratio,ph] = bode(T,freq);
% (4) --- convert the magnitude ratios to a 2D array of dB
% values ----
temp = 20*log10(mag_ratio);
mag_db = reshape(temp,[1 nn]);
% (5) ---- convert 3D array of phase values to a 2D array
phase = reshape(ph,[1 nn]);
% (6A) ---- plot magnitude in dB vs logarithmic frequency
subplot(211)
semilogx(freq,mag_db); grid
xlabel('Frequency (rad/s)'); ylabel('Magnitude (dB)')
% (6B) ---- plot phase angle versus logarithmic frequency
subplot(212)
semilogx(freq,phase);grid
xlabel('Frequency (rad/s)'); ylabel('Phase (deg)')
```

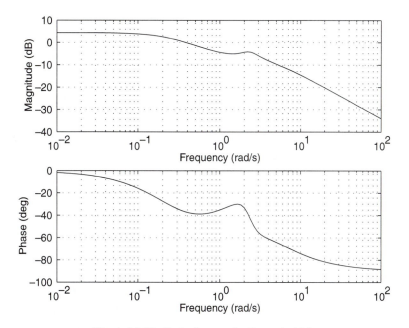

Figure 14.10 Bode diagram for Example 14.9.

Because the outputs of the `bode` command are cell arrays, we use the `reshape` command to obtain `mag_db` and `phase` as row vectors for plotting. The desired Bode diagram is shown in Figure 14.10. From the upper part of Figure 14.10, we see that for low frequencies, $|T(j\omega)|$ is asymptotic to a value of just less than 5 dB. The actual value is $4.437 \text{ dB} = 20\log_{10}(10/6)$, which can be obtained by taking the limit as $\omega \to 0$ of $T(s)$ as given by (3). The lower part of Figure 14.10 shows that for low frequencies the phase is asymptotic to zero. As ω increases, the magnitude of the frequency response remains flat until about $\omega = 0.1$ rad/s and then begins to fall off, with a slight peak near $\omega = 2$ rad/s. It continues to decrease at a slope of -20 dB/decade as the frequency increases. The phase plot exhibits a similar behavior as the frequency increases but becomes asymptotic to $-90°$ as $\omega \to \infty$.

Using MATLAB to Construct Linear Plots

In the discussion of frequency response in Section 8.4, we showed plots of the magnitude ratio $M(\omega)$ and the phase angle $\theta(\omega)$ in radians, plotted versus a linear frequency scale. In the following example, we illustrate the MATLAB commands that will produce frequency-response plots in this alternate form.

▶ **EXAMPLE 14.10**

In Example 8.18, we showed that the transfer function for the notch filter in Figure 8.19 is

$$T(s) = \frac{6s^2 + 1}{6s^2 + 6s + 1}$$

The magnitude $M(\omega)$ was plotted versus linear frequency in Figure 8.20. Use MATLAB to verify this curve and also plot the corresponding phase angle $\theta(\omega)$.

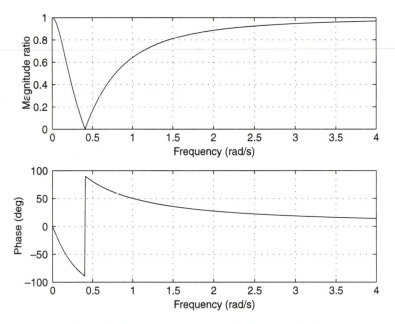

Figure 14.11 Frequency response for Example 14.10.

SOLUTION The following commands will produce the specified plots of magnitude and phase angle versus linear frequency.

```
num = [6 0 1];
den = [6 6 1];
H = tf(num,den)
freq = [0:0.01:4];
nn = length(freq);
[mag_ratio,ph] = bode(H,freq);
mag = reshape(mag_ratio,[1 nn]);
phase = reshape(ph,[1 nn]);
subplot(211)
plot(freq,mag);grid
subplot(212)
plot(freq,phase);grid
```

The plot that results is shown in Figure 14.11. In Example 8.18, we saw that $M(\omega) = 0$ at $\omega = 1/\sqrt{6} = 0.4082$ rad/s, as can be verified by replacing s by $j(1/\sqrt{6})$ in the expression for $T(s)$. We can use MATLAB's find command to determine the frequency at which the minimum of $M(\omega)$ occurs by entering the following:

```
ii=find(mag==min(mag))
freq(ii)
```

The first command returns the index ii corresponding to the minimum magnitude. Then the second command displays the frequency corresponding to the minimum magnitude: 0.4100 rad/s, which is very close to the analytical value of $1/\sqrt{6}$. For the frequency vector used, which has 401 points between 0 and 4.0 rad/s, the minimum magnitude turns out to be 0.0035, whereas the analytical value is 0. Should we wish to know this minimum value and the frequency at which it occurs more accurately, we might select the frequency

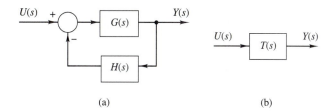

Figure 14.12 (a) Feedback connection of two subsystems. (b) Equivalent system.

vector to have 1001 points between $\omega = 0.30$ and 0.50 rad/s by using the command `freq` = `[0.3:0.001:0.5]` and repeating the calculations. In this case, we obtain a minimum magnitude of 0.00050 at $\omega = 0.4080$, which are very close to the analytical values.

14.4 MODELS FOR FEEDBACK SYSTEMS

In the remainder of this book, we shall apply the results of earlier sections and earlier chapters to the design of feedback control systems. This will afford us an opportunity to illustrate how s-domain, time-domain, and frequency-domain analysis can be used for one important class of systems. For the basic feedback configuration shown in Figure 14.12(a), we know from Chapter 13 that

$$T(s) = \frac{Y(s)}{U(s)} = \frac{G(s)}{1 + G(s)H(s)} \tag{14}$$

If the sign of the feedback signal at the summing junction is positive, the effect is the same as replacing $H(s)$ by $-H(s)$ in (14).

In this section, we first show how to use MATLAB to find $T(s)$ without any intermediate calculations, as soon as $G(s)$ and $H(s)$ are known. However, we frequently would then want to be able to modify the original $G(s)$ and $H(s)$ in order to improve the performance of the overall system. The second part of this section will lay the groundwork for doing this.

Constructing MATLAB Models

When the subsystems G and H exist in the MATLAB workspace as single-input, single-output LTI objects, and when the feedback signal has a negative sign where it enters the summing junction, the command `T = feedback(G,H)` will create an LTI object called T, having the desired transfer function for the feedback interconnection.

We can handle the feedback interconnection of two multi-input, multi-output subsystems also as LTI objects, provided that we take care to ensure that the numbers of inputs and outputs are consistent at each connection point. The default condition is for a negative sign on the feedback signal where it enters the summing junction, although it is possible to obtain a positive sign by using an additional argument of $+1$ in the `feedback` command.

▷ EXAMPLE 14.11

Find the closed-loop transfer function of the feedback system shown in Figure 14.13. Give the resulting LTI object in TF form and determine its zeros, poles, and gain.

SOLUTION   The two subsystems can be created by entering the commands

```
kG = 2;
zerG = -1;
```

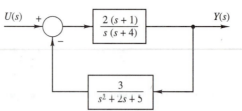

Figure 14.13 Feedback connection for Example 14.11.

```
polG = [0; -4];
G = zpk(zerG,polG,kG)
numH = 3;
denH = [1 2 5];
H = tf(numH, denH)
```

Once G and H have been defined, they can be connected in a feedback configuration (with a negative sign where the feedback signal enters the summing junction) by issuing the command

```
T = feedback(G,H)
```

The result displayed on the screen will be

```
Zero/pole/gain:
         2 (s+1) (s^2 + 2s + 5)
-------------------------------------------
(s+4.308) (s+0.2609) (s^2 + 1.431s + 5.338)
```

which is the closed-loop transfer function $T(s)$ given in ZPK form. Recall that this form will be used because $G(s)$ was entered in TF form and $H(s)$ in ZPK form, which takes precedence over the TF form for the result.

We can extract the numerator and denominator polynomials of the closed-loop transfer function by entering

```
[numT, denT] = tfdata(T, 'v')
```

which results in the display

```
numT =        0    2.0000    6.0000    14.0000    10.0000

denT =   1.0000    6.0000   13.0000    26.0000     6.0000
```

Thus, we can write the closed-loop transfer function as

$$T(s) = \frac{2s^3 + 6s^2 + 14s + 10}{s^4 + 6s^3 + 13s^2 + 26s + 6} \tag{15}$$

To determine the closed-loop poles and zeros, we enter

```
[zerT,polT,kT] = zpkdata(T, 'v')
```

and obtain

```
zerT =      -1.0000 + 2.0000i
            -1.0000 - 2.0000i
            -1.0000
```

```
polT =        -4.3083
              -0.2609
              -0.7154 + 2.1969i
              -0.7154 - 2.1969i


kT  =      2
```

Note that all four of the closed-loop poles differ from those of $G(s)$ and $H(s)$. This is usually the case when two blocks are joined in a feedback configuration. In contrast, the poles and zeros of a series connection are the same as those of the individual blocks. Also, we have seen that the poles of a parallel connection are the same as those of the individual blocks, but the zeros will be different.

We can see that the closed-loop zeros are $s = -1$, which is the zero of $G(s)$, and $s = -1 + j2$ and $s = -1 - j2$, which are the poles of $H(s)$. It is a general property of feedback systems that the closed-loop zeros are the *zeros* of the *forward*-path transfer function and the *poles* of the *feedback*-path transfer function.

Stability

In the design of a feedback control system, the engineer must satisfy a number of specifications, including those on the transient and steady-state response. The most fundamental criterion is to make certain that the system is stable, even when allowing for imperfections in the physical devices. Any study of stability is strongly linked to the form of the transient response.

We shall see in the next section that even when $G(s)$ and $H(s)$ are both stable, the closed-loop system might still be unstable. Conversely, even when $G(s)$ or $H(s)$ is unstable, the closed-loop transfer function $T(s)$ might be stable. It is important to be able to predict the stability of the closed-loop system knowing the open-loop transfer function $G(s)H(s)$, in order to be able to modify $G(s)$ or $H(s)$ in ways that will improve the overall system performance. In the next section, we construct diagrams in the s-plane to do this. In the final section, we accomplish this using the Bode diagrams introduced in Section 14.3.

▶ 14.5 ROOT-LOCUS PLOTS

The poles of a system's overall transfer function determine whether the system is stable and, for a stable system, determine the nature of the transient response. Because the positions of the poles in the complex s-plane constitute one of the key design considerations, it is important to know how these positions change when one of the parameters of the system is varied.

The effect of the pole positions on stability and on the transient response was discussed in detail in Chapter 8. Keep in mind that the poles of the transfer function are roots of the characteristic polynomial used in Chapter 8. If all the poles are inside the left half of the s-plane, then the zero-input response decays to zero and the system is stable. If the transfer function has a pole inside the right half-plane or repeated poles on the imaginary axis, the zero-input response increases without limit and the system is unstable. Finally, if all the poles are inside the left half-plane except for first-order poles on the imaginary axis (possibly including the origin), a nonzero but finite component of the zero-input response remains for large values of t, and the system is said to be marginally stable.

The nature of the zero-input response corresponding to various pole positions was illustrated in Figures 8.5 through 8.8. For a single pole on the negative real axis of the s-plane, the distance from the vertical axis is the reciprocal of the time constant. For a pair of first-order poles at $s = -\alpha \pm j\beta$ in the left half-plane, the zero-input response has the form

$$y(t) = K\epsilon^{-\alpha t}\cos(\beta t + \phi)$$

where α, the distance from the vertical axis, is again the reciprocal of the time constant for the exponential factor.

The movement of the pole positions when a particular parameter is varied can be shown by drawing the path traced out in the s-plane as the parameter is increased from very small to very large values. Because poles of the transfer function are roots of the characteristic equation, such a path is called a **root locus**. It can help us select appropriate values for some of the elements.

▶ **EXAMPLE 14.12**

Consider the transfer function $T(s) = X(s)/F_a(s)$ for the second-order mechanical system shown in Figure 14.14(a). Asume that M and K are fixed, while the friction coefficient B is varied from zero toward infinity.

SOLUTION   The input-output differential equation is easily shown to be $M\ddot{x} + B\dot{x} + Kx = f_a(t)$, so the transfer function is

$$T(s) = \frac{1}{Ms^2 + Bs + K} = \frac{\dfrac{1}{M}}{s^2 + \dfrac{B}{M}s + \dfrac{K}{M}}$$

There are two poles, which can be found by applying the quadratic formula to the denominator of $T(s)$. When $B = 0$, the poles are at $s = \pm j\sqrt{K/M}$, as indicated by the crosses in Figure 14.14(b). For $0 < B < 2\sqrt{KM}$, we have a pair of complex conjugate poles in the left half-plane at

$$s = -\frac{B}{2M} \pm j\sqrt{\frac{K}{M} - \frac{B^2}{4M^2}} = \frac{1}{2M}\left[-B \pm j\sqrt{4KM - B^2}\right]$$

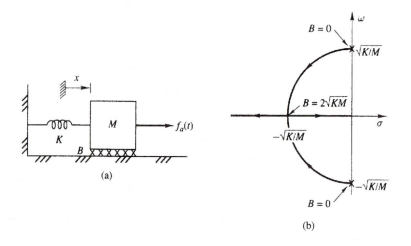

(a)

(b)

Figure 14.14   (a) Mechanical system for Example 14.12. (b) Locus of pole positions when B is varied.

The distance of these poles from the origin of the s-plane is

$$\frac{1}{2M}[B^2 + (4KM - B^2)]^{1/2} = \sqrt{K/M}$$

Thus for constant values of K and M, this part of the locus is the arc of a circle of radius $\sqrt{K/M}$. When $B = 2\sqrt{KM}$, the denominator of $T(s)$ becomes

$$M(s^2 + 2\sqrt{K/M}\,s + K/M) = M\left(s + \sqrt{K/M}\right)^2$$

corresponding to a double pole on the negative real axis at $s = -\sqrt{K/M}$. For values of B larger than $2\sqrt{KM}$, there are two distinct poles on the negative real axis. The complete locus traced out by the pole positions is shown by the heavy lines. The arrows indicate increasing values of B.

We see from Equation (8.32) that the damping ratio for this system is

$$\zeta = \frac{B}{2\sqrt{MK}}$$

so the locus in part (b) of the figure agrees with Figure 8.10(b).

Although a locus of possible pole positions can be drawn for any system in which only one of the parameters is varied, it is particularly important for us to be able to do this for the feedback configuration shown in Figure 14.12(a). From (14), the closed-loop transfer function is

$$T(s) = \frac{G(s)}{1 + G(s)H(s)} \tag{16}$$

We shall assume that the open-loop transfer function $G(s)H(s)$ can be expressed in factored form as

$$G(s)H(s) = K\frac{(s - z_1)\cdots(s - z_m)}{(s - p_1)\cdots(s - p_n)} \tag{17a}$$

$$= KF(s) \tag{17b}$$

Note from this definition that $F(s)$ contains all the factors of $G(s)H(s)$ except the multiplying constant K. Parts of $G(s)H(s)$ are normally under the control of the designer, who typically takes a particular $F(s)$ and then constructs a locus of the poles of $T(s)$ as K is varied. The designer can do this *without* having first to write $T(s)$ as an explicit rational function and without having to factor high-order polynomials.

The quantities z_1 through z_m in (17) are the zeros of $F(s)$ in the finite s-plane, which are referred to as the **open-loop zeros**. The quantities p_1 through p_n are the poles, referred to as the **open-loop poles**. For all practical physical systems $m \le n$. If $m < n$, then $F(s)$ is said to have a zero of order $n - m$ at infinity. We seek a locus of the poles of $T(s)$ as the positive constant K is varied. The poles of the closed-loop transfer function $T(s)$ will lie on the root locus and will be called *roots*. In the ensuing discussion, the terms *poles* and *zeros* refer to p_i and z_j, respectively, in the open-loop transfer function $KF(s)$.

We will describe shortly how to use MATLAB to construct the root locus when $F(s)$ is given. However, it is helpful to be able to predict the general shape of the locus, even when we plan to rely on a computer program for an exact final plot. We present here, without proof, a number of characteristics that any root locus must have for the case where $K \ge 0$ and where the poles and zeros are distinct. We let n denote the number of poles of $F(s)$, and m the number of finite-plane zeros.

1. The locus is symmetrical with respect to the real axis of the s-plane.

2. The locus has n branches.

3. A point on the real axis is part of the locus if and only if the total number of real poles and zeros to the right of that point is odd.

4. As K increases from zero, one branch of the locus departs from each of the poles of $F(s)$.

5. As K approaches infinity, the branches of the locus approach the zeros of $F(s)$. There will be m branches approaching the finite-plane zeros. If $m \neq n$, the remaining $n - m$ branches will approach infinity.

6. The branches approaching infinity will be asymptotic to equally spaced straight lines emanating from the center of mass for the poles and zeros. This center can be found by regarding each pole as a positive unit mass and each zero as a negative unit mass, and the center of mass is a point on the real axis. The angles of the equally spaced asymptotes can be determined by remembering that the branches of the locus must be symmetrical about the real axis of the s-plane. Whenever $n - m$ is an odd integer, one of the asymptotes is the negative real axis. Whenever $n - m$ exceeds two, some of the asymptotes will extend into the right half of the complex plane.

We now present two examples of root-locus diagrams for feedback systems and make some remarks about each of them. Following that, we shall show how to use MATLAB to obtain such plots.

▶ **EXAMPLE 14.13**

Show and discuss the locus for the poles of the closed-loop transfer function when

$$KF(s) = \frac{K}{s^3 + 7s^2 + 14s + 8} = \frac{K}{(s+1)(s+2)(s+4)}$$

Under what conditions is the system stable?

SOLUTION   The parts of the real axis that belong to the locus are indicated by heavy lines. Because $F(s)$ has no finite-plane zeros, all three branches of the locus eventually go to infinity. The complete plot is shown in Figure 14.15. Arrows indicate the directions in which the closed-loop poles move as K is increased. Note that two branches of the locus pass into the right half-plane when K becomes sufficiently large, and the system becomes unstable.

It is often necessary to calibrate the locus by showing the values of K that correspond to specific points, and we shall describe shortly a MATLAB command that can easily do this. Of particular interest is the value of K for which the branches cross the imaginary axis of the s-plane. For this example, we can show that these points correspond to $K = 90$. Thus in order for all the branches of the locus to be inside the left half-plane, as is required for a stable system, the positive constant K must be restricted to $K < 90$.

▶ **EXAMPLE 14.14**

Show the root locus and find the values of K for which the overall system is stable when

$$KF(s) = \frac{K(s+2)}{s(s-1)}$$

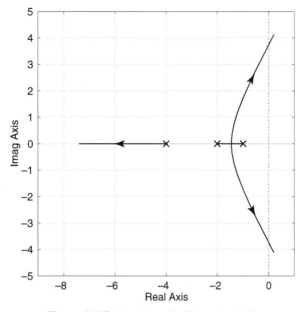

Figure 14.15 Root locus for Example 14.13.

The open-loop transfer function $KF(s)$ is unstable because of the pole in the right half-plane. This might be an unavoidable characteristic of one part of the system that cannot be changed. However, others factors in $KF(s)$ might allow both branches of the locus to be brought inside the left half-plane for some range of K.

SOLUTION The complete locus is shown in Figure 14.16. Notice that the zero close to the open-loop poles tends to attract the locus toward it as K increases. This is a general

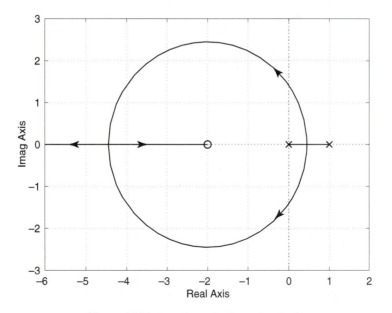

Figure 14.16 Root locus for Example 14.14.

property that is helpful when the designer is adding additional zeros and poles to a specified $F(s)$. When K is sufficiently large, the locus moves into the left half-plane, so that the closed-loop system becomes stable. We can show that the right-hand branches pass into the left half-plane when $K = 1$, so the closed-loop system is stable for all $K > 1$.

Using MATLAB to Construct the Locus

A root-locus plot can be used to determine how the closed-loop poles of a feedback system will vary as the open-loop gain is varied. As with the other MATLAB commands that we are discussing, the user has several options available. The model can be given in any of the three LTI forms (transfer-function, zero-pole-gain, or state-space). A list of gain values can be specified, or MATLAB will select a set that it considers appropriate. In Figure 14.15, for example, the maximum value of K selected by MATLAB causes the branches of the locus to stop where they do.

The user can have MATLAB either draw the plot without returning the numerical values of the roots or return the values without automatically drawing the plot. The region within the s-plane covered by the plot can also be specified or left to MATLAB. If we wish to ensure that the scales used for the horizontal and vertical axes are identical, we can insert into the file the command `axis equal`. There is an additional command called `rlocfind` that allows the user to select any point on the locus once it has been drawn and to obtain both the value of the gain parameter corresponding to that point and the values of all n of the closed-loop characteristic roots for that value of gain.

The following example illustrates how root-locus plots can be created with MATLAB and how the gain can be determined for specific points on the locus. The reader should keep in mind that the use of the `rlocfind` command requires the user to position the cursor at or near a specific point on the root-locus diagram. Also, the root-locus plot is created by computing a set of closed-loop roots for a discrete set of gains, which normally are not selected by the user, and then connecting these points with straight lines.

Hence, the results obtained with the `rlocfind` command should be viewed as *approximations* to the actual gain, rather than the exact value. It is possible for the user to obtain better approximations by specifying more gain values for the plot and using the `axis` command to zoom in on the region of interest, but this is beyond the scope of our discussion.

▶ **EXAMPLE 14.15**

Generate a root-locus plot for the feedback system shown in Figure 14.17. Use the plot to determine the value of the gain K for which the closed-loop system will become marginally stable.

SOLUTION We recognize that the forward and feedback transfer functions, excluding the gain K, are the functions $G(s)$ and $H(s)$ that were built and connected in series in Example 14.11. We can therefore create the open-loop model as the LTI object GH by giving the command GH = H*G. The gain K in the forward path is included automatically when we construct the root locus. To generate the root-locus plot with the plotting region and values of K used to plot the locus determined by MATLAB, we enter the command `rlocus(GH)`. The plot shown in Figure 14.18(a) will appear on the screen. Note that the symbol x has been placed at each of the four open-loop poles and the symbol o at the single open-loop zero.

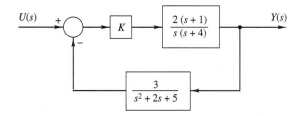

Figure 14.17 Feedback system for Example 14.15.

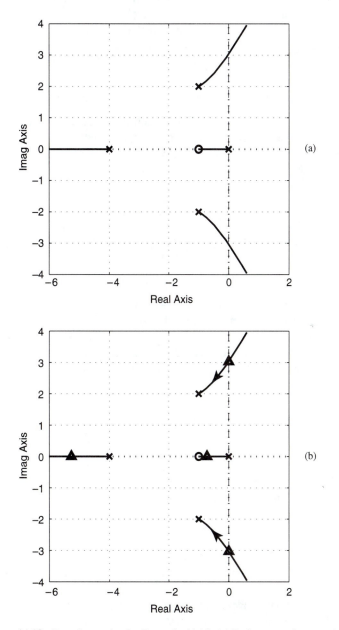

Figure 14.18 Root-locus plot for Example 14.15. (a) Before use of `rlocfind` command. (b) After use of `rlocfind` command, with four closed-loop poles marked.

To determine an approximation to the value of K for which the complex branches cross into the right half of the s-plane, we use the rlocfind command by entering [K,CLpoles] = rlocfind(GH). MATLAB will prompt the user to select the point for which the gain is to be found by clicking with the mouse directly on the root-locus plot that has just been drawn in the graphics window.

We get $K^* = 5.83$ and the corresponding closed-loop poles are $s = -5.273, -0.724$, and $-0.0015 \pm j3.028$. Part (b) of Figure 14.18 shows the root-locus plot after the rlocfind command has computed the values of K^* and the corresponding closed-loop poles. MATLAB normally adds plus signs at the locations of the four computed poles, so the user can easily see where all of the closed-loop system poles will be for that specific value of the gain. We have used triangles instead for increased visibility.

Other Applications of the Root Locus

The concepts that we have developed for the construction of a root locus can be extended to other situations. We consider two of them now. First, suppose that the sign of the signal fed back to the summer in Figure 14.12(a) is changed from minus to plus. Then the closed-loop transfer function is

$$T(s) = \frac{G(s)}{1 - G(s)H(s)}$$

By comparing this equation with (16) and (17), we see that the only change has been to replace K by $-K$. In such a case, we can start with the pole-zero pattern of $G(s)H(s)$ and then ask MATLAB to draw the root locus for negative values of K by issuing the command rlocus(-GH).

Second, root-locus diagrams can be used even when the system model is not presented in the standard feedback configuration shown in Figure 14.12(a). We require only that the denominator of the overall transfer function be written in the form $1 + KF(s)$. As one simple example, recall the mechanical system that is shown in Figure 14.14(a) and considered in Example 14.12 and that has the transfer function

$$T(s) = \frac{\dfrac{1}{M}}{s^2 + \dfrac{B}{M}s + \dfrac{K}{M}}$$

We want the locus of the poles of $T(s)$ when B/M is varied. To put the expression into a suitable form, we divide both the numerator and the denominator of the fraction by those denominator terms that do *not* contain the parameter to be varied. Thus we write

$$T(s) = \frac{\dfrac{1/M}{s^2 + K/M}}{1 + \dfrac{(B/M)s}{s^2 + K/M}}$$

so that the denominator now has has the desired form, namely $1 + (B/M)F(s)$, where $F(s)$ does not depend on the variable gain B/M. We regard the open-loop transfer function as having poles at $s = \pm j\sqrt{K/M}$ and a zero at the origin. Drawing the locus as B/M increases from zero to infinity gives Figure 14.14(b).

14.6 STABILITY CRITERIA

Consider again the standard feedback configuration that is shown in Figure 14.12(a). We discussed in Section 14.4 the importance of being able to predict the stability of the overall system from a knowledge of the open-loop transfer function $G(s)H(s)$. If the behavior of the overall system needs to be modified, we can consider changing the gain constant associated with $G(s)H(s)$ or can even arrange to give it additional poles and zeros.

The most fundamental way of relating the frequency-response function $G(j\omega)H(j\omega)$ to the stability of the overall system is to construct a polar plot. We would regard $G(j\omega)H(j\omega)$ as a vector drawn from the origin of a new complex plane (different from the s-plane) and would look at the path traced out by the tip of that vector as ω increases. A test known as the **Nyquist stability criterion** can be applied to determine whether the system is stable. Furthermore, that test can give considerable insight into whether the system will remain stable even if some of the characteristics of the open-loop transfer function undergo moderate changes.

Knowledge of the Nyquist criterion is necessary for a complete understanding of how the Bode diagram for $G(s)H(s)$ can be related to the stability of the overall system. Unfortunately, developing the Nyquist criterion is fairly involved, so we must refer those interested in it to the references in Appendix F. In the following discussion, we assume that all the poles of the open-loop transfer function are inside the left half-plane, except for a possible first-order pole at the origin. Then we can state, without proof, the following definitions and properties.

The Bode diagram in Figure 14.19 represents a typical open-loop transfer function $G(s)H(s)$. The frequency at which the magnitude curve crosses the zero-dB line is denoted by ω_{pm} and is sometimes called the **magnitude crossover frequency**. The **phase margin**

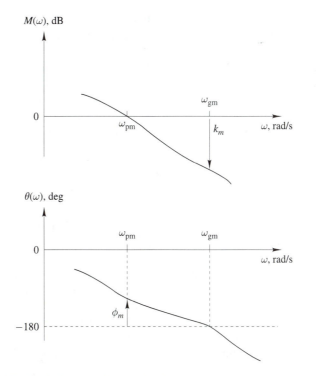

Figure 14.19 Bode diagram showing the phase margin ϕ_m and the gain margin k_m.

ϕ_m is the amount by which the angle curve would have to be moved down in order to make the angle $\theta(\omega_{\text{pm}})$ be $-180°$.

The frequency at which the angle curve crosses $-180°$, sometimes called the **phase crossover frequency**, is denoted by ω_{gm}. The **gain margin** k_m is the number of decibels by which the magnitude curve would have to be moved up in order to make $M(\omega_{\text{gm}})|_{\text{dB}} = 0$. The gain and phase margins are labeled in Figure 14.19. For the overall system to be stable, both the gain margin and the phase margin must be positive, as is assumed in Figure 14.19.

There are no standard symbols for the gain and phase margins or for the frequencies at which these quantities are measured. The subscripts we have used in the symbols ω_{pm} and ω_{gm} indicate the frequencies at which the phase margin and gain margin, respectively, are measured. Although the nature of the plots in Figure 14.19 is fairly typical, cases exist for which our definitions of k_m and ϕ_m do not apply. For example, it is possible for the curve of $\theta(\omega)$ never to cross the $-180°$ axis or to cross it more than once. Similarly, the curve of $M(\omega)|_{\text{dB}}$ could cross the zero-dB axis more than once.

In order for the closed-loop system not to be too lightly damped and in order to be sure that it will remain stable even if there are some variations in the parameters, a designer would insist on certain minimum values for the gain and phase margins. The ranges of desired gain and phase margins depend on the particular system being considered and on the intended applications. However, a gain margin of about 10 dB and a phase margin of $45°$ are typical for many systems.

Increasing the phase margin generally increases the damping ratio ζ associated with a pair of complex poles. For specific second-order systems, explicit relationships between ϕ_m and ζ can be derived. Some of the references in Appendix F contain curves of ζ versus ϕ_m, but it is important to note the specific open-loop transfer function for which they apply.

In order to give some plausibility to the foregoing statements about the gain and phase margin, let us look at the steady-state behavior of the feedback system in Figure 14.12(a) when the input is sinusoidal. At the frequency ω_{gm}, the angle of the open-loop transfer function is $-180°$. Because of the signs shown on the summer, the signal fed back from the $G(s)H(s)$ path is then in phase with the input and adds directly to it.

Suppose the input is now removed. If the magnitude of the open-loop gain is less than unity at the frequency ω_{gm}, then the signal fed back to the summer continually diminishes, and the output goes to zero. If $|M(\omega_{\text{gm}})| = 1$, corresponding to 0 dB, the signal fed back to the summer is just large enough to sustain the sinusoidal oscillations at a constant amplitude. Such a condition describes a marginally stable system. If $|M(\omega_{\text{gm}})| > 1$, then the oscillations continually grow in magnitude, and the system is unstable.

A positive gain margin means that $M(\omega_{\text{gm}})|_{\text{dB}} < 0$, corresponding to a stable system. A gain margin of zero occurs when $M(\omega_{\text{gm}})|_{\text{dB}} = 0$. A negative gain margin means that $M(\omega_{\text{gm}})|_{\text{dB}} > 0$, corresponding to an unstable system.

The MATLAB computer program can not only yield the Bode diagram for a given open-loop transfer function but can also list the gain and phase margins and the frequencies at which they are measured. The gain margin can be expressed both in decibels and as the additional multiplying factor needed to make $M(\omega_{\text{gm}}) = 1$. As in any computer program that produces a smooth curve from discrete numerical calculations, values corresponding to special points on the curve are only approximations to the exact numbers. Although a designer would require only reasonably close answers, more accurate values can always be achieved by increasing the number of points at which calculations are made. We now make use of MATLAB to draw and interpret the Bode diagram for a third-order system.

► **EXAMPLE 14.16**

Draw the Bode diagram for the open-loop transfer function

$$G(s)H(s) = \frac{K}{(s+1)(s+2)(s+4)}$$

when $K = 20$. Find the gain and phase margins and determine the value of the positive constant K for which the closed-loop transfer function is marginally stable.

SOLUTION   The following MATLAB commands will build the open-loop system GH and produce the plot shown in Figure 14.20.

```
zerGH = [ ]
polGH = [-1; -2; -4]
kGH = 20
GH = zpk(zerGH,polGH,kGH)
margin(GH)
```

The `margin` command draws the Bode plot of the open-loop system and displays the values of the gain and phase margins, along with the frequencies at which the margins are defined. For the specified value of $K = 20$, the gain margin is $k_m = 13.064$ dB. This value is computed at the frequency at which the phase angle of the open-loop system equals $-180°$, namely $\omega_{gm} = 3.7417$ rad/s. To convert the gain margin k_m from decibels to the multiplicative gain A, we compute $A = 10^{13.064/20} = 10^{0.6532} = 4.4999$. Hence, this analysis predicts that the closed-loop system should become marginally stable when the gain $K = 20 \times 4.4999 = 89.998$. The phase margin $\phi_m = 63.939°$ and is computed at the frequency $\omega_{pm} = 1.5486$ rad/s, which is where the open-loop magnitude plot crosses the 0-dB line. MATLAB prints the numerical values above the magnitude plot and also adds vertical lines to show where the gain and phase margins are measured.

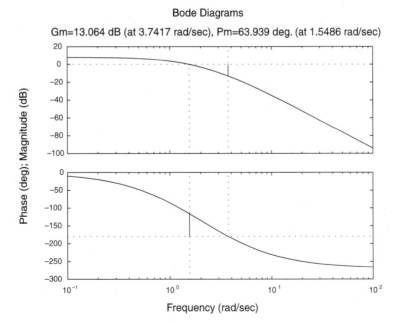

Figure 14.20   Bode diagram for Example 14.16.

A root-locus plot for the transfer function used in this example was drawn in Example 14.13. That method showed that the closed-loop system is unstable for $K > 90$, which agrees with the results of this example.

▶ SUMMARY

We have used MATLAB, with the Control System Toolbox, to build models of linear dynamic systems as linear time-invariant objects in three different forms: transfer-function, zero-pole-gain, and state-space. Then we showed how individual models can be interconnected in series and in parallel. We saw how a variety of results can be obtained from these models and plotted in the time and frequency domains.

For the feedback configuration, we need to be able to predict the stability of the closed-loop system from knowledge of the open-loop transfer function. To do this, the final two sections developed two techniques, using root-locus plots and Bode diagrams. Examples showed that the closed-loop system might be unstable even when each of the subsystems is stable. Furthermore, an example was presented that demonstrated that the closed-loop system could be stable, even if one of its subsystems was unstable.

In the following chapter, we will use the Bode and root-locus plots to design feedback controllers for linear systems that will result in stable closed-loop systems and that will achieve specified performance measures in the time and frequency domains.

▶ PROBLEMS

Use MATLAB or another suitable software package for all of the following problems. You should produce a diary file that shows the input, the numerical results, and (where appropriate) computer-generated plots.

14.1. The transfer function of a linear system has zeros at $s = -1, -4 + j2, -4 - j2$, poles at $s = -0.6, -1 + j8, -1 - j8, -15$, and a gain of 2.

   a. Obtain the transfer function as a ratio of polynomials.

   b. Find the four matrices that describe the state-space model.

   c. Compute the zeros, poles, and gain of the transfer function from the state-space matrices.

14.2. The transfer function of a linear system is

$$T(s) = \frac{2s^2 + 4s + 1}{3s^4 + 8s^3 + 9s^2 + 10s}$$

   a. Determine the zeros, poles, and gain of the transfer function.

   b. Find the four matrices that describe the state-space model.

   c. Compute the numerator and denominator polynomials of the transfer function from the state-space matrices.

*14.3. The state-space matrices for a linear system are

$$\mathbf{A} = \begin{bmatrix} 0 & 1 & 2 & 0 \\ -3 & -2 & 0 & 1 \\ -2 & 0 & 0 & 1 \\ 0 & 0 & -1 & -5 \end{bmatrix} \qquad \mathbf{B} = \begin{bmatrix} 0 \\ 1 \\ 0 \\ 2 \end{bmatrix} \qquad \mathbf{C} = \begin{bmatrix} 1 & 0 & -2 & 3 \end{bmatrix} \qquad \mathbf{D} = 0$$

    a.   Obtain the transfer function as a ratio of polynomials.

    b.   Determine the zeros, poles, and gain of the transfer function.

14.4.

    a.   Use Equation (13.38) to develop a state-space model of the system that was studied in Example 13.11, whose block diagram is shown in Figure 13.20(b).

    b.   Convert this model to transfer-function form and compare the result with the differential equation given in the example statement.

*14.5.

    a.   Use Equations (13.42a), (13.42b), (13.42c), and (13.43) to develop a state-space model of the system that was studied in Example 13.12, whose block diagram is shown in Figure 13.21(c).

    b.   Convert this model to transfer-function form and compare the result with the differential equation given in the example statement.

14.6. Obtain state-space models for the individual subsystems defined in Example 14.2. Then, for the series combination, obtain both the zero-pole-gain and transfer-function forms of the overall transfer function $G_{12s}(s)$. Verify that the results agree with those found in the example.

*14.7. For the series connection shown in Figure P14.7, obtain the zero-pole-gain model and the transfer-function model.

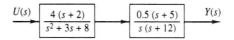

$U(s)$ $\dfrac{4\,(s+2)}{s^2+3s+8}$ $\dfrac{0.5\,(s+5)}{s\,(s+12)}$ $Y(s)$

Figure P14.7

14.8.

    a.   Use the series and parallel commands of MATLAB to build the model represented in Figure 13.3 in transfer-function form.

    b.   Determine the zeros, poles, and gain of $T(s) = Z(s)/U(s)$.

14.9. Repeat Problem 14.8 for the block diagram shown in Figure P14.9.

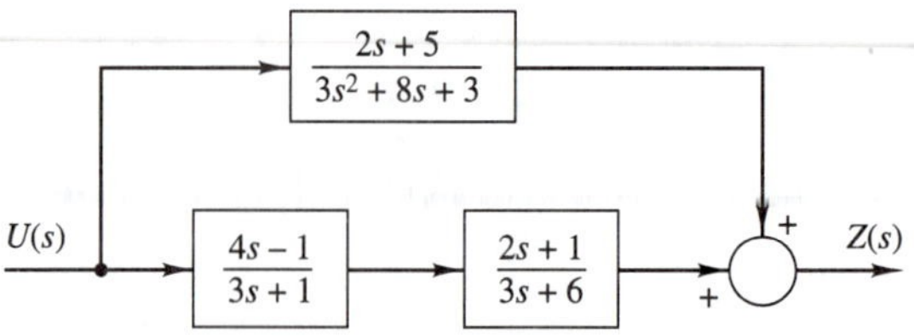

Figure P14.9

14.10. Obtain plots of the unit step and unit impulse responses for the linear system given in Problem 14.2.

14.11. Obtain plots of the unit step and unit impulse responses for the linear system given in Problem 14.3.

*14.12. Obtain the zero-state response of the system described in Problem 14.2 to the input shown in Figure P14.12 over the interval $0 \leq t \leq 50$ s. Plot the input and the output on the same axes.

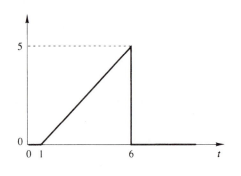

Figure P14.12

14.13. Obtain the zero-state response of the system described in Problem 14.3 to the input shown in Figure P14.13 over the interval $0 \leq t \leq 20$ s. Plot the input and the output on the same axes.

Figure P14.13

14.14. Obtain the Bode plots shown in Figure 14.8.

In the next three problems, obtain a computer-generated Bode plot for the magnitude and phase angle curves. Explain the low- and high-frequency asymptotes.

14.15.

$$T(s) = \frac{25(s+2)}{s+50}$$

14.16.

$$T(s) = \frac{320(s+5)}{(s+40)^2}$$

14.17.

$$T(s) = \frac{50}{s(s+50)}$$

14.18. Obtain a computer-generated frequency-response plot similar to Figure 14.11 for the three transfer functions given in Problem 8.47. In many filtering applications, we want the system to respond relatively strongly to sinusoidal components within a certain frequency range, but to attenuate components outside of that range. In such cases, the bandwidth is often defined as the frequency range over which the magnitude of the output is at least $1/\sqrt{2}$ times the maximum output. Find the bandwidth for each transfer function.

14.19. Obtain a computer-generated frequency-response plot similar to Figure 14.11 for the three transfer functions given in Problem 8.48.

14.20. For the feedback connection shown in Figure P14.20, obtain the model for the closed-loop system in zero-pole-gain form, transfer-function form, and state-space form.

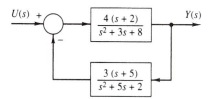

Figure P14.20

14.21. Repeat Problem 14.20 when the negative sign at the summing junction is changed to a plus sign—that is, when the feedback is positive rather than negative.

*14.22. Repeat Problem 14.20 for the feedback system shown in Figure P14.22.

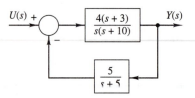

Figure P14.22

14.23.

a. Obtain a root-locus plot similar to the one shown in Figure 14.15 for the feedback system studied in Example 14.13.

b. Determine the value of K that will result in a complex closed-loop pole having an imaginary part of +2. Find the other closed-loop poles that correspond to this value of gain.

14.24.

a. Obtain a root-locus plot for the system described by

$$KF(s) = \frac{K}{(s+1)(s+6-j2)(s+6+j2)}$$

b. Determine the value of K that will result in a complex closed-loop pole having a real part of -2. Find the other closed-loop poles that correspond to this value of gain.

14.25. Find the root locus for

$$KF(s) = \frac{K(s+\alpha)}{(s+1)(s+2)(s+10)}$$

when $\alpha = 20$. Determine the maximum value of K for which the closed-loop system is stable.

*14.26.

a. Obtain a root-locus plot for the system described by

$$KF(s) = \frac{K(s+2)}{s(s+1)(s+4)(s+6)}$$

b. Determine the value of K that will result in a complex closed-loop pole having an imaginary part of +2. Find the other closed-loop poles that correspond to this value of gain.

14.27. Find the root locus for

$$KF(s) = \frac{K(s+1)}{s^2(s+10)}$$

14.28. Repeat Example 14.13 for negative values of K.

14.29. Find the root locus for the system described in Problem 14.24 for negative values of K. For what values of K is the closed-loop system stable?

14.30. Repeat Problem 14.29 for the system described in Problem 14.25.

14.31.

    a. Obtain the Bode plot for the open-loop transfer function that was considered in Example 14.13 when $K = 10$.

    b. Determine the gain and phase margins and the frequencies at which they are defined.

    c. Use the gain margin to determine the maximum value of K for which the closed-loop system is stable.

*14.32. Repeat Problem 14.31 for the open-loop transfer function described in Problem 14.24 when $K = 150$.

14.33. Repeat Problem 14.31 for the open-loop transfer function described in Problem 14.26 when $K = 40$.

14.34.

    a. Obtain a computer-generated Bode diagram for the open-loop transfer function in Problem 14.24, with $K = 100$. Determine the gain margin and the phase margin.

    b. Find the value of K that will give a gain margin of 10 dB.

    c. Find K^*, the maximum value of K for which the closed-loop transfer function will be stable.

14.35. Repeat Problem 14.34 for the open-loop transfer function in Problem 14.26 with $K = 40$.

15

FEEDBACK DESIGN WITH MATLAB

In Section 14.6 we explained how certain characteristics of the frequency response of an open-loop system can be used to infer the stability properties of the closed-loop system. This is a useful tool for the design of feedback systems.

In this chapter we introduce and illustrate some of the practical design criteria. We consider ways of meeting these criteria using Bode diagrams and the root-locus technique. We include stability considerations, the transient response, the steady-state errors corresponding to particular inputs, the need for satisfactory performance even if the system parameters change somewhat, minimizing the response to unwanted disturbance inputs, and rejecting unwanted noise.

▶ 15.1 DESIGN GUIDELINES

The block diagram for a basic feedback configuration that includes both reference and disturbance inputs is shown in Figure 15.1(a). Particular systems may have more complicated diagrams with a number of additional feedback and feedforward paths. However, many diagrams can be reduced to the one in the figure, and analyzing it will provide an introduction to some of the design techniques most often used.

The **plant** is the system to be controlled. It may contain almost any collection of linearized components or processes, and its parameters are generally already fixed and beyond the designer's control. The purpose of the **sensor** is to measure the output and feed a signal back to the input summing device. The engineer may also choose to add other components within the sensor block to improve the system performance.

The **controller** provides the excitation for the plant and can be designed to meet the specifications for the overall system behavior. Its characteristics and implementation are chosen by the engineer. Piecewise linear systems can sometimes be linearized about more than one operating point, in which case we might be able to design a single controller that is satisfactory for all such points. In other cases, we may need more than one controller, with provision for switching smoothly among them as the operating point changes.

The notation that we shall use is shown in Figure 15.1(b). The transfer functions for the controller, plant, and sensor are denoted by $G_c(s)$, $G_p(s)$, and $H(s)$, respectively. Because the diagram has two inputs, we distinguish their Laplace transforms by using $R(s)$ for the reference input and $D(s)$ for the disturbance input. We assume that we want the output $y(t)$ to follow closely any changes in the reference input $r(t)$. For the system examined in Section 13.4, the input and output variables were the angular displacements of mechanical components. However, they might equally well be any other type of variable.

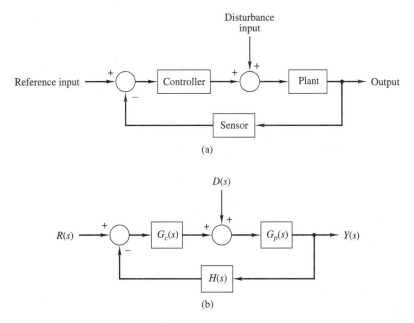

Figure 15.1 Block diagrams for a basic feedback system.

A system may be subjected to unpredictable disturbances that tend to affect the output adversely. Examples include wind gusts on a moving vehicle and a load torque exerted on a rotating shaft to which a cutting tool is attached. Such unwanted inputs are usually exerted directly on the plant and are represented in the figure by the transformed variable $D(s)$. We want the output to be relatively insensitive to a disturbance input. An additional concern, which we shall discuss only in a qualitative way, is the need for the system's behavior to remain acceptable even if some of its parameters undergo moderate changes or if random noise signals arise within the system.

We define the following two transfer functions for the system shown in Figure 15.1(b): $T_R(s) = Y(s)/R(s)$ when the disturbance input $D(s)$ is zero, and $T_D(s) = Y(s)/D(s)$ when the reference input $R(s)$ is zero. For $T_R(s)$, we let $G(s) = G_c(s)G_p(s)$ in Equation (13.3). For $T_D(s)$, we use Equation (13.4) with $G(s) = G_p(s)$ and with $H(s)$ replaced by $-G_c(s)H(s)$. Then

$$T_R(s) = \frac{Y(s)}{R(s)} = \frac{G_c(s)G_p(s)}{1 + G_c(s)G_p(s)H(s)} \tag{1a}$$

$$T_D(s) = \frac{Y(s)}{D(s)} = \frac{G_p(s)}{1 + G_c(s)G_p(s)H(s)} \tag{1b}$$

We would of course like $T_R(s)$ to approach unity and $T_D(s)$ to approach zero. Because both transfer functions have the same denominator, they will have the same poles. Thus consideration of stability and consideration of the form of the transient response will yield the same results, no matter which of the two functions is used. Information about these items can be obtained from root-locus or Bode plots corresponding to the open-loop transfer function $G_c(s)G_p(s)H(s)$.

One design criterion is the steady-state response to common reference inputs, such as the unit step function. Because we want the output to follow the reference input, we define the error function

$$e(t) = r(t) - y(t) \tag{2}$$

which, when transformed, gives

$$E(s) = R(s) - Y(s)$$

For the special case where $H(s) = 1$, $E(s)$ is the transformed output of the first summer and the input to the controller. For the general case, however, $E(s)$ does not appear in the block diagram.

When examining steady-state responses, we can apply the final-value theorem, which was presented in Equation (7.100) and is repeated here:

$$f(\infty) = \lim_{s \to 0} sF(s) \tag{3}$$

provided that $sF(s)$ has no poles on the imaginary axis or in the right half of the complex plane. We can also write a partial-fraction expansion, omitting those terms that do not contribute to the steady-state response.

In order to investigate the effects of different types of controllers, we shall use for purposes of illustration the following second-order transfer function for the plant:

$$G_p(s) = \frac{K_p}{(s+a)(s+b)} \tag{4}$$

where a and b are nonnegative real constants. For the servomechanism treated in Section 13.4, the transfer function of the motor had this form with $a = 0, b = 1/\tau_m$, and $K_p = K_m$. If we include an input for the hydraulic system shown in Figure 12.13, it will have a transfer function similar to (4) with real positive values for a and b.

In the ensuing discussion, we shall also let $H(s) = 1$. Then the system diagram in Figure 15.1(b) reduces to the one in Figure 15.2, for which

$$T_R(s) = \frac{G_c(s)\dfrac{K_p}{(s+a)(s+b)}}{1 + G_c(s)\dfrac{K_p}{(s+a)(s+b)}} \tag{5a}$$

$$T_D(s) = \frac{\dfrac{K_p}{(s+a)(s+b)}}{1 + G_c(s)\dfrac{K_p}{(s+a)(s+b)}} \tag{5b}$$

where the open-loop transfer function is

$$\frac{K_p G_c(s)}{(s+a)(s+b)}$$

Proportional Control

We first let $G_c(s)$ be the positive constant K_c. Then (5) reduces to

$$T_R(s) = \frac{\dfrac{K_c K_p}{(s+a)(s+b)}}{1 + \dfrac{K_c K_p}{(s+a)(s+b)}} = \frac{K_c K_p}{s^2 + (a+b)s + (ab + K_c K_p)} \tag{6a}$$

$$T_D(s) = \frac{\dfrac{K_p}{(s+a)(s+b)}}{1 + \dfrac{K_c K_p}{(s+a)(s+b)}} = \frac{K_p}{s^2 + (a+b)s + (ab + K_c K_p)} \tag{6b}$$

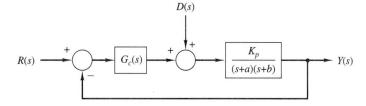

Figure 15.2 Block diagram for second-order plant and unity feedback.

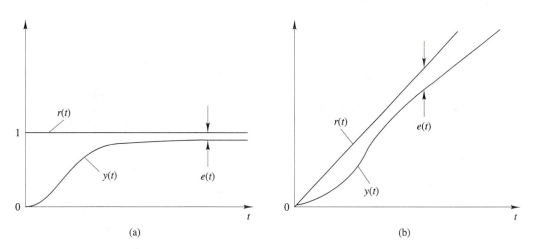

Figure 15.3 Response to the reference input when $G_c(s) = K_c$. (a) Unit step response. (b) Unit ramp response.

When the disturbance input is zero and the reference input is the unit step function, $R(s) = 1/s$ and

$$Y(s) = \frac{K_c K_p}{s[s^2 + (a+b)s + (ab + K_c K_p)]}$$

By the final-value theorem, the steady-state response is

$$y_{ss} = \frac{K_c K_p}{ab + K_c K_p} \tag{7}$$

The steady-state error is

$$e_{ss} = 1 - \frac{K_c K_p}{ab + K_c K_p} = \frac{ab}{ab + K_c K_p} \tag{8}$$

Although this is not zero, it can be made small by choosing K_c such that $K_c K_p \gg ab$.

The step input $r(t)$ and a possible curve for the output $y(t)$ are shown in Figure 15.3(a). The nature of the transient response depends on the poles of the closed-loop transfer function $T_R(s)$, which for the plot shown have been assumed to be on the negative real axis.

We next let the reference input be the unit ramp function

$$r(t) = \begin{cases} 0 & \text{for } t \leq 0 \\ t & \text{for } t > 0 \end{cases} \tag{9}$$

If $r(t)$ and $y(t)$ are angular displacements of rotating mechanical components, then the input in (9) corresponds to a constant angular velocity of 1 for all $t > 0$. Then $R(s) = 1/s^2$ and

$$Y(s) = \frac{K_c K_p}{s^2[s^2 + (a+b)s + (ab + K_c K_p)]}$$

for which the beginning of the partial-fraction expansion is

$$Y(s) = \frac{K_c K_p/(ab + K_c K_p)}{s^2} + \cdots$$

so that

$$y(t) = \frac{K_c K_p}{ab + K_c K_p}t + \cdots \tag{10}$$

The dots represent terms that are either constant or decaying with time. Thus for large values of time, $y(t)$ approaches a straight line with a slope of $K_c K_p/(ab + K_c K_p)$. For any finite value of $K_c K_p$, this slope will be less than 1, and the error will continually increase, as shown in Figure 15.3(b). For a rotating mechanical system, this means that the steady-state output angular velocity is less than that for the input, so the output angular displacement lags further and further behind the input. This lag can be reduced by making K_c very large.

We consider next the response to an unwanted disturbance input. If $r(t) = 0$ and $d(t)$ is the unit step function, then $D(s) = 1/s$, so

$$Y(s) = \frac{K_p}{s[s^2 + (a+b)s + (ab + K_c K_p)]}$$

and

$$y_{ss} = \frac{K_p}{ab + K_c K_p} \tag{11}$$

For the disturbance response to be small, we require $K_c \gg K_p$ in addition to $K_c K_p \gg ab$. Once again, we see that a large controller gain will improve the steady-state response. The reader may wish to show that if $d(t)$ is the unit ramp function, then

$$Y(s) = \frac{K_p}{s^2[s^2 + (a+b)s + (ab + K_c K_p)]}$$

and

$$y(t) = \frac{K_p}{ab + K_c K_p}t + \cdots \tag{12}$$

which again indicates the need for a large value of K_c.

We now look at the nature of the transient response, which is governed by the poles of the closed-loop transfer function. These are points on the root locus that corresponds to the open-loop transfer function

$$\frac{K_c K_p}{(s+a)(s+b)}$$

and that is shown in Figure 15.4. As long as a, b, and $K_c K_p$ are positive numbers, the system is always stable because the locus remains in the left half-plane. For large values of $K_c K_p$, however, the roots will be complex and far from the real axis, corresponding to a small damping ratio ζ. The transient response will then contain stronger oscillations than would normally be desired.

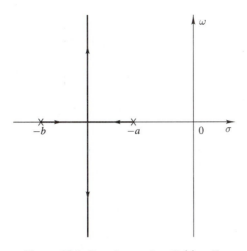

Figure 15.4 Root locus when $G_c(s) = K_c$.

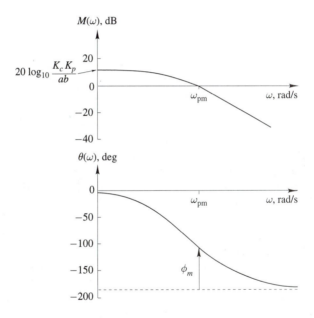

Figure 15.5 Bode diagram when $G_c(s) = K_c$.

Note that there are conflicting priorities concerning the choice of the controller gain K_c. For the best steady-state behavior, the gain should be very large. However, this can result in an undesirable transient response.

A typical Bode diagram is sketched in Figure 15.5. The magnitude curve has an initial slope of zero and a final slope of −40 dB per decade. As expected, the phase margin ϕ_m is always positive. Increasing $K_c K_p$ raises the magnitude curve without affecting the phase angle. This improves the steady-state behavior of the system but decreases ϕ_m. The gain margin k_m is not defined, because the angle curve never crosses −180°.

Proportional-Plus-Derivative Control

In Section 13.4, we showed that including a tachometer, which provided a feedback signal proportional to the derivative of the output angle, could improve the dynamic response of the overall system. We now generalize that result. We take the **proportional-plus-derivative** (PD) controller's transfer function to be

$$G_c(s) = K_c(1 + K_D s) \tag{13}$$

Because the transfer function for a component described by $y(t) = K_D \dot{x}$ is $K_D s$, the controller includes both proportional and derivative action. Equations (5) become

$$T_R(s) = \frac{\dfrac{K_c K_p (1 + K_D s)}{(s+a)(s+b)}}{1 + \dfrac{K_c K_p (1 + K_D s)}{(s+a)(s+b)}}$$

$$= \frac{K_c K_p (1 + K_D s)}{s^2 + (a+b+K_c K_p K_D)s + (ab + K_c K_p)} \tag{14}$$

and

$$T_D(s) = \frac{\dfrac{K_p}{(s+a)(s+b)}}{1 + \dfrac{K_c K_p (1 + K_D s)}{(s+a)(s+b)}}$$

$$= \frac{K_p}{s^2 + (a+b+K_c K_p K_D)s + (ab + K_c K_p)} \tag{15}$$

We can examine the steady-state responses to the reference and disturbance inputs by the procedure used for the proportional controller. Because K_D does not affect the limit of $T_R(s)$ or $T_D(s)$ as $s \to 0$, the key expressions turn out to be the same as before. For large values of time, $y(t)$ is again given by (7) and (10) when $r(t)$ is the unit step and unit ramp, respectively. It is also given by (11) and (12) when $d(t)$ is the unit step and unit ramp, respectively.

Although K_D does not influence the steady-state response, it can change the transient behavior significantly. The open-loop transfer function

$$K_c K_p \frac{1 + K_D s}{(s+a)(s+b)}$$

has a zero at $s = -1/K_D$. Root-locus plots for three different zero positions are shown in Figure 15.6. In all three cases, the locus is confined to the left half-plane and the system is always stable. Recall, however, that the distances of the roots from the imaginary axis are the reciprocals of the time constants in the corresponding transient terms. The time constants should be reasonably small so that the transient response dies out quickly. Thus we usually want to move the roots away from the vertical axis. In order to accomplish this, we choose K_D such that the zero at $s = -1/K_D$ is to the left of both open-loop poles, as in part (a) of Figure 15.6. A graphical comparison of Figures 15.4 and 15.6(a) shows that the added zero in Figure 15.6(a) has caused the vertical parts of the locus to curve to the left and intersect again with the negative real axis.

Figure 15.7 shows a typical Bode diagram for the open-loop pole-zero pattern in Figure 15.6(a). The phase angle approaches zero degrees at low frequency and is asymptotic to $-90°$ at high frequency. The phase margin ϕ_m is always positive, but the gain margin k_m is again undefined because the angle curve does not cross or even approach the $-180°$ line.

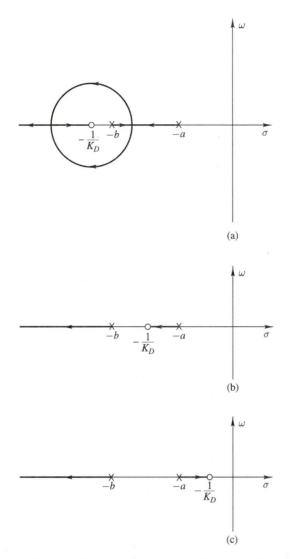

(a)

(b)

(c)

Figure 15.6 Possible root-locus plots when $G_c(s) = K_c(1 + K_D s)$.

There is a potential problem in constructing a PD controller. The numerator of the transfer function for a practical controller cannot be of higher order than the denominator— that is, the number of finite-plane zeros cannot exceed the number of finite-plane poles. The expression in (13) has a zero at $s = -1/K_D$ but no poles. Viewed as the ratio of two polynomials, the denominator of $G_c(s)$ is 1 (a polynomial of order zero), whereas its numerator is of order one.

It is true that one of the problems at the end of this chapter shows how a circuit using an ideal operational amplifier can have a transfer function similar to (13). However, an amplifier that has a finite, nonzero gain at arbitrarily high frequencies cannot be built. All realizable amplifiers have at least one pole, which may be at a very high frequency. If this frequency is much higher than any frequency of interest in the system at hand, then the "unrealizability" constraint is of no practical importance. A controller with a single zero and no pole of consequence could then be built and used.

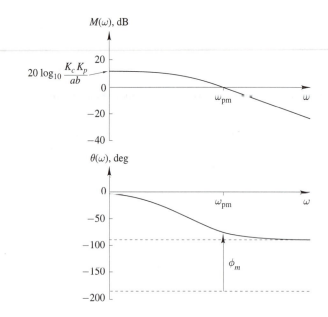

Figure 15.7 Bode diagram when $G_c(s) = K_c(1 + K_D s)$.

These problems are not confined to analog systems. When such systems are simulated or built digitally, as with MATLAB, similar related problems occur. In particular, if a differentiating block is required in a system, then the integration step size needed to obtain stable, accurate simulations may become very small, and the time required to calculate a response may become excessive. This problem is most severe for input functions containing high frequencies, as is the case for a step function.

This realizability problem occurs in systems in which the only variable available to the controller is the system output, which must be differentiated by the controller. In some systems, such as the servomechanism studied in Section 13.4, a signal is available that is proportional to the derivative of the output variable. That system contained a tachometer that produced a signal proportional to the derivative of the angle that was to be controlled. If such a signal is not available, then in order to make the controller realizable, we must add a pole to it. Since we do not want to influence the behavior of the root locus in the vicinity of the zero of $G_c(s)$, we place this pole far to the left of the zero. We follow the rule of thumb that a factor of ten is sufficiently large, and define a realizable version of $G_c(s)$ as

$$G_c(s) = \frac{K_c(1 + K_D s)}{1 + K_D s / 10}$$

Proportional-Plus-Integral Control

The steady-state error in the step response for a proportional controller can be eliminated by including a feedback term proportional to the integral of the error. If it were possible for the steady-state error to have a nonzero constant value, then the signal fed back by this new component would increase without limit. To obtain a **proportional-plus-integral** (PI) controller, we let

$$G_c(s) = K_c \left(1 + \frac{K_I}{s}\right) = \frac{K_c(s + K_I)}{s} \tag{16}$$

where K_I/s is the transfer function for a component described by $y(t) = K_I \int_0^t x(\lambda)d\lambda$. An implementation of this equation using an ideal op-amp was studied in Example 6.14. Inserting (16) into (5) gives

$$T_R(s) = \frac{\dfrac{K_c K_p(s+K_I)}{s(s+a)(s+b)}}{1 + \dfrac{K_c K_p(s+K_I)}{s(s+a)(s+b)}}$$

$$= \frac{K_c K_p(s+K_I)}{s^3 + (a+b)s^2 + (ab+K_c K_p)s + K_c K_p K_I} \tag{17}$$

and

$$T_D(s) = \frac{\dfrac{K_p}{(s+a)(s+b)}}{1 + \dfrac{K_c K_p(s+K_I)}{s(s+a)(s+b)}}$$

$$= \frac{K_p s}{s^3 + (a+b)s^2 + (ab+K_c K_p)s + K_c K_p K_I} \tag{18}$$

We first look at the steady-state response when the disturbance input is zero. When $r(t) = U(t)$, $Y(s) = T_R(s)/s$ and

$$y_{ss} = T_R(0) = \frac{K_c K_p K_I}{K_c K_p K_I} = 1$$

In contrast to the results for the previous types of controllers, the steady-state error is zero for all nonzero values of $K_c K_p K_I$. When $r(t)$ is the unit ramp function,

$$Y(s) = \frac{K_c K_p(s+K_I)}{s^2[s^3 + (a+b)s^2 + (ab+K_c K_p)s + K_c K_p K_I]}$$

$$= \frac{1}{s^2} - \frac{ab/(K_c K_p K_I)}{s} + \cdots$$

and

$$y(t) = t - \frac{ab}{K_c K_p K_I} + \cdots$$

The dots represent those terms that decay to zero as t becomes large. Thus the steady-state error to the unit ramp is

$$e_{ss} = \frac{ab}{K_c K_p K_I}$$

Again in contrast to the controllers previously considered, the error for large values of t does not continually increase. In fact, it can be made arbitrarily small by making $K_c K_p K_I$ sufficiently large.

The response to a unit step disturbance input, when $r(t) = 0$, is $Y(s) = T_D(s)/s$. From (18) and the final-value theorem, we see that

$$y_{ss} = 0$$

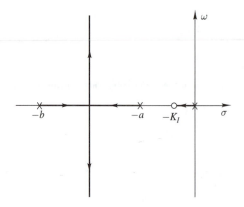

Figure 15.8 Root-locus diagram for $G_c(s) = K_c(1 + K_I/s)$.

When $d(t)$ is the unit ramp function, $Y(s) = T_D(s)/s^2$ and

$$y_{ss} = \frac{K_p}{K_c K_p K_I} = \frac{1}{K_c K_I}$$

which becomes small for large values of $K_c K_I$.

For the transient response, we look at the root-locus and Bode diagrams corresponding to the open-loop transfer function

$$\frac{K_c K_p (s + K_I)}{s(s + a)(s + b)}$$

The location of the open-loop zero at $s = -K_I$ is usually chosen close to the pole at the origin, as shown in Figure 15.8. Then the main part of the root-locus diagram will not differ greatly from that for proportional control. There will be another branch of the locus near the origin, and this will result in an additional term in the transient response. Although this term will decay relatively slowly, its magnitude will be small because of the short distance between the pole and the zero of $G_c(s)$.

The Bode diagram will be similar to the one for proportional control, except at low frequencies. For very small values of ω, the magnitude plot will have a slope of -20 dB per decade, and the angle curve will approach $-90°$.

Other Types of Control

We have seen that proportional-plus-integral control gives very good steady-state behavior for both reference and disturbance inputs. The dynamic response, however, will be slower than with proportional-plus-derivative control. In order to get both good steady-state and good dynamic characteristics, we can use a controller that combines derivative and integral action. A general **proportional-plus-integral-plus-derivative** (PID) controller has the transfer function

$$G_c(s) = K_c \left(1 + K_D s + \frac{K_I}{s} \right) \tag{19}$$

An ideal op-amp circuit for (19) was examined in Problem 8.62 at the end of Chapter 8. A practical disadvantage arises, however, with both a PI and PID controller. It is very difficult to construct an electronic controller with a pole of $G_c(s)$ right at the origin of the s-plane. It is more realistic to move such a pole slightly to the left of $s = 0$.

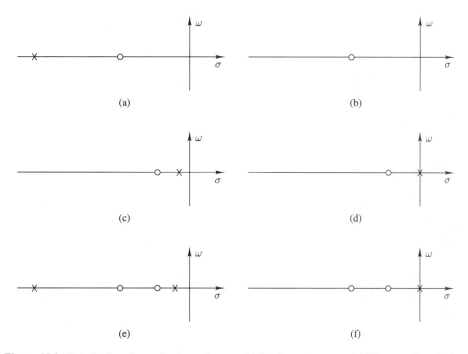

Figure 15.9 Relative locations of poles and zeros. (a) Lead compensator. (b) PD controller. (c) Lag compensator. (d) PI controller. (e) Lag-lead compensator. (f) PID controller.

Pole-zero plots for the idealized PD, PI, and PID controllers are shown in parts (b), (d), and (f), respectively, of Figure 15.9. We have pointed out, however, the practical problems of implementing $G_c(s)$ when there are more finite-plane zeros than poles or when there is a pole exactly at $s = 0$.

Three other basic types of controllers avoid both of these difficulties. These are represented by lead, lag, and lag-lead transfer functions, whose pole-zero patterns are displayed in Figure 15.9(a), (c), and (e). Because some of the poles or zeros may be quite close to the origin and others far away, it can be difficult to indicate their true positions in a single diagram. In Figure 15.9, therefore, the diagrams are not to scale but show only the relative pole and zero locations. From Figure 15.9, we see how the lead, lag, and lag-lead characteristics might be approximated by those for the PD, PI, and PID controllers that are shown to their right.

In Figure 15.10, we show Bode plots for typical **lead** and **lag** compensators. The Bode diagram for the **lag-lead** case will be included in the next section. The names for these transfer functions refer to their phase-angle characteristics. Positive and negative values of $\theta(\omega)$ are called leading and lagging angles, respectively. For a lag network, the majority of the changes in the curves normally take place at lower frequencies than for a lead network. Keep in mind that Figures 15.9 and 15.10 show the characteristics only of the controller and not of the entire open-loop transfer function.

Many useful transfer functions can be implemented with electrical, mechanical, hydraulic, or pneumatic components. Two of the problems at the end of Chapter 8 presented op-amp networks that can be used for lead and PID control. Some of the problems at the end of this chapter examine op-amp networks for achieving other types of transfer functions. Because such devices can be incorporated into the controller to compensate for deficiencies that would otherwise exist in the system's behavior, they are often called **compensators**.

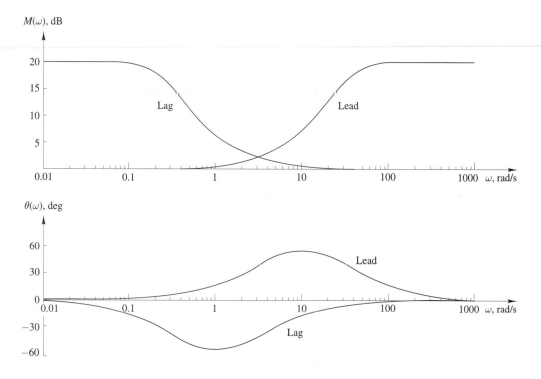

Figure 15.10 Bode diagrams for lead and lag compensators.

In some cases, appropriate compensators are put in the sensor block of Figure 15.1(a) or in an inner feedback loop, rather than in the controller block.

Additional Concerns

A mathematical model provides only an approximate description of the physical system. In the construction of the components, some deviations from the nominal parameter values are to be expected. Wear and environmental factors might further modify their characteristics over time. In any event, not all the secondary features can be incorporated into the model. In fact, there may be some uncertainty in the complete description of the plant to be controlled.

It is important to be sure that the system's behavior will remain acceptable when the parameters undergo moderate changes. It is possible to calculate the effects of individual parameter changes on the responses of the model. Fortunately, the very structure of feedback systems tends to reduce these effects. Even then, however, a safety margin is needed because of uncertainty about how closely the model describes the actual system. Often, factors that may be relatively unimportant under most conditions play a more significant role as the open-loop gain becomes large. Even though the original model might have indicated that the system is always stable, additional poles and zeros may need to be included and may even drive the root locus into the right half-plane for large values of the gain constant K. Nonlinear effects also become increasingly important. More sophisticated models, extensive computer simulation, and testing may be necessary.

The operation of practical sensors and other system components often produces unwanted high-frequency signals in addition to the desired responses. It is important that such internally generated signals, referred to as noise, be attenuated rapidly as they travel around the feedback loop. Thus the amplitude curve should fall off sufficiently fast at high frequencies. This is another disadvantage of the PD and PID controllers, whose Bode plots rise at 20 dB per decade at high frequencies. It may be necessary to specify a cut-off frequency, above which the open-loop gain must be sufficiently small. If the designer is willing to use more costly high-performance components, then this cut-off requirement can be relaxed somewhat.

▶ 15.2 APPLICATIONS

We shall illustrate the concepts discussed in Section 15.1 by considering two relatively simple systems. In each case, we shall assume that the system can be represented by Figure 15.1(b) so that $T_R(s)$ and $T_D(s)$ are given by (1). The transfer function of the plant will be specified, will have distinct poles only on the negative real axis, and will have no zeros in the finite s-plane. We assume that we may choose the transfer function $G_c(s)$ for the controller, with no restriction on the values of its coefficients.

In the solutions, we shall assume that there are no constraints other than those given in the problem statements. We shall, however, comment on some of the practical difficulties that one should consider when choosing $G_c(s)$. The design choices that we make will not represent an approach that should always be used, nor will our results be unique. The purpose of this section is simply to introduce some of the important design techniques and to show where design choices are required.

▶ EXAMPLE 15.1

Let the transfer function of the plant in Figure 15.1(b) be

$$G_p(s) = \frac{10}{(s+1)(s+10)} \tag{20}$$

Assume a unity-feedback system, so that $H(s) = 1$. The following specifications must be satisfied:

1. The steady-state error for a reference step input should not exceed 2 percent of the input.

2. The steady-state response to a disturbance step input should not exceed 2 percent of the disturbance.

3. The closed-loop transfer function $T_R(s)$ should have poles with time constants less than 0.1 s for the exponential factors in the transient response.

4. For any pair of complex conjugate poles, the damping ratio ζ should be at least 0.8.

Determine whether it is possible to satisfy these design constraints by using proportional control. If it is not, use proportional-plus-derivative control to do so.

SOLUTION   The transfer function for the plant is given by (4) with $K_p = 10$, $a = 1$, and $b = 10$. For proportional control, where $G_c(s) = K_c$, we can substitute these numerical values into (6) through (12). From (6), the closed-loop transfer functions for a reference or disturbance input are

$$T_R(s) = \frac{10K_c}{s^2 + 11s + 10(1 + K_c)}$$

$$T_D(s) = \frac{10}{s^2 + 11s + 10(1 + K_c)} \tag{21}$$

When $r(t)$ is the unit step function, the steady-state error is, from (8),

$$e_{ss} = \frac{10}{10 + 10K_c} = \frac{1}{1 + K_c} \tag{22}$$

which should not exceed 0.02. Because we find it generally unwise to choose the controller gain K_c larger than necessary, we let $K_c = 49$, the lowest value it can have and still keep e_{ss} as low as 0.02. This is a design choice, and larger values of K_c could have been chosen that would still produce a satisfactory controller.

When $d(t)$ is the unit step function, we see from (11) that the steady-state response is $y_{ss} = 1/(1 + K_c)$. Thus selecting $K_c = 49$ also limits the steady-state response to a constant disturbance input to 2 percent.

Once K_c has been chosen to meet the steady-state requirements, we have no further means of affecting the dynamic behavior. The denominator in (21) is the closed-loop characteristic polynomial. Thus with $K_c = 49$, the characteristic equation is

$$s^2 + 11s + 500 = 0 \tag{23}$$

from which the closed-loop poles are at $s = -5.50 \pm j21.67$. The time constant of the exponential factor associated with these poles is $1/5.50 = 0.182$ s. To find the damping ratio, we can compare (23) with $s^2 + 2\zeta\omega_n s + \omega_n^2 = 0$ or we can use Equation (8.30) to write $\zeta = \cos[\tan^{-1}(21.67/5.50)] = 0.246$. The dynamic behavior does not come close to meeting the specifications.

In order to improve the transient response without adversely affecting the steady-state characteristics, we can use a PD controller that has the transfer function

$$G_c(s) = K_c(1 + K_D s)$$

Then the open-loop transfer function becomes

$$G_c(s)G_p(s) = \frac{10K_c(1 + K_D s)}{(s + 1)(s + 10)} \tag{24}$$

The expressions for $T_R(s)$ and $T_D(s)$ are those in (14) and (15) with $K_p = 10$, $a = 1$, and $b = 10$. As explained in Section 15.1, the parameter K_D has no effect on $T_R(0)$ and $T_D(0)$. Thus we again take $K_c = 49$ in order to satisfy the steady-state specifications.

The extra zero at $s = -1/K_D$ in the open-loop transfer function in (24) should be placed to the left of both the poles, as shown in Figure 15.6(a). Suppose we choose this zero at $s = -15$, plot the root-locus diagram, and then find the point on the locus corresponding to $K_c = 49$. We use MATLAB to do this, with the results shown in Figure 15.11. The pair of closed-loop poles turn out to be at $s = -21.83 \pm j4.83$, corresponding to a time constant $\tau = 0.0458$ s and a damping ratio $\zeta = 0.976$. This is more than satisfactory, so we let $K_D = 1/15 = 0.06667$. We have made another design choice. As Figure 15.11 suggests, moving the open-loop zero from $s = -15$ to other nearby locations would not displace the root locus very far or substantially change the closed-loop poles for $K_c = 49$.

If the dynamic response were not satisfactory, we could reposition the zero at $s = -1/K_D$ and plot a new root-locus diagram. Actually, because the overall transfer function in this example is only second order, we could achieve the necessary results without plotting an accurate root locus. However, constructing such diagrams is generally a key step in the design of a controller.

▶ EXAMPLE 15.2

For the plant specified in Example 15.1, we want zero steady-state error when $r(t) = U(t)$ and also zero steady-state response when $d(t) = U(t)$. The conditions on the closed-loop

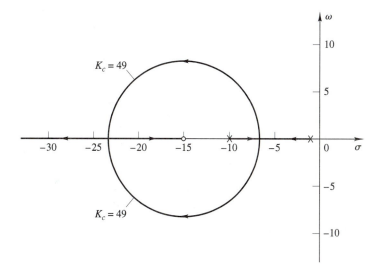

Figure 15.11 Root locus for Example 15.1 with PD control.

poles are the same as before. The new steady-state requirements could be met with a PI controller, as in (16) through (18). However, the transient response would be no better than that achieved with proportional control. In order to get the improved dynamic character-istics associated with derivative action and the excellent steady-state behavior associated with integral action, we use a PID controller that has the transfer function given in (19), which is repeated here.

$$G_c(s) = K_c\left(1 + K_D s + \frac{K_I}{s}\right) = \frac{K_c(K_D s^2 + s + K_I)}{s} \tag{25}$$

SOLUTION The open-loop transfer function is

$$G_c(s)G_p(s) = \frac{10K_c(K_D s^2 + s + K_I)}{s(s+1)(s+10)} \tag{26}$$

Inserting this expression into (1), with $H(s) = 1$, gives

$$T_R(s) = \frac{10K_c(K_D s^2 + s + K_I)}{s^3 + (11 + 10K_c K_D)s^2 + 10(1 + K_c)s + 10K_c K_I}$$

$$T_D(s) = \frac{10s}{s^3 + (11 + 10K_c K_D)s^2 + 10(1 + K_c)s + 10K_c K_I} \tag{27}$$

Because $T_R(0) = 1$, the steady-state error when $r(t) = U(t)$ is zero for all nonzero values of $K_c K_I$. And because $T_D(0) = 0$, the steady-state response to a constant disturbance input is also zero.

The controller has contributed to the open-loop transfer function in (26) a pole at the origin and two zeros. We shall let one of these zeros be at $s = -15$, as for the PD con-troller. The other zero is normally placed close to the pole at the origin, so that the dynamic behavior of the overall system will not be too different from that exhibited when the PD controller is used. We find it convenient to select this zero position at $s = -1$, which is the location of one of the open-loop poles.

This is another design choice which could have been made somewhat differently. The problem is easier to handle analytically if the zero of the controller cancels the pole of the

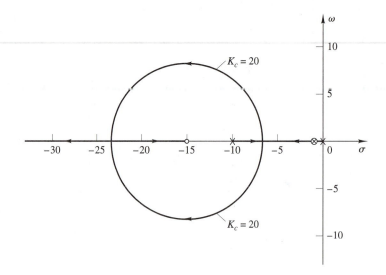

Figure 15.12 Root locus for Example 15.2 with PID control.

plant at $s = -1$. However, the ease with which MATLAB can obtain the desired responses numerically makes this consideration relatively unimportant. As noted below, it is nearly impossible to exactly cancel a pole in actual practice. Thus, the zero could be placed anywhere near $s = -1$ and the response would be little changed.

Specifying the two zeros of $G_c(s)$ indirectly determines the values of K_D and K_I. For our choice, we find that $K_D = 1/16$ and $K_I = 15/16$. Then (26) can be rewritten as

$$G_c(s)G_p(s) = \frac{(10/16)K_c\,(s^2 + 16s + 15)}{s(s+1)(s+10)} = \frac{(5/8)K_c\,(s+1)(s+15)}{s(s+1)(s+10)} \tag{28}$$

We again use MATLAB to plot the root locus for this open-loop transfer function as shown in Figure 15.12. Because the steady-state specification is satisfied for all values of K_c, we are now free to select this parameter to place the closed-loop poles anywhere on the locus. The computer program can easily calculate the points on the locus for a given value of K_c and can also determine the value of K_c that corresponds to a specified point. If, for example, $K_c = 20$, the closed-loop poles are at $s = -11.25 \pm j7.81$. The time constant and damping ratio for this pair of poles are $\tau = 1/11.25 = 0.889$ s and $\zeta = \cos[\tan^{-1}(7.81/11.25)] = 0.822$.

Although the foregoing choice of K_c satisfies the problem specifications, we should point out two possible problems. First, because of normal tolerances in the physical system, the pole and zero at $s = -1$ in $G_c(s)G_p(s)$ cannot be expected to cancel exactly. In addition to modifying the existing branches of the locus somewhat, this creates a third branch, which in turn yields another term in the transient response, with a time constant of about 1 s. However, the amplitude of this transient term is small because of the short distance between the pole and zero.

The second possible problem becomes apparent when we carefully examine the expression for the closed-loop disturbance transfer function. It turns out that, unlike $T_R(s)$, $T_D(s)$ contains a pole at $s = -1$ even if there is an exact cancellation of the pole and zero of $G_c(s)G_p(s)$ at that point. We shall not take the time to investigate the effects of this, but the reader should be aware of it.

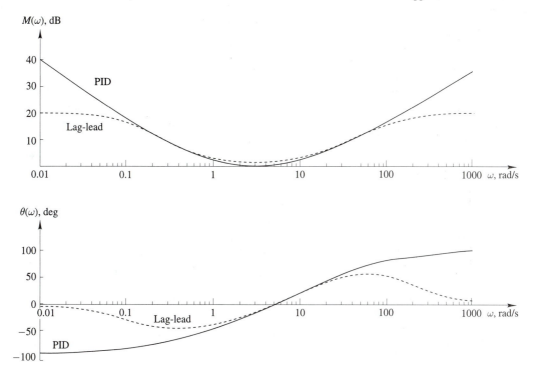

Figure 15.13 Bode diagrams for PID and lag-lead controllers.

We have already alluded to three practical difficulties that arise with a PID controller. First, it is difficult to achieve a pole right at $s = 0$. Second, $G_c(s)$ contains more finite-plane zeros than poles. This relates to the third problem, where unwanted noise might be amplified if $|G_c(j\omega)|$ increases at high frequencies.

Consider the Bode diagram shown in Figure 15.13 for the PID controller described by (25), with K_c chosen to be unity and with the values of K_D and K_I that we used in the last example. For very low and very high frequencies, $|G_c(j\omega)|$ becomes K_I/ω and $K_D\omega$, respectively. Thus the low- and high-frequency asymptotes for the magnitude curve have slopes of -20 dB per decade and $+20$ dB per decade, respectively. The low- and high-frequency asymptotes for the angle curve are $-90°$ and $+90°$. The fact that the magnitude approaches infinity as $\omega \to 0$ and as $\omega \to \infty$ results in the difficulties described in the previous paragraph. To avoid this, we can use a lag-lead transfer function with the pole-zero pattern shown in part (e) of Figure 15.9.

In order to compare the characteristics of the two controllers, we assume that their multiplying constants have been adjusted so that for both cases, $M(\omega)$ approaches unity in the mid-frequency range. For any controller, increasing the value of K_c just raises the entire magnitude curve by a constant amount without affecting the angle curve. A Bode diagram for the lag-lead controller used in the next example is also shown in Figure 15.13, again for $K_c = 1$. Note that except at low and high frequencies, the two diagrams are quite similar.

▷ **EXAMPLE 15.3**

Determine a suitable lag-lead compensator for the plant considered in the last two examples. The steady-state and transient requirements are the same as those given in Example 15.1.

SOLUTION   The transfer function for the PID controller used in Example 15.2 was

$$\frac{K_c(s+1)(s+15)}{16s}$$

For the lag-lead controller, we keep the same zero positions but move the pole at the origin to $s = -0.1$. We place the new pole at $s = -150$. This is a design choice, made to obtain a rapid decay of the transient response by keeping the poles of the closed-loop transfer function far to the left. Then

$$G_c(s) = \frac{10K_c(s+1)(s+15)}{(s+0.1)(s+150)} \tag{29}$$

The multiplying factor of 10 has been added to the numerator so that in the mid-frequency range (say, $3 < \omega < 5$), $M(\omega)$ is approximately the same for the PID and lag-lead controllers when the values of K_c are the same. This makes possible a more meaningful comparison of the values of K_c needed for the two cases.

Using (29), we have for the open-loop transfer function

$$G_c(s)G_p(s) = \frac{100K_c\,(s+1)(s+15)}{(s+0.1)(s+1)(s+10)(s+150)} \tag{30}$$

To find the closed-loop transfer functions, we substitute this expression into (1), with $H(s) = 1$, giving

$$T_R(s) = \frac{100K_c(s+15)}{s^3 + 160.1s^2 + (1516 + 100K_c)s + (150 + 1500K_c)}$$

$$T_D(s) = \frac{10(s+0.1)(s+150)}{(s+1)[s^3 + 160.1s^2 + (1516 + 100K_c)s + (150 + 1500K_c)]}$$

We see that $T_R(0) = 10K_c/(1 + 10K_c)$. When $r(t) = U(t)$, the steady-state error is

$$e_{ss} = 1 - \frac{10K_c}{1 + 10K_c} = \frac{1}{1 + 10K_c} \tag{31}$$

Because $T_D(0) = 1/(1 + 10K_c)$, the steady-state response when $d(t) = U(t)$ is

$$y_{ss} = \frac{1}{1 + 10K_c} \tag{32}$$

The expressions in (31) and (32) are specified not to exceed 0.02, so we require $K_c \geq 4.9$.

The root-locus plot for the open-loop transfer function in (30) could be obtained from MATLAB by entering its zero ($s = -15$), its poles ($s = -0.1$, $s = -10$, and $s = -150$), and its gain ($k = 100$) to create a ZPK object, as described in Section 14.1. An alternative solution can be obtained using Simulink to model the plant and the controller. MATLAB can then be used to create and study the resulting linear time-invariant object. In order to use this approach, we draw the Simulink diagram shown in Figure 15.14.

The icons labeled `Controller` and `Plant` are both `Zero-Pole` blocks from the `Linear` library. The sensor is represented by a block with a gain of unity. Figure 15.14 shows `step` inputs for both the reference and disturbance inputs.

In order to use the MATLAB function `rlocus`, we need only the open-loop transfer function given by (30). This can be obtained by simplifying the original Simulink diagram to the one shown in Figure 15.15. The root locus will be the same whether the input is $r(t)$ or $d(t)$, so only one input is included. We have also removed the factor K_c from the controller's transfer function. This is because the root-locus command will automatically introduce the gain parameter, whose numerical value has not yet been chosen.

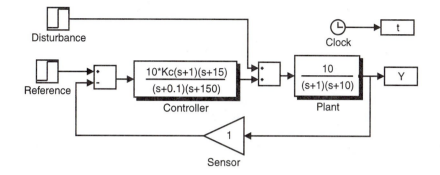

Figure 15.14 Simulink diagram for Example 15.3.

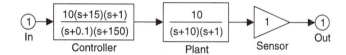

Figure 15.15 Simplified Simulink diagram for Example 15.3.

The In and Out blocks are found in the Connections library. They identify the input and output of the open-loop transfer function for which we want to produce the root locus. Note that the loop has not been closed. In order to use the rlocus command in MATLAB, we must first produce the open-loop transfer function using the linmod command. If the Simulink model in Figure 15.15 is called ex15_3.mdl, then the command linmod('ex15_3') will form the required model. If this is followed by rlocus, a plot similar to Figure 15.16(a) is produced. The following M-file shows the required commands:

```
[a,b,c,d] = linmod('ex15_3');
T = ss(a,b,c,d);
k=[0:0.05:200];
rlocus(T,k)
axis([-160 1 -60 60])
```

The presence of a pole at $s = -150$ causes the default choice of the gains for which points on the locus are plotted to be too widely spaced. To overcome this, we specify a vector k of 4001 gain values between zero and 200. Figure 15.16(a) shows all the branches of the root locus, but with little detail. Figure 15.16(b) shows the detail around the origin.

The values of k (which is the same as K_c) corresponding to any points of interest can be determined using rlocfind. The points marked with triangles correspond to $K_c = 49$. Keep in mind that we may select any value of K_c greater than 4.9 and still meet the steady-state error specifications.

If a pair of complex closed-loop poles has a real part of -10, the exponential factor in the corresponding terms in the transient response has a time constant of 0.1 s. This just satisfies one of the conditions on the dynamic behavior. For our locus, the real part is -10 when $K_c = 14.41$. This is determined using the rlocfind command, as described in Example 14.15. With this value of K_c, the pair of complex poles are then at $s = -10 \pm j7.443$, which corresponds to a damping ratio $\zeta = \cos[\tan^{-1}(7.443/10)] = 0.802$. Because this value is greater than the minimum value allowed by the problem statement, we may choose $K_c = 14.41$. By substituting this value into (31) and (32), we find that in the steady state,

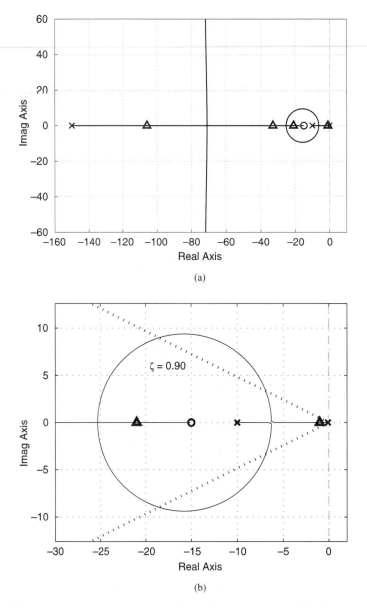

Figure 15.16 (a) Root locus for Example 15.3 with lag-lead control. (b) Enlargement of locus near the origin.

the error to a reference step input and the response to a constant disturbance input have both been reduced to 0.69 percent.

It is instructive to consider what could be done if the problem statement had required a damping ratio $\zeta = 0.9$. For a pair of complex poles, we see from Figure 8.10(a) and Equation (8.30) that all points having a damping ratio ζ lie on a line that makes an angle $\theta = \cos^{-1}\zeta$ with the negative real axis. We sketch such a line on Figure 15.16(b) and note that it intersects the root locus at two places. One of these has a real part of s about -7, so the transient response would not decay fast enough to meet the specifications. The second intersection occurs when the real part of s is about -18. Using rlocfind, we identify a

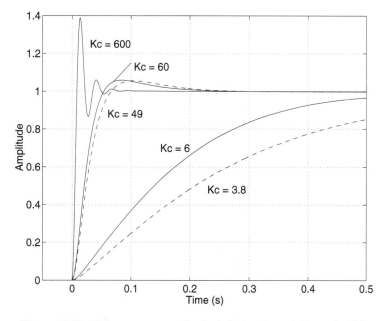

Figure 15.17 Unit step responses for the reference input in Example 15.3.

point at $K_c = 35.0$ where $s = -18.72 \pm j8.92$ and $s = -122.67$, so that $\zeta = 0.90$ and the time constant is $\tau = 1/18.72 = 0.053$ s. The steady-state error for a constant input, found from (31), is $1/351 = 0.28$ percent, and the steady-state response to a unit step disturbance, found from (32), is also 0.28 percent.

Another useful insight into the characteristics of this system can be gained by changing K_c over a wide range and plotting the resulting step responses on the same axes. This is done in Figure 15.17 for $K_c = 3.8, 6, 49, 60,$ and 600. We plotted the response when $K_c = 3.8$ because at that value the rightmost branches of the root locus first leave the real axis.

The system is also subject to disturbances applied directly to the plant. The response of the closed-loop system to disturbance inputs depends strongly on K_c. Figure 15.18 shows these responses for the same values of K_c used for Figure 15.17. The maximum amplitude of the response to a unit step disturbance is reduced by higher values of controller gain.

For the plant characteristics given by (20), we can use MATLAB to plot the unit step response for a reference input for different choices of $G_c(s)$. In Figure 15.19, the curves for proportional control and PD control are for Example 15.1, both with $K_c = 49$. The curve for lag-lead compensation is for Example 15.3 with $K_c = 49$.

For proportional control, we could have anticipated the large oscillations and the relatively long time to approach the steady state. Recall that the closed-loop poles had a damping ratio $\zeta = 0.246$ and a time constant $\tau = 0.182$ s. The curve for proportional-plus-derivative control exhibits excellent dynamic characteristics (corresponding to closed-loop poles with $\zeta = 0.976$ and $\tau = 0.046$ s). However, a PD controller would be more sensitive to unwanted noise and in practice would be implemented with a lead compensator.

For the lag-lead design, the pair of dominant complex poles were quite close to those for the PD controller. Because the damping ratio was slightly smaller and the time constant slightly larger, the overshoot and the time to approach the steady state have increased

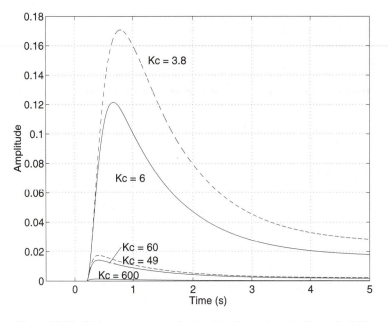

Figure 15.18 Unit step responses for the disturbance input in Example 15.3.

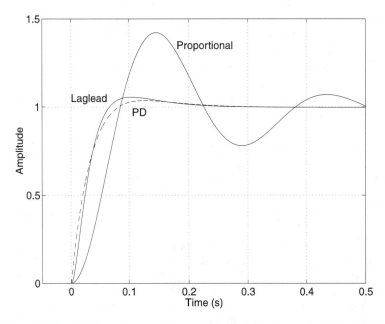

Figure 15.19 Unit step responses for a reference input for Examples 15.1 and 15.3.

slightly. However, the lag-lead design has a far better steady-state behavior, with an error of only 0.20 percent instead of 2.0 percent.

For any of the designs, a large value of K_c may cause practical difficulties. At some point, the linear model will no longer represent the physical system with reasonable accuracy. As the open-loop gain is increased, unmodeled modes and nonlinearities generally become important. Before selecting a final value of K_c, the designer might run additional computer simulations using a more complex but more accurate model. In Example 15.3, it may be better to choose $K_c = 15$, which will meet all the specifications given in Example 15.1, than to seek a faster response by choosing $K_c = 49$.

In the final two examples, the transfer function $G_p(s)$ for the plant will have three poles and no finite-plane zeros. Increasing the number of poles of $G_p(s)$ without increasing the number of zeros tends to make the system's transient response more difficult to control. Consider the case of a unity-feedback system with proportional control. The poles and zeros of the open-loop transfer function $K_c G_p(s)$ are the same as those of $G_p(s)$. From Section 14.5, the number of branches in the root locus that approach infinity is $n - m$, where n and m are the numbers of open-loop poles and zeros, respectively, in the finite s-plane. If $n - m \geq 3$, some of these branches pass into the right half-plane as the gain increases, making the overall system unstable.

Even if a more complicated function were to be used for $G_c(s)$, it would be difficult in practice to add more zeros than poles to the open-loop transfer function. This is because we want to attenuate any high-frequency noise. Nevertheless, we can still choose the poles and zeros of $G_c(s)$ in such a way as to try to pull the root locus to the left in the s-plane.

When the transfer function of the plant has a pole at the origin, the steady-state error for a reference step input is zero. However, there can still be a nonzero steady-state response to a constant disturbance input. In order to illustrate this, let

$$G_p(s) = \frac{K_p}{s(s+a)(s+b)}$$

and $G_c(s) = K_c$. With $H(s) = 1$, we obtain from (1)

$$T_R(s) = \frac{K_c K_p}{s^3 + (a+b)s^2 + abs + K_c K_p}$$

$$T_D(s) = \frac{K_p}{s^3 + (a+b)s^2 + abs + K_c K_p} \tag{33}$$

We see that $T_R(0) = 1$, so there is no steady-state error when $r(t)$ is constant. Because $T_D(0) = 1/K_c$, the steady-state response to $d(t) = U(t)$ is $1/K_c$.

▶ **EXAMPLE 15.4**

A particular feedback system can be represented by Figure 15.1(b), with $H(s) = 1$ and with

$$G_p(s) = \frac{500}{s(s+10)(s+50)} \tag{34}$$

The unit step response to $r(t)$ should have a steady-state value of unity, should have an overshoot of less than 15 percent, and should be within 4 percent of the steady-state value for all $t > 0.5$ s. The steady-state response to a constant disturbance input should be limited to 1 percent of the input. Determine a transfer function for the controller that meets these specifications. Because of the practical considerations discussed earlier in this section, the number of finite-plane zeros of $G_c(s)$ should not exceed the number of poles, and none of its poles should be at the origin.

SOLUTION If $G_c(s) = K_c$, the only poles of the open-loop transfer function will be at $s = 0$, -10, and -50, which leads to the root-locus plot shown in Figure 15.20. The expressions for $T_R(s)$ and $T_D(s)$ are given by (33) with $K_p = 500$, $a = 10$, and $b = 50$. In order to limit to 1 percent the steady-state response to a constant disturbance input, we must choose K_c to be at least 100. With $K_c = 100$, the closed-loop poles are at $s = 2.16 \pm j27.80$ and -64.31, and the system is unstable.

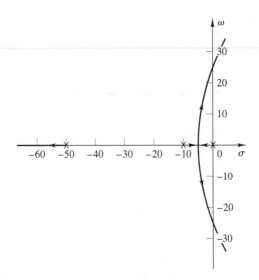

Figure 15.20 Root-locus plots for Example 15.4 with $G_c(s) = K_c$.

In order to meet the requirements on both the steady-state and transient behavior, we shall use a lag-lead compensator. We put one zero of $G_c(s)$ at $s = -12$, which is a little to the left of the pole of $G_p(s)$ at -10, in order to pull to the left the two right-hand branches of the locus. We also put a pole of $G_c(s)$ close to the origin (specifically, at $s = -0.1$) to reduce the steady-state error to a disturbance input.

Associated with the pole at $s = -0.1$ we include a zero at $s = -1$, so that the shape of the root-locus branches moving toward the right half-plane will not be greatly altered. Associated with the zero at $s = -12$ we add a pole at $s = -120$, so that $G_c(s)$ will not have more zeros than poles in the finite s-plane. Once we have selected the zero at $s = -12$ and the pole at $s = -0.1$, it is not uncommon to use a multiplying factor of 10 to locate the additional zero at $-(10)(0.1) = -1$ and the additional pole at $-(10)(12) = -120$. Thus we let

$$G_c(s) = \frac{10K_c(s+1)(s+12)}{(s+0.1)(s+120)}$$

Then the open-loop transfer function becomes

$$G_c(s)G_p(s) = \frac{5000K_c(s+1)(s+12)}{s(s+0.1)(s+10)(s+50)(s+120)} \tag{35}$$

We find that

$$T_R(s) = 5000K_c(s+1)(s+12)/P(s)$$
$$T_D(s) = 500(s+0.1)(s+120)/P(s)$$

where

$$P(s) = s^5 + 180.1s^4 + 7718s^3 + (60{,}770 + 5000K_c)s^2 + (6000 + 52{,}000K_c)s + 60{,}000K_c$$

We see that $T_R(0) = 1$ and $T_D(0) = 0.1/K_c$. In order to satisfy the steady-state error specification, we require $K_c \geq 10$.

The root locus for (35) is plotted in Figure 15.21(a). It is difficult to show the entire locus clearly in a single diagram, so the portion near the origin is enlarged in part (b) of

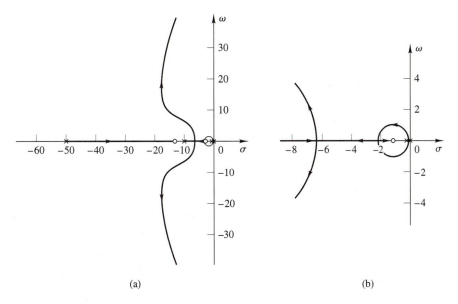

(a)   (b)

Figure 15.21 (a) Root locus for Example 15.4 with lag-lead control. (b) Enlargement of locus near the origin.

Figure 15.21. If we choose $K_c = 16$, the closed-loop poles are at $s = -1.96, -19.16$, $-16.01 \pm j10.63$, and -127.85. Note that the time constant corresponding to the pole at $s = -1.96$ is $\tau = 1/1.96 = 0.51$ s. This has the effect of delaying the approach to the steady-state. However, the magnitude of this term in the transient response will be relatively small because of the presence of the closed-loop zero at $s = -1$.

The response to $r(t) = U(t)$ is shown in Figure 15.22. There is some overshoot because of the complex poles. Because of the pole at $s = -1.96$, the response takes some time to approach its final steady-state value. However, it does satisfy all the conditions in the problem statement.

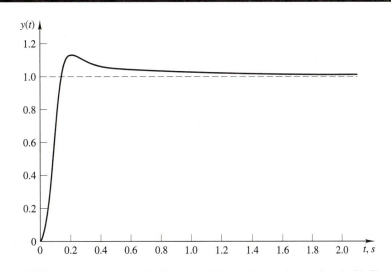

Figure 15.22 Unit step response for Example 15.4 with lag-lead control and with $K_c = 16$.

We next consider a design for this system to meet specifications on the gain and phase margins, rather than on the damping ratios or time constants.

▶ EXAMPLE 15.5

Design a lag-lead controller for the system in Example 15.4 which meets the following specifications:

1. The steady-state error specifications are the same as for Example 15.4. This requires $K_c \geq 10$.

2. The phase margin must be at least 45 degrees.

3. The gain margin must be 20 dB.

SOLUTION   We could enter the model directly in MATLAB, as described in Example 15.3, but we will use the Simulink diagram in Figure 15.23 instead.

We begin by setting $K_c = 1$ and using the command `margin`, after using `linmod` to create the model, as was shown in the M-file in Example 15.3. The result is Figure 15.24 where the gain margin is reported as 45.6 dB. The specified margin is 20.0 dB. We therefore choose K_c to be 25.6 dB, which corresponds to a multiplicative constant of 19.0, in order

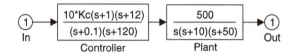

Figure 15.23 Simplified Simulink diagram for Example 15.5.

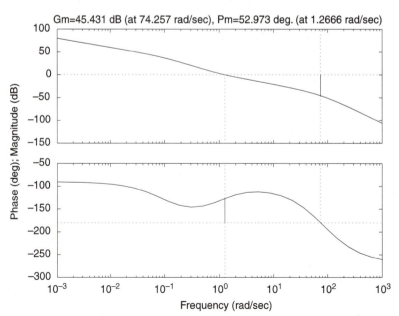

Figure 15.24 Bode diagram for Example 15.5 when $K_c = 1$.

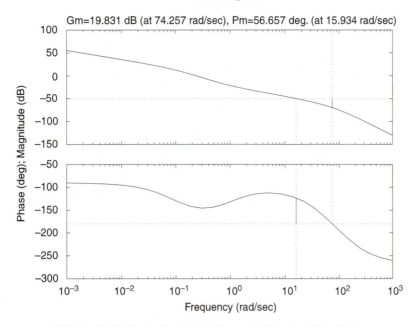

Figure 15.25 Bode diagram for Example 15.5 when $K_c = 19.0$.

to raise the magnitude curve by this constant amount. Setting $K_c = 19.0$ and re-running `margin` results in Figure 15.25.

We can use `zpkdata` directly in MATLAB, or we can refer to the root-locus diagram and use `rlocfind` for the point where $K_c = 19.0$. Either method tells us that the closed-loop poles are located at $s = -129.5$, $s = -17.38 \pm j17.24$, $s = -14.8$, and $s = -1.05$.

We have chosen the examples in this section to illustrate some of the techniques for handling feedback systems and some of the concerns encountered in practice. The availability of computer programs such as MATLAB for quickly plotting root-locus diagrams, Bode diagrams, and time responses is an important asset. An actual design would normally include many more diagrams than we have plotted, showing the effect of varying the controller's characteristics and allowing for the tolerances in physical devices.

In all our examples, we assumed a simple unity-feedback system, with the compensator placed only in the controller block in the forward path. Had space permitted, we could also have discussed the merits of inserting compensators in the sensor block or in separate inner feedback loops. Note, for example, that the servomechanism examined in Section 13.4 does not exactly fit the configuration assumed in this section.

▶ SUMMARY

Root-locus and Bode diagrams are important tools for the analysis and design of feedback systems. When applied to a feedback system, a root locus shows how the poles of the closed-loop transfer function move in the s-plane as the open-loop gain is increased. Once the locus is drawn, a program such as MATLAB can locate the points corresponding to a particular gain. It can also determine the gain needed to place a closed-loop pole at a particular point on the locus. If the general shape of the locus is not satisfactory, changes can be made in those parameters that are under the designer's control, and a new locus can be plotted.

A Bode diagram for the open-loop transfer function can give important information about the behavior of the closed-loop system. An indirect indication of the form of the free response is given by the gain margin and phase margin, which were defined in Section 14.6. Increasing the open-loop gain raises the magnitude curve by a constant amount without affecting the angle curve.

Among the typical constraints given to the designer of a feedback system are conditions on the steady-state responses to reference and disturbance inputs, as well as conditions on the nature of the transient response. In simple systems, increasing the open-loop gain tends to improve the steady-state response but adversely affects the dynamic behavior. In order to satisfy all the constraints, a more complicated controller may be needed. Root-locus and Bode diagrams help the designer investigate the effects of possible changes in the poles and zeros of the open-loop transfer function.

▶ PROBLEMS

15.1. This problem involves the servomechanism considered in Section 13.4, and the numerical values of the parameters given in Table 13.1 should be used where they are needed.

    a. Calculate the value of K_A for a proportional-only controller that will result in a damping ratio of $\zeta = 0.7071$ for the closed-loop system. Give its transfer function $T(s)$ as a ratio of polynomials.

    b. Assuming that a tachometer signal $e_3 = K_T\dot{\phi}$ is available for proportional-plus-derivative control, calculate the values of K_A and K_T that will yield a closed-loop system with $\zeta = 0.7071$ and $\omega_n = 8$ rad/s. Determine the closed-loop transfer function and verify that its poles have the specified values for ζ and ω_n.

*15.2. Repeat Problem 15.1 when the desired damping ratio is $\zeta = 1.0$ and, for part (b), the desired undamped natural frequency is $\omega_n = 10$ rad/s.

*15.3.

    a. Find the closed-loop transfer function $Y(s)/U(s)$ in terms of the parameter K for the feedback system shown in Figure P15.3.

    b. Write an expression for the closed-loop poles in terms of K, and sketch the locus of these poles in the complex plane for $K \geq 0$. Indicate the pole locations for $K = 0, 1, 2,$ and 3 on the locus.

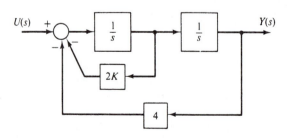

Figure P15.3

15.4.

a. Obtain a Bode plot for the open-loop transfer function

$$G(s) = \frac{K}{(s^2 + 12s + 40)(s + 1)}$$

Determine the gain margin and the phase margin when $K = 100$.

b. Find the value of K that will give a gain margin of 10 dB.

c. Find K^*, the maximum value of K for which the closed-loop transfer function will be stable.

*15.5. Repeat Problem 15.4 for the open-loop transfer function

$$G(s) = \frac{K(s+2)}{s^4 + 11s^3 + 34s^2 + 24s}$$

where $K = 40$.

15.6.

a. Use the result of Example 8.23 to verify that the transfer function for the circuit shown in Figure P15.6 is

$$\frac{E_o(s)}{E_i(s)} = -\frac{R_3(R_4C_4s + 1)[(R_1 + R_2)C_2s + 1]}{R_1(R_2C_2s + 1)[(R_3 + R_4)C_4s + 1]}$$

b. Express the poles and zeros in terms of the resistances and capacitances. Plot the pole-zero pattern, assuming that $R_4C_4 > (R_1 + R_2)C_2$.

c. By referring to Figure 15.9, identify the type of controller or compensator that the circuit implements.

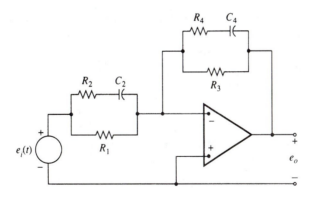

Figure P15.6

*15.7.

a. Use the result of Example 8.23 to determine the transfer function $E_o(s)/E_i(s)$ for the circuit shown in Figure P15.7.

b. Express the poles and zeros in terms of R_1, R_2, and C. Plot the pole-zero pattern.

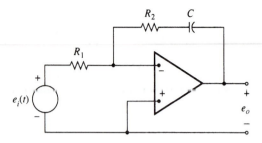

Figure P15.7

c. By referring to Figure 15.9, identify the type of controller or compensator that the circuit implements.

15.8. Repeat Problem 15.7 for the circuit shown in Figure P15.8.

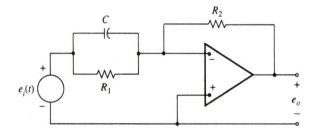

Figure P15.8

*15.9. The servomechanism discussed in Section 13.4 has a disturbance torque $\tau_d(t)$ applied to the shaft to which the output potentiometer is attached. In Figure 13.22, the positive sense of this torque is clockwise.

a. Show that the block diagram of Figure 13.23 must be modified to appear as in Figure P15.9(a).

b. When $R = 8\ \Omega$, $\alpha = 5.0$ V·s/rad, and the two potentiometer gain blocks K_θ are moved to the output of the summing junction and combined with the amplifier gain block K_A, verify that the model can be represented as shown in Figure P15.9(b). Use the numerical values in Table 13.1.

c. Assume that proportional control is used with $K_A = 0.08727$ V/V. Determine the steady-state errors to

 (i) A unit ramp input in the reference angle $\theta_i(t)$
 (ii) A unit step in the disturbance torque $\tau_d(t)$

d. Draw a root-locus plot and determine the value of K_A that will result in a closed-loop system having $\zeta = 0.8$. Find the corresponding value of ω_n.

e. Draw a Bode plot for the open-loop system when $K_A = 0.4$ and determine the phase margin. Explain why the gain margin is not defined.

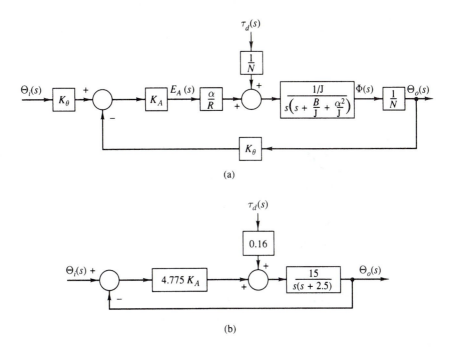

(a)

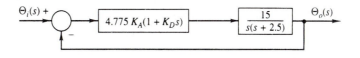

(b)

Figure P15.9

15.10. The servomechanism discussed in Section 13.4 is to be controlled by inserting a proportional-plus-derivative (PD) compensator immediately after the amplifier.

  a. Verify that the system can be represented by the block diagram shown in Figure P15.10, where K_D is the derivative gain in seconds.

  b. Show analytically that when the derivative gain is $K_D = 0.1$ s, it is possible to have closed-loop poles with $\zeta = 0.5$ and $\omega_n = 5.0$ rad/s. Determine the required value of the amplifier gain K_A.

  c. Draw a root-locus plot and verify the locations of the closed-loop poles for the value of K_A found in part (b).

  d. Using MATLAB or another computer program, compute and plot the response of the closed-loop system to a unit step in the reference input $\theta_i(t)$.

Figure P15.10

15.11. The servomechanism discussed in Section 13.4 is to be controlled by inserting a lead compensator immediately after the amplifier.

  a. Verify that when the magnitude of the lead pole is 10 times that of the lead zero, the system can be represented by the block diagram shown in Figure P15.11.

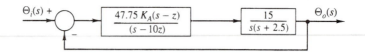

Figure P15.11

b. Draw the root-locus plot with the lead zero placed at $s = -10$, and show that the upper branch of the locus passes close to the point in the s-plane that corresponds to $\zeta = 0.5$ and $\omega_n = 5.0$ rad/s. Select a value for the amplifier gain K_A that will result in closed-loop poles close to this point, and determine the locations of all the closed-loop poles for this value of K_A.

c. Using MATLAB or another computer program, compute and plot the response of the closed-loop system to a unit step in the reference input $\theta_i(t)$.

15.12. The servomechanism discussed in Section 13.4 has a disturbance torque $\tau_d(t)$ as described in Problem 15.9 and is to be controlled with a lag-lead compensator.

a. Adapt the block diagram in Figure P15.9(b) to have the controller transfer function

$$G_c(s) = \frac{10K_A(s+1)(s+10)}{(s+0.1)(s+100)}$$

in place of the gain K_A.

b. Evaluate the closed-loop transfer functions from each of the inputs to the output.

c. Solve for the steady-state error in θ_o that is due to a unit step in the disturbance torque $\tau_d(t)$.

d. Draw the root-locus plot and determine the value of K_A that will result in closed-loop poles with $\zeta = 0.7$. Determine the locations of all the closed-loop poles for this value of K_A.

e. Using MATLAB or another computer program, compute and plot the response of the closed-loop system to a unit step in the reference input $\theta_i(t)$.

*15.13. Figure P15.13 shows a thermal process with a feedback temperature controller. The controller receives the desired temperature $\theta_d(t)$ and the measured temperature θ_m as inputs and determines the heat q_h supplied by the heater. The uncontrolled process was modeled in Example 11.7, where we derived transfer functions for $\hat{\Theta}(s)/\hat{Q}_h(s)$ and $\hat{\Theta}(s)/\hat{\Theta}_i(s)$. Assume that the thermal resistance R is infinite and that the liquid flow rate is $\bar{w}$, a constant. In terms of incremental variables, the controller is modeled by the relationship

$$\hat{Q}_h(s) = G_c(s)[\hat{\Theta}_d(s) - \hat{\Theta}_m(s)]$$

where $G_c(s)$ is the controller transfer function and the quantity $\hat{\Theta}_d(s) - \hat{\Theta}_m(s)$ is the transform of the measured temperature error. It is assumed that the sensor measures the actual temperature exactly—that is, $\hat{\theta}_m = \hat{\theta}$.

a. Draw a block diagram representing the closed-loop system that has the incremental inputs $\hat{\Theta}_d(s)$ and $\hat{\Theta}_i(s)$ and the incremental output $\hat{\Theta}(s)$.

b. Evaluate the closed-loop transfer functions $\hat{\Theta}(s)/\hat{\Theta}_d(s)$ and $\hat{\Theta}(s)/\hat{\Theta}_i(s)$ in terms of the unspecified controller transfer function $G_c(s)$.

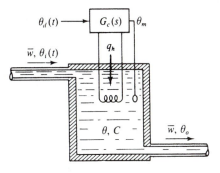

Figure P15.13

c. Using proportional control with $G_c(s) = K_c$, evaluate the two closed-loop transfer functions $\hat{\Theta}(s)/\hat{\Theta}_d(s)$ and $\hat{\Theta}(s)/\hat{\Theta}_i(s)$. Show that for step inputs in $\hat{\theta}_d(t)$ and $\hat{\theta}_i(t)$, taken separately, the steady-state temperature error is not zero for finite values of K_c.

d. Using proportional-plus-integral control with $G_c(s) = K_c(1 + K_I/s)$, reevaluate the closed-loop transfer functions. Show that for step inputs in $\hat{\theta}_d(t)$ and $\hat{\theta}_i(t)$, taken separately, the steady-state temperature error is zero for all positive values of K_c and K_I.

15.14. Repeat the analysis of the temperature-control system described in Problem 15.13 when a dynamic model for the temperature sensor is included. Ignore possible fluctuations in the inlet temperature so that $\hat{\theta}_i(t)$ can be taken to be zero. In terms of the incremental variables, the sensor is assumed to have the transfer function

$$\frac{\hat{\Theta}_m(s)}{\hat{\Theta}(s)} = \frac{1}{as + 1}$$

where a is the sensor time constant.

a. Find the closed-loop transfer function $\hat{\Theta}(s)/\hat{\Theta}_d(s)$ for proportional control, where $G_c(s) = K_c$. Determine the steady-state error to a step input in $\hat{\theta}_d(t)$.

b. Repeat part (a) for PI control, where $G_c(s) = K_c(1 + K_I/s)$.

15.15. Figure P15.15 shows a liquid-level control system such as might be found in a typical chemical process. The sensed level signal h_s is obtained by measuring the gauge pressure $p_1{}^*$ at the bottom of the tank. The controller also receives a signal $h_d(t)$ indicating the desired level. The controller output x is used to position a linear control valve in a bypass line connected around a centrifugal pump. The bypass flow rate w_b is given by

$$w_b = k\sqrt{\Delta p}\left(\frac{x}{x_m}\right)$$

where x_m is the maximum valve opening, x is the actual valve opening, Δp is the pressure difference developed by the pump, and k is the valve coefficient. The pump is driven at a constant speed, and at the operating point, the slope of the curve of Δp versus w is $-K$.

a. Derive the linearized model of the control valve by finding the coefficients α and β in the expression $\hat{w}_b = \alpha\widehat{\Delta p} + \beta\hat{x}$.

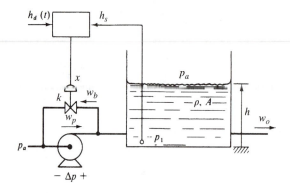

Figure P15.15

b. Write the linearized system equations in terms of the incremental variables $\hat{h}, \hat{w}_o$, $\hat{w}_b, \hat{w}_p, \widehat{\Delta p}$, and $\hat{p}_1$. Then draw the block diagram of the open-loop system with $\hat{X}(s)$ and $\hat{W}_o(s)$ as the inputs and $\hat{H}(s)$ as the output. Evaluate the transfer functions $\hat{H}(s)/\hat{X}(s)$ and $\hat{H}(s)/\hat{W}_o(s)$.

c. Taking $\hat{W}_o(s) = 0$, draw a block diagram of the closed-loop system when the controller is described by $\hat{X}(s) = K_c[\hat{H}_d(s) - \hat{H}_s(s)]$ and the sensor is described by $\hat{H}_s(s) = \hat{H}(s)$. Find the closed-loop transfer function $\hat{H}(s)/\hat{H}_d(s)$. Explain why the controller gain K_c should be negative.

d. Repeat part (c), using a dynamic model of the sensor such that $\hat{H}_s(s)/\hat{H}(s) = 1/(\tau s + 1)$.

15.16. Find the phase margin for the solution given for Example 15.4.

15.17. Find the unit step response to a disturbance applied directly to the plant in Example 15.4 when $K_c = 16$.

15.18. Find the value of K_c required to produce a closed-loop system in Example 15.4 having a damping ratio $\zeta = 0.9$.

A

UNITS

We use the **International System of Units**, abbreviated as SI for **Système International d'Unités**. In Table A.1, we list the seven basic SI units, of which we use only the first five in this book. A supplementary unit for plane angles is the radian (rad).

In Table A.2, we give those derived SI units that we use. In addition to the physical quantity, the unit, and its symbol, there is a fourth column that expresses the unit in terms of units previously given. For example, $1\ N = 1\ kg \cdot m/s^2$.

Basic Units

TABLE A.1   Names and Symbols of the Basic Units

Physical quantity	Name	Symbol
Length	meter	m
Mass	kilogram	kg
Time	second	s
Electrical current	ampere	A
Thermodynamic temperature	kelvin	K
Luminous intensity	candela	cd
Amount of substance	mole	mol

Derived Units

TABLE A.2   Names, Symbols, and Equivalents of the Derived Units

Physical quantity	Name	Symbol	In terms of other units
Force	newton	N	$kg \cdot m/s^2$
Energy	joule	J	$kg \cdot m^2/s^2$
Power	watt	W	J/s
Electrical charge	coulomb	C	A·s
Voltage	volt	V	W/A
Electrical resistance	ohm	Ω	V/A
Electrical capacitance	farad	F	A·s/V
Inductance	henry	H	V·s/A
Magnetic flux	weber	Wb	V·s

Prefixes

The standard prefixes for decimal multiples and submultiples of a unit are given in elementary physics books. The only ones used in this book are kilo (k), milli (m), and micro (μ). The terms in which they appear are as follows: $1\ k\Omega = 10^3\ \Omega$, $1\ mV = 10^{-3}\ V$, $1\ \mu F = 10^{-6}\ F$.

Constants

The following three constants are used in the numerical solution of many examples and problems.

Atmospheric pressure at sea level: $p_a = 1.013 \times 10^5\ N/m^2$

Base of natural logarithms: $\epsilon = 2.718$

Gravitational constant at the surface of the earth: $g = 9.807\ m/s^2$

B

MATRICES

This appendix is intended as a refresher for the reader who has had an introductory course in linear algebra and is studying those sections of the book that use matrix methods. It is not a suitable introduction for someone who has not had formal exposure to matrices. Only those aspects of the subject that are used in Sections 3.3 and 14.1 are emphasized. References to more complete treatments appear in Appendix F.

Definitions

A **matrix** is a rectangular array of elements that are either constants or functions of time. We refer to a matrix having m rows and n columns as being of **order** $m \times n$. The element in the ith row and the jth column of the matrix **A** is denoted by a_{ij}, such that

$$\mathbf{A} = \begin{bmatrix} a_{11} & a_{12} & \cdots & a_{1n} \\ a_{21} & a_{22} & \cdots & a_{2n} \\ \vdots & \vdots & & \vdots \\ a_{m1} & a_{m2} & \cdots & a_{mn} \end{bmatrix}$$

A matrix having the same number of rows as columns, in which case $m = n$, is a **square matrix** of order n. A matrix with a single column is a **column vector**. Examples of column vectors are

$$\mathbf{b} = \begin{bmatrix} b_1 \\ b_2 \\ \vdots \\ b_m \end{bmatrix} \qquad \mathbf{c} = \begin{bmatrix} 4 \\ -1 \\ 3 \end{bmatrix} \qquad \mathbf{x}(t) = \begin{bmatrix} x_1(t) \\ x_2(t) \\ \vdots \\ x_n(t) \end{bmatrix}$$

A matrix with a single row is called a **row vector**.

A matrix that consists of a single element—that is, one that has only one row and one column—is referred to as a **scalar**. A square matrix of n rows and n columns that has unity for each of the elements on its main diagonal and zero for the remaining elements is the **identity matrix** of order n, denoted by **I**. For example,

$$\begin{bmatrix} 1 & 0 \\ 0 & 1 \end{bmatrix} \quad \text{and} \quad \begin{bmatrix} 1 & 0 & 0 \\ 0 & 1 & 0 \\ 0 & 0 & 1 \end{bmatrix}$$

are the identity matrices of order 2 and 3, respectively. A matrix all of whose elements are zero is the **null matrix**, denoted by **0**.

We obtain the **transpose** of a matrix by interchanging its rows and columns such that the ith row of the original matrix becomes the ith column of the transpose, for all rows. Hence if A is of order $m \times n$ with elements a_{ij}, then its transpose, A^T, is of order $n \times m$ and has elements $(A^T)_{ij} = a_{ji}$.

Operations

We can add or subtract two matrices that have the same order by adding or subtracting their respective elements. For example, if A and B are both $m \times n$, the ijth element of $A + B$ is $a_{ij} + b_{ij}$. We can form the product $C = AB$ only if the number of rows of B, the right matrix, is equal to the number of columns of A, the left matrix. If this condition is met, we find the $i\ell$th element of C by summing the products of the elements in the ith row of A with the corresponding elements in the ℓth column of B. If A is $m \times n$ and B is $n \times p$, then C is $m \times p$ and

$$c_{i\ell} = \sum_{j=1}^{n} a_{ij} b_{j\ell} \qquad \text{for } i = 1, 2, \ldots, m \qquad \text{and} \qquad \ell = 1, 2, \ldots, p \qquad (1)$$

For example, if

$$A = \begin{bmatrix} 2 & -1 & 3 \\ 1 & 0 & 4 \end{bmatrix} \qquad \text{and} \qquad B = \begin{bmatrix} 1 & -1 \\ 2 & 3 \\ -2 & 1 \end{bmatrix}$$

then

$$c_{11} = (2)(1) + (-1)(2) + (3)(-2) = -6$$
$$c_{12} = (2)(-1) + (-1)(3) + (3)(1) = -2$$
$$c_{21} = (1)(1) + (0)(2) + (4)(-2) = -7$$
$$c_{22} = (1)(-1) + (0)(3) + (4)(1) = 3$$

giving

$$C = \begin{bmatrix} -6 & -2 \\ -7 & 3 \end{bmatrix}$$

We obtain the product of a matrix and a scalar by multiplying each element of the matrix by the scalar. For example, if k is a scalar, the ijth element of kA is ka_{ij}.

Some properties of matrices are summarized in Table B.1, where A, B, and C denote general matrices and where I and 0 are the identity matrix and the null matrix, respectively. Some of the properties that hold for scalars are not always valid for matrices. For example, except in special cases, $AB \neq BA$. Also, the equation $AB = AC$ does not necessarily imply that $B = C$. Furthermore, the equation $AB = 0$ does not necessarily imply that either A or B is zero.

TABLE B.1 Matrix Properties

$$A + B = B + A$$
$$A + (B + C) = (A + B) + C$$
$$A(B + C) = AB + AC$$
$$AI = IA = A$$
$$0A = 0$$
$$A0 = 0$$
$$A + 0 = A$$

C

COMPLEX ALGEBRA

A complex quantity z is represented in **rectangular form** as

$$z = x + jy \tag{1}$$

where the real quantities x and y are called the real and imaginary parts of z, respectively. The common algebraic operations are similar to those for real numbers, with $j^2 = -1$.[1] The operations of **addition** and **subtraction** are defined by

$$\begin{aligned} z_1 + z_2 &= (x_1 + x_2) + j(y_1 + y_2) \\ z_1 - z_2 &= (x_1 - x_2) + j(y_1 - y_2) \end{aligned} \tag{2}$$

The **multiplication** of complex quantities is given by

$$\begin{aligned} z_1 z_2 &= (x_1 + jy_1)(x_2 + jy_2) \\ &= (x_1 x_2 + j^2 y_1 y_2) + j(x_1 y_2 + x_2 y_1) \\ &= (x_1 x_2 - y_1 y_2) + j(x_1 y_2 + x_2 y_1) \end{aligned} \tag{3}$$

The **division** of two complex quantities, $z_3 = z_1/z_2$, is defined by $z_1 = z_2 z_3$ if $z_2 \neq 0$. The division process can be carried out by **rationalization** as follows:

$$\begin{aligned} \frac{z_1}{z_2} &= \frac{x_1 + jy_1}{x_2 + jy_2} \\ &= \left(\frac{x_1 + jy_1}{x_2 + jy_2} \right) \left(\frac{x_2 - jy_2}{x_2 - jy_2} \right) \\ &= \left(\frac{x_1 x_2 + y_1 y_2}{x_2{}^2 + y_2{}^2} \right) + j \left(\frac{x_2 y_1 - x_1 y_2}{x_2{}^2 + y_2{}^2} \right) \end{aligned} \tag{4}$$

The **complex conjugate** of $z = x + jy$ is denoted and defined by

$$z^* = x - jy \tag{5}$$

and can be formed from z by replacing j by $-j$. Thus, the rationalization process used in (4) consists of multiplying the numerator and denominator of the given fraction by the complex conjugate of the denominator.

---

[1] The symbol i is also used for $\sqrt{-1}$, but we reserve i for electric current.

It can be easily shown that the above definitions result in the usual commutative, associative, and distributive laws, namely,

$$
\begin{aligned}
z_1 z_2 &= z_2 z_1 \\
z_1 + z_2 &= z_2 + z_1 \\
z_1(z_2 z_3) &= (z_1 z_2) z_3 \\
z_1 + (z_2 + z_3) &= (z_1 + z_2) + z_3 \\
z_1(z_2 + z_3) &= z_1 z_2 + z_1 z_3 \\
(z_1 + z_2) z_3 &= z_1 z_3 + z_2 z_3
\end{aligned}
\tag{6}
$$

A complex number z can be represented geometrically by a point in a plane, as shown in Figure C.1(a). Every combination of x and y determines a unique point in the plane. In the equivalent geometrical interpretation shown in Figure C.1(b), z is represented by a directed line or vector drawn from the origin to the point $x + jy$. Then the addition or subtraction of two complex quantities can be carried out graphically, as in parts (c) and (d) of the figure, using the usual rules for vector addition and subtraction. Figure C.1(e) shows the graphical addition of z and its complex conjugate z^*. Notice that $z + z^* = 2x$.

The symbol $|z|$ denotes the length of the directed line representing z and is called the **magnitude** (or **absolute value** or **modulus**) of z. The angle of the directed line (measured

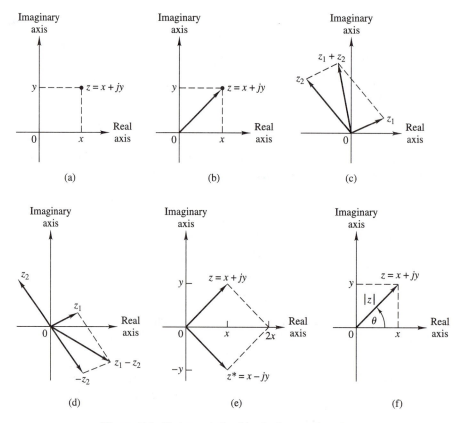

Figure C.1 Various relationships in the complex plane.

counterclockwise from the positive real axis) is denoted by θ and is called the **angle** or **argument** of z. From Figure C.1(f),

$$x = |z| \cos \theta$$
$$y = |z| \sin \theta$$
$$|z| = \sqrt{x_2 + y_2} \tag{7}$$
$$\theta = \arctan \frac{y}{x}$$

When the last equation is used, the quadrant of θ must be determined by inspection.

Euler's formulas

$$\epsilon^{j\theta} = \cos \theta + j \sin \theta$$
$$\epsilon^{-j\theta} = \cos \theta - j \sin \theta \tag{8}$$

and the two corollaries

$$\cos \theta = \frac{\epsilon^{j\theta} + \epsilon^{-j\theta}}{2}$$
$$\sin \theta = \frac{\epsilon^{j\theta} - \epsilon^{-j\theta}}{j2} \tag{9}$$

are frequently needed when handling complex quantities. From (8) and (7)

$$|z| \epsilon^{j\theta} = |z|(\cos \theta + j \sin \theta) = x + jy = z$$

Thus,

$$z = |z| \epsilon^{j\theta} \tag{10}$$

which is called the **polar form** of z. The angle θ is measured in radians (rad), where 1 radian $= 180/\pi = 57.296$ degrees.

Equations (8) and (10) lead to another interpretation of j, which was previously defined by $j^2 = -1$. If $\theta = \pi/2$ in these equations, $\epsilon^{j\pi/2} = j$, so j is a complex number with a magnitude of unity and an angle of $\pi/2$ radians. Notice also that $1/j = j/j^2 = -j$.

It is frequently necessary to convert a complex number from the rectangular to polar form, or polar to rectangular form, which can be done using (7). For example,

$$4 + j3 = \sqrt{4^2 + 3^2} \, \epsilon^{j\arctan \frac{3}{4}} = 5\epsilon^{j0.6435}$$

or

$$5\epsilon^{j0.6435} = 5\cos 0.6435 + j\sin 0.6435 = 4 + j3$$

Many hand calculators can easily convert complex numbers between rectangular and polar form and can use complex numbers in algebraic operations.

The multiplication and division of complex quantities are normally carried out in polar form. If $z_1 = |z_1|\epsilon^{j\theta_1}$ and $z_2 = |z_2|\epsilon^{j\theta_2}$,

$$z_1 z_2 = |z_1||z_2|\epsilon^{j(\theta_1 + \theta_2)}$$
$$\frac{z_1}{z_2} = \frac{|z_1|}{|z_2|}\epsilon^{j(\theta_1 - \theta_2)} \tag{11}$$

In forming a product, therefore, the magnitudes are multiplied, and the angles added. In forming a quotient, the magnitudes are divided, and the angles subtracted.

It should be noticed that multiplying a complex quantity by $j = 1\epsilon^{j\pi/2}$ does not change the magnitude, but adds $\pi/2$ radians to the angle. If the original quantity was represented by a directed line in the complex plane, the multiplication by j rotates the line 90 degrees in the counterclockwise direction. Notice also that the complex conjugate of $z = x + y = |z|\epsilon^{j\theta}$ is $z^* = x - jy = |z|\epsilon^{-j\theta}$, so $zz^* = |z|^2$.

Raising a complex number to a power is also easily handled with the number in polar form. If $z = |z|\epsilon^{j\theta}$,

$$z^n = |z|^n \epsilon^{jn\theta} \tag{12}$$

In MATLAB, all numbers are stored as complex numbers, so the user does not have to take any special action when dealing with them, such as declaring them to be complex. MATLAB commands are available for determining the magnitude, angle, real part, and imaginary parts of a complex number. Also, the variables $i = j = \sqrt{-1}$ are predefined, so they can be used without explicit definition. These four commands are summarized in Table C.1 and their use is illustrated in the example that follows.

TABLE C.1 MATLAB Commands for Complex Numbers as Applied to $z = x + jy = re^{j\theta}$

Purpose	Syntax	Result
real part	`real(z)`	x
imaginary part	`imag(z)`	y
magnitude	`abs(z)`	r
angle	`angle(z)`	θ

▶ **EXAMPLE C.1**

Find the rectangular and polar forms of

$$z = \frac{j100(1 - j2)^2}{(-3 + j4)(-1 - j)}$$

SOLUTION   If we express each of the factors in polar form,

$$z = \frac{(100\epsilon^{j\pi/2})(\sqrt{5}\,\epsilon^{-j1.1071})^2}{(5\epsilon^{j2.2143})(\sqrt{2}\,\epsilon^{-j2.3562})} = 70.711\epsilon^{-j0.5016} = 62 - j34$$

Alternatively, if the factors are left in rectangular form, and the resulting fraction is rationalized,

$$z = \left[\frac{100(4 - j3)}{7 - j1}\right]\left(\frac{7 + j1}{7 + j1}\right) = \frac{100(31 - j17)}{49 + 1} = 62 - j34$$

To create the complex variable z in MATLAB, we can enter the following commands:

```
a = j*100;
b = 1 - j*2;
c = -3 + j*4;
d = -1 -j;
z = (a*b^2)/(c*d)
```

Having created z in the MATLAB workspace, we compute its real and imaginary parts, magnitude, and angle by entering the following commands:

```
r = abs(z)
theta = angle(z)
x = real(z)
y = imag(z)
```

The results are

```
r =    70.7107
theta =   -0.5016
x =    62.0000
y =   -34
```

The fact that MATLAB returns the number 62.0000 for x indicates that the result is not exactly the integer 62, but rather something very close to it.

D

CLASSICAL SOLUTION OF DIFFERENTIAL EQUATIONS

This appendix summarizes the basic procedures that can be used for solving differential equations without the use of the Laplace transform. For many readers, this will be a review. We assume that we have fixed linear differential equations with real coefficients. Although the methods we describe are not sufficient for all possible inputs, they are satisfactory for the inputs that are commonly encountered. Proofs and detailed justifications are omitted, but some general references are included in Appendix F.

We assume that the system model has been put into input-output form with all other variables eliminated, as discussed in Section 3.2. The model for an nth-order system with input $u(t)$ and output $y(t)$ can be written, as in Equation (3.25), as

$$a_n y^{(n)} + \cdots + a_2 \ddot{y} + a_1 \dot{y} + a_0 y = b_m u^{(m)} + \cdots + b_1 \dot{u} + b_0 u(t) \tag{1}$$

where $y^{(n)} = d^n y / dt^n$, etc. The collection of terms on the right side of (1), which involves the input and its derivatives, is represented by

$$F(t) = b_m u^{(m)} + \cdots + b_1 \dot{u} + b_0 u(t) \tag{2}$$

where $F(t)$ is called the **forcing function**. With this definition, we may rewrite (1) as

$$a_n y^{(n)} + \cdots + a_2 \ddot{y} + a_1 \dot{y} + a_0 y = F(t) \tag{3}$$

The desired solution, $y(t)$ for $t \geq 0$, must satisfy the differential equation for $t \geq 0$ and also n specified initial conditions, which are usually $y(0), \dot{y}(0), \ldots, y^{(n-1)}(0)$.

▶ D.1 HOMOGENEOUS DIFFERENTIAL EQUATIONS

If $F(t)$ is replaced by zero, (3) reduces to the **homogeneous differential equation**

$$a_n y_H^{(n)} + \cdots + a_2 \ddot{y}_H + a_1 \dot{y}_H + a_0 y_H = 0 \tag{4}$$

where the subscript H has been added to emphasize that $y_H(t)$ is the solution to the homogeneous equation. Assume that a solution of (4) has the form

$$y_H(t) = K \epsilon^{rt}$$

where we must determine the constant r such that (4) is satisfied. Multiplying y_H by a constant just multiplies the entire left side of (4) by that constant, so any nonzero value will be satisfactory for the constant K. Substituting the assumed solution into (4) gives

$$(a_n r^n + \cdots + a_2 r^2 + a_1 r + a_0) K \epsilon^{rt} = 0$$

Because $K \epsilon^{rt} \neq 0$ for a nontrivial solution, we must have

$$a_n r^n + \cdots + a_2 r^2 + a_1 r + a_0 = 0 \tag{5}$$

which is called the **characteristic equation**. Note that the coefficients in this algebraic equation are identical to those on the left side of the differential equation, so we can write (5) by inspection. An nth-order algebraic equation has n roots, which we denote by r_1, r_2, $\ldots$, r_n. Not only are $K_1 \epsilon^{r_1 t}$, $K_2 \epsilon^{r_2 t}$, $\ldots$, $K_n \epsilon^{r_n t}$ all solutions of (4) for arbitrary K_i, but

$$y_H(t) = K_1 \epsilon^{r_1 t} + K_2 \epsilon^{r_2 t} + \cdots + K_n \epsilon^{r_n t} \tag{6}$$

is a solution for any values of the arbitrary constants K_1 through K_n. If all the roots of the characteristic equation are different, then (6) is the most general solution.

If two or more of the characteristic roots are identical—that is, if there is a repeated root of (5)—then we must modify the form of (6) to contain n independent terms and to represent the most general solution. If $r_1 = r_2$ with the remaining roots being distinct, the solution is

$$y_H(t) = K_1 \epsilon^{r_1 t} + K_2 t \epsilon^{r_1 t} + K_3 \epsilon^{r_3 t} + \cdots + K_n \epsilon^{r_n t}$$

If $r_1 = r_2 = r_3$, the most general solution to the homogeneous differential equation is

$$y_H(t) = (K_1 + K_2 t + K_3 t^2) \epsilon^{r_1 t} + K_4 \epsilon^{r_4 t} + \cdots + K_n \epsilon^{r_n t}$$

If any of the roots of the characteristic equation are complex numbers, they must occur in complex conjugate pairs because the coefficients in the characteristic equation are real. Suppose that $r_1 = \alpha + j\beta$ and $r_2 = \alpha - j\beta$, where $j = \sqrt{-1}$, [2] and that the characteristic equation has no other roots. Then

$$y_H(t) = K_1 \epsilon^{(\alpha + j\beta)t} + K_2 \epsilon^{(\alpha - j\beta)t}$$

However, because the solution is a real function of time, we should rewrite this equation by using suitable trigonometric identities. Factoring out $\epsilon^{\alpha t}$ from the right side, then using the first two entries in Table D.1 with $\theta = \beta t$, and finally collecting like terms, we have

$$y_H(t) = \epsilon^{\alpha t} [K_1 \epsilon^{j\beta t} + K_2 \epsilon^{-j\beta t}]$$
$$= \epsilon^{\alpha t} [K_1(\cos\beta t + j\sin\beta t) + K_2(\cos\beta t - j\sin\beta t)]$$
$$= \epsilon^{\alpha t} [(K_1 + K_2)\cos\beta t + j(K_1 - K_2)\sin\beta t]$$

With $K_3 = K_1 + K_2$ and $K_4 = j(K_1 - K_2)$, the equation becomes

$$y_H(t) = \epsilon^{\alpha t}(K_3 \cos\beta t + K_4 \sin\beta t) \tag{7}$$

By using the last entry in Table D.1, we can rewrite (7) in the alternative form

$$y_H(t) = K \epsilon^{\alpha t} \cos(\beta t + \phi) \tag{8}$$

where the constants K and ϕ depend on K_3 and K_4. In these expressions involving complex roots, K_1 and K_2 are complex numbers, with K_2 the complex conjugate of K_1; but K_3, K_4, K, and ϕ are real constants. Either of the two forms given by (7) and (8) should be used.

---

[2] The symbol i is also used for $\sqrt{-1}$, but we reserve i for electrical current.

TABLE D.1   Useful Trigonometric Identities

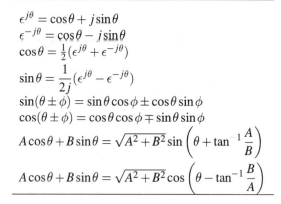

We treat repeated complex characteristic roots much like repeated real roots. For a fourth-order characteristic equation with $r_1 = r_3 = \alpha + j\beta$ and $r_2 = r_4 = \alpha - j\beta$, the most general solution to the homogeneous differential equation has the form

$$y_H(t) = K_A \epsilon^{\alpha t} \cos(\beta t + \phi_A) + K_B t \epsilon^{\alpha t} \cos(\beta t + \phi_B)$$

▶ D.2 NONHOMOGENEOUS DIFFERENTIAL EQUATIONS

We now consider the nonhomogeneous differential equation, where the forcing function $F(t)$ in (3) is nonzero. For mathematical convenience, we express the solution as the sum of two parts, namely

$$y(t) = y_H(t) + y_P(t) \tag{9}$$

where $y_H(t)$ and $y_P(t)$ are known as the complementary and the particular solutions, respectively. The **complementary solution** $y_H(t)$ is the solution to the homogeneous differential equation in (4); examples are given by (6) through (8). The **particular solution** $y_P(t)$ must satisfy the entire original differential equation, so

$$a_n y_P^{(n)} + \cdots + a_2 \ddot{y}_P + a_1 \dot{y}_P + a_0 y_P = F(t) \tag{10}$$

A general procedure for finding $y_P(t)$ is the variation-of-parameters method. When the forcing function $F(t)$ has only a finite number of different derivatives, however, the method of undetermined coefficients is satisfactory. Some examples of functions possessing only a finite number of independent derivatives are given in the left-hand column of Table D.2. Others include $\cosh \omega t$ and $t^2 \epsilon^{\alpha t} \cos(\omega t + \phi)$. An example of a forcing function that does *not* fall into this category is $1/t$.

In using the **method of undetermined coefficients**, we normally assume that the form of $y_P(t)$ consists of terms similar to those in $F(t)$ and its derivatives, with each term multiplied by a constant to be determined. We find the values of these constants by substituting the assumed $y_P(t)$ into the differential equation and then equating corresponding coefficients. The particular solution to be assumed for some common forcing functions is shown in Table D.2.

**TABLE D.2   Usual Form of the
Particular Solution**

$F(t)$	$y_P(t)$
α	A
$\alpha_1 t + \alpha_0$	$At + B$
$\epsilon^{\alpha t}$	$A\epsilon^{\alpha t}$
$\cos \omega t$	$A\cos \omega t + B\sin \omega t$
$\sin \omega t$	$A\cos \omega t + B\sin \omega t$

If $F(t)$ or one of its derivatives contains a term identical to a term in $y_H(t)$, the corresponding terms in the right-hand column of Table 2 should be multiplied by t. Thus if $y_H(t) = K_1\epsilon^{-t} + K_2\epsilon^{-2t}$ and $F(t) = 3\epsilon^{-t}$, then $y_P(t) = At\epsilon^{-t}$ should be used. If a term in $F(t)$ corresponds to a double root of the characteristic equation, the normal form for $y_P(t)$ is multiplied by t^2. If, for example, $y_H(t) = K_1\epsilon^{-t} + K_2 t\epsilon^{-t}$ and $F(t) = \epsilon^{-t}$, then $y_P(t) = At^2\epsilon^{-t}$.

In the **general solution** $y(t) = y_H(t) + y_P(t)$ for an nth-order nonhomogeneous differential equation, the complementary solution $y_H(t)$ contains n arbitrary constants, which we earlier denoted by K_1 through K_n. The original differential equation in (3) will be satisfied regardless of the values of these constants, so their evaluation requires n separately specified initial conditions: $y(0), \dot{y}(0), \ldots, y^{(n-1)}(0)$. Because these initial values are associated with the entire solution and not just with the complementary solution, we cannot evaluate the arbitrary constants until we have found both $y_H(t)$ and $y_P(t)$.

▶ **EXAMPLE D.1**

Find the solution to the differential equation $\ddot{y} + 2\dot{y} + 5y = t$ for $t \geq 0$ if $y(0) = 0$ and $\dot{y}(0) = 2$.

SOLUTION   The characteristic equation is $r^2 + 2r + 5 = 0$, which by the quadratic formula has roots at

$$r = \frac{-2 \pm \sqrt{4 - 20}}{2} = -1 \pm j2$$

As in (7), we can write the complementary solution as

$$y_H(t) = \epsilon^{-t}[K_1 \cos 2t + K_2 \sin 2t]$$

The particular solution is assumed to have the form $y_P(t) = At + B$. Substituting this expression into the differential equation gives

$$0 + 2A + 5(At + B) = t$$

or

$$(5A - 1)t + (2A + 5B) = 0$$

which requires that $5A = 1$ and $2A + 5B = 0$. Thus $A = \frac{1}{5}, B = -\frac{2}{25}$, and the general solution for $t \geq 0$ is

$$y(t) = \epsilon^{-t}[K_1 \cos 2t + K_2 \sin 2t] + \frac{1}{5}t - \frac{2}{25}$$

The derivative of this general solution is

$$\dot{y}(t) = \epsilon^{-t}[(2K_2 - K_1)\cos 2t - (K_2 + 2K_1)\sin 2t] + \frac{1}{5}$$

At $t = 0$ and with the given initial conditions, the last two equations reduce to

$$0 = K_1 - \frac{2}{25}$$

$$2 = 2K_2 - K_1 + \frac{1}{5}$$

from which $K_1 = \frac{2}{25}$ and $K_2 = \frac{47}{50}$. For $t \geq 0$,

$$y(t) = \epsilon^{-t}\left[\frac{2}{25}\cos 2t + \frac{47}{50}\sin 2t\right] + \frac{1}{5}t - \frac{2}{25}$$

E

LAPLACE TRANSFORMS

In Table E.1, we list the Laplace transforms for common functions of time. When using this table to take inverse transforms, the reader should keep in mind that the time functions are valid only for positive values of t.

Table E.2 contains the most important properties that are needed in the application of the transform method to dynamic models. The restrictions on the use of these properties are not included in Table E.2 but may be found in Chapter 7. In that chapter, we discuss the proper interpretation of the initial-condition terms and the conditions on the initial-value and final-value theorems. The functions of time are again valid only for $t > 0$.

TABLE E.1 Transforms of Functions

Time functions	Transformed functions
$\delta(t)$	1
$U(t)$	$\dfrac{1}{s}$
A	$\dfrac{A}{s}$
t	$\dfrac{1}{s^2}$
t^2	$\dfrac{2!}{s^3}$
t^n for $n = 1, 2, 3, \ldots$	$\dfrac{n!}{s^{n+1}}$
ϵ^{-at}	$\dfrac{1}{s+a}$
$t\epsilon^{-at}$	$\dfrac{1}{(s+a)^2}$
$t^2\epsilon^{-at}$	$\dfrac{2!}{(s+a)^3}$
$\sin \omega t$	$\dfrac{\omega}{s^2 + \omega^2}$
$\cos \omega t$	$\dfrac{s}{s^2 + \omega^2}$
$\epsilon^{-at} \sin \omega t$	$\dfrac{\omega}{(s+a)^2 + \omega^2}$
$\epsilon^{-at} \cos \omega t$	$\dfrac{s+a}{(s+a)^2 + \omega^2}$
$\epsilon^{-at}\left[B\cos\omega t + \left(\dfrac{C - aB}{\omega} \right) \sin\omega t \right]$	$\dfrac{Bs + C}{(s+a)^2 + \omega^2}$
$2K\epsilon^{-at} \cos(\omega t + \phi)$	$\dfrac{K\epsilon^{j\phi}}{s+a-j\omega} + \dfrac{K\epsilon^{-j\phi}}{s+a+j\omega}$

TABLE E.2   Transform Properties

Time functions	Transformed functions
$f(t)$	$F(s)$
$af(t)$	$aF(s)$
$f(t) + g(t)$	$F(s) + G(s)$
$\epsilon^{-at}f(t)$	$F(s+a)$
$tf(t)$	$-\dfrac{d}{ds}F(s)$
$f(t/a)$	$aF(as)$
$[f(t-a)]U(t-a)$	$\epsilon^{-sa}F(s)$
$\dot{f}(t)$	$sF(s) - f(0)$
$\ddot{f}(t)$	$s^2F(s) - sf(0) - \dot{f}(0)$
$\dddot{f}(t)$	$s^3F(s) - s^2f(0) - s\dot{f}(0) - \ddot{f}(0)$
$\dfrac{d^n f}{dt^n}$ for $n = 1,2,3,\ldots$	$s^nF(s) - s^{n-1}f(0) - \cdots - sf^{(n-2)}(0) - f^{(n-1)}(0)$
$\displaystyle\int_0^t f(\lambda)d\lambda$	$\dfrac{1}{s}F(s)$
$f(0+)$	$\displaystyle\lim_{s\to\infty} sF(s)$
$f(\infty)$	$\displaystyle\lim_{s\to 0} sF(s)$

F

SELECTED READING

This appendix suggests a number of books that are suitable for undergraduates. Some provide background for specific topics or can be used as collateral reading. Others extend the topics we treat to a more advanced level or offer an introduction to additional areas. The works cited here are included in the References. This appendix and the references provide typical starting points for further reading, but these lists are not intended to be comprehensive.

Background Sources

Introductory college physics textbooks describe the basic mechanical, electrical, and electromechanical elements and some other components as well. They also illustrate simple applications of physical laws to such systems. One good reference is Halliday, *et al.* Comprehensive lists of units and conversion factors are given in Wildi.

A good example of a book on differential equations is Boyce and DiPrima. For a background in linear algebra and matrices, see Anton.

Books at a Comparable Level

The following three introductory systems books discuss a variety of physical components, as well as analysis techniques: Rosenberg and Karnopp; Ogata (1997); and Palm. Included are electrical, mechanical, hydraulic, pneumatic, and thermal systems. The references include several books, easily recognized by their titles, that illustrate the application of MAT-LAB to the analysis of dynamic systems.

Most books dealing with modeling and analysis are restricted to a particular discipline. Kimbrell; Meriam and Kraige; and Thomson and Dahleh deal with mechanical components. Textbooks on chemical processes include Bequette; Seborg *et al.*; and Luyben.

Two of the standard books on electrical circuits are Johnson *et al.* and Nilsson and Riedel. Sedra and Smith covers electronic circuits in detail. Books such as Smith and Dorf, and also Carlson and Gisser, treat devices and applications in both electrical and electromechanical systems. The books by Guru and Hiziroglu and by McPherson and Laramore are confined to electromechanical machinery.

Books That Extend the Analytical Techniques

It is natural to follow up your study of this book with more advanced books on feedback control systems, such as Raven; Ogata (1996); and some others in the list of references. Some of these include discrete-time systems in addition to continuous systems. Two books that emphasize discrete-time control systems are Phillips and Nagle; and Franklin *et al.*

A number of books broaden the study of transform methods to include both the Fourier and z transforms, as well as the Laplace transform. Good examples are Chen and Kamen.

The text by Bolton will introduce the reader to the new specialty of mechatronics. This field integrates electronics, electrical engineering, computer technology, and mechanical engineering.

MATLAB-based books

An extensive bibliography of over 400 MATLAB-based books can be found at the Web site of The MathWorks, at http://www.mathworks.com/support/books/index.php3. The books by Frederick and Chow; Ogata (1994); and Strum and Kirk may be of particular interest.

▶ F.1 REFERENCES

Anton, H. and C. Rorres. *Elementary Linear Algebra*, 8th ed. Wiley, New York, 2000.

Bequette, B. W. *Process Dynamics: Modeling, Analysis and Simulation*. Prentice-Hall, Upper Saddle River, NJ, 1998.

Bolton, W. *Mechatronics: Electronic Control Systems in Mechanical and Electrical Engineering*, 2nd ed. Addison-Wesley Longman, New York, 1999.

Boyce, W. E. and R. C. DiPrima. *Elementary Differential Equations*, 7th ed. Wiley, New York, 2000.

Carlson, A. B. and D. G. Gisser. *Electrical Engineering, Concepts and Applications*, 2nd ed. Addison-Wesley, Reading, MA, 1990.

Chen, C. T. *System and Signal Analysis*. Oxford University Press, New York, 1994.

Doebelin, E. O. *Measurement Systems*, 4th ed. McGraw-Hill, New York, 1990.

Dorf, R. C. and R. H. Bishop. *Modern Control Systems*, 9th ed. Prentice-Hall, Upper Saddle River, NJ, 2001.

Franklin, G. F., J. D. Powell, and A. Emami-Naeini. *Feedback Control of Dynamic Systems*, 3rd ed. Addison-Wesley, Reading, MA, 1993.

Franklin, G. F., J. D. Powell, and M. L. Workman. *Digital Control of Dynamic Systems*, 3rd ed. Addison-Wesley, Reading, MA, 1997.

Frederick, D. K. and J. H. Chow. *Feedback Control Problems Using MATLAB and The Control System Toolbox*. Brooks/Cole, Pacific Grove, CA, 1999.

Guru, B. S. and H. R. Hiziroglu. *Electrical Machinery and Transformers*, 3rd ed. Oxford University Press, New York, 2000.

Halliday, D., R. Resnick, and V. Walker. *Fundamentals of Physics*, 6th ed. Wiley, New York, 2000.

Hayt, W. H., Jr. and J. E. Kemmerly. *Engineering Circuit Analysis*, 5th ed. McGraw-Hill, New York, 1993.

Johnson, D. E., J. L. Hilburn, J. R. Johnson, and P. D. Scott. *Electric Circuit Analysis*, 3rd ed. Prentice-Hall, Englewood Cliffs, NJ, 1998.

Kamen, E. W. *Introduction to Signals and Systems*, 2nd ed. Prentice-Hall, Englewood Cliffs, NJ, 1990.

Kimbrell, J. T. *Kinematics Analysis and Synthesis*. McGraw-Hill, New York, 1991.

Kuo, B. C. *Automatic Control Systems*, 7th ed. Wiley, New York, 1999.

Luyben, W. L. *Process Modeling, Simulation and Control for Chemical Engineers*, 2nd ed. McGraw-Hill, New York, 1989.

The MathWorks. *Using MATLAB*. The MathWorks, Inc., Natick, MA, 2000.

The MathWorks. *Control System Toolbox*. The MathWorks, Inc., Natick, MA, 2000.

The MathWorks. *Using Simulink*. The MathWorks, Inc., Natick, MA, 2000.

McPherson, G. and R. D. Laramore. *An Introduction to Electrical Machines and Transformers*, 2nd ed. Wiley, New York, 1990.

Meriam, J. L. and L. G. Kraige. *Engineering Mechanics Dynamics/Statics*, 4th ed. Wiley, New York, 1996.

Nilsson, J. W. and S. A. Riedel. *Electric Circuits*, 6th ed. Prentice-Hall, Englewood Cliffs, NJ, 1999.

Ogata, K. *Designing Linear Control Systems with MATLAB*. Prentice-Hall, Englewood Cliffs, NJ, 1994.

Ogata, K. *Modern Control Engineering*, 3rd ed. Prentice-Hall, Englewood Cliffs, NJ, 1996.

Ogata, K. *System Dynamics*, 3rd ed. Prentice-Hall, Englewood Cliffs, NJ, 1997.

Palm, W. J. *Modeling, Analysis, and Control of Dynamic Systems*, 2nd ed. Wiley, New York, 1999.

Phillips, C. L. and H. T. Nagle. *Digital Control System Analysis and Design*, 3rd ed. Prentice-Hall, Englewood Cliffs, NJ, 1994.

Raven, F. H. *Automatic Control Engineering*, 4th ed. McGraw-Hill, New York, 1987.

Rosenberg, R. C. and D. C. Karnopp. *Introduction to Physical Systems Dynamics*. McGraw-Hill, New York, 1983.

Seborg, D. E. E., T. F. Edgar, and D. A. Mellichamp. *Process Dynamics and Control*. Wiley, New York, 1989.

Sedra, O. S. and K. C. Smith. *Microelectronic Circuits*, 4th ed. Oxford University Press, New York, 1997.

Smith, R. J. and R. C. Dorf. *Circuits, Systems, and Devices*, 5th ed. Wiley, New York, 1992.

Strum, R. D. and D. E. Kirk. *Contemporary Linear Systems with MATLAB*. Brooks/Cole, Pacific Grove, CA, 1999.

Thomson, W. T. and M. D. Dahleh. *Theory of Vibrations with Applications*, 5th ed. Prentice-Hall, Englewood Cliffs, NJ, 1997.

Wildi, T. *Metric Units and Conversion Charts*, 2nd ed. IEEE Press, Piscataway, NJ, 1994.

ANSWERS TO SELECTED PROBLEMS

Chapter 2—Translational Mechanical Systems

2.1 $M_1\ddot{x}_1 + B_1\dot{x}_1 + (K_1 + K_2)x_1 - K_2 x_2 = f_a(t)$,
$-K_2 x_1 + M_2\ddot{x}_2 + B_2\dot{x}_2 + K_2 x_2 = 0$

2.4 $M_1\ddot{x}_1 + (B_1 + B_3)\dot{x}_1 + (K_1 + K_3)x_1 - B_3\dot{x}_2 - K_3 x_2 = f_a(t)$,
$-B_3\dot{x}_1 - K_3 x_1 + M_2\ddot{x}_2 + (B_2 + B_3)\dot{x}_2 + (K_2 + K_3)x_2 = 0$

2.7
    a. $M_1\ddot{x}_1 + B_1\dot{x}_1 + (K_1 + K_2)x_1 = K_2 x_2(t)$
    b. $f_2 = -K_2 x_1 + M_2\ddot{x}_2 + B_2\dot{x}_2 + K_2 x_2(t)$

2.10 $M_1\ddot{x}_1 + K_1 x_1 - K_2 x_2 = 0$
$M_2\ddot{x}_1 + B\dot{x}_1 + M_2\ddot{x}_2 + B\dot{x}_2 + K_2 x_2 = f_a(t)$

2.12 $M_1\ddot{x}_1 + K_1 x_1 + B\dot{x}_2 = K_1 x_3(t)$
$-M_2\ddot{x}_1 - K_2 x_1 + M_2\ddot{x}_2 + B\dot{x}_2 + K_2 x_2 = 0$

2.15
    a. $M_1\ddot{x}_1 + B\dot{x}_1 + K_1 x_1 - B\dot{x}_2 - K_1 x_2 = M_1 g$
        $-B\dot{x}_1 - K_1 x_1 + M_2\ddot{x}_2 + B\dot{x}_2 + (K_1 + K_2)x_2 = M_2 g + f_a(t)$
    b. $x_{1_0} = [(M_1 + M_2)/K_2 + (M_1/K_1)]g$
        $x_{2_0} = [(M_1 + M_2)/K_2]g$
    c. $M_1\ddot{z}_1 + B\dot{z}_1 + K_1 z_1 - B\dot{z}_2 - K_1 z_2 = 0$
        $-B\dot{z}_1 - K_1 z_1 + M_2\ddot{z}_2 + B\dot{z}_2 + (K_1 + K_2)z_2 = f_a(t)$

2.18
    a. $M_1\ddot{x}_1 + B\dot{x}_1 + 2K x_1 - B\dot{x}_2 - K x_2 = M_1 g$
        $-B\dot{x}_1 - K x_1 + M_2\ddot{x}_2 + B\dot{x}_2 + 3K x_2 - K x_3 = M_2 g$
        $-K x_2 + M_3\ddot{x}_3 + K x_3 = M_3 g - f_a(t)$
    b. $x_{1_0} = (2M_1 + M_2 + M_3)g/(3K)$
        $x_{2_0} = (M_1 + 2M_2 + 2M_3)g/(3K)$
        $x_{2_0} - x_{1_0} = (-M_1 + M_2 + M_3)g/(3K)$
        $x_{3_0} - x_{2_0} = M_3 g/K$

2.22 $M_1\ddot{x}_1 + (K_1 + K_2)x_1 - K_2x_2 = M_1g$
$-K_2x_1 + M_2\ddot{x}_2 + B\dot{x}_2 + (K_2 + K_3)x_2 = -M_2g$

2.25 $\dot{x}_1 = [B_2/(B_1 + B_2)]\dot{x}_2$
$B_{eq} = B_1B_2/(B_1 + B_2)$

2.27 $M\ddot{x}_1 + (B_1 + B_2)\dot{x}_1 + (K_1 + K_2)x_1 - B_2\dot{x}_2 - K_2x_2 = 0$
$-B_2\dot{x}_1 - K_2x_1 + B_2\dot{x}_2 + (K_2 + K_3)x_2 = K_3x_3(t)$

Chapter 3—Standard Forms for System Models

3.3 $\dot{y} = v$
$\dot{v} = -2y - 4v + x$
$\dot{x} = -y - x + f_a(t)$

3.5 $\dot{x}_1 = v_1$
$\dot{v}_1 = [-K_1x_1 - (B_1 + B_2 + B_3)v_1 + B_2v_2]/M_1$
$\dot{x}_2 = v_2$
$\dot{v}_2 = [B_2v_1 - K_2x_2 - B_2v_2 + f_a(t)]/M_2$
$y_1 = B_2(v_2 - v_1)$
$y_2 = K_2x_2$

3.9 $\dot{x}_1 = v_1$
$\dot{v}_1 = [-K_1x_1 - Bv_1 + K_1x_2 + Bv_2 + M_1g]/M_1$
$\dot{x}_2 = v_2$
$\dot{v}_2 = [K_1x_1 + Bv_1 - (K_1 + K_2)x_2 - Bv_2 + M_2g + f_a(t)]/M_2$
$y_1 = x_1 - x_2$
$a_1 = [-K_1x_1 - Bv_1 + K_1x_2 + Bv_2 + M_1g]/M_1$

3.11 $\dot{x}_1 = v_1$
$\dot{v}_1 = [-3Kx_1 - Bv_1 + Kx_2 + M_1g]/M_1$
$\dot{x}_2 = v_2$
$\dot{v}_2 = [Kx_1 - Kx_2 + M_2g + f_a(t)]/M_2$
$y = x_2 - x_1$

3.13 $\dot{x}_1 = v_1$
$\dot{v}_1 = [-K_1x_1 - B_1v_1 - K_3x_2 + K_3x_3(t)]/M$
$\dot{x}_2 = [K_2x_1 + B_2v_1 - (K_2 + K_3)x_2 + K_3x_3(t)]/B_2$
$y = K_2x_1 - (K_2 + K_3)x_2 + K_3x_3(t)$

3.15 $\dot{q}_1 = -3q_1 + 2q_2 + 3u_1(t) - 6u_2(t)$
$\dot{q}_2 = 2q_1 + q_2 + u_1(t) + 4u_2(t)$
$y = q_1 - q_2 - u_1(t) + 3u_2(t)$
where $q_1 = x_1 - 2u_2(t)$ and $q_2 = x_2 - u_1(t)$

3.18 $\dot{x}_1 = q_1 + (B_1/M_1)x_3(t)$
$\dot{q}_1 = [-(K_1 + K_2)x_1 - B_1q_1 + K_2x_2 + (K_1 - B_1^2/M_1)x_3(t) + f_a(t)]/M_1$
$\dot{x}_2 = q_2 + (B_2/M_2)x_3(t)$
$\dot{q}_2 = [K_2x_1 - K_2x_2 - B_2q_2 - (B_2^2/M_2)x_3(t)]/M_2$
$v_1 = q_1 + (B_1/M_1)x_3(t)$
$v_2 = q_2 + (B_2/M_2)x_3(t)$

3.21 $\dot{x}_1 = v_1$
$\dot{v}_1 = [-Kx_1 - Bv_1 + Bv_2 - f_a(t)]/M_1$
$\dot{x}_2 = v_2$
$\dot{v}_2 = [Bv_1 - Bv_2 + M_2g]/M_2$
$m_1 = M_1v_1$
$m_2 = M_2v_2$
$f_c = B(v_2 - v_1)$

3.24 $M_1\ddot{x}_1 + B\dot{x}_1 + (K_1 + K_2)x_1 = B\dot{x}_2 + K_2x_2(t)$

3.26 $x_1^{(iv)} + 2x_1^{(iii)} + 5\ddot{x}_1 + \dot{x}_1 + 2x_1 = 4g + \ddot{f}_a + \dot{f}_a + 2f_a(t)$

3.29 $\dot{\mathbf{q}} = \begin{bmatrix} 0 & -1 & 1 \\ K_2/M_1 & -B/M_1 & B/M_1 \\ -K_2/M_2 & B/M_2 & -B/M_2 \end{bmatrix} \mathbf{q} + \begin{bmatrix} 0 \\ 0 \\ 1/M_2 \end{bmatrix} f_a(t),$

$\mathbf{y} = \begin{bmatrix} K_2 & 0 & 0 \\ 0 & M_1 & M_2 \end{bmatrix} \mathbf{q} + \begin{bmatrix} 0 \\ 0 \end{bmatrix} f_a(t)$

Chapter 4—Block Diagrams and Computer Simulation

4.2

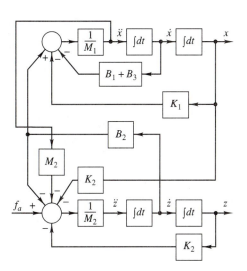

4.5

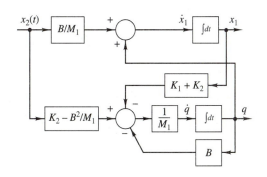

4.8

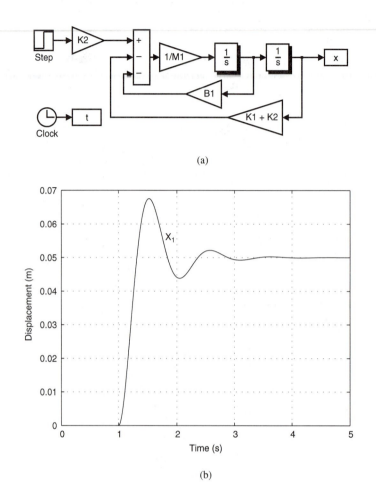

(a)

(b)

4.10

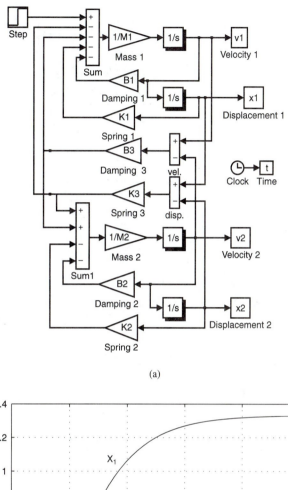

(a)

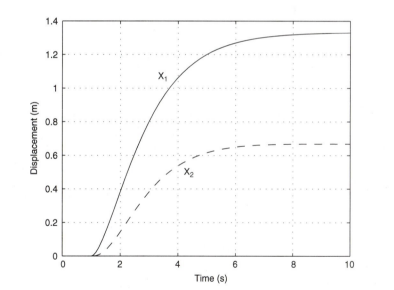

(b)

4.13

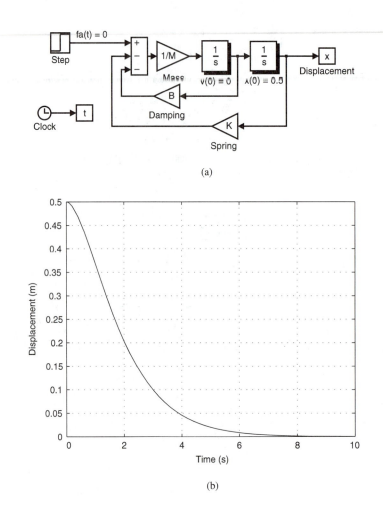

(a)

(b)

4.15 a.–d.

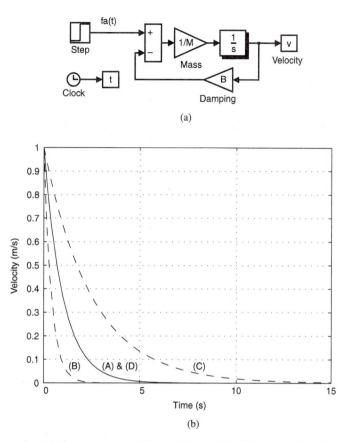

(a)

(b)

e. Increasing M slows the rate of decay. Increasing B increases the rate of decay.

Chapter 5—Rotational Mechanical Systems

5.2 $J_1 J_2 \ddot{\omega}_2 + B J_2 \ddot{\omega}_2 + K(J_1 + J_2)\dot{\omega}_2 + KB\omega_2 = BK\omega_a(t)$

5.6

    a. With $\theta_1, \omega_1, \theta_2, \omega_2$ as state variables:

$$\dot{\theta}_1 = \omega_1$$
$$\dot{\omega}_1 = [-K_1\theta_1 - B\omega_1 + B\omega_2 + \tau_a(t)]/J_1$$
$$\dot{\theta}_2 = \omega_2$$
$$\dot{\omega}_2 = (B\omega_1 - K_2\theta_2 - B\omega_2)/J_2$$
$$\tau_B = B(\omega_2 - \omega_1)$$

    b. $\theta_2^{(iv)} + 2\theta_2^{(iii)} + 2\ddot{\theta}_2 + 2\dot{\theta}_2 + \theta_2 = \dot{\tau}_a$

5.7 $J\ddot{\theta} + K_{eq}\theta = \tau_a(t)$ where $K_{eq} = K_1 + [K_2 K_3/(K_2 + K_3)]$

5.10

    a. $\dot{x}_1 = v_1$
$$\dot{v}_1 = [-(K_1 + K_2)x_1 - B_1 v_1 + (K_2 L/4)\theta]/M$$
$$\dot{\theta} = (16/9B_2 L)[(K_2/4)x_1 - (K_2 L/16)\theta - (3/4)f_a(t)]$$

    b. $f_r = (4K_2/3)x_1 - (K_2 L/3)\theta$

5.13 With x_1, v_1, θ, ω as state variables:

$$\dot{x}_1 = v_1$$
$$\dot{v}_1 = [-(K_1 + K_2)x_1 - B_1 v_1 + (1/4)K_2 L\theta]/M$$
$$\dot{\theta} = \omega$$
$$\dot{\omega} = (3/7ML)[4K_2 x_1 - LK_2 \theta - 9LD_2 \omega - 12f_a(t)]$$
$$f_{K_2} = -K_2 x_1 + (1/4)LK_2 \theta$$

5.18 a. $J_1 \ddot{\theta}_1 + B\dot{\theta}_1 - R_1 f_c = \tau_a(t)$
     $J_2 \ddot{\theta}_2 + K\theta_2 + R_2 f_c = 0$

  b. With θ_1, ω_1 as state variables:
     $$\dot{\theta}_1 = \omega_1$$
     $$\dot{\omega}_1 = [-(K/N^2)\theta_1 - B\omega_1 + \tau_a(t)]/[J_1 + (J_2/N^2)]$$
     $$m_T = [J_1 + (J_2/N)]\omega_1$$
     $$\tau_B = -B\omega_1$$

  c. $[J_1 + (J_2/N^2)]\ddot{\theta}_1 + B\dot{\theta}_1 + (K/N^2)\theta_1 = \tau_a(t)$

5.19 $\dot{\theta}_1 = \omega_1$
$$\dot{\omega}_1 = [-K_1 \theta_1 - B\omega_1 + K_1 \theta_2 + \tau_a(t)]/J_1$$
$$\dot{\theta}_2 = \omega_2$$
$$\dot{\omega}_2 = [K_1 \theta_1 - (K_1 + N^2 K_2)\theta_2]/(J_2 + N^2 J_3)$$
$$\theta_3 = -N\theta_2$$

5.26 With θ, ω as state variables:
$$\dot{\theta} = \omega$$
$$\dot{\omega} = [-KR\theta - (B/R)\omega + Mg + f_a(t)]/[(J/R) + MR]$$
$$z = R\theta - Mg/K$$

5.29 With θ, ω, x_2, v_2 as state variables:
$$\dot{\theta} = \omega$$
$$\dot{\omega} = [-(R_1^2 K_1 + K_2)\theta + (K_2/R_2)x_2 + R_1 K_1 x_1(t)]/J_1$$
$$\dot{x}_2 = v_2$$
$$\dot{v}_2 = [(K_2/R_2)\theta - (K_2/R_2^2 + 2K_3)x_2 - Bv_2 + M_2 g]/M_2$$
$$y = K_2 \theta - (K_2/R_2)x_2$$

5.32

$$\frac{d}{dt}\begin{bmatrix} \theta \\ \omega \\ x \\ v \end{bmatrix} = \begin{bmatrix} 0 & 1 & 0 & 0 \\ -(K_1 + K_2 R^2)/J & -B/J & K_2 R/J & 0 \\ 0 & 0 & 0 & 1 \\ K_2 R/M & 0 & -K_2/M & 0 \end{bmatrix}\begin{bmatrix} \theta \\ \omega \\ x \\ v \end{bmatrix}$$
$$+ \begin{bmatrix} 0 & 0 \\ 0 & 0 \\ 0 & 0 \\ 1/M & 1/M \end{bmatrix}\begin{bmatrix} Mg \\ f_a(t) \end{bmatrix}$$

Chapter 6—Electrical Systems

6.3 $\dot{e}_o + e_o = (7/24)\dot{e}_i + (2/3)e_i(t)$

6.7 $2\ddot{e}_o + 2\dot{e}_o + 4e_o = 3e_i(t)$

6.9 $\ddot{e}_o + \dot{e}_o + (1/2)e_o = 2di_i/dt$

6.12 $\ddot{e}_o + 2\dot{e}_o + 2e_o = \ddot{e}_i + (1/2)\dot{e}_i$

6.17 $e_o = 12$ V

6.18

    a.   With i_L (positive to right), e_C (positive at upper node) as state variables:

$$\dot{e}_C = [-e_C - R_1 i_L + R_1 i_i(t) - 6]/(R_1 C)$$
$$di_L/dt = (e_C - R_2 i_L)/L$$

    b. $i_o = i_i(t) - (e_C + 6)/R_1$

6.21 With i_L (positive downward), e_o as state variables:

$$\dot{e}_o = -(1/2)e_o - 2i_L + 2i_i(t)$$
$$di_L/dt = (1/8)e_o - (1/2)i_L + (1/2)i_i(t)$$

6.24 With i_L (positive to right), e_C (positive at upper node) as state variables:

$$\dot{e}_C = [-e_C + 2i_L + e_i(t)]/3$$
$$di_L/dt = [-e_C - 4i_L + e_i(t)]/6$$
$$e_o = (2/3)[-e_C + 2i_L + e_i(t)]$$

6.27 $\dot{e}_{C_1} = -i_L + i_i(t)$
$$\dot{e}_{C_2} = [-e_{C_2} - 2i_L + 6i_i(t)]/12$$
$$di_L/dt = (3e_{C_1} + e_{C_2} - 4i_L)/9$$
$$i_o = (1/6)e_{C_2} + (1/3)i_L$$

6.31 $e_o = -(R_3/R_1)e_1(t) - (R_3/R_2)e_2(t)$; summer with gains determined by $R_1, R_2,$ and R_3.

6.34 $C_2 \dot{e}_o + (1/R_2)e_o = -C_1 \dot{e}_i - (1/R_1)e_i(t)$

6.38 $\dfrac{d}{dt}\begin{bmatrix} e_C \\ i_L \end{bmatrix} = \begin{bmatrix} -0.6 & 0.3 \\ -1.2 & -0.8 \end{bmatrix} \begin{bmatrix} e_C \\ i_L \end{bmatrix} + \begin{bmatrix} 0.2 \\ 0.4 \end{bmatrix} e_i(t)$

$$e_o = \begin{bmatrix} -0.6 & -0.2 \end{bmatrix} \begin{bmatrix} e_C \\ i_L \end{bmatrix} + \begin{bmatrix} 0.2 \end{bmatrix} e_i(t)$$

Chapter 7—Transform Solutions of Linear Models

Note: Expressions for time functions are valid for $t > 0$.

7.3

    a. $F_1(s) = (s^2 + 4s - 5)/[(s^2 + 4s + 13)^2]$
    b. $F_2(s) = (6\omega s^2 - 2\omega^3)/[(s^2 + \omega^2)^3]$
    c. $F_3(s) = 2s/[(s + 1)^3]$
    d. $F_4(s) = 2/[s(s + 1)^3]$

7.7

    a. $\tau = 3/2$ s
    b. $e_o(t) = 2 - (4/3)\epsilon^{-2t/3}$
    c. $(e_o)_{ss} = 2$ V

7.11

   a. $y_U(t) = 2(1 - \epsilon^{-t/2})$
   b. $y(t) = 4 - 5\epsilon^{-t/2}$
   c. $y(t) = 2(1 - \epsilon^{-t/2})$ for $0 \le t \le 2$ and $y(t) = 2(\epsilon - 1)\epsilon^{-t/2}$ for $t \ge 2$
   d. $h(t) = \epsilon^{-t/2}$

7.17

   a. $f(t) = 2 - (5 + t)\epsilon^{-t} + 4\epsilon^{-2t}$
   b. $f(t) = 2 + 2\sqrt{2}\,\epsilon^{-t}\cos(2t - \pi/4)$
   c. $f(t) = 3\delta(t) + (1/3) - [(37/3) - 20t]\epsilon^{-3t}$
   d. $f(t) = \delta(t) + 3\epsilon^{-t} - 8\epsilon^{-2t}$

7.18

   a. $f(t) = 2\epsilon^{-2t} + 2.692\epsilon^{-t}\cos(2t + 1.190)$
   b. $f(t) = 2\epsilon^{-2t} + \epsilon^{-t}\cos 2t - 2.5\epsilon^{-t}\sin 2t$

7.20 $f(t) = 1.25 - 2\epsilon^{-t} + 0.75\epsilon^{-4t}$

7.23

   a. $F_1(s) = A(1 - 2\epsilon^{-s} + \epsilon^{-2s})/s$
   b. $F_2(s) = A(1 - \epsilon^{-s} - \epsilon^{-2s} + \epsilon^{-3s})/s^2$

7.26

   a. $f(0+) = 1, \quad f(\infty) = 2$
   b. $f(0+) = 4, \quad f(\infty) = 2$
   c. The initial-value theorem is not applicable, $f(\infty) = 1/3$
   d. The initial-value theorem is not applicable, $f(\infty) = 0$

Chapter 8—Transfer Function Analysis

   Note: Expressions for time functions are valid for $t > 0$.

8.3 $y(t) = 2.5t - 3.75 + 6\epsilon^{-t} - 1.25\epsilon^{-2t}$

8.10

   b. $\zeta = [(B_1 + B_2)/2]/\sqrt{K_2(M_1 + M_2)}$ and $\omega_n = \sqrt{K_2/(M_1 + M_2)}$
   c. $x_{ss} = (1 - M_2 g)/K_2$

8.13

   a. $\theta(t) = [3\theta(0) + \dot{\theta}(0)]\epsilon^{-2t} - [2\theta(0) + \dot{\theta}(0)]\epsilon^{-3t}$
   b. The mode functions are ϵ^{-2t} and ϵ^{-3t}. To eliminate the first of these, let $\dot{\theta}(0) = -3\theta(0)$. To eliminate the second, let $\dot{\theta}(0) = -2\theta(0)$.

8.20 $X_R(s)/F_a(s) = M_1/[M_1 M_2 s^2 + (M_1 + M_2)Bs + (M_1 + M_2)K_2]$

8.23 $X(s)/\tau_a(s) = (K/R)/P(s)$ where
     $P(s) = s[JMs^3 + (B_1 M + JB_2)s^2 + (KM + B_1 B_2 + JK/R^2)s + KB_2 + (KB_1/R^2)]$

8.25 $E_o(s)/E_i(s) = (10s^2 + 4s)/P(s)$ where $P(s) = 50s^2 + 70s + 42$

8.28

   a. Double pole at $s = -2$, zeros at $s = -1 \pm j1$
   b. $y_{zi}(t) = (K_1 + K_2 t)\epsilon^{-2t}, \quad \ddot{y} + 4\dot{y} + 4y = \ddot{u} + 2\dot{u} + 2u(t)$
   c. $y_U(0+) = 1, \quad y_U(\infty) = 0.5$
   d. $y_U(t) = 0.5 + (0.5 - t)\epsilon^{-2t}, \quad h(t) = \delta(t) - 2(1 - t)\epsilon^{-2t}$

8.31

    a. Poles at $s = -0.5, \pm j2$, double zero at $s = 0$

    b. $y_{zi}(t) = K_1\epsilon^{-0.5t} + K_2\cos(2t + \phi)$ or

        $y_{zi}(t) = K_1\epsilon^{-0.5t} + K_3\cos 2t + K_4\sin 2t$, $2\ddot{y} + \ddot{y} + 8\dot{y} + 4y = \ddot{u}$

    c. $y_U(0+) = 0$, $y_U(\infty) = 0$

    d. $y_U(t) = -0.05882\epsilon^{-0.5t} + 0.2426\cos(2t - 1.326)$,

        $h(t) = 0.02941\epsilon^{-0.5t} + 0.4850\cos(2t + 0.2450)$

8.34

    a. $\zeta = 1/\sqrt{2}, \omega_n = \sqrt{2}$

    b. $y_U(t) = -0.5\epsilon^{-t}(\cos t + \sin t) + 0.5$, $h(t) = \epsilon^{-t}\sin t$

8.36

    a. $\zeta = 1/\sqrt{5} = 0.4472, \omega_n = \sqrt{5} = 2.236$

    b. $h(t) = 3.414\delta(t) - 4.467\epsilon^{-t}[\sin(2t + 1.249)]U(t)$

8.37 $h(t) = \delta(t) + te^{-2t} - 2e^{-2t}; y_U(t) = 0.25 - 0.5te^{-2t} + 0.75\epsilon^{-2t}$

8.41 $e_o(t) = 1 + (1/3)\epsilon^{-t} + (2/3)\epsilon^{-4t}$

8.48

    a. Pole at $s = -1$, zero at $s = +1$; $M(\omega) = 1$ for all ω

        $\theta(\omega) = \pi - 2\tan^{-1}\omega$

    b. Poles at $s = 0, -5$; double zero at $s = -1$

        $M(\omega) = (1 + \omega^2)/(\omega\sqrt{25 + \omega^2})$

        $\theta(\omega) = 2\tan^{-1}\omega - \pi/2 - \tan^{-1}(\omega/5)$

    c. Double poles at $s = -0.1 \pm j10$; zero at $s = 0$

        $M(\omega) = \omega/[(100 - \omega^2)^2 + 0.04\omega^2]$

        $\theta(\omega) = \pi/2 - 2\tan^{-1}[0.2\omega/(100 - \omega^2)]$

8.49 $A = 0.0667, B = 5.0, C = 0.120, A/B = 0.01333, C/B = 0.0240$

8.51

    a. $H(s) = Cs/(LCs^2 + RCs + 1)$

    b. $\omega = 1/\sqrt{LC}$

8.61

    b. $E_o(s)/E_i(s) = (s^2 + 0.5s)/(s^2 + 2s + 2)$

    c. $\ddot{e}_o + 2\dot{e}_o + 2e_o = \ddot{e}_i + 0.5\dot{e}_i$

Chapter 9—Developing a Linear Model

9.2 For $\bar{x} = -1$, $f(x) \simeq A(0.6321 + 0.3679\hat{x})$

    for $\bar{x} = 0$, $f(x) \simeq A\hat{x}$

    for $\bar{x} = +1$, $f(x) \simeq A(-0.6321 + 0.3679\hat{x})$

9.5 $f(y) \simeq 2 - 4\hat{y}$

9.8

    b. $\bar{x} = 0.39$ m

    c. $1.5\ddot{\hat{x}} + 0.5\dot{\hat{x}} + 55\hat{x} = 0$

    d. Satisfactory for at least $0.27 < x < 0.60$

9.10   a. $\bar{x} = 4, \ddot{\hat{x}} + 2\dot{\hat{x}} + \hat{x} = B\sin 3t$
 b. For $A = 4$, $\bar{x} = 1$, $k = 2$;   for $A = -4$, $\bar{x} = -1$, $k = 2$
 c. $\hat{x}(0) = 0.5$, $\dot{\hat{x}}(0) = 0.5$

9.12

 a. $\bar{y} = -1$, $\ddot{\hat{y}} + 2\dot{\hat{y}} + 4\hat{y} = B\cos t$
 b. $\bar{y} = 3$, $\ddot{\hat{y}} + 2\dot{\hat{y}} + 8\hat{y} = B\cos t$

9.18

 a. $\bar{\theta} = 2$
 b. $\ddot{\hat{\theta}} + 2\dot{\hat{\theta}} + 12\hat{\theta} = \hat{\tau}_a(t)$
 c. $\hat{\theta}(0) = -1.5$ rad, $\dot{\hat{\theta}}(0) = -0.5$ rad/s

9.21

 b. $\bar{x} = \sqrt[3]{Mg}, M\ddot{\hat{x}} + B\dot{\hat{x}} + 3(Mg)^{(2/3)}\hat{x} = \hat{f}_a(t)$
 c. $\bar{x} = 1.260\sqrt[3]{Mg}, M\ddot{\hat{x}} + B\dot{\hat{x}} + 4.762(Mg)^{(2/3)}\hat{x} = \hat{f}_a(t)$

9.23

 a. $\bar{x} = -4, \bar{y} = -2$
 b. $\dot{\hat{x}} = -2\hat{x} + 12\hat{y}, \dot{\hat{y}} = \hat{x} + \cos t$
 c. $\ddot{\hat{x}} + 2\dot{\hat{x}} - 12\hat{x} = 12\cos t$

9.27

 b. $\bar{i}_L = 0.3622$ A; $\bar{e}_o = 0.3936$ V; $d\hat{i}_L/dt = -2.515\hat{i}_L + 0.05714\cos t; \hat{e}_o = 2.173\hat{i}_L$
 c. $\tau = 0.3976$ s

9.28

 b. $\bar{e}_o = 4$ V, $\bar{i}_o = 16$ A
 c. $\ddot{\hat{e}}_o + 7\dot{\hat{e}}_o + \hat{e}_o = 2\dot{\hat{e}}_i + \hat{e}_i(t)$

Chapter 10—Electromechanical Systems

10.1   a. $L_T\dot{e}_o + R_T e_o = L_2(x)\dot{e}_i + R_2(x)e_i(t)$
 b. $L_T\dot{e}_o + R_T e_o = [x(t)/x_{max}][L_T\dot{e}_i + R_T e_i(t)]$

10.6 $E_o(s)/F_a(s) = dBR_o/P(s)$ where
  $P(s) = MLs^2 + (BL + MR + MR_o)s + B(R + R_o) + (dB)^2$

10.8

 a. $di/dt = [-Ri + Bdv + e_i(t)]/L$, $\dot{v} = (-Bdi + Mg)/M$
 b. $\bar{e}_i = RMg/Bd$
 c. $\bar{v} = RMg/(Bd)^2$

10.11

 a. $(J_R + J_L)\dot{\omega} + (B_R + B_L)\omega = \gamma k_\phi i_F(t)i_A(t)$
 b. $\bar{\omega} = [\gamma k_\phi/(B_R + B_L)]\bar{i}_F\bar{i}_A$
 c. $(J_R + J_L)\dot{\hat{\omega}} + (B_R + B_L)\hat{\omega} = \gamma k_\phi[\bar{i}_F\hat{i}_A(t) + \bar{i}_A\hat{i}_F(t)]$

10.14

 a. $di_A/dt = (-R_A i_A - \gamma k_\phi i_F\omega + E_A)/L_A$
  $di_F/dt = [-R_F i_F + e_F(t)]/L_F$
  $\dot{\omega} = [\gamma k_\phi i_F i_A - B\omega - \tau_L(t)]/J$
 b. $\bar{\omega} = \gamma k_\phi \bar{i}_F \bar{i}_A/B$ where $\bar{i}_F = \bar{e}_F/R_F$ and
  $\bar{i}_A = BR_F^2 E_A/[BR_A R_F^2 + (\gamma k_\phi \bar{e}_F)^2]$

c. $d\hat{i}_A/dt = (-R_A\hat{i}_A - \gamma k_\phi \bar{\omega}\hat{i}_F - \gamma k_\phi \bar{i}_F \hat{\omega})/L_A$
$d\hat{i}_F/dt = [-R_F\hat{i}_F + \hat{e}_F(t)]/L_F$
$\dot{\hat{\omega}} = [\gamma k_\phi \bar{i}_F \hat{i}_A + \gamma k_\phi \bar{i}_A \hat{i}_F - B\hat{\omega} - \hat{\tau}_L(t)]/J$

10.18 $\omega_{ss} = 37.5$ rad/s

Chapter 11—Thermal Systems

11.2 $R_{eq} = R_a R_b/(2R_a + 4R_b); \quad q = [(4/R_a) + (2/R_b)](\theta_1 - \theta_a)$

11.5

a. $\dot{\theta}_1 = [-(1/R_{eq})\theta_1 + (1/R_1)\theta_a + (1/R_2)\theta_2(t) + q_i(t)]/C$ where
$1/R_{eq} = (1/R_1) + (1/R_2)$
b. $H_1(s) = (1/R_2)/[Cs + (1/R_{eq})]; \quad H_2(s) = 1/[Cs + (1/R_{eq})]$

11.7

a. $\dot{\theta}_1 = [-(1/R_1 + 1/R_2)\theta_1 + (1/R_2)\theta_2 + (1/R_1)\theta_a + q_1(t)]/C_1$
$\dot{\theta}_2 = [(1/R_2)\theta_1 - (1/R_2 + 1/R_3)\theta_2 + (1/R_3)\theta_3(t) + q_2(t)]/C_2$
b. $\dot{\hat{\theta}}_1 = [-(1/R_1 + 1/R_2)\hat{\theta}_1 + (1/R_2)\hat{\theta}_2 + q_1(t)]/C_1$
$\dot{\hat{\theta}}_2 = [(1/R_2)\hat{\theta}_1 - (1/R_2 + 1/R_3)\hat{\theta}_2 + (1/R_3)\hat{\theta}_3(t) + q_2(t)]/C_2$
$H_1(s) = R_1/P(s)$ and $H_2(s) = (R_1 R_2 C_1 s + R_1 + R_2)/P(s)$ where
$P(s) = R_1 R_2 R_3 C_1 C_2 s^2 + [R_1 C_1(R_2 + R_3) + R_3 C_2(R_1 + R_2)]s + R_1 + R_2 + R_3$

11.9

b. $\theta_m = \{C_w \theta_w(0) + C_m \theta_m(0) + C_w[\theta_m(0) - \theta_w(0)]\epsilon^{-t/\tau}\}/(C_m + C_w)$
$\theta_w = \{C_w \theta_w(0) + C_m \theta_m(0) - C_m[\theta_m(0) - \theta_w(0)]\epsilon^{-t/\tau}\}/(C_m + C_w)$
for $t \geq 0$ where $\tau = RC_m C_w/(C_m + C_w)$
c. $\dot{\theta}_m$ is unchanged ; $\dot{\theta}_w = [(1/R)\theta_m - (1/R + 1/R_a)\theta_w + (1/R_a)\theta_a]/C_w$

11.11 $\dot{\hat{\theta}}_1 = (9/RC)[-2\hat{\theta}_1 + \hat{\theta}_2 + \hat{\theta}_i(t)]$
$\dot{\hat{\theta}}_2 = (9/RC)(\hat{\theta}_1 - 2\hat{\theta}_2 + \hat{\theta}_3)$
$\dot{\hat{\theta}}_3 = (9/RC)(\hat{\theta}_2 - \hat{\theta}_3)$
$P(s) = s^3 + (45/RC)s^2 + [486/(RC)^2]s + [729/(RC)^3]$

Chapter 12—Fluid Systems

12.1 $C = \pi h^2/\rho g, \quad p^* = \rho g \sqrt[3]{3v/\pi}$

12.5 $\dot{\hat{p}}_1 = [-(1/K)\hat{p}_1 + (1/K)\hat{p}_2 + \hat{w}_i(t)]/C_1$
$\dot{\hat{p}}_2 = [(1/K)\hat{p}_1 - (1/K + 1/R)\hat{p}_2]/C_2$

12.7

b. $\hat{W}_o(s)/\hat{P}_i(s) = [(R_3 C_1 s + 1)/R_2 + (1/R_1)]/[(R_4 C_2 s + 1)(R_3 C_1 s + 1)]$
c. $(\hat{w}_o)_{ss} = (1/R_1) + (1/R_2)$

12.10

b. $\tau = \beta A/\alpha \rho g, \quad h_{ss} = \beta/\rho g$
c. $h(t) = (\beta/\rho g)(1 - \epsilon^{-t/\tau})$ for $t \geq 0$

Chapter 13—Block Diagrams for Dynamic Systems

13.3 $T_1(s) = (3s - 6)/(s^3 + 7s^2 + 14s + 8)$

$T_2(s) - (2s^4 + 14s^3 + 28s^2 + 13s + 6)/(s^4 + 7s^3 + 14s^2 + 8s)$

13.5

a. $\dot{q}_1 = q_2,\quad \dot{q}_2 = -6q_1 - 5q_2 + 12u(t),\quad y = q_1$

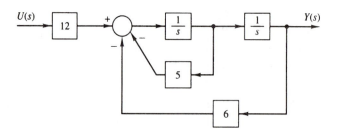

b. $\dot{q}_1 = q_2,\quad \dot{q}_2 = -6q_1 - 5q_2 + u(t),\quad y = 10q_1 + 9q_2 + 2u(t)$

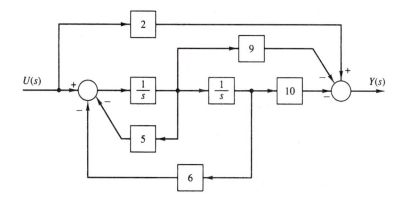

c. $\dot{q}_1 = q_2,\quad \dot{q}_2 = q_3,\quad \dot{q}_3 = -4q_1 - 3q_2 - 2q_3 + u(t)\quad y = 4q_1 - q_2 + 2q_3$

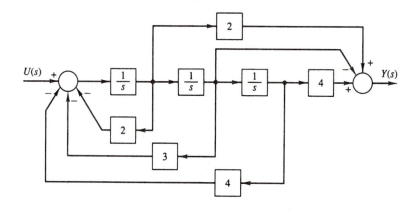

13.7 $Y(s)/U(s) = 2/(s^2 + 8s + 9)$

13.9 $X(s)/\tau_a(s) = (K/R)/P(s)$ where
$P(s) = JMs^4 + (MB_1 + JB_2)s^3 + [MK + B_1B_2 + (JK/R^2)]s^2 + [B_2K + (B_1K/R^2)]s$

13.12

    a. $T(s) = 2/(s^2 + 9s + 2K)$
    b. $y_{ss} = 1/K$
    c. $\zeta = 9/(2\sqrt{2K}), \quad \omega_n = \sqrt{2K}, \quad K = 20.25, \quad \omega_n = 6.364$ rad/s

13.14

    a. $T_1(s) = 4/(s^2 + 4Ks + 4)$
    b. $T_2(s) = s/(s^2 + 4Ks + 4)$
    c. $\omega_n = 2$ rad/s, $\zeta = K$, so use $K = 1$
    d. For $u(t) = U(t)$, $y_{ss} = 1$;   for $v(t) = U(t)$, $y_{ss} = 0$.

13.16

    a. $T(s) = K_1/[s^2 + (K_2 + 1)s + K_3]$
    b. $\zeta = (K_2 + 1)/(2\sqrt{K_3}), \omega_n = \sqrt{K_3}$
    c. $K_2 > -1$
    d. $-1 < K_2 < 2\sqrt{K_3} - 1$
    e. $K_2 = 1, K_3 = 4$

13.18 $T_1(s) = (AB + BDE + D)/(1 + BE + AC)$
$T_2(s) = A/(1 + BE + AC)$

Chapter 14—Modeling, Analysis, and Design Tools

14.3

    a. $T(s) = (6s^3 + 9s^2 + 49s + 79)/(s^4 + 7s^3 + 18s^2 + 45s + 41)$
    b. Zeros are $s = 0.0429 \pm j2.881, -1.586$; poles are $s = -0.4218 \pm j2.502, -1.316, -4.840$;
       gain = 6

14.5

    a. $\mathbf{A} = \begin{bmatrix} 0 & 1 & 0 \\ 0 & 0 & 1 \\ -2 & -2 & -4 \end{bmatrix}, \quad \mathbf{B} = \begin{bmatrix} 0 \\ 0 \\ 2 \end{bmatrix}$

       $\mathbf{C} = \begin{bmatrix} -3 & -7 & -8 \end{bmatrix}, \quad \mathbf{D} = 4$
    b. $T(s) = (4s^3 - 6s + 2)/(s^3 + 4s^2 + 2s + 2)$

14.7 Zeros are $s = -2, -5$;
    poles are $s = 0, -1.500 \pm j2.398, -12.00$;   gain = 2;
    $T(s) = (2s^2 + 14s + 20)/(s^4 + 15s^3 + 44s^2 + 96s)$

14.12

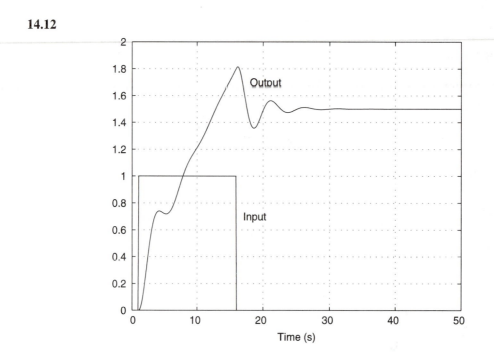

14.22 Zeros are $s = -3, -5$; poles are $s = -1.0956, -6.952 \pm j2.536$,

$T(s) = (4s^2 + 32s + 60)/(s^3 + 15s^2 + 70s + 60)$,

$$\mathbf{A} = \begin{bmatrix} -15 & -70 & -60 \\ 1 & 0 & 0 \\ 0 & 1 & 0 \end{bmatrix}, \quad \mathbf{B} = \begin{bmatrix} 1 \\ 0 \\ 0 \end{bmatrix}, \mathbf{C} = \begin{bmatrix} 4 & 32 & 60 \end{bmatrix}, \quad \mathbf{D} = 0$$

14.26

  b. $K = 39.7, s = -0.574 \pm j2.00, -2.50, -7.36$

14.32   a.

Bode Diagrams

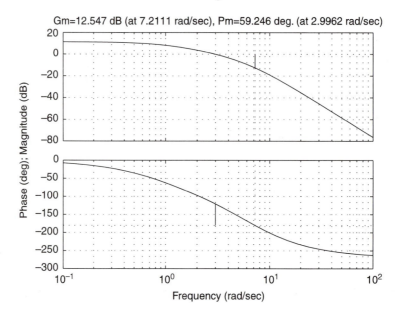

b. $k_m = 4.240 = 12.547$ dB, $\omega_{gm} = 7.211$ rad/s, $\phi_m = 59.25°$, $\omega_{pm} = 2.996$ rad/s
c. $K = 636$

Chapter 15—Feedback Design with MATLAB

15.2

    a. $K_A = 0.0218$, $T(s) = 1.563/(s^2 + 2.50s + 1.563)$
    b. $K_A = 1.396$, $K_T = 0.0836$ V·s/rad, $T(s) = 100/(s^2 + 20s + 100)$

15.3

    a. $T(s) = 1/(s^2 + 2Ks + 4)$
    b. Poles at $s = -K \pm \sqrt{K^2 - 4}$;   for $K = 0$, $s = \pm j2$;
       for $K = 1$, $s = -1 \pm j\sqrt{3}$; for $K = 2$, $s = -2, -2$;
       for $K = 3, s = -0.764, -5.236$

15.5

    a. $k_m = 11.05$ dB, $\phi_m = 28.95°$

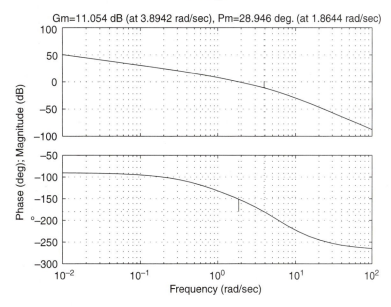

Bode Diagrams

Gm=11.054 dB (at 3.8942 rad/sec), Pm=28.946 deg. (at 1.8644 rad/sec)

    b. $K = 45.16$
    c. $K^* = 142.8$

15.7

    a. $H(s) = -(R_2Cs + 1)/(R_1Cs)$
    b. Pole at $s = 0$, zero at $s = -1/(R_2C)$
    c. PI compensator

15.9 c. (i) $\theta_{ss} = 0.400$ rad; (ii) $\theta_{ss} = -0.384$ rad

d. $K_A = 0.034$ V/V, $\omega_n = 1.55$ rad/s

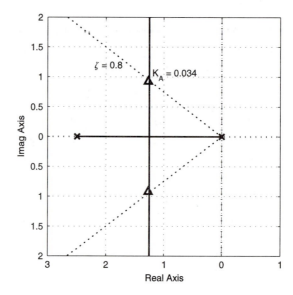

e. $\phi_m = 26.25°$, phase is always $\geq -180°$

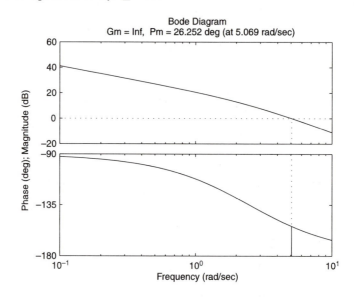

15.13

a. The block diagram is based on the equation
$$\hat{\Theta}(s) = \tau\hat{Q}_h(s)/[C(\tau s + 1)] + \hat{Q}_i(s)/(\tau s + 1)$$

b. $\hat{\Theta}(s)/\hat{\Theta}_d(s) = \tau G_c(s)/P(s)$
$\hat{\Theta}(s)/\hat{\Theta}_i(s) = C/P(s)$ where $P(s) = \tau Cs + C + \tau G_c(s)$

c. For unit step in $\hat{\theta}_d(t)$, $\hat{\theta}_{ss} = \tau K_c/(C + \tau K_c)$
for unit step in $\hat{\theta}_i(t)$, $\hat{\theta}_{ss} = C/(C + \tau K_c)$

d. $\hat{\Theta}(s)/\hat{\Theta}_d(s) = \tau K_c(s + K_I)/P(s)$ and $\hat{\Theta}(s)/\hat{\Theta}_i(s) = Cs/P(s)$
where $P(s) = \tau Cs^2 + (C + \tau K_c)s + \tau K_c K_I$

INDEX